The Homebrew Industrial Revolution

A Low-Overhead Manifesto

KEVIN A. CARSON
Center for a Stateless Society

BOOKSURGE

Published by
BookSurge

ISBN 978-1439266991

Carson, Kevin A.
The Homebrew Industrial Revolution: A Low-Overhead Manifesto
Includes bibliographic references and index
1. Technology—social aspects. 2. Production management.
3. Reengineering (Management) 4. Anarchism. I. Title

To my mother, Ruth Emma Rickert,
and the memory of my father, Amos Morgan Carson.

Contents

How?

Preface

In researching and writing my last book, *Organization Theory: A Libertarian Perspective,* I was probably more engaged and enthusiastic about working on material related to micromanufacturing, the microenterprise, the informal economy, and the singularity resulting from them, than on just about any other part of the book. When the book went to press, I didn't feel that I was done writing about those things. As I completed that book, I was focused on several themes that, while they recurred throughout the book, were imperfectly tied together and developed.

In my first paper as research associate at Center for a Stateless Society,[1] I attempted to tie these themes together and develop them in greater detail in the form of a short monograph. I soon found that it wasn't going to stop there, as I elaborated on the same theme in a series of C4SS papers on industrial history.[2] And as I wrote those papers, I began to see them as the building blocks for a standalone book.

One of the implicit themes in *Organization Theory* which I have attempted to develop since, and which is central to this book, is the central role of fixed costs—initial capital outlays and other overhead—in economics. The higher the fixed costs of an enterprise, the larger the income stream required to service them. That's as true for the household microenterprise, and for the "enterprise" of the household itself, as for more conventional businesses. Regulations that impose artificial capitalization and other overhead costs, the purchase of unnecessarily expensive equipment of a sort that requires large batch production to amortize, the use of stand-alone buildings, etc., increase the size of the minimum revenue stream required to stay in business, and effectively rule out part-time or intermittent self-employment. When such restrictions impose artificially high fixed costs on the means of basic subsistence (housing and feeding oneself, etc.), their effect is to make cheap and comfortable subsistence impossible, and to mandate ongoing external sources of income just to survive. As Charles Johnson has argued,

[1]Kevin Carson, "Industrial Policy: New Wine in Old Bottles," C4SS Paper No. 1 (1st Quarter 2009) <http://c4ss.org/content/78>.

[2]Carson, "MOLOCH: Mass Production Industry as a Statist Construct," C4SS Paper No. 3 (July 2009) <http://c4ss.org/ content/888>; "The Decline and Fall of Sloanism," C4SS Paper No. 4 (August 2009) <http://c4ss.org/content/category/ studies>; "The Homebrew Industrial Revolution," C4SS Paper No. 5 (September 2009) <http://c4ss.org/content/1148>; "Resilient Communities and Local Economies," C4SS Paper No. 6 (4th Quarter 2009) <http://c4ss.org/content/1415>; "The Alternative Economy as a Singularity," C4SS Paper No. 7 (4th Quarter 2009) <http://c4ss.org/content/1523>.

> If it is true (as Kevin has argued, and as I argued in Scratching By[1]) that, absent the state, most ordinary workers would experience a dramatic decline in the fixed costs of living, including (among other things) considerably better access to individual ownership of small plots of land, no income or property tax to pay, and no zoning, licensing, or other government restraints on small-scale neighborhood home-based crafts, cottage industry, or light farming/heavy gardening, I think you'd see a lot more people in a position to begin edging out or to drop out of low-income wage labor entirely—in favor of making a modest living in the informal sector, by growing their own food, or both . . . [2]

On the other hand, innovation in the technologies of small-scale production and of daily living reduce the worker's need for a continuing income stream. It enables the microenterprise to function intermittently and to enter the market incrementally, with no overhead to be serviced when business is slow. The result is enterprises that are lean and agile, and can survive long periods of slow business, at virtually no cost; likewise, such increased efficiencies, by minimizing the ongoing income stream required for comfortable subsistence, have the same liberating effect on ordinary people that access to land on the common did for their ancestors three hundred years ago.

The more I thought about it, the more central the concept of overhead became to my analysis of the two competing economies. Along with setup time, fixed costs and overhead are central to the difference between agility and its lack. Hence the subtitle of this book: "A Low Overhead Manifesto."

Agility and Resilience are at the heart of the alternative economy's differences with its conventional predecessor. Its superiorities are summed up by a photograph I found at Wikimedia Commons, which I considered using as a cover image; a tiny teenage Viet Cong girl leading an enormous captured American soldier. I'm obliged to Jerry Brown (via *Reason* magazine's Jesse Walker) for the metaphor: guerrillas in black pajamas, starting out with captured Japanese and French arms, with a bicycle-based supply train, kicking the living shit out of the best-trained and highest-technology military force in human history.

> But Governor Brown was much more of a fiscal conservative than Governor Reagan, even if he made arguments for austerity that the Republican would never use. (At one point, to get across the idea that a lean organization could outperform a bloated bureaucracy, he offered the example of the Viet Cong.)[3]

I since decided to go with the picture of the Rep-Rap 3-D printer which you see on the cover now, but a guerrilla soldier is still an appropriate symbol for all the characteristics of the alternative economy I'm trying to get across. As I write in the concluding chapter of the book:

> Running throughout this book, as a central theme, has been the superior efficiency of the alternative economy: its lower burdens of overhead, its more intensive use of inputs, and its avoidance of idle capacity.
>
> Two economies are fighting to the death: one of them a highly-capitalized, high-overhead, and bureaucratically ossified conventional economy, the subsidized and protected product of one and a half century's collusion between big

[1]Charles Johnson, "Scratching By: How Government Creates Poverty as We Know It," *The Freeman: Ideas on Liberty*, December 2007 <http://www.thefreemanonline.org/featured/scratching-by-how-government-creates-poverty-as-we-know- it/>.

[2]Johnson comment under Roderick Long, "Amazon versus the Market," *Austro-Athenian Empire*, December 13, 2009 <http://aaeblog.com/2009/12/13/amazon-versus-the-market/comment-page-1/#comment-354091>.

[3]Jesse Walker, "Five Faces of Jerry Brown," *The American Conservative*, November 1, 2009 <http://www.amconmag.com/article/2009/nov/01/00012/>.

government and big business; the other a low capital, low-overhead, agile and resilient alternative economy, outperforming the state capitalist economy despite being hobbled and driven underground.

The alternative economy is developing within the interstices of the old one, preparing to supplant it. The Wobbly phrase "building the structure of the new society within the shell of the old" is one of the most fitting phrases ever conceived for summing up the concept.

I'd like to thank Brad Spangler and Roderick Long for providing me the venue, at Center for a Stateless Society, where I wrote the series of essays this book is based on. I couldn't have written this without all the valuable information I gathered as a participant in the P2P Research email list and the Open Manufacturing list at Google Groups. My participation (no doubt often clueless) was entirely that of a fanboy and enthusiastic layman, since I can't write a line of code and can barely hammer a nail straight. But I thank them for allowing me to play the role of Jane Goodall. And finally, thanks to Professor Gary Chartier of La Sierra University, for his beautiful job formatting the text and designing the cover, as well as his feedback and kind promotion of this work in progress.

1

A Wrong Turn

A. Preface: Mumford's Periodization of Technological History

Lewis Mumford, in *Technics and Civilization*, divided the progress of technological development since late medieval times into three considerably overlapping periods (or phases): the eotechnic, paleotechnic, and neotechnic.

The original technological revolution of the late Middle Ages, the eotechnic, was associated with the skilled craftsmen of the free towns, and eventually incorporated the fruits of investigation by the early scientists. It began with agricultural innovations like the horse collar, horseshoe and crop rotation. It achieved great advances in the use of wood and glass, masonry, and paper (the latter including the printing press). The agricultural advances of the early second millennium were further built on by the innovations of market gardeners in the sixteenth and seventeenth centuries—like, for example, raised bed horticulture, composting and intensive soil development, and the hotbeds and greenhouses made possible by advances in cheap production of glass.

In mechanics, in particular, its greatest achievements were clockwork machinery and the intensive application of water and wind power. The first and most important prerequisite of machine production was the transmission of power and control of movement by use of meshed gears. Clockwork, Mumford argued, was "the key-machine of the modern industrial age." It was

> a new kind of power-machine, in which the source of power and the transmission were of such a nature as to ensure the even flow of energy throughout the works and to make possible regular production and a standardized product. In its relationship to determinable quantities of energy, to standardization, to automatic action, and finally to its own special product, accurate timing, the clock has been the foremost machine in modern technics. . . . The clock, moreover, served as a model for many other kinds of mechanical works, and the analysis of motion that accompanied the perfection of the clock, with the various types of gearing and transmission that were elaborated, contributed to the success of quite different kinds of machine.[1]
>
> If power machinery be a criterion, the modern industrial revolution began in the twelfth century and was in full swing by the fifteenth.[2]

With this first and largest hurdle cleared, Renaissance tinkerers like DaVinci quickly turned to the application of clockwork machinery to specific processes.[3] Given the existence of clockwork, the development of machine processes for every

[1]Lewis Mumford, *Technics and Civilization* (New York: Harcourt, Brace, and Company, 1934), pp. 14-15.

[2]*Ibid.*, p. 112.

[3]*Ibid.*, p. 68.

imaginable specific task was inevitable. Regardless of the prime mover at one end, or the specific process at the other, clockwork transmission of power was the defining feature of automatic machinery.

> In solving the problems of transmitting and regulating motion, the makers of clockwork helped the general development of fine mechanisms. To quote Usher once more: "The primary development of the fundamental principles of applied mechanics was . . . largely based upon the problems of the clock." Clockmakers, along with blacksmiths and locksmiths, were among the first machinists:
>
> Nicholas Forq, the Frenchman who invented the planer in 1751, was a clockmaker: Arkwright, in 1768, had the help of a Warrington clockmaker; it was Huntsman, another clockmaker, desirous of a more finely tempered steel for the watchspring, who invented the process of producing crucible steel: these are only a few of the more outstanding names. In sum, the clock was the most influential of machines, mechanically as well as socially; and by the middle of the eighteenth century it had become the most perfect: indeed, its inception and its perfection pretty well delimit the eotechnic phase. To this day, it is the pattern of fine automatism.[1]

With the use of clockwork to harness the power of prime movers and transmit it to machine production processes, eotechnic industry proliferated wherever wind or running water was abundant. The heartland of eotechnic industry was the river country of the Rhineland and northern Italy, and the windy areas of the North and Baltic seas.[2]

> Grinding grain and pumping water were not the only operations for which the water-mill was used: it furnished power for pulping rags for paper (Ravensburg: 1290) : it ran the hammering and cutting machines of an ironworks (near Dobrilugk, Lausitz, 1320) : it sawed wood (Augsburg: 1322) : it beat hides in the tannery, it furnished power for spinning silk, it was used in fulling-mills to work up the felts, and it turned the grinding machines of the armorers. The wire-pulling machine invented by Rudolph of Nürnberg in 1400 was worked by water-power. In the mining and metal working operations Dr. Georg Bauer described the great convenience of water-power for pumping purposes in the mine, and suggested that if it could be utilized conveniently, it should be used instead of horses or man-power to turn the underground machinery. As early as the fifteenth century, water-mills were used for crushing ore. The importance of water-power in relation to the iron industries cannot be over-estimated: for by utilizing this power it was possible to make more powerful bellows, attain higher heats, use larger furnaces, and therefore increase the production of iron.
>
> The extent of all these operations, compared with those undertaken today in Essen or Gary, was naturally small: but so was the society. The diffusion of power was an aid to the diffusion of population: as long as industrial power was represented directly by the utilization of energy, rather than by financial investment, the balance between the various regions of Europe and between town and country within a region was pretty evenly maintained. It was only with the swift concentration of financial and political power in the sixteenth and seventeenth centuries, that the excessive growth of Antwerp, London, Amsterdam, Paris, Rome, Lyons, Naples, took place.[3]

With the "excessive growth of Antwerp, London, Amsterdam, Paris, Rome, Lyons, Naples," came the triumph of a new form of industry associated with the concentrated power of those cities. The eotechnic phase was supplanted or

[1] *Ibid.*, p. 134.
[2] *Ibid.*, p. 113.
[3] *Ibid.*, pp. 114-115.

crowded out in the early modern period by the paleotechnic—or what is referred to, wrongly, in most conventional histories simply as "the Industrial Revolution."

Paleotechnic had its origins in the new centralized state and the industries closely associated with it (most notably mining and armaments), and centered on mining, iron, coal, and steam power. To give some indication of the loci of the paleotechnic institutional complex, the steam engine was first introduced for pumping water out of mines, and its need for fuel in turn reinforced the significance of the coal industry[1]; the first appearance of large-scale factory production was in the armaments industry.[2] The paleotechnic culminated in the "dark satanic mills" of the nineteenth century and the giant corporations of the late nineteenth and early twentieth.

The so-called "Industrial Revolution," in conventional parlance, conflates two distinct phenomena: the development of mechanized processes for specific kinds of production (spinning and weaving, in particular), and the harnessing of the steam engine as a prime mover. The former was a direct outgrowth of the mechanical science of the eotechnic phase, and would have been fully compatible with production in the small shop if not for the practical issues raised by steam power. The imperative to concentrate machine production in large factories resulted, not from the requirements of machine production as such, but from the need to economize on steam power.

Although the paleotechnic incorporated some contributions from the eotechnic period, it was a fundamental departure in direction, and involved the abandonment of a rival path of development. Technology was developed in the interests of the new royal absolutists, mercantilist industry and the factory system that grew out of it, and the new capitalist agriculturists (especially the Whig oligarchy of England); it incorporated only those eotechnic contributions that were compatible with the new tyrannies, and abandoned the rest.

But its successor, the neotechnic, is what concerns us here.

B. The Neotechnic Phase

Much of the centralization of paleotechnic industry resulted, in addition to the authoritarian institutional culture associated with its origins, from the need (which we saw above) to economize on power.

> .... the steam engine tended toward monopoly and concentration.... Twenty-four hour operations, which characterized the mine and the blast furnace, now came into other industries which had heretofore respected the limitations of day and night. Moved by a desire to earn every possible sum on their investments, the textile manufacturers lengthened the working day.... The steam engine was pacemaker. Since the steam engine requires constant care on the part of the stoker and engineer, steam power was more efficient in large units than in small ones: instead of a score of small units, working when required, one large engine was kept in constant motion. Thus steam power fostered the tendency toward large industrial plants already present in the subdivision of the manufacturing process. Great size, forced by the nature of the steam engine, became in turn a symbol of efficiency. The industrial leaders not only accepted concentration and magnitude as a fact of operation, conditioned by the steam engine: they came to believe in it by itself, as a mark of progress. With the big steam engine, the big factory, the big bonanza farm, the big blast furnace, efficiency was supposed to exist in direct ratio to size. Bigger was an-

[1] *Ibid.*, pp. 159, 161.
[2] *Ibid.*, p. 90.

ciency was supposed to exist in direct ratio to size. Bigger was another way of saying better.

> [Gigantism] was . . . abetted by the difficulties of economic power production with small steam engines: so the engineers tended to crowd as many productive units as possible on the same shaft, or within the range of steam pressure through pipes limited enough to avoid excessive condensation losses. The driving of the individual machines in the plant from a single shaft made it necessary to spot the machines along the shafting, without close adjustment to the topographical needs of the work itself. . . .[1]

Steam power meant that machinery had to be concentrated in one place, in order to get the maximum use out of a single prime mover. The typical paleotechnic factory, through the early 20th century, had machines lined up in long rows, "a forest of leather belts one arising from each machine, looping around a long metal shaft running the length of the shop," all dependent on the factory's central power plant.[2]

The neotechnic revolution of the late nineteenth century put an end to all these imperatives.

If the paleotechnic was a "coal-and-iron complex," in Mumford's terminology, the neotechic was an "electricity-and-alloy complex."[3] The defining features of the neotechnic were the decentralized production made possible by electricity, and the light weight and ephemeralization (to borrow a term from Buckminster Fuller) made possible by the light metals.

The beginning of the neotechnic period was associated, most importantly, with the invention of the prerequisites for electrical power—the dynamo, the alternator, the storage cell, the electric motor—and the resulting possibility of scaling electrically powered production machinery to the small shop, or even scaling power tools to household production.

Electricity made possible the use of virtually any form of energy, indirectly, as a prime mover for production: combustibles of all kinds, sun, wind, water, even temperature differentials.[4] As it became possible to run free-standing machines with small electric motors, the central rationale for the factory system disappeared. "In general," as Paul Goodman wrote, "the change from coal and steam to electricity and oil has relaxed one of the greatest causes for concentration of machinery around a single driving shaft."[5]

The decentralizing potential of small-scale, electrically powered machinery was a common theme among many writers from the late 19th century on. That, and the merging of town and village it made possible, were the central themes of Kropotkin's *Fields, Factories and Workshops*. With electricity "distributed in the houses for bringing into motion small motors of from one-quarter to twelve horsepower," it was possible to produce in small workshops and even homes. Freeing machinery up from a single prime mover ended all limits on the location of machine production. The primary basis for economy of scale, as it existed in the nineteenth century, was the need to economize on horsepower—a justification that

[1] *Ibid.*, p. 224.

[2] William Waddell and Norman Bodek, *The Rebirth of American Industry: A Study of Lean Management* (Vancouver, WA: PCS Press, 2005), pp. 119-121.

[3] Mumford, *Technics and Civilization*, p. 110.

[4] *Ibid.*, pp. 214, 221.

[5] Paul and Percival Goodman, *Communitas: Means of Livelihood and Ways of Life* (New York: Vintage Books, 1947, 1960), p. 156.

vanished when the distribution of electrical power eliminated reliance on a single source of power.[1]

William Morris seems to have made some Kropotkinian technological assumptions in his depiction of a future libertarian communist society in *News From Nowhere*:

> "What building is that?" said I, eagerly; for it was a pleasure to see something a little like what I was used to: "it seems to be a factory."
>
> "Yes, he said," "I think I know what you mean, and that's what it is; but we don't call them factories now, but Banded-workshops; that is, places where people collect who want to work together."
>
> "I suppose," said I, "power of some sort is used there?"
>
> "No, no," said he. "Why should people collect together to use power, when they can have it at the places where they live or hard by, any two or three of them, or any one, for the matter of that? . . . "[2]

The introduction of electrical power, in short, put small-scale machine production on an equal footing with machine production in the factory.

> The introduction of the electric motor worked a transformation within the plant itself. For the electric motor created flexibility in the design of the factory: not merely could individual units be placed where they were wanted, and not merely could they be designed for the particular work needed: but the direct drive, which increased the efficiency of the motor, also made it possible to alter the layout of the plant itself as needed. The installation of motors removed the belts which cut off light and lowered efficiency, and opened the way for the rearrangement of machines in functional units without regard for the shafts and aisles of the old-fashioned factory: each unit could work at its own rate of speed, and start and stop to suit its own needs, without power losses through the operation of the plant as a whole.
>
> . . . [T]he efficiency of small units worked by electric motors utilizing current either from local turbines or from a central power plant has given small-scale industry a new lease on life: on a purely technical basis it can, for the first time since the introduction of the steam engine, compete on even terms with the larger unit. Even domestic production has become possible again through the use of electricity: for if the domestic grain grinder is less efficient, from a purely mechanical standpoint, than the huge flour mills of Minneapolis, it permits a nicer timing of production to need, so that it is no longer necessary to consume bolted white flours because whole wheat flours deteriorate more quickly and spoil if they are ground too long before they are sold and used. To be efficient, the small plant need not remain in continuous operation nor need it produce gigantic quantities of foodstuffs and goods for a distant market: it can respond to local demand and supply; it can operate on an irregular basis, since the overhead for permanent staff and equipment is proportionately smaller; it can take advantage of smaller wastes of time and energy in transportation, and by face to face contact it can cut out the inevitable red-tape of even efficient large organizations.[3]

Mumford's comments on flour milling also anticipated the significance of small-scale powered machinery in making possible what later became known as "lean production"; its central principle is that overall flow is more important to cost-cutting than maximizing the efficiency of any particular stage in isolation.

[1]Peter Kropotkin, *Fields, Factories and Workshops: or Industry Combined with Agriculture and Brain Work with Manual Work* (New York: Greenwood Press, Publishers, 1968 [1898]), pp. 154., 179-180.

[2]William Morris, *News From Nowhere: or, An Epoch of Rest* (1890). Marxists.Org online text <http://www.marxists.org/archive/morris/works/1890/nowhere/nowhere.htm>.

[3]Mumford, *Technics and Civilization*, pp. 224-225.

The modest increases in unit production cost at each separate stage are offset not only by greatly reduced transportation costs, but by avoiding the large eddies in overall production flow (buffer stocks of goods-in-process, warehouses full of goods "sold" to inventory without any orders, etc.) that result when production is not geared to demand.[1]

Neotechnic methods, which could be reproduced anywhere, made possible a society where "the advantages of modern industry [would] be spread, not by transport—as in the nineteenth century—but by local development." The spread of technical knowledge and standardized methods would make transportation far less important.[2]

Mumford also described, in quite Kropotkinian terms, the "marriage of town and country, of industry and agriculture," that could result from the application of further refined eotechnic horticultural techniques and the decentralization of manufacturing in the neotechnic age.[3]

Mumford saw the neotechnic phase as a continuation of the principles of the eotechnic, with industrial organization taking the form it would have done if allowed to develop directly from the eotechnic without interruption.

> The neotechnic, in a sense, is a resumption of the lines of development of the original eotechnic revolution, following the paleotechnic interruption. The neotechnic differs from the paleotechnic phase almost as white differs from black. But on the other hand, it bears the same relation to the eotechnic phase as the adult form does to the baby.
>
> .... The first hasty sketches of the fifteenth century were now turned into working drawings: the first guesses were now re-enforced with a technique of verification: the first crude machines were at last carried to perfection in the exquisite mechanical technology of the new age, which gave to motors and turbines properties that had but a century earlier belonged almost exclusively to the clock.[4]

Or as Ralph Borsodi put it, "[t]he steam engine put the water-wheel out of business. But now the gasoline engine and the electric motor have been developed to a point where they are putting the steam engine out of business."

> The modern factory came in with steam. Steam is a source of power that almost necessitates factory production. But electricity does not. It would be poetic justice if electricity drawn from the myriads of long neglected small streams of the country should provide the power for an industrial counter-revolution.[5]

Mumford suggested that, absent the abrupt break created by the new centralized states and their state capitalist clients, the eotechnic might have evolved directly into the neotechnic. Had not the eotechnic been aborted by the paleotechnic, a full-scale modern industrial revolution would still almost certainly have come about "had not a ton of coal been dug in England, and had not a new iron mine been opened."[6]

[1]In the case of flour, according to Borsodi, the cost of custom-milled flour from a local mill was about half that of flour from a giant mill in Minneapolis, and flour from a small electric household mill was cheaper still. *Prosperity and Security: A Study in Realistic Economics* (New York and London: Harper & Brothers Publishers, 1938), pp. 178-181.

[2]*Ibid.*, pp. 388-389.

[3]Mumford, *Technics and Civilization*, pp. 258-259.

[4]Mumford, *Technics and Civilization*, p. 212.

[5]Ralph Borsodi, *This Ugly Civilization* (Philadelphia: Porcupine Press, 1929, 1975), p. 65.

[6]Mumford, *Technics and Civilization*, p. 118.

Jefferson – enough gov't to assert sov of citizens over corp.s

The amount of work accomplished by wind and water power compared quite favorably with that of the steam-powered industrial revolution. Indeed, the great advances in textile output of the eighteenth century were made with water-powered factories; steam power was adopted only later. The Fourneyron water-turbine, perfected in 1832, was the first prime-mover to exceed the poor 5% or 10% efficiencies of the early steam engine, and was a logical development of earlier water-power technology that would likely have followed much earlier in due course, had not the development of water-power been sidetracked by the paleotechnic revolution.[1]

Had the spoonwheel of the seventeenth century developed more rapidly into Fourneyron's efficient water-turbine, water might have remained the backbone of the power system until electricity had developed sufficiently to give it a wider area of use.[2]

The eotechnic phase survived longest in America, according to Mumford. Had it survived a bit longer, it might have passed directly into the neotechnic. In *The City in History*, he mentioned abortive applications of eotechnic means to decentralized organization, unfortunately forestalled by the paleotechnic revolution, and speculated at greater length on the Kropotkinian direction social evolution might have taken had the eotechnic passed directly into the neotechnic. Of the societies of seventeenth century New England and New Netherlands, he wrote:

> This eotechnic culture was incorporated in a multitude of small towns and villages, connected by a network of canals and dirt roads, supplemented after the middle of the nineteenth century by short line railroads, not yet connected up into a few trunk systems meant only to augment the power of the big cities. With wind and water power for local production needs, this was a balanced economy; and had its balance been maintained, had balance indeed been consciously sought, a new general pattern of urban development might have emerged. . . .
>
> In 'Technics and Civilization' I pointed out how the earlier invention of more efficient prime movers, Fourneyron's water turbine and the turbine windmill, could perhaps have provided the coal mine and the iron mine with serious technical competitors that might have kept this decentralized regime long enough in existence to take advantage of the discovery of electricity and the production of the light metals. With the coordinate development of science, this might have led directly into the more humane integration of 'Fields, Factories, and Workshops' that Peter Kropotkin was to outline, once more, in the eighteen-nineties.[3]

Borsodi speculated, along lines similar to Mumford's, on the different direction things might have taken had the eotechnic phase been developed to its full potential without being aborted by the paleotechnic:

> It is impossible to form a sound conclusion as to the value to mankind of this institution which the Arkwrights, the Watts, and the Stephensons had brought into being if we confine ourselves to a comparison of the efficiency of the factory system of production with the efficiency of the processes of production which prevailed before the factory appeared.
>
> 
>
> A very different comparison must be made.
>
> We must suppose that the inventive and scientific discoveries of the past two centuries had not been used to destroy the methods of production which prevailed before the factory.

[1]*Ibid.*, p. 118.

[2]*Ibid.*, p. 143.

[3]Lewis Mumford, *The City in History: Its Transformations, and Its Prospects* (New York: Harcourt, Brace, & World, Inc., 1961), pp. 333-34.

> We must suppose that an amount of thought and ingenuity precisely equal to that used in developing the factory had been devoted to the development of domestic, custom, and guild production.
>
> We must suppose that the primitive domestic spinning wheel had been gradually developed into more and more efficient domestic machines; that primitive looms, churns, cheese presses, candle molds, and primitive productive apparatus of all kinds had been perfected step by step without sacrifice of the characteristic "domesticity" which they possessed.
>
> In short, we must suppose that science and invention had devoted itself to making domestic and handicraft production efficient and economical, instead of devoting itself almost exclusively to the development of factory machines and factory production.
>
> The factory-dominated civilization of today would never have developed. Factories would not have invaded those fields of manufacture where other methods of production could be utilized. Only the essential factory would have been developed. Instead of great cities, lined with factories and tenements, we should have innumerable small towns filled with the homes and workshops of neighborhood craftsmen. Cities would be political, commercial, educational, and entertainment centers. . . . Efficient domestic implements and machines developed by centuries of scientific improvement would have eliminated drudgery from the home and the farm.[1]

And, we might add, the home production machinery itself would have been manufactured, not in Sloanist mass-production factories, but mainly in small factories and shops integrating power machinery into craft production.

C. A Funny Thing Happened on the Way to the Neotechnic Revolution

The natural course of things, according to Borsodi, was that the "process of shifting production from the home and neighborhood to the distantly located factory" would have peaked with "the perfection of the reciprocating steam-engine," and then leveled off until the invention of the electric motor reversed the process and enabled families and local producers to utilize the powered machinery previously restricted to the factory.[2] But it didn't happen that way. Instead, electricity was incorporated into manufacturing in an utterly perverse way.

Michael Piore and Charles Sabel described a fork in the road, based on which of two possible alternative ways were chosen for incorporating electrical power into manufacturing. The first, more in keeping with the unique potential of the new technology, was to integrate electrically powered machinery into small-scale craft production: "a combination of craft skill and flexible equipment," or "mechanized craft production."

> Its foundation was the idea that machines and processes could augment the craftsman's skill, allowing the worker to embody his or her knowledge in ever more varied products: the more flexible the machine, the more widely applicable the process, the more it expanded the craftsman's capacity for productive expression.

The other was to adapt electrical machinery to the preexisting framework of paleotechnic industrial organization—in other words, what was to become twentieth century mass-production industry. This latter alternative entailed breaking the production process down into its separate steps, and then substituting extremely

[1] Borsodi, *This Ugly Civilization*, pp. 60-61.
[2] Borsodi, Prosperity and Security, p. 182.

expensive and specialized machinery for human skill. "The more specialized the machine—the faster it worked and the less specialized its operator needed to be—the greater its contribution to cutting production costs.[1]

The first path, unfortunately, was for the most part the one not taken; it has been followed only in isolated enclaves, particularly in assorted industrial districts in Europe. The most famous current example is Italy's Emilia-Romagna region, which we will examine in a later chapter.

The second, mass-production model became the dominant form of industrial organization. Neotechnic advances like electrically powered machinery, which offered the potential for decentralized production and were ideally suited to a fundamentally different kind of society, have so far been integrated into the framework of mass production industry.

Mumford argued that the neotechnic advances, rather than being used to their full potential as the basis for a new kind of economy, were instead incorporated into a paleotechnic framework. Neotechnic had not "displaced the older regime" with "speed and decisiveness," and had not yet "developed its own form and organization."

> Emerging from the paleotechnic order, the neotechnic institutions have nevertheless in many cases compromised with it, given way before it, lost their identity by reason of the weight of vested interests that continued to support the obsolete instruments and the anti-social aims of the middle industrial era. *Paleotechnic ideals still largely dominate the industry and the politics of the Western World*. . . . To the extent that neotechnic industry has failed to transform the coal-and-iron complex, to the extent that it has failed to secure an adequate foundation for its humaner technology in the community as a whole, to the extent that it has lent its heightened powers to the miner, the financier, the militarist, the possibilities of disruption and chaos have increased.[2]
>
> True: the industrial world produced during the nineteenth century is either technologically obsolete or socially dead. But unfortunately, its maggoty corpse has produced organisms which in turn may debilitate or possibly kill the new order that should take its place: perhaps leave it a hopeless cripple.[3]
>
> The new machines followed, not their own pattern, but the pattern laid down by previous economic and technical structures.[4]
>
> The fact is that in the great industrial areas of Western Europe and America . . . , the paleotechnic phase is still intact and all its essential characteristics are uppermost, even though many of the machines it uses are neotechnic ones or have been made over—as in the electrification of railroad systems—by neotechnic methods. In this persistence of paleotechnics . . . we continue to worship the twin deities, Mammon and Moloch. . . .[5]
>
> We have merely used our new machines and energies to further processes which were begun under the auspices of capitalist and military enterprise: we have not yet utilized them to conquer these forms of enterprise and subdue them to more vital and humane purposes. . . .[6]
>
> Not alone have the older forms of technics served to constrain the development of the neotechnic economy: but the new inventions and devices have been

[1]Michael S. Piore and Charles F. Sabel, *The Second Industrial Divide: Possibilities for Prosperity* (New York: HarperCollins, 1984), pp. 4-6, 19.

[2]Mumford, *Technics and Civilization*, pp. 212-13.

[3]*Ibid.*, p. 215.

[4]*Ibid.*, p. 236.

[5]*Ibid.*, p. 264.

[6]*Ibid.*, p. 265.

> frequently used to maintain, renew, stabilize the structure of the old social order. . . .[1]
>
> The present pseudomorph is, socially and technically, third-rate. It has only a fraction of the efficiency that the neotechnic civilization as a whole may possess, provided it finally produces its own institutional forms and controls and directions and patterns. At present, instead of finding these forms, we have applied our skill and invention in such a manner as to give a fresh lease of life to many of the obsolete capitalist and militarist institutions of the older period. Paleotechnic purposes with neotechnic means: that is the most obvious characteristic of the present order.[2]

Mumford used Spengler's idea of the "cultural pseudomorph" to illustrate the process: " . . . in geology . . . a rock may retain its structure after certain elements have been leached out of it and been replaced by an entirely different kind of material. Since the apparent structure of the old rock remains, the new product is called a pseudomorph."

> A similar metamorphosis is possible in culture: new forces, activities, institutions, instead of crystallizing independently into their own appropriate forms, may creep into the structure of an existing civilization. . . . As a civilization, we have not yet entered the neotechnic phase. . . . [W]e are still living, in Matthew Arnold's words, between two worlds, one dead, the other powerless to be born.[3]

For Mumford, Soviet Russia was a mirror image of the capitalist West in shoehorning neotechnic technology into a paleotechnic institutional framework. Despite the neotechnic promise of Lenin's "electrification plus Soviet power," the Soviet aesthetic ideal was that of the Western mass-production factory: "the worship of size and crude mechanical power, and the introduction of a militarist technique in both government and industry. . . . "[4] That Lenin's vision of "communism" entailed a wholesale borrowing of the mass-production model, under state ownership, is suggested for his infatuation with Taylorism and his suppression of worker self-management in the factories. The Stalinist fetish for gigantism, with its boasts of having the biggest factory, power plant, etc. in the world, followed as a matter of course.

How were existing institutional interests able to thwart the revolutionary potential of electrical power, and divert neotechnic technologies into paleotechnic channels? The answer is that the state tipped the balance.

The state played a central role in the triumph of mass-production industry in the United States.

The state's subsidies to long-distance transportation were first and most important. There never would have been large manufacturing firms producing for a national market, had not the federal government first created a national market with the national railroad network. A high-volume national transportation system was an indispensable prerequisite for big business.

We quoted Mumford's observation above, that the neotechnic revolution offered to substitute industrialization by local economic development for reliance on long-distance transport. State policies, however, tipped the balance in the other direction: they artificially shifted the competitive advantage toward industrial concentration and long-distance distribution.

[1] *Ibid.*, p. 266.
[2] *Ibid.*, p. 267.
[3] *Ibid.*, p. 265.
[4] *Ibid.*, p. 264.

Alfred Chandler, the chief apostle of the large mass-production corporation, himself admitted as much: all the advantages he claimed for mass production *presupposed* a high-volume, high-speed, high-turnover distribution system on a national scale, without regard to whether the costs of the latter exceeded the alleged benefits of the former.

> ... [M]odern business enterprise appeared for the first time in history when the volume of economic activities reached a level that made administrative coordination more efficient and more profitable than market coordination.[1]
>
> ... [The rise of administrative coordination first] occurred in only a few sectors or industries where technological innovation and market growth created high-speed and high-volume throughput.[2]

William Lazonick, a disciple of Chandler, described the process as obtaining "a large market share in order to transform the high fixed costs into low unit costs. . . . "[3]

The railroad and telegraph, "so essential to high-volume production and distribution," were in Chandler's view what made possible this steady flow of goods through the distribution pipeline.[4]

The primacy of such state-subsidized infrastructure is indicated by the very structure of Chandler's book. He begins with the railroads and telegraph system, themselves the first modern, multi-unit enterprises.[5] And in subsequent chapters, he recounts the successive evolution of a national wholesale network piggybacking on the centralized transportation system, followed by a national retail system, and only then by large-scale manufacturing for the national market. A national long-distance transportation system led to mass distribution, which in turn led to mass production.

> The revolution in the processes of distribution and production rested in large part on the new transportation and communications infrastructure. Modern mass production and mass distribution depend on the speed, volume, and regularity in the movement of goods and messages made possible by the coming of the railroad, telegraph and steamship.[6]
>
> The coming of mass distribution and the rise of the modern mass marketers represented an organizational revolution made possible by the new speed and regularity of transportation and communication.[7]
>
> ... The new methods of transportation and communication, by permitting a large and steady flow of raw materials into and finished products out of a factory, made possible unprecedented levels of production. The realization of this potential required, however, the invention of new machinery and processes.[8]

In other words, the so-called "internal economies of scale" in manufacturing could come about only when the offsetting external diseconomies of long-distance distribution were artificially nullified by corporate welfare. Such "economies" can only occur given an artificial set of circumstances which permit the reduced unit

[1] Alfred D. Chandler, Jr., *The Visible Hand: The Managerial Revolution in American Business* (Cambridge and London: The Belknap Press of Harvard University Press, 1977), p. 8.

[2] *Ibid.*, p. 11.

[3] William Lazonick, *Business Organization and the Myth of the Market Economy* (Cambridge, 1991), pp. 198-226.

[4] Chandler, *The Visible Hand*, p. 79.

[5] *Ibid.*, pp. 79, 96-121.

[6] *Ibid.*, p. 209.

[7] *Ibid.*, p. 235.

[8] *Ibid.*, p. 240.

costs of expensive, product-specific machinery to be considered in isolation, because the indirect costs entailed are all externalized on society. And if the real costs of long-distance shipping, high-pressure marketing, etc., do in fact exceed the savings from faster and more specialized machinery, then the "efficiency" is a false one.

It's an example of what Ivan Illich called "counterproductivity": the adoption of a technology beyond the point, not only of diminishing returns, but of *negative* returns. Illich also used the term "second watershed" to describe the same concept: e.g., in the case of medicine, the first watershed included such basic things as public sanitation, the extermination of rats, water purification, and the adoption of antibiotics; the second watershed was the adoption of skill- and capital-intensive methods to the point that iatrogenic (hospital- or doctor-induced) illness exceeded the health benefits. In other areas, the introduction of motorized transportation, beyond a certain point, produces artificial distance between things and generates congestion faster than it can be relieved.[1]

Where Illich went wrong was in seeing counterproductivity as inevitable, if adoption of technologies wasn't restrained by regulation. In fact, when all costs and benefits of a technology are internalized by the adopter, adoption beyond the point of counterproductivity will not occur. Adoption beyond the point of counterproductivity is profitable only when the costs are externalized on society or on the taxpayer, and the benefits are appropriated by a privileged class.

As Chandler himself admitted, the greater "efficiency" of national wholesale organizations lay in their "even more effective exploitation of the existing railroad and telegraph systems."[2] That is, they were more efficient parasites. But the "efficiencies" of a parasite are usually of a zero-sum nature.

Chandler also admitted, perhaps inadvertently, that the "more efficient" new production methods were adopted almost as an afterthought, given the artificially large market areas and subsidized distribution:

> . . . the nature of the market was more important than the methods of production in determining the size and defining the activities of the modern industrial corporation.[3]

And finally, Chandler admitted that the new mass-production industry was not more efficient at producing in response to autonomous market demand. He himself helpfully pointed out, as we shall see in the next chapter, that the first large industrialists only integrated mass-production with mass-distribution because they were forced to: "They did so because existing marketers were unable to sell and distribute products in the volume they were produced."[4]

Despite all this, Chandler—astonishingly—minimized the role of the state in creating the system he so admired:

> The rise of modern business enterprise in American industry between the 1880s and World War I was little affected by public policy, capital markets, or entrepreneurial talents because it was part of a more fundamental economic development. Modern business enterprise . . . was the organizational response to fun-

[1]Ivan Illich, "The Three Dimensions of Public Opinion," in *The Mirror of the Past: Lectures and Addresses, 1978-1990* (New York and London: Marion Boyars, 1992), p. 84; *Tools for Conviviality* (New York, Evanston, San Francisco, London: Harper & Row, 1973), pp. xxii-xxiii, 1-2, 3, 6-7, 84-85; *Disabling Professions* (New York and London: Marion Boyars, 1977), p. 28.

[2]Chandler, *The Visible Hand*, p. 215.

[3]*Ibid.*, p. 363.

[4]*Ibid.*, p. 287.

> damental changes in processes of production and distribution made possible by the availability of new sources of energy and by the increasing application of scientific knowledge to industrial technology. The coming of the railroad and telegraph and the perfection of new high-volume processes . . . made possible a historically unprecedented volume of production.[1]

"The coming of the railroad"? In Chandler's language, the railroads seem to be an inevitable force of nature rather than the result of deliberate actions by policy makers.

We can't let Chandler get by without challenging his implicit assumption (shared by many technocratic liberals) that paleotechnic industry was more efficient than the decentralized, small-scale production methods of Kropotkin and Borsodi. The possibility never occurred to him that massive state intervention, at the same time as it enabled the revolutions in corporate size and capital-intensiveness, might also have tipped the balance between alternative forms of production technology.

The national railroad system simply never would have come into existence on such a scale, with a centralized network of trunk lines of such capacity, had not the state rammed the project through.

Piore and Sabel describe the enormous capital outlays, and the enormous transaction costs to be overcome, in creating a national railroad system. Not only the startup costs of actual physical capital, but those of securing rights of way, were "huge":

> It is unlikely that railroads would have been built as quickly and extensively as they were but for the availability of massive government subsidies.

Other transaction costs overcome by government, in creating the railroad system, included the revision of tort and contract law (e.g., to exempt common carriers from liability for many kinds of physical damage caused by their operation).[2]

According to Matthew Josephson, for ten years or more before 1861, "the railroads, especially in the West, were 'land companies' which acquired their principal raw material through pure grants in return for their promise to build, and whose directors . . . did a rushing land business in farm lands and town sites at rising prices." For example, under the terms of the Pacific Railroad bill, the Union Pacific (which built from the Mississippi westward) was granted twelve million acres of land and $27 million worth of thirty-year government bonds. The Central Pacific (built from the West Coast eastward) received nine million acres and $24 million worth of bonds.[3]

The federal railroad land grants, according to Murray Rothbard, included fifteen mile tracts of land on either side of the actual right of way. As the railroads were completed, this land skyrocketed in value. And as new towns were built along the railroad routes, every house and business was built on land sold by the railroads. The tracts included valuable timber land, as well.[4]

Theodore Judah, chief engineer for what became the Central Pacific, assured potential investors "that it could be done—*if government aid were obtained*. For the cost would be terrible." Collis Huntington, the leading promoter for the project, engaged in a sordid combination of strategically placed bribes and appeals to

[1] *Ibid.*, p. 376.

[2] Piore and Sabel, pp. 66-67.

[3] Matthew Josephson, *The Robber Barons: The Great American Capitalists 1861-1901* (New York: Harcourt, Brace & World, Inc., 1934, 1962), pp. 77-78.

[4] Murray N. Rothbard, *Power and Market: Government and the Economy* (Menlo Park, Calif.: Institute for Humane Studies, Inc., 1970), p. 70.

communities' fears of being bypassed, in order to extort grants of "rights of way, terminal and harbor sites, and . . . stock or bond subscriptions ranging from $150,000 to $1,000,000" from a long string of local governments that included San Francisco, Stockton, and Sacramento.[1]

Absent the land grants and government purchases of railroad bonds, the railroads would likely have developed instead along the initial lines described by Mumford: many local rail networks linking communities into local industrial economies. The regional and national interlinkages of local networks, when they did occur, would have been far fewer and far smaller in capacity. The comparative costs of local and national distribution, accordingly, would have been quite different. In a nation of hundreds of local industrial economies, with long-distance rail transport much more costly than at present, the natural pattern of industrialization would have been to integrate small-scale power machinery into flexible manufacturing for local markets.

Instead, the state artificially aggregated the demand for manufactured goods into a single national market, and artificially lowered the costs of distribution for those serving that market. In effect, it created an artificial ecosystem to which large-scale, mass-production industry was best "adapted."

The first organisms to adapt themselves to this artificial ecosystem, as recounted by Chandler, were the national wholesale and retail networks, with their dependence on high turnover and dependability. Then, piggybacked on them, were the large manufacturers serving the national market. But they were only "more efficient" in terms of their more efficient exploitation of an artificial environment which itself was characterized by the concealment and externalization of costs. With all the concealed and externalized costs fully subsumed into the price of mass-produced goods, rather than shifted onto society or the taxpayer, it is likely that the overall cost of goods produced flexibly on general-purpose machinery for local markets would have been less than that of mass-produced goods.

Besides almost single-handedly creating the artificially unified and cheap national market without which national manufacturers could not have existed, the railroad companies also actively promoted the concentration of industry through their rate policies. Piore and Sabel argue that "the railroads' policy of favoring their largest customers, through rebates," was a central factor in the rise of the large corporation. Once in place, the railroads—being a high fixed-cost industry—had

> a tremendous incentive to use their capacity in a continuous, stable way. This incentive meant, in turn, that they had an interest in stabilizing the output of their principal customers—an interest that extended to protecting their customers from competitors who were served by other railroads. It is therefore not surprising that the railroads promoted merger schemes that had this effect, nor that they favored the resulting corporations or trusts with rebates.

"Indeed, seen in this light, the rise of the American corporation can be interpreted more as the result of complex alliances among Gilded Age robber barons than as a first solution to the problem of market stabilization faced by a mass-production economy."[2] According to Josephson,

> while the tillers of the soil felt themselves subject to extortion, they saw also that certain interests among those who handled the grain or cattle they produced, the elevators, millers and stockyards, or those from whom they purchased their necessities, the refiners of oil, the great merchant-houses, were encouraged by the railroads to combine against the consumer. In the hearings before the Hep-

[1]Josephson, pp. 83-84.
[2]Piore and Sabel, pp. 66-67.

> burn Committee in 1879 it was revealed that the New York Central, like railways all over the country, had some 6,000 secret rebate agreements, such as it had made with the South Improvement Company. . . .[1]
>
> . . . [T]he secret tactics of the rebate gave certain producing groups (as in petroleum, beef, steel) those advantages which permitted them to outstrip competitors and soon to conduct their business upon as large a scale as the railways themselves.[2]
>
> . . . Upon the refined oil [Rockefeller] shipped from Cleveland he received a rebate of 50 cents a barrel, giving him an advantage of 25 per cent over his competitors.[3]
>
> In the meantime the political representatives whom the disabused settlers sent forth to Washington or to the state legislatures seemed not only helpless to aid them, but were seen after a time riding about the country wherever they listed by virtue of free passes generously distributed to them.[4]
>
> The railroads also captured the state legislatures and railroad commissions.[5]

Among certain Objectivists and vulgar libertarians of the Right, this is commonly transformed into a morality play in which men of innovative genius built large businesses through sheer effort and entrepreneurship, and the power of superior efficiency. These heroic John Galts then charged rates based on the new railroad's benefits to customers, and were forced into political lobbying only as a matter of self-defense against government extortion. This is a lie.

What happened was nothing to do with a free market, unless one belongs to the right-wing strain of libertarianism for which "free market" equates to "beneficial to big business." It was, rather, a case of the government intervening to create an industry almost from scratch, and by the same act putting it in a commanding height from which it could extort monopoly profits from the public. The closest modern analogy is the drug companies, which use unlimited patent monopolies granted by the state to charge extortionate prices for drugs developed entirely or almost entirely with government research funds. But then the Randroids and vulgar libertarians are also fond of Big Pharma.

Of course, the railroads were only the first of many centralizing infrastructure projects. The process continued through the twentieth century, with the development of the subsidized highway system and the civil aviation system. But unlike the railroads, whose chief significance was their role in creating the national market in the first place, civil aviation and the automobile-industrial complex were arguably most important as sinks for surplus capital and output. They will be treated in the next chapter, accordingly, as examples of a phenomenon described by Paul Baran and Paul Sweezy in *Monopoly Capitalism*: government creation of new industries to absorb the surplus resulting from corporate capitalism's chronic tendencies toward overinvestment and overproduction.

Second, the American legal framework was transformed in the mid-nineteenth century in ways that made a more hospitable environment for large corporations operating on a national scale. Among the changes were the rise of a general federal commercial law, general incorporation laws, and the status of the corporation as a person under the Fourteenth Amendment. The functional significance of these changes on a national scale was analogous to the later effect, on a global scale, of the Bretton Woods agencies and the GATT process: a centralized

[1]Josephson, pp. 250-251.
[2]*Ibid*., p. 253.
[3]*Ibid*., p. 265.
[4]*Ibid*., p. 251.
[5]*Ibid*., p. 252.

legal order was created, prerequisite for their stable functioning, coextensive with the market areas of large corporations.

The federalization of the legal regime is associated, in particular, with the recognition of a general body of federal commercial law in *Swift v. Tyson* (1842), and with the application of the Fourteenth Amendment to corporate persons in *Santa Clara County v. Southern Pacific Railroad Company* (1886).

The Santa Clara decision was followed by an era of federal judicial activism, in which state laws were overturned on the basis of "substantive due process." The role of the federal courts in the national economy was similar to the global role of the contemporary World Trade Organization, with higher tribunals empowered to override the laws of local jurisdictions which were injurious to corporate interests.

In the federal courts, the "due process" and "equal protection" rights of corporations as "juristic persons" have been made the basis of protections against legal action aimed at protecting the older common law rights of flesh and blood persons. For example local ordinances to protect groundwater and local populations against toxic pollution and contagion from hog farms, to protect property owners from undermining and land subsidance caused by coal extraction—surely indistinguishable in practice from the tort liability provisions of any just market anarchy's libertarian law code—have been overturned as violations of the "equal protection" rights of hog factory farms and mining companies.

Still another component of the corporate legal revolution was the increased ease, under general incorporation laws, of forming limited liability corporations with permanent entity status apart (severally or collectively) from the shareholders.

Arguably, as Robert Hessen and others have made a case, corporate entity status and limited liability against creditors could be achieved entirely through private contract. Whether or not that is so, the government has tilted the playing field decisively toward the corporate form by providing a ready-made and automatic procedure for incorporation. In so doing, it has made the corporation the standard or default form of organization, reduced the transaction costs of establishing it relative to what would prevail were it negotiated entirely from scratch, and thereby reduced the bargaining power of other parties in negotiating the terms on which it operates.

Third, not only did the government indirectly promote the concentration and cartelization of industry through the railroads it had created, but it did so directly through patent law. As we shall see in the next chapter, mass-production requires large business organizations capable of exercising sufficient power over their external environment to guarantee the consumption of their output. Patents promoted the stable control of markets by oligopoly firms through the control, exchange and pooling of patents.

According to David Noble, two essentially new science-based industries (those that "grew out of the soil of scientific rather than traditional craft knowledge") emerged in the late 19th century: the electrical and chemical industries.[1]

In the electrical industry, General Electric had its origins first in a merger between Edison Electric (which controlled all of Edison's electrical patents) and the Sprague Electric Railway and Motor Company, and then in an 1892 merger between Edison General Electric and Thomas-Houston—both of them motivated primarily by patent considerations. In the latter case, in particular, Edison General Electric and Thomas-Houston each needed patents owned by the others and could

[1]David F. Noble, *America by Design: Science, Technology, and the Rise of Corporate Capitalism* (New York: Alfred A. Knopf, 1977), p. 5.

not "develop lighting, railway or power equipment without fear of infringement suits and injunctions."[1] From the 1890s on, the electrical industry was dominated by two large firms: GE and Westinghouse, both of which owed their market shares largely to patent control. In addition to the patents which they originally owned, they acquired control over patents (and hence over much of the electrical manufacturing market) through "acquisition of the patent rights of individual inventors, acquisition of competing firms, mergers with competitors, and the systematic and strategic development of their own patentable inventions. As GE and Westinghouse together secured a deadlock on the electrical industry through patent acquisition, competition between them became increasingly intense and disruptive. By 1896 the litigation cost from some three hundred pending patent suits was enormous, and the two companies agreed to form a joint Board of Patent Control. General Electric and Westinghouse pooled their patents, with GE handling 62.5% of the combined business.[2]

The structure of the telephone industry had similar origins, with the Bell Patent Association forming "the nucleus of the first Bell industrial organization" (and eventually of AT&T) The National Bell Telephone Company, from the 1880s on, fought vigorously to "occupy the field" (in the words of general manager Theodore N. Vail) through patent control. As Vail described the process, the company surrounded itself

> with everything that would protect the business, that is the knowledge of the business, all the auxiliary apparatus; a thousand and one little patents and inventions with which to do the business which was necessary, that is what we wanted to control and get possession of.
>
> To achieve this, the company early on established an engineering department whose business it was to study the patents, study the development and study these devices that either were originated by our own people or came in to us from the outside. Then early in 1879 we started our patent department, whose business was entirely to study the question of patents that came out with a view to acquiring them, because . . . we recognized that if we did not control these devices, somebody else would.[3]

This approach strengthened the company's position of control over the market not only during the seventeen year period of the main patents, but (as Frederick Fish put it in an address to the American Institute of Electrical Engineers) during the subsequent seventeen years of

> each and every one of the patents taken out on subsidiary methods and devices invented during the progress of commercial development. [Therefore] one of the first steps taken was to organize a corps of inventive engineers to perfect and improve the telephone system in all directions . . . that by securing accessory inventions, possession of the field might be retained as far as possible and for as long a time as possible.[4]

This method, preemptive occupation of the market through strategic patent acquisition and control, was also used by GE and Westinghouse.

Even with the intensified competition resulting from the expiration of the original Bell patents in 1894, and before government favoritism in the grants of rights-of-way and regulated monopoly status, the legacy effect of AT&T's control of the secondary patents was sufficient to secure it half the telephone market thir-

[1] *Ibid.*, p. 9.
[2] *Ibid.*, pp. 9-10.
[3] *Ibid.*, pp. 11-12.
[4] *Ibid.*, p. 12.

teen years later, in 1907.[1] AT&T, anticipating the expiration of its original patents, had (to quote Vail again) "surrounded the business with all the auxiliary protection that was possible." For example, the company in 1900 purchased Michael Pupin's patent on loading coils and in 1907 acquired exclusive domestic rights for Cooper-Hewitt's patents on the mercury-arc repeater—essential technologies underlying AT&T's monopoly on long-distance telephony.[2]

By the time the FCC was formed in 1935, the Bell System had acquired patents to "some of the most important inventions in telephony and radio," and "through various radio-patent pool agreements in the 1920s . . . had effectively consolidated its position relative to the other giants in the industry." In so doing, according to an FCC investigation, AT&T had gained control of "the exploitation of potentially competitive and emerging forms of communication" and "pre-empt[ed] for itself new frontiers of technology for exploitation in the future. . . . "[3]

The radio-patent pools included AT&T, GE and Westinghouse, RCA (itself formed as a subsidiary of GE after the latter acquired American Marconi), and American Marconi.[4] Alfred Chandler's history of the origins of the consumer electronics industry is little more than an extended account of which patents were held, and subsequently acquired, by which companies.[5] This should give us some indication, by the way, of what he meant by "organizational capability," a term of his that will come under more scrutiny in the next chapter. In an age where the required capital outlays for actual physical plant and equipment are rapidly diminishing in many forms of manufacturing, one of the chief functions of "intellectual property" is to create artificial "comparative advantage" by giving a particular firm a monopoly on technologies and techniques, and prevent their diffusion throughout the market.

The American chemical industry, in its modern form, was made possible by the Justice Department's seizure of German chemical patents in WWI. Until the war, some 98% of patent applications in chemical industry came from German firms, and were never worked in the U.S. As a result the American chemical industry was technically second-rate, largely limited to final processing of intermediate goods imported from Germany. Attorney General A. Mitchell Palmer, as "Alien Property Custodian" during the war, held the patents in trust and licensed 735 of them to American firms; Du Pont alone received three hundred.[6]

More generally, patents are an effective tool for cartelizing markets in industry at large. They were used in the automobile and steel industries among others, according to Noble.[7] In a 1906 article, mechanical engineer and patent lawyer Edwin Prindle described patents as "the best and most effective means of controlling competition."

> Patents are the only legal form of absolute monopoly. In a recent court decision the court said, "within his domain, the patentee is czar. . . . cries of restraint of trade and impairment of the freedom of sales are unavailing, because for the promotion of the useful arts the constitution and statutes authorize this very monopoly."

[1]*Ibid.*, p. 12.
[2]*Ibid.*, p. 91.
[3]*Ibid.*, p. 92.
[4]*Ibid.*, pp. 93-94.
[5]Alfred Chandler, Jr., *Inventing the Electronic Century* (New York: The Free Press, 2001).
[6]Noble, *America by Design*, p. 16.
[7]*Ibid.*, p. 91.

> The power which a patentee has to dictate the conditions under which his monopoly may be exercised has been used to form trade agreements throughout practically entire industries, and if the purpose of the combination is primarily to secure benefit from the patent monopoly, the combination is legitimate. Under such combinations there can be effective agreements as to prices to be maintained . . . ; the output for each member of the combination can be specified and enforced . . . and many other benefits which were sought to be secured by trade combinations made by simple agreements can be added. Such trade combinations under patents are the only valid and enforceable trade combinations that can be made in the United States.[1]

And unlike purely private cartels, which tend toward defection and instability, patent control cartels—being based on a state-granted privilege—carry a credible and effective punishment for defection.

Through ttangible propertyheir "Napoleonic concept of industrial warfare, with inventions and patents as the soldiers of fortune," and through "the research arm of the 'patent offensive,'" manufacturing corporations were able to secure stable control of markets in their respective industries.[2]

These were the conditions present at the outset of the mass production revolution, in which the development of the corporate industrial economy began. In the absence of these necessary preconditions, there simply would not have been a single national market or large industrial corporations serving it. Rather than being adopted into the framework of the paleotechnic factory system, the introduction of electrical machinery would likely have followed its natural course and lived up to its unique potential: powered machinery would have been incorporated into small-scale production for local markets, and the national economy would have developed as "a hundred Emilia-Romagnas."

But these were only the necessary conditions at the outset. As we shall see in the next chapter, the growth of big government continued to parallel that of big business, introducing newer and larger-scale forms of political intervention to address the corporate economy's increasing tendencies toward destabilization, and to insulate the giant corporation from the market forces that would otherwise have destroyed it.

[1] *Ibid.*, p. 89.
[2] *Ibid.*, p. 95.

2

Moloch: The Sloanist Mass Production Model

Introduction

The mass-production model carried some strong imperatives: first, it required large-batch production, running the enormously expensive product-specific machinery at full capacity, to minimize unit costs (in Amory Lovins' words, "ever-faster once-through flow of materials from depletion to pollution"[1]); and second, it required social control and predictability to ensure that the output would be consumed, lest growing inventories and glutted markets cause the wheels of industry to stop turning. Utilize capacity, utilize capacity, that is Moses and the prophets. Here's Lewis Mumford on the principle:

> As mechanical methods have become more productive, the notion has grown up that consumption should become more voracious. In back of this lies an anxiety lest the productivity of the machine create a glut in the market. . . .

This threat is overcome by "the devices of competitive waste, of shoddy workmanship, and of fashion . . . "[2]

As described by Michael Piore and Charles Sabel, the problem was that product-specific resources could not be reallocated when the market shifted; under such conditions, the cost of market unpredictability was unacceptably high. Markets for the output of mass-production industry had to be guaranteed because highly specialized machinery could not be reallocated to other uses with changes in demand. "A piece of modern machinery dedicated to the production of a single part cannot be turned to another use, no matter how low the price of that part falls, or how high the price of other goods rises."[3]

> Mass production required large investments in highly specialized equipment and narrowly trained workers. In the language of manufacturing, these resources were "dedicated": suited to the manufacture of a particular product—often, in fact, to just one make or model. When the market for that particular product declined, the resources had no place to go. Mass production was therefore profitable only with markets that were large enough to absorb an enormous output of a single, standardized commodity, and stable enough to keep the re-

[1]Paul Hawken, Amory Lovins, and L. Hunter Lovins, *Natural Capitalism: Creating the Next Industrial Revolution* (Boston, New York, London: Little, Brown, and Company, 1999), p. 81

[2]Lewis Mumford, *Technics and Civilization* (New York: Harcourt, Brace, and Company, 1934), pp. 396-397.

[3]Michael J. Piore and Charles F. Sabel, *The Second Industrial Divide: Possibilities for Prosperity* (New York: HarperCollins, 1984), p. 50.

> sources involved in the production of that commodity continuously employed. Markets of this kind . . . did not occur naturally. They had to be created.[1]
>
> . . . . It became necessary for firms to organize the market so as to avoid fluctuations in demand and create a stable atmosphere for profitable, long-term investment.[2]
>
> . . . [There were] two consequences of the Americans' discovery that the profitability of investment in mass-production equipment depends on the stabilization of markets. The first of these consequences was the construction, from the 1870s to the 1920s, of giant corporations, which could balance demand and supply within their industries. The second consequence was the creation, two decades later, of a Keynesian system for matching production and consumption in the national economy as a whole.[3]

Ralph Borsodi argued that "[w]ith serial production, . . . man has ventured into a topsy-turvy world in which

> goods that wear out rapidly or that go out of style before they have a chance to be worn out seem more desirable than goods which are durable and endurable. Goods now have to be consumed quickly or discarded quickly so that the buying of goods to take their place will keep the factory busy.
>
> By the old system production was merely the means to an end.
>
> By the new system production itself has become the end.[4]
>
> With continuous operation of [the factory's] machinery, much larger quantities of its products must be sold to the public. The public buys normally only as fast as it consumes the product. The factory is therefore confronted by a dilemma; if it makes things well, its products will be consumed but slowly, while if it makes them poorly, its products will be consumed rapidly.
>
> It naturally makes its products as poorly as it dares.
>
> It encourages premature depreciation.[5]

(In a free market, of course, firms that made stuff well would have a competitive advantage. But in our unfree market, the state's subsidies to inefficiency cost, "intellectual property" laws, and other restraints on competition insulate firms from the full competitive disadvantage of offering inferior products.)

Because of the imperative for overcapitalized industry to operate at full capacity, on round-the-clock shifts, in order to spread the cost of its expensive machinery over the greatest possible number of units of output, the imperative of guaranteeing consumption of the output was equally great. As Benjamin Barber puts it, capitalism manufactures needs for the goods it's producing rather than producing goods in response to needs.[6]

This is not just a caricature by the enemies of Sloanist mass-production. It has been a constant theme of the model's most enthusiastic advocates and defenders. They disagree with economic decentralists, not on the systemic requirements of the mass-production model, but only on whether or not it has on the whole been a good thing, and whether there is any viable alternative.

In *The New Industrial State*, Galbraith wrote about the connection between capital intensiveness and the "technostructure's" need for predictability and control:

[1]*Ibid.*, p. 49.

[2]*Ibid.*, p. 54.

[3]*Ibid.*, p. 15.

[4]Ralph Borsodi, *This Ugly Civilization* (Philadelphia: Porcupine Press, 1929, 1975), pp. 64-65.

[5]*Ibid.*, p. 126.

[6]"Manufacture Goods, Not Needs," E. F. Schumacher Society Blog, October 11, 2009 <http://efssociety.blogspot.com/2009/10/manufacture-goods-not-needs_11.html>.

> . . . [Machines and sophisticated technology] require . . . heavy investment of capital. They are designed and guided by technically sophisticated men. They involve, also, a greatly increased lapse of time between any decision to produce and the emergence of a salable product.
>
> From these changes come the need and the opportunity for the large organization. It alone can deploy the requisite capital; it alone can mobilize the requisite skills. . . . The large commitment of capital and organization well in advance of result requires that there be foresight and also that all feasible steps be taken to insure that what is foreseen will transpire.[1]
>
> . . . From the time and capital that must be committed, the inflexibility of this commitment, the needs of large organization and the problems of market performance under conditions of advanced technology, comes the necessity for planning.[2]
>
> The need for planning . . . arises from the long period of time that elapses during the production process, the high investment that is involved and the inflexible commitment of that investment to the particular task.[3]
>
> Planning exists because [the market] process has ceased to be reliable. Technology, with its companion commitment of time and capital, means that the needs of the consumer must be anticipated—by months or years. . . . [I]n addition to deciding what the consumer will want and will pay, the firm must make every feasible step to see that what it decides to produce is wanted by the consumer at a remunerative price. . . . It must exercise control over what is sold. . . . It must replace the market with planning.[4]
>
> . . . The need to control consumer behavior is a requirement of planning. Planning, in turn, is made necessary by extensive use of advanced technology and capital and by the relative scale and complexity of organization. These produce goods efficiently; the result is a very large volume of production. As a further consequence, goods that are related only to elementary physical sensation—that merely prevent hunger, protect against cold, provide shelter, suppress pain—have come to comprise a small and diminishing part of all production. Most goods serve needs that are discovered to the individual not by the palpable discomfort that accompanies deprivation, but by some psychic response to their possession. . . .[5]

For Galbraith, the "accepted sequence" of consumer sovereignty (what Mises called "dollar democracy"), in which consumer demand determines what is produced, was replaced by a "revised sequence" in which oligopoly corporations determine what is produced and then dispose of it by managing consumer behavior. In contemporary terms, the demand-pull economy is replaced by a supply-push model.

Alfred Chandler, like Galbraith, was thoroughly sold on the greater efficiencies of the large corporation. He argued that the modern multi-unit enterprise arose when administrative coordination "permitted" greater efficiencies.[6]

> By linking the administration of producing units with buying and distributing units, costs for information on markets and sources of supply were reduced. Of much greater significance, the internalization of many units permitted the flow of goods from one unit to another to be administratively coordinated. More

[1]John Kenneth Galbraith, *The New Industrial State* (New York: Signet Books, 1967), p. 16

[2]*Ibid.*, p. 28.

[3]*Ibid.*, p. 31.

[4]*Ibid.*, pp. 34-35.

[5]*Ibid.*, pp. 210-212.

[6]Alfred D.Chandler, Jr., *The Visible Hand: The Managerial Revolution in American Business* (Cambridge and London: The Belknap Press of Harvard University Press, 1977), p. 6.

> effective scheduling of flows achieved a more intensive use of facilities and personnel employed in the processes of production and so increased productivity and reduced costs.[1]
>
> Organizationally, output was expanded through improved design of manufacturing or processing plants and by innovations in managerial practices and procedures required to synchronize flaws and supervise the work force. Increases in productivity also depend on the skills and abilities of the managers and the workers and the continuing improvement of their skills over time. Each of these factors or any combination of them helped to increase the speed and volume of the flow, or what some processors call the "throughput," of materials within a single plant or works. . . .[2]
>
> Integration of mass production with mass distribution afforded an opportunity for manufacturers to lower costs and increase productivity through more effective administration of the processes of production and distribution and coordination of the flow of goods through them. Yet the first industrialists to integrate the two basic sets of processes did not do so to exploit such economies. They did so because existing marketers were unable to sell and distribute products in the volume they were produced.[3]

The mass-production factory achieved "economies of speed" from "greatly increasing the daily use of equipment and personnel."[4] (Of course, Chandler starts by assuming the greater inherent efficiency of capital-intensive modes of production, which *then* require "economies of speed" to reduce unit costs from the expensive capital assets).

What Chandler meant by "economies of speed" was entirely different from lean production's understanding of flow. Chandler's meaning is suggested by his celebration of the new corporate managers who "developed techniques to purchase, store, and move huge stocks of raw and semifinished materials. In order to maintain a more certain flow of goods, they often operated fleets of railroad cars and transportation equipment."[5] In other words, both the standard Sloanist model of enormous buffer stocks of unfinished goods, and warehouses full of finished goods awaiting orders—and the faux "lean" model in which inventory is swept under the rug and moved into warehouses on wheels and in container-ships.

(The reader may be puzzled or even annoyed by my constant use of the term "Sloanism." I got it from the insightful commentary of Eric Husman at *GrimReader* blog, in which he treats the production and accounting methods of General Motors as paradigmatic of 20th century American mass-production industry, and contrasts them with the lean methods popularly identified with Taichi Ohno's Toyota production system.)

"Sloanism" refers, in particular, to the management accounting system identified with General Motors. It was first developed by Brown at DuPont, and brought to GM when DuPont acquired a controlling share of the company and put Alfred Sloan in charge. Brown's management accounting system, whose perverse incentives are dissected in detail by William Waddell and Norman Bodek in *Rebirth of American Industry*, became the prevailing standard throughout American corporate management.

In Sloanist management accounting, inventory is counted as an asset "with the same liquidity as cash." Regardless of whether a current output is needed to fill an order, the producing department sends it to inventory and is credited for it.

[1] *Ibid.*, pp. 6-7.
[2] *Ibid.*, p. 241.
[3] *Ibid.*, p. 287.
[4] *Ibid.*, p. 244.
[5] *Ibid.*, p. 412.

Under the practice of "overhead absorption," all production costs are fully incorporated into the price of goods "sold" to inventory, at which point they count as an asset on the balance sheet.

> With inventory declared to be an asset with the same liquidity as cash, it did not really matter whether the next 'cost center,' department, plant, or division actually needed the output right away in order to consummate one of these paper sales. The producing department put the output into inventory and took credit.[1]
>
> . . . Expenses go down . . . , while inventory goes up, simply by moving a skid full of material a few operations down the stream. In fact, expenses can go down and ROI can improve even when the plant pays an overtime premium to work on material that is not needed; or if the plant uses defective material in production and a large percentage of the output from production must be scrapped.[2]

In other words, by the Sloanist accounting principles predominant in American industry, the expenditure of money on inputs is by definition the creation of value. As Waddell described it at his blog,

> companies can make a bunch of stuff, assign huge buckets of fixed overhead to it and move those overheads over to the balance sheet, making themselves look more profitable.

In other words, "they accept cost as a fait accompli. . . . " Paul Goodman's idea of the culture of cost-plus (about which more below) sums it up perfectly. And as Waddell points out, the GDP as a metric depends on the same assumptions as the management accounting system used by American industry: it counts expenditure on inputs, by definition, as the creation of wealth.[3]

American factories frequently have warehouses filled with millions of dollars worth of obsolete inventory, which is still there "to avoid having to reduce profits this quarter by writing it off." When the corporation finally does have to adjust to reality, the result is costly write-downs of inventory.

> It did not take much of a mathematician to figure out that, if all you really care about is the cost of performing one operation to a part, and you were allowed to make money by doing that single operation as cheaply as possible and then calling the partially complete product an asset, it would be cheaper to make them a bunch at a time.
>
> It stood to reason that spreading set-up costs over many parts was cheaper than having to set-up for just a few even if it meant making more parts than you needed for a long time. It also made sense, if you could make enough parts all at once, to just make them cheaply, and then sort out the bad ones later.
>
> Across the board, batches became the norm because the direct cost of batches was cheap and they could be immediately turned into money—at least as far as Mr. DuPont was concerned—by classifying them as work-in-process inventory.[4]

[1]William H. Waddell and Norman Bodek, *Rebirth of American Industry: A Study of Lean Management* (Vancouver, WA: PCS Press, 2005), p. 75.

[2]*Ibid.*, p. 140.

[3]William Waddell, "The Irrelevance of the Economists," *Evolving Excellence*, May 6, 2009 <http://www.evolvingexcellence.com/blog/2009/05/the-irrelevance-of-the-economists.html>. Paul T. Kidd anticipated much of Waddell's and Bodek's criticism in *Agile Manufacturing: Forging New Frontiers* (Wokingham, England; Reading, Mass.; Menlo Park, Calif.; New York; Don Mills, Ontario; Amsterdam; Bonn; Sydney; Singapore; Tokyo; Madrid; San Juan; Paris; Mexico City; Seoul; Taipei: Addison-Wesley Publishing Company, 1994), especially Chapter Four.

[4]Waddell and Bodek, p. 98.

And the effect of these inventories on cost is enormous. In the garment industry, making to forecast rather than to order, and maintaining large enough inventory to avoid idle machines, is estimated to account for some 25% of retail price.[1] That means your clothes cost about a third more because of the "efficiencies" of Sloanist mass production.

Under the Sloan system, if a machine can be run at a certain speed, it *must* be run at that speed to maximize efficiency. And the only way to increase efficiency is to increase the speed at which individual machines can be run.[2] The Sloan system focuses, exclusively, on labor savings "perceived to be attainable only through faster machines. Never mind that faster machines build inventory faster, as well."[3]

The incredible bureaucratic inefficiencies resulting from these inventories is suggested by GM's "brilliant innovation" of MRP software in the 1960s—a central planning system that surely would have made the folks at Gosplan green with envy. Of course, as Toyota Production System father Taichi Ohno pointed out, MRP would be useless to a company operating on zero lead time and lot sizes of one.[4] The point of MRP is that it "allows each cost center to operate at its individual optimum without regard to the performance of the other cost centers."

> If the machining department is having a good week, that supervisor can claim credit for his production—perhaps even exceeding the schedule.
>
> It does not affect him at all that the next department upstream—assembly, for example—is having major problems and will not come close to making schedule. . . .
>
> . . . [MRP's] core is the logic and a set of algorithms to eanble each component of a product to be produced at different volumes and speeds; and, in fact, the same components of a product going through different operations to be produced at different volumes and speeds, in order to optimum efficiency at each operation. It is based on the assumption that manufacturing is best performed in such a disjointed manner, and it assures adequate inventory to buffer all of this unbalanced production.[5]

The lean approach has its own "economies of speed," but they are the direct opposite of the Sloanist approach. The Sloanist approach focuses on maximizing economies of speed in terms of the unit cost of a particular machine, without regard to the inventories of unfinished goods that must accumulate as buffer stocks as a result, and all the other enormous eddies in the flow of production. As the authors of *Natural Capitalism* put it, it attempts to optimize each step of the production process in isolation, "thereby pessimizing the entire system." A machine can reduce the labor cost of one step by running at enormous speeds, and yet be out of sync with the overall process.[6] Waddell and Bodek give the example of Ernie Breech, sent from GM to "save" Ford, demanding a plant manager tell him the cost of manufacturing the steering wheel so he could calculate ROI for that step of the process. The plant manager was at a loss trying to figure out what Breech wanted: did he think steering wheel production was a bottleneck in production flow, or what? But for Breech, if the unit cost of that machine and the direct cost of the labor working it were low enough compared to the "value" of the steering wheels

[1] Raphael Kaplinsky, "From Mass Production to Flexible Specialization: A Case Study of Microeconomic Change in a Semi-Industrialized Economy," *World Development* 22:3 (March 1994), p. 346.

[2] Waddell and Bodek, p. 122.

[3] *Ibid.*, p. 119.

[4] *Ibid.*, p. xx.

[5] *Ibid.*, pp. 112-114.

[6] Lovins et al, *Natural Capitalism*, pp. 129-30.

"sold" to inventory, that was all that mattered. Under the Sloan accounting system, producing a steering wheel—even in isolation, and regardless of what was done with it or whether there was an order for the car it was a part of—was a money-making proposition. "Credit for that work—it looks like a payment on the manufacturing budget—is given for performing that simple task because it moves money from expenses to assets.[1]

"Selling to inventory," under standard management accounting rules, is equivalent to the incentive systems for production under a Five-Year Plan: there is no incentive to produce goods that will actually work or be consumed. Hence the carloads of refrigerators, for which Soviet factories were credited toward their 5YP quotas, thrown off trains with no regard to whether they were damaged beyond repair in the process.

The lean approach, in contrast, gears production flow to orders, and then sizes individual machines and steps in the production process to the volume of overall flow. Under lean thinking, it's better to have a less specialized machine with a lower rate of output, in order to avoid an individual step out of proportion to the overall production flow. This is what the Toyota Production System calls *takt*: pacing the output of each stage of production to meet the needs of the next stage, and pacing the overall flow of all the stages in accordance with current orders.[2] In a Sloan factory, the management would select machinery to produce the entire production run "as fast as they humanly could, then sort out the pieces and put things together later."[3]

To quote the authors of *Natural Capitalism* again: "The essence of the lean approach is that in almost all modern manufacturing,

> the combined and often synergistic benefits of the lower capital investment, greater flexibility, often higher reliability, lower inventory cost, and lower shipping cost of much smaller and more localized production equipment will far outweigh any modest decreases in its narrowly defined "efficiency" per process step. It's more efficient overall, in resources and time and money, to scale production properly, using flexible machines that can quickly shift between products. By doing so, all the different processing steps can be carred out immediately adjacent to one another with the product kept in continuous flow. The goal is to have no stops, no delays, no backflows, no inventories, no expediting, no bottlenecks, no buffer stocks, and no *muda* [waste].[4]

The contrast is illustrated by a couple of examples from *Natural Capitalism*: an overly "efficient" grinding machine at Pratt & Whitney, and a cola bottling machine likewise oversized in relation to its task:

> The world's largest maker of jet engines for aircraft had paid $80 million for a "monument"—state-of-the-art German robotic grinders to make turbine blades. The grinders were wonderfully fast, but their complex computer controls required about as many technicians as the old manual production system had required machinists. Moreover, the fast grinders required supporting processes that were costly and polluting. Since the fast grinders were meant to produce big, uniform batches of product, but Pratt & Whitney needed agile production of small, diverse batches, the twelve fancy grinders were replaced with eight simple ones costing one-fourth as much. Grinding time increased from 3 to 75 minutes, but the throughput time for the entire process decreased from 10 days to 75 minutes because the nasty supporting processes were eliminated. Viewed from the

[1]Waddell and Bodek, pp. 89, 92.
[2]*Ibid.*, pp. 122-123.
[3]*Ibid.*, p. 39.
[4]Hawken et al, pp. 129-130.

> whole-system perspective of the complete production process, not just the grinding step, the big machines had been so fast that they slowed down the process too much, and so automated that they required too many workers. The revised production system, using a high-wage traditional workforce and simple machines, produced $1 billion of annual value in a single room easily surveyable from a doorway. It cost half as much, worked 100 times faster, cut changeover time from 8 hours to 100 seconds, and would have repaid its conversion costs in a year even if the sophisticated grinders were simply scrapped.[1]

In the cola industry, the problem is "the mismatch between a very small-scale operation—drinking a can of cola—and a very large-scale one, producing it." The most "efficient" large-scale bottling machine creates enormous batches that are out of scale with the distribution system, and result in higher unit costs overall than would modest-sized local machines that could immediately scale production to demand-pull. The reason is the excess inventories that glut the system, and the "pervasive costs and losses of handling, transport, and storage between all the elephantine parts of the production process." As a result, "the giant cola-canning machine may well cost *more* per delivered can than a small, slow, unsophisticated machine that produces the cans of cola locally and immediately on receiving an order from the retailer."[2]

As Womack and Jones put it in *Lean Thinking*, "machines rapidly making unwanted parts during one hundred percent of their available hours and employees earnestly performing unneeded tasks during every available minute are only producing *muda*."[3] Lovins et al sum it up more broadly:

> Their basic conclusion, from scores of practical case studies, is that specialized, large-scale, high-speed, highly production departments and equipment are the key to *in*efficiency and *un*competitiveness, and that maximizing the utilization of productive capacity, the pride of MBAs, is nearly always a mistake.[4]

Rather, it's better to scale productive capacity to demand.

In a genuine lean factory, managers are hounded in daily meetings about meeting the numbers for inventory reduction and reduction of cycle time, in the same way that they're hounded on a daily basis to reduce direct labor hours and increase ROI in a Sloanist factory (including the American experiments with "lean production" in firms still governed by Donaldson Brown's accounting principles). James Womack et al., in *The Machine That Changed the World*, recount an amusing anecdote about a delegation of lean production students from Corporate America touring a Toyota plant. Reading a question on their survey form as to how many days of inventory were in the plant, the Toyota manager politely asked whether the translator could have meant *minutes* of inventory.[5]

As Mumford put it, "Measured by effective work, that is, human effort transformed into direct subsistence or into durable works of art and technics, the relative gains of the new industry were pitifully small."[6] The amount of wasted resources and crystallized labor embodied in the enormous warehouses of Sloanist factories and the enormous stocks of goods in process, the mushrooming cost of

[1] *Ibid.*, pp. 128-129.

[2] *Ibid.*, p. 129.

[3] James P. Womack and Daniel T. Jones, *Lean Thinking: Banish Waste and Create Wealth in Your Corporation* (New York: Simon and Schuster, 1996), p. 60.

[4] Lovins et al., *Natural Capitalism*, p. 127.

[5] James P. Womack, Daniel T. Jones, Daniel Roos, *The Machine That Changed the World* (New York: Macmillian Publishing Company, 1990), p. 80.

[6] Mumford, *Technics and Civilization*, p. 196.

marketing, the "warehouses on wheels," and the mountains of discarded goods in the landfills that could have been repaired for a tiny fraction of the cost of replacing them, easily outweigh the savings in unit costs from mass production itself. As Michael Parenti put it, the essence of corporate capitalism is "the transformation of living nature into mountains of commodities and commodities into heaps of dead capital."[1] The cost savings from mass production are more than offset by the costs of mass distribution.

Chandler's model of production resulted in the adoption of increasingly specialized, asset-specific production machinery:

> The large industrial enterprise continued to flourish when it used capital-intensive, energy-consuming, continuous or large-batch production technology to produce for mass markets.[2]
>
> The ratio of capital to labor, materials to labor, energy to labor, and managers to labor for each unit of output became higher. Such high-volume industries soon became capital-intensive, energy-intensive, and manager-intensive.[3]

Of course this view is fundamentally wrong-headed. To regard a particular machine as "more efficient" based on its unit costs taken in isolation is sheer idiocy. If the costs of idle capacity are so great as to elevate unit costs above those of less specialized machinery, at the levels of spontaneous demand occurring without push marketing, and if the market area required for full utilization of capacity results in distribution costs greater than the unit cost savings from specialized machinery, then the expensive product-specific machinery is, in fact, *less* efficient. The basic principle was stated by F. M. Scherer:

> Ball bearing manufacturing provides a good illustration of several *product-specific* economies. If only a few bearings are to be custom-made, the ring machining will be done on general-purpose lathes by a skilled operator who hand-positions the stock and tools and makes measurements for each cut. With this method, machining a single ring requires from five minutes to more than an hour, depending on the part's size and complexity and the operator's skill. If a sizable batch is to be produced, a more specialized automatic screw machine will be used instead. Once it is loaded with a steel tube, it automatically feeds the tube, sets the tools and adjusts its speed to make the necessary cuts, and spits out machined parts into a hopper at a rate of from eighty to one hundred forty parts per hour. A substantial saving of machine running and operator attendance time per unit is achieved, but setting up the screw machine to perform these operations takes about eight hours. If only one hundred bearing rings are to be made, setup time greatly exceeds total running time, and it may be cheaper to do the job on an ordinary lathe.[4]

The Sloanist approach is to choose the specialized automatic machine and find a way to make people buy more bearing rings.

Galbraith and Chandler write as though the adoption of the machinery were enough to automatically increase efficiency, in and of itself, regardless of how much money had to be spent elsewhere to "save" that money.

But if we approach things from the opposite direction, we can see that flexible manufacturing with easily redeployable assets makes it feasible to shift quickly

[1]Michael Parenti, "Capitalism's Self-Inflicted Apocalypse," *Common Dreams*, January 21, 2009 <http://www.commondreams.org/view/2009/01/20-9>.

[2]Mumford, *Technics and Civilization*, p. 347.

[3]*Ibid.*, p. 241.

[4]F.M. Scherer and David Ross, *Industrial Market Structure and Economic Performance*. 3rd ed (Boston: Houghton Mifflin, 1990), p. 97.

from product to product in the face of changing demand, and thus eliminates the imperative of controlling the market. As Barry Stein said,

> if firms could respond to local conditions, they would not *need* to control them. If they must control markets, then it is a reflection of their lack of ability to be adequately responsive.[1]
>
> . . . Consumer needs, if they are to be supplied efficiently, call increasingly for organizations that are more flexibly arranged and in more direct contact with those customers. The essence of planning, under conditions of increasing uncertainty, is to seek better ways for those who have the needs to influence or control the productive apparatus more effectively, not less.
>
> Under conditions of rapid environmental change, implementing such planning is possible only if the "distance" between those supplied and the locus of decision-making on the part of those producing is reduced. . . . But it can be shown easily in information theory that the feedback—information linking the environment and the organization attempting to service that environment—necessarily becomes less accurate *or* less complete as the rate of change of data increases, *or* as the number of steps in the information transfer process continues.

Stein suggested that Galbraith's solution was to suppress the turbulence: "to control the changes, in kind and extent, that the society will undergo."[2] But far better, he argues, would be "a value shift that integrates the organization and the environment it serves."

> This problem is to be solved not by the hope of better planning on a large scale . . . , but by the better integration of productive enterprises with the elements of society needing that production.
>
> Under conditions of rapid change in an affluent and complex society, the only means available for meeting differentiated and fluid needs is an array of producing units small enough to be in close contact with *their* customers, flexible enough to produce for *their* demands, and able to do so in a relatively short time. . . . It is a contradiction in terms to speak of the necessity for units large enough to control their environment, but producing products which in fact no one may want![3]
>
> As to the problem of planning—large firms are said to be needed here because the requirements of sophisticated technology and increasingly specialized knowledge call for long lead times to develop, design, and produce products. Firms must therefore have enough control over the market to assure that the demand needed to justify that time-consuming and costly investment will exist. This argument rests on a foundation of sand; first, because the needs of society should *precede*, not follow, decisions about what to produce, and second, because the data do not substantiate the need for large production organizations except in rare and unusual instances, like space flight. On the contrary, planning for social needs requires organizations and decision-making capabilities in which the feedback and interplay between productive enterprises and the market in question is accurate and timely—conditions more consistent with smaller organizations than large ones.[4]

[1]Barry Stein, *Size, Efficiency, and Community Enterprise* (Cambridge: Center for Community Economic Development, 1974), p. 41.

[2]*Ibid.*, p. 43.

[3]*Ibid.*, p. 44.

[4]*Ibid.*, p. 58.

A. Institutional Forms to Provide Stability

In keeping with the need for stability and control Galbraith described above, the technostructure resorted to organizational expedients within the corporate enterprise to guarantee reliable outlets for production and provide long-term predictability in the availability and price of inputs. These expedients can be summed up as replacing the market price mechanism with planning.

> A firm cannot usefully foresee and schedule future action or prepare for contingencies if it does not know what its prices will be, what its sales will be, what its costs including labor and capital costs will be and what will be available at these costs. . . . Much of what the firm regards as planning consists in minimizing or getting rid of market influences.[1]

There's a reason for twentieth century liberalism's strong affinity for mass-production industry (e.g. Michael Moore's nostalgia for the consensus capitalism of the '50s, when the predominant mode of employment was a factory job with lifetime security). Twentieth century liberalism had its origins as the ideology of the managerial and professional classes, particularly the managers and engineers who ran the giant manufacturing corporations. And the centerpiece of their ideology was to extend to society outside the corporation the same planning and control, the same government by disinterested experts, that prevailed inside it. And this ideological affinity for social planning dovetailed exactly with mass-production industry's need to reshape society as a whole to guarantee consumption of its output.[2]

Galbraith describes three institutional expedients taken by the technostructure to control the uncertainties of the market and permit long-term predictability: vertical integration, the use of market power to control suppliers and outlets, and long-term contractual arrangements with suppliers and outlets.[3]

In vertical integration, "[t]he planning unit takes over the source of supply or the outlet; a transaction that is subject to bargaining over prices and amounts is thus replaced with a transfer within the planning unit."[4]

One of the most important forms of "vertical integration" is the choice to "make" rather than "buy" credit—replacing the external credit markets with internal finance through retained earnings.[5] The theory that management is controlled by outside capital markets assumes a high degree of dependence on outside finance. But in fact management's first line of defense, in maintaining its autonomy from shareholders and other outside interests, is to *minimize* its reliance on outside finance. Management tends to finance new investments as much as possible with retained earnings, followed by debt, with new issues of shares only as a last resort.[6] Issues of stock are important sources of investment capital only for start-ups and small firms undertaking major expansions.[7] Most corporations finance a majority of their new investment from retained earnings, and tend to limit invest-

[1]Galbraith, *The New Industrial State*, p. 37.

[2]See Kevin Carson, *Organization Theory: A Libertarian Perspective* (Booksurge, 2008), Chapter Four.

[3]Galbraith, *New Industrial State*, p. 38.

[4]*Ibid.*, p. 39.

[5]*Ibid.*, pp. 50-51.

[6]Martin Hellwig, "On the Economics and Politics of Corporate Finance and Corporate Control," in Xavier Vives, ed., *Corporate Governance: Theoretical and Empirical Perspectives* (Cambridge: Cambridge University Press, 2000),pp. 100-101.

[7]Ralph Estes, *Tyranny of the Bottom Line: Why Corporations Make Good People Do Bad Things* (San Francisco: Berrett-Koehler Publishers, 1996), p. 51.

ment to the highest priorities when retained earnings are scarce.[1] As Doug Henwood says, in the long run "almost all corporate capital expenditures are internally financed, through profits and depreciation allowances." Between 1952 and 1995, almost 90% of investment was funded from retained earnings.[2]

The prevailing reliance on internal financing tends to promote concentration. Internally generated funds that exceed internal requirements are used to expand or diversify internal operations, or for horizontal and vertical integration, rather than "lending it or making other kinds of arm's-length investments."[3] Martin Hellwig, in his discussion of the primacy of finance by retained earnings, makes one especially intriguing observation, in particular. He denies that reliance primarily on retained earnings necessarily leads to a "rationing" of investment, in the sense of underinvestment; internal financing, he says, can just as easily result in overinvestment, if the amount of retained earnings exceeds available opportunities for rational capital investment.[4] This confirms Schumpeter's argument that double taxation of corporate profits promoted excessive size and centralization, by encouraging reinvestment in preference to the issue of dividends. Of course it may result in structural misallocations and irrationality, to the extent that retention of earnings prevents dividends from returning to the household sector to be invested in other firms, so that overaccumulation in the sectors with excessive retained earnings comes at the expense of a capital shortage in other sectors.[5] Doug Henwood contrasts the glut of retained earnings, under the control of corporate bureaucracies with a shortage of investment opportunities, to the constraints the capital markets place on small, innovative firms that need capital the most.[6]

Market control "consists in reducing or eliminating the independence of action of those to whom the planning unit sells or from whom it buys," while preserving "the outward form of the market." Market power follows from large size in relation to the market. A decision to buy or not to buy, as in the case of General Motors and its suppliers, can determine the life or death of a firm. What's more, large manufacturers always have the option of vertical integration—making a part themselves instead of buying it—to discipline suppliers. "The option of eliminating a market is an important source of power for controlling it."[7]

Long-term contracting can reduce uncertainty by "specifying prices and amounts to be provided or bought for substantial periods of time." Each large firm creates a "matrix of contracts" in which market uncertainty is eliminated.[8]

Piore and Sabel mention Edison Electric as an example of using long-term contracts to guarantee stability,

> inducing its customers to sign long-term "future delivery" contracts, under which they had to buy specified quantities of Edison products at regular intervals over ten years. By assuring the demand for output, these contracts enabled the company to invest in large plants. . . . As one Edison executive explained:
>
> *It is essential in order to make lamps at a minimum cost that the factory should be run constantly at as uniform an output as possible. Our future delivery plan in lamps has been very successful* [in this regard]. . . . *It is very expensive*

[1]Hellwig, pp. 101-102, 113.

[2]Doug Henwood, *Wall Street: How it Works and for Whom* (London and New York: Verso, 1997), p. 3.

[3]Piore and Sabel, pp. 70-71.

[4]Hellwig, pp. 114-115.

[5]*Ibid.*, p. 117.

[6]Henwood, *Wall Street*, pp. 154-155.

[7]Galbraith, *The New Industrial State*, pp. 39-40.

[8]*Ibid.*, pp. 41-42.

> *work changing from one rate of production to another in factories. . . . The benefit of the future delivery plan is apparent since we can manufacture to stock knowing that all the stock is to be taken within a certain time.*[1]

Unlike lean, demand-pull production, which minimizes inventory costs by producing only in response to orders, mass production requires supply-push distribution (guaranteeing a market before production takes place).

The use of contracts to stabilize input availability and price is exemplified, in particular, by the organizational expedients to stabilize wages and reduce labor turnover. After mixed success with a variety of experiments with company unions, the "American Plan," and other forms of welfare capitalism, employers finally turned to the official organized labor regime under the Wagner Act to establish long-term predictability in the supply and price of labor inputs, and to secure management's control of production. Under the terms of "consensus capitalism," the comparatively small profile of labor costs in the total cost package of capital-intensive industry meant that management was willing to pay comparatively high wages and benefits (up to the point of gearing wages to productivity), to provide more or less neutral grievance procedures, etc., so long as management's prerogatives were recognized for directing production. But the same had been true in many cases of the American Plan: it allowed for formalized grievance procedures and progressive discipline, and in some cases negotiation over rates of pay. The common goal of all these various attempts, however much they disagreed in their particulars, was "by stabilizing wages and employment, to insulate the cost of a major element of production from the flux of a market economy."[2] From management's perspective, the sort of bureaucratized industrial union established under Wagner had the primary purposes of enforcing contracts on the rank and file and suppressing wildcat strikes. The corporate liberal managers who were most open to industrial unionism in the 1930s were, in many cases, the same people who had previously relied on company unions and works councils. Their motivation, in both cases, was the same. For example, GE's Gerard Swope, one of the most "progressive" of corporate liberals and the living personification of the kinds of corporate interests that backed FDR, had attempted in 1926 to get the AFL's William Green to run GE's works council system.[3]

Another institutional expedient of Galbraith's technostructure is to regulate the pace of technical change, with the oligopoly firms in an industry colluding to introduce innovation at a rate that maximizes returns. Baran and Sweezy described the regulation of technical change, as it occurs in oligopoly markets under corporate capitalism:

> Here innovations are typically introduced (or soon taken over) by giant corporations which act not under the compulsion of competitive pressures but in accordance with careful calculations of the profit-maximizing course. Whereas in the competitive case no one, not even the innovating firms themselves, can control the rate at which new technologies are generally adopted, this ceases to be true in the monopolistic case. It is clear that the giant corporation will be guided not by the profitability of the new method considered in isolation, but by the net effect of the new method on the overall profitability of the firm. And this means

[1]Piore and Sabel, p. 58.
[2]*Ibid.*, p. 65.
[3]*Ibid.*, p. 132.

> that in general there will be a slower rate of introduction of innovation than under competitive criteria.[1]

Or as Paul Goodman put it, a handful of manufacturers control the market, "competing with fixed prices and slowly spooned-out improvements."[2]

Besides these microeconomic structures created by the nominally private corporation to provide stability, the state engaged in the policies described by Gabriel Kolko as "political capitalism."

> *Political capitalism* is the utilization of political outlets to attain conditions of stability, predictability, and security—to attain rationalization—in the economy. *Stability* is the elimination of internecine competition and erratic fluctuations in the economy. *Predictability* is the ability, on the basis of politically stabilized and secured means, to plan future economic action on the basis of fairly calculable expectations. By *security* I mean protection from the political attacks latent in any formally democratic political structure. I do not give to *rationalization* its frequent definition as the improvement of efficiency, output, or internal organization of a company; I mean by the term, rather, the organization of the economy and the larger political and social spheres in a manner that will allow corporations to function in a predictable and secure environment permitting reasonable profits over the long run.[3]

The state played a major role in cartelizing the economy, to protect the large corporation from the destructive effects of price competition. At first the effort was mainly private, reflected in the trust movement at the turn of the 20th century. Chandler celebrated the first, private efforts toward consolidation of markets as a step toward rationality:

> American manufacturers began in the 1870s to take the initial step to growth by way of merger—that is, to set up nationwide associations to control price and production. They did so primarily as a response to the continuing price decline, which became increasingly impressive after the panic of 1873 ushered in a prolonged economic depression.[4]

The process was further accelerated by the Depression of the 1890s, with mergers and trusts being formed through the beginning of the next century in order to control price and output: "the motive for merger changed. Many more were created to replace the association of small manufacturing firms as the instrument to maintain price and production schedules."[5]

From the turn of the twentieth century on, there was a series of attempts by J.P. Morgan and other promoters to create some institutional structure for the corporate economy by which price competition could be regulated and their respective market shares stabilized. "It was then," Paul Sweezy wrote,

> that U.S. businessmen learned the self-defeating nature of price-cutting as a competitive weapon and started the process of banning it through a complex

[1]Paul Baran and Paul Sweezy, *Monopoly Capitalism: An Essay in the American Economic and So-*
cial Order (New York: Monthly Review Press, 1966), pp. 93-94.

[2]Paul Goodman, *People or Personnel*, in *People or Personnel* and *Like a Conquered Province* (New York: Vintage Books, 1964, 1966), p. 58.

[3]Gabriel Kolko. *The Triumph of Conservatism: A Reinterpretation of American History 1900-1916* (New York: The Free Press of Glencoe, 1963), p. 3.

[4]Chandler, The Visible Hand, p. 316.

[5]*Ibid.*, p. 331.

network of laws (corporate and regulatory), institutions (e.g., trade associations), and conventions (e.g., price leadership) from normal business practice.[1]

Chandler's celebratory account of the trust movement, as a progressive force, ignores one central fact: the trusts were less efficient than their smaller competitors. They immediately began losing market share to less leveraged firms outside the trusts. The trust movement was an unqualified failure, as big business quickly recognized. Subsequent attempts to cartelize the economy, therefore, enlisted the state. As recounted by Gabriel Kolko,[2] the main force behind the Progressive Era regulatory agenda was big business itself, the goal being to restrict price and quality competition and to reestablish the trusts under the aegis of government. His thesis was that, "contrary to the consensus of historians, it was not the existence of monopoly that caused the federal government to intervene in the economy, but the lack of it."

Merely private attempts at cartelization (i.e., collusive price stabilization) before the Progressive Era—namely the so-called "trusts"—were miserable failures, according to Kolko. The dominant trend at the turn of the century—despite the effects of tariffs, patents, railroad subsidies, and other existing forms of statism—was competition. The trust movement was an attempt to cartelize the economy through such voluntary and private means as mergers, acquisitions, and price collusion. But the over-leveraged and over-capitalized trusts were even less efficient than before, and steadily lost market share to their smaller, more efficient competitors. Standard Oil and U.S. Steel, immediately after their formation, began to lose market share.

In the face of this resounding failure, big business acted through the state to cartelize itself—hence, the Progressive regulatory agenda.

> Ironically, contrary to the consensus of historians, it was not the existence of monopoly that caused the federal government to intervene in the economy, but the lack of it."[3]
>
> If economic rationalization could not be attained by mergers and voluntary economic methods, a growing number of important businessmen reasoned, perhaps political means might succeed."[4]

The rationale of the Progressive Era regulatory state was stated in 1908 by George Perkins, whom Kolko described as "the functional architect . . . of political capitalism during Roosevelt's presidency. . . . " The modern corporation

> must welcome federal supervision, administered by practical businessmen, that "should say to stockholders and the public from time to time that the management's reports and methods of business are correct." With federal regulation, which would free business from the many states, industrial cooperation could replace competition.[5]

Kolko provided considerable evidence that the main force behind the Progressive Era legislative agenda was big business. The Meat Inspection Act, for instance, was passed primarily at the behest of the big meat packers.[6] This pattern was re-

[1]Paul Sweezy. "Competition and Monopoly," *Monthly Review* (May 1981), pp. 1-16.

[2]Kolko, *Triumph of Conservatism*.

[3]*Ibid.*, p. 5.

[4]*Ibid.*, p. 58.

[5]*Ibid.*, p. 129.

[6]*Ibid.*, pp. 98-108. In the 1880s, repeated scandals involving tainted meat had resulted in U.S. firms being shut out of several European markets. The big packers had turned to the government to inspect exported meat. By organizing this function jointly, through the state, they removed quality inspection as a competitive issue between them, and the gov-

peated, in its essential form, in virtually every component of the "Progressive" regulatory agenda.

The various safety and quality regulations introduced during this period also worked to cartelize the market. They served essentially the same purpose as attempts in the Wilson war economy to reduce the variety of styles and features available in product lines, in the name of "efficiency." Any action by the state to impose a uniform standard of quality (e.g. safety), across the board, necessarily eliminates that feature as a competitive issue between firms. As Butler Shaffer put it, the purpose of "wage, working condition, or product standards" is to "universalize cost factors and thus restrict price competition."[1] Thus, the industry is partially cartelized, to the very same extent that would have happened had all the firms in it adopted a uniform quality standard, and agreed to stop competing in that area. A regulation, in essence, is a state-enforced cartel in which the members agree to cease competing in a particular area of quality or safety, and instead agree on a uniform standard which they establish through the state. And unlike private cartels, which are unstable, no member can seek an advantage by defecting.

Although theoretically the regulations might simply put a floor on quality competition and leave firms free to compete by exceeding the standard, in practice corporations often take a harsh view of competitors that exceed regulatory safety or quality requirements. A good example is Monsanto's (often successful) attempts to secure regulatory suppression of commercial speech by competitors who label their milk rBGH-free; more generally, the frankenfoods industry relies on FDA regulations to prohibit the labeling of food as GMO-free. Another example is the beef industry's success at getting the government to prohibit competitors from voluntarily testing their cattle for mad cow disease more frequently than required by law.[2] So the regulatory floor frequently becomes a ceiling.

More importantly, the FTC and Clayton Acts reversed the long trend toward competition and loss of market share and made stability possible.

> The provisions of the new laws attacking unfair competitors and price discrimination meant that the government would now make it possible for many trade associations to stabilize, for the first time, prices within their industries, and to make effective oligopoly a new phase of the economy.[3]

The Federal Trade Commission created a hospitable atmosphere for trade associations and their efforts to prevent price cutting.[4] Butler Shaffer, in *In Restraint*

ernment provided a seal of approval in much the same way a trade association would. The problem with this early inspection regime was that only the largest packers were involved in the export trade, which gave a competitive advantage to the small firms that supplied only the domestic market. The main effect of Roosevelt's Meat Inspection Act was to bring the small packers into the inspection regime, and thereby end the competitive disability it imposed on large firms. Upton Sinclair simply served as an unwitting shill for the meatpacking industry.

[1]Butler Shaffer, *Calculated Chaos: Institutional Threats to Peace and Human Survival* (San Francisco: Alchemy Books, 1985), p. 143.

[2]Associated Press, "U.S. government fights to keep meatpackers from testing all slaughtered cattle for mad cow," *International Herald-Tribune*, May 29, 2007 <http://www.iht.com/articles/ap/2007/05/29/america/NA-GEN-US-Mad-Cow.php>. "Monsanto Declares War on 'rBGH-free' Dairies," April 3, 2007 (reprint of Monsanto press release by Organic Consumers Association) <http://www.organicconsumers.org/articles/article_4698 .cfm>. "Pa. bars hormone-free milk labels," *USA Today*, November 13, 2007 <http://www.usatoday.com/news/nation/2007-11-13-milk-labels_N.htm>.

[3]Kolko, *The Triumph of Conservatism*, p. 268.

[4]*Ibid.*, p. 275.

of Trade, provides a detailed account of the functioning of these trade associations, and their attempts to stabilize prices and restrict "predatory price cutting," through assorted codes of ethics.[1] Specifically, the trade associations established codes of ethics directly under FTC auspices that had the force of law: "[A]s early as 1919 the FTC began inviting members of specific industries to participate in conferences designed to identify trade practices that were felt by "the practically unanimous opinion" of industry members to be unfair." The standard procedure, through the 1920s, was for the FTC to invite members of a particular industry to a conference, and solicit their opinions on trade practice problems and recommended solutions.

> The rules that came out of the conferences and were approved by the FTC fell into two categories: Group I rules and Group II rules. Group I rules were considered by the commission as expressions of the prevailing law for the industry developing them, and a violation of such rules by any member of that industry—whether that member had agreed to the rules or not—would subject the offender to prosecution under Section 5 of the Federal Trade Commission Act as an "unfair method of competition." . . .
>
> Contained within Group I were rules that dealt with practices considered by most business organizations to be the more "disruptive" of stable economic conditions. Generally included were prohibitions against inducing "breach of contract; . . . commercial bribery; . . . price discrimination by secret rebates, excessive adjustments, or unearned discounts; . . . *selling of goods below cost or below published list of prices for purpose of injuring competitor*; misrepresentation of goods; . . . use of inferior materials or deviation from standards; [and] falsification of weights, tests, or certificates of manufacture [emphasis added]."[2]

The two pieces of legislation accomplished what the trusts had been unable to: they enabled a handful of firms in each industry to stabilize their market share and to maintain an oligopoly structure between them.

> It was during the war that effective, working oligopoly and price and market agreements became operational in the dominant sectors of the American economy. The rapid diffusion of power in the economy and relatively easy entry virtually ceased. Despite the cessation of important new legislative enactments, the unity of business and the federal government continued throughout the 1920s and thereafter, using the foundations laid in the Progressive Era to stabilize and consolidate conditions within various industries. And, on the same progressive foundations and exploiting the experience with the war agencies, Herbert Hoover and Franklin Roosevelt later formulated programs for saving American capitalism. The principle of utilizing the federal government to stabilize the economy, established in the context of modern industrialism during the Progressive Era, became the basis of political capitalism in its many later ramifications.[3]

The regulatory state provided "rationality" in two other ways: first, by the use of federal regulation to preempt potentially harsher action by populist governments at the state and local level; and second, by preempting and overriding older common law standards of liability, replacing the potentially harsh damages imposed by local juries with a least common denominator of regulatory standards based on "sound science" (as determined by industry, of course). Regarding the first, whatever view one takes of the validity of the local regulations in and of themselves, it is hardly legitimate for a centralized state to act on behalf of corpo-

[1]Butler Shaffer, *In Restraint of Trade: The Business Campaign Against Competition, 1918-1938* (Lewisburg: Bucknell University Press, 1997).

[2]*Ibid.*, pp. 82-84.

[3]Kolko, *Triumph of Conservatism*, p. 287.

rate interests, in suppressing unfriendly local regulations and overcoming the transaction costs of operating in a large number of conflicting jurisdictions, all at taxpayer expense. "Free trade" simply means the state does not hinder those under its own jurisdiction from trading with anyone else on whatever terms they can obtain on their own—not that the state actually opens up markets. Regarding the second, it is interesting that so many self-described "libertarians" support what they call "tort reform," when civil liability for damages is in fact the libertarian *alternative* to the regulatory state. Much of such "tort reform" amounts to indemnifying business firms from liability for reckless fraud, pollution, and other externalities imposed on the public.

There is also the regulatory state's function, which we will examine below in more depth, of imposing mandatory minimum overhead costs and thus erecting barriers to competition from low-overhead producers.

State spending serves to cartelize the economy in much the same way as regulation. Just as regulation removes significant areas of quality and safety as issues in cost competition, the socialization of operating costs on the state (e.g. R&D subsidies, government-funded technical education, etc.) allows monopoly capital to remove them as components of price in cost competition between firms, and places them in the realm of guaranteed income to all firms in a market alike. Transportation subsidies reduce the competitive advantage of locating close to one's market. Farm price support subsidies turn idle land into an extremely lucrative real estate investment. Whether through regulations or direct state subsidies to various forms of accumulation, the corporations act through the state to carry out some activities jointly, and to restrict competition to selected areas.

An ever-growing portion of the functions of the capitalist economy have been carried out through the state. According to James O'Connor, state expenditures under monopoly capitalism can be divided into "social capital" and "social expenses."

> *Social capital* is expenditures required for profitable private accumulation; it is indirectly productive (in Marxist terms, social capital indirectly expands surplus value). There are two kinds of social capital: social investment and social consumption (in Marxist terms, social constant capital and social variable capital). . . . *Social investment* consist of projects and services that increase the productivity of a given amount of laborpower and, other factors being equal, increase the rate of profit. . . . *Social consumption* consists of projects and services that lower the reproduction costs of labor and, other factors being equal, increase the rate of profit. An example of this is social insurance, which expands the productive powers of the work force while simultaneously lowering labor costs. The second category, social expenses, consists of projects and services which are required to maintain social harmony—to fulfill the state's "legitimization" function. . . . The best example is the welfare system, which is designed chiefly to keep social peace among unemployed workers.[1]

According to O'Connor, such state expenditures counteract the falling direct rate of profit that Marx predicted in volume 3 of *Capital*. Monopoly capital is able to externalize many of its operating expenses on the state; and since the state's expenditures indirectly increase the productivity of labor and capital at taxpayer expense, the apparent rate of profit is increased. "In short, monopoly capital socializes more and more costs of production."[2]

[1]James O'Connor, *The Fiscal Crisis of the State* (New York: St. Martin's Press, 1973), pp. 6-7.

[2]*Ibid.*, p. 24.

(In fact, O'Connor makes the unwarranted assumption that the subsidized increase in capital-intensiveness actually increases productivity, rather than simply subsidizing the cost of increasing the ratio of capital to unit of output and despite the inefficiency of more capital-intensive methods. The subsidized capital-intensive production methods are, in fact, as surely a means of destroying surplus capital as sinking it in the ocean would be.)

O'Connor listed several ways in which monopoly capital externalizes its operating costs on the political system:

> Capitalist production has become more interdependent—more dependent on science and technology, labor functions more specialized, and the division of labor more extensive. Consequently, the monopoly sector (and to a much lesser degree the competitive sector) requires increasing numbers of technical and administrative workers. It also requires increasing amounts of infrastructure (physical overhead capital)—transportation, communication, R&D, education, and other facilities. In short, the monopoly sector requires more and more social investment in relation to private capital. . . . The costs of social investment (or social constant capital) are not borne by monopoly capital but rather are socialized and fall on the state.[1]

The general effect of the state's intervention in the economy, then, is to remove ever increasing spheres of economic activity from the realm of competition in price or quality, and to organize them collectively through organized capital as a whole.

B. Mass Consumption and Push Distribution to Absorb Surplus

As we have already seen, the use of expensive product-specific machinery requires large-batch production to achieve high throughput and thus spread production costs out over as many units as possible. And to do this, in turn, requires enormous exercises of power to ensure that a market existed for this output.

First of all, it required the prior forms of intervention described in the last chapter and in the previous section of this chapter: state intervention to create a unified national market and transportation system, and state intervention to promote the formation of stable oligopoly cartels.

But despite all the state intervention up front to make the centralized corporate economy possible, state intervention is required *afterward* as well as before in order to keep the system running. Large, mass-production industry is unable to survive without the government guaranteeing an outlet for its overproduction, and insulating it from a considerable amount of market competition. As Paul Baran and Paul Sweezy put it, monopoly capitalism

> tends to generate ever more surplus, yet it fails to provide the consumption and investment outlets required for the absorption of a rising surplus and hence for the smooth working of the system. Since surplus which cannot be absorbed will not be produced, it follows that the *normal* state of the monopoly capitalist economy is stagnation. With a given stock of capital and a given cost and price structure, the system's operating rate cannot rise above the point at which the amount of surplus produced can find the necessary outlets. And this means chronic underutilization of available human and material resources. Or, to put the point in slightly different terms, the system must operate at a point low enough on its profitability schedule not to generate more surplus than can be absorbed. Since the profitability schedule is always moving upward, there is a corresponding downdrift of the "equilibrium" operating rate. Left to itself—that

[1] *Ibid.*, p. 24.

is to say, in the absence of counteracting forces which are no part of what may be called the "elementary logic" of the system—monopoly capitalism would sink deeper and deeper into a bog of chronic depression.[1]

Mass production divorces production from consumption. The rate of production is driven by the imperative of keeping the machines running at full capacity so as to minimize unit costs, rather than by customer orders. So in addition to contractual control of inputs, mass-production industry faces the imperative of guaranteeing consumption of its output by managing the consumer. It does this through push distribution, high-pressure marketing, planned obsolescence, and consumer credit.

Mass advertising serves as a tool for managing aggregate demand. According to Baran and Sweezy, the main function of advertising is "waging, on behalf of the producers and sellers of consumer goods, a relentless war against saving and in favor of consumption." And that function is integrally related to planned obsolescence:

> The strategy of the advertiser is to hammer into the heads of people the unquestioned desirability, indeed the imperative necessity, of owning the newest product that comes on the market. For this strategy to work, however, producers have to pour on the market a steady stream of "new" products, with none daring to lag behind for fear his customers will turn to his rivals for newness.
>
> Genuinely new or different products, however, are not easy to come by, even in our age of rapid scientific and technological advance. Hence much of the newness with which the consumer is systematically bombarded is either fraudulent or related trivially and in many cases even negatively to the function and serviceability of the product.[2]
>
> .... In a society with a large stock of consumer durable goods like the United States, an important component of the total demand for goods and services rests on the need to replace a part of this stock as it wears out or is discarded. Built-in obsolescence increases the rate of wearing out, and frequent style changes increase the rate of discarding. . . . The net result is a stepping up in the rate of replacement demand and a general boost to income and employment. In this respect, as in others, the sales effort turns out to be a powerful antidote to monopoly capitalism's tendency to sink into a state of chronic depression.[3]

Although somewhat less state-dependent than the expedients discussed later in this chapter, mass advertising had a large state component. For one thing, the founders of the mass advertising and public relations industries were, in large part, also the founders of the science of "manufacturing consent" used to manipulate Anglo-American populations into support for St. Woodrow's crusade. Edward Bernays and Harold Lasswell, who played a central role in the Creel Commission and other formative prowar propaganda efforts in WWI, went on to play similarly prominent roles in the development of public relations and mass consumer advertising.

For another, the state's own organs of propaganda (through the USDA, school home economics classes, etc.) reinforced the message of advertising, placing great emphasis on discrediting "old-fashioned" atavisms like home-baked bread and home-grown and -canned vegetables, and promoting in their place the "up-to-

[1]Paul Baran and Paul Sweezy, *Monopoly Capitalism : An Essay in the American Economic and So-*
cial Order (New York: Monthly Review Press, 1966), p. 108.
[2]*Ibid.*, pp. 128-129.
[3]*Ibid.*, p. 131.

date" housewifely practice of heating stuff up out of cans from the market.[1] Jeffrey Kaplan described this, in a recent article, as the "gospel of consumption":

> [Industrialists] feared that the frugal habits maintained by most American families would be difficult to break. Perhaps even more threatening was the fact that the industrial capacity for turning out goods seemed to be increasing at a pace greater than people's sense that they needed them.
>
> It was this latter concern that led Charles Kettering, director of General Motors Research, to write a 1929 magazine article called "Keep the Consumer Dissatisfied." . . . Along with many of his corporate cohorts, he was defining a strategic shift for American industry—from fulfilling basic human needs to creating new ones.
>
> In a 1927 interview with the magazine *Nation's Business*, Secretary of Labor James J. Davis provided some numbers to illustrate a problem that the *New York Times* called "need saturation." Davis noted that "the textile mills of this country can produce all the cloth needed in six months' operation each year" and that 14 percent of the American shoe factories could produce a year's supply of footwear. The magazine went on to suggest, "It may be that the world's needs ultimately will be produced by three days' work a week."
>
> Business leaders were less than enthusiastic about the prospect of a society no longer centered on the production of goods. For them, the new "labor-saving" machinery presented not a vision of liberation but a threat to their position at the center of power. John E. Edgerton, president of the National Association of Manufacturers, typified their response when he declared: "Nothing . . . breeds radicalism more than unhappiness unless it is leisure."
>
> By the late 1920s, America's business and political elite had found a way to defuse the dual threat of stagnating economic growth and a radicalized working class in what one industrial consultant called "the gospel of consumption"—the notion that people could be convinced that however much they have, it isn't enough. President Herbert Hoover's 1929 Committee on Recent Economic Changes observed in glowing terms the results: "By advertising and other promotional devices . . . a measurable pull on production has been created which releases capital otherwise tied up." They celebrated the conceptual breakthrough: "Economically we have a boundless field before us; that there are new wants which will make way endlessly for newer wants, as fast as they are satisfied."[2]

Right-wing libertarians like Murray Rothbard answer critiques of mass advertising by saying they downplay the role of the audience as an active moral agent in deciding what to accept and what to reject, and fail to recognize that information has a cost and that there's such a thing as "rational ignorance." Interestingly, however, many of Rothbard's followers at Mises.Org and Lew Rockwell.Com show no hesitancy whatsoever in attributing a cumulative sleeper effect to statist propaganda in the public schools and state-allied media. No doubt they would argue that, in the latter case, both the volume and the content of the propaganda are artificially shifted in the direction of a certain message, thus artificially raising the cost of defending against the propaganda message. But that is exactly my point concerning mass advertising. The state capitalist system makes mass-production industry for the national market artificially prevalent, and makes its need to dispose of surplus output artificially urgent, thus subjecting the consumer to a barrage of pro-consumption propaganda far greater in volume than would be experienced in a decentralized, free market society of small-scale local commodity production.

[1]Stuart Ewen, *Captains of Consciousness: Advertising and the Social Roots of Consumer Culture* (New York: McGraw-Hill, 1976), pp. 163, 171-172.

[2]Jeffrey Kaplan, "The Gospel of Consumption: And the better future we left behind," *Orion*, May/June 2008 <http://www.orionmagazine.org/index.php/articles/article/2962>.

Chandler's model of "high-speed, high-throughput, turning high fixed costs into low unit costs," and Galbraith's "technostructure," presuppose a "push" model of distribution. The push paradigm, according to, is characterized by the following assumptions:

- There's not enough to go around
- Elites do the deciding
- Organisations must be hierarchical
- People must be molded
- Bigger is better
- Demand can be forecast
- Resources can be allocated centrally
- Demand can be met[1]

Here's how push distribution was described by Paul and Percival Goodman not long after World War II:

> ... in recent decades ... the center of economic concern has gradually shifted from either providing goods for the consumer or gaining wealth for the enterpriser, to keeping the capital machines at work and running at full capacity; for the social arrangements have become so complicated that, unless the machines are running at full capacity, all wealth and subsistence are jeopardized, investment is withdrawn, men are unemployed. That is, when the system depends on all the machines running, unless every kind of good is produced and sold, it is also impossible to produce bread.[2]

The same imperative was at the root of the hypnopaedic socialization in Huxley's Brave New World: "ending is better than mending"; "the more stitches, the less riches." Or as GM designer Harley Earl said in the 1950s,

> My job is to hasten obsolescence. I've got it down to two years; now when I get it down to one year, I'll have a perfect score.[3]

Along the same lines, Baran and Sweezy cite a New York investment banker on the disaster that would befall capitalism without planned obsolescence or branding: "Clothing would be purchased for its utility value; food would be bought on the basis of economy and nutritional value; automobiles would be stripped to essentials and held by the same owners for the full ten to fifteen years of their useful lives; homes would be built and maintained for their characteristics of shelter. . . . "[4]

The older economy that the "push" distribution system replaced was one in which most foods and drugs were what we would today call "generic." Flour, cereal, and similar products were commonly sold in bulk and weighed and packaged by the grocer (the ratio had gone from roughly 95% bulk to 75% package goods during the twenty years before Borsodi wrote in 1927); the producers geared production to the level of demand that was relayed to them by the retailers' orders. Drugs, likewise, were typically compounded by the druggist on-premises to the

[1]John Hagel III, John Seely Brown, and Lang Davison, *The Power of Pull: How Small Moves, Smartly Made, Can Set Big Things in Motion*, quoted in JP Rangaswami, "Thinking about predictability: More musings about Push and Pull," *Confused of Calcutta*, May 4, 2010 <http://confusedofcalcutta.com/2010/05/04/thinking-about-predictability-more-musings-about-push-and-pull/>.

[2]Paul and Percival Goodman, *Communitas: Means of Livelihood and Ways of Life* (New York: Vintage Books, 1947, 1960), pp. 188-89.

[3]Eric Rumble, "Toxic Shocker," *Up! Magazine*, January 1, 2007 <http://www.up-magazine.com/magazine/exclusives/Toxic_Shocker_3.shtml>.

[4]Baran and Sweezy, *Monopoly Capital*, p. 124.

physician's specifications, from generic components.[1] Production was driven by orders from the grocer, as customers used up his stock of bulk goods.

Under the new "push" system, the producers appealed directly to the consumer through brand-name advertising, and relied on pressure on the grocer to create demand for what they chose to produce. Brand loyalty helps to stabilize demand for a particular manufacturer's product, and eliminate the fluctuation of demand that accompanies price competition in pure commodities.

> It is possible to roughly classify a manufacturer as belonging either to those who "make" products to meet requirements of the market, or as belonging to those who "distribute" brands which they decide to make. The manufacturer in the first class relies upon the natural demand for his product to absorb his output. He relies upon competition among wholesalers and retailers in maintaining attractive stocks to absorb his production. The manufacturer in the second class creates a demand for his brand and forces wholesalers and retailers to buy and "stock" it. In order to market what he has decided to manufacture, he figuratively has to make water run uphill.[2]

The problem was that the consumer, under the new regime of Efficiency, paid about four times as much for trademarked flour, sugar, etc., as he had paid for bulk goods under the old "inefficient" system.[3] Under the old regime, the grocer was a purchasing agent for the customer; under the new, he was a marketing agent for the producer.

Distribution costs are increased still further by the fact that larger-scale production and greater levels of capital intensiveness increase the unit costs resulting from idle capacity, and thereby (as we saw in the last chapter) greatly increase the resources devoted to high-pressure, "push" forms of marketing.

Borsodi's book *The Distribution Age* was an elaboration of the fact that, as he stated in the Preface, production costs fell by perhaps a fifth between 1870 and 1920, even as the cost of marketing and distribution nearly tripled.[4] The modest reduction in unit production cost was more than offset by the increased costs of distribution and high-pressure marketing. "[E]very part of our economic structure," he wrote, was "being strained by the strenuous effort to market profitably what modern industry can produce."[5]

Distribution costs are far lower under a demand-pull regime, in which production is geared to demand. As Borsodi argued,

> ... [I]t is still a fact ... that the factory which sells only in its natural field because that is where it can serve best, meets little sales-resistance in marketing through the normal channels of distribution. The consumers of such a factory are so "close" to the manufacturer, their relations are so intimate, that buying from that factory has the force of tradition. Such a factory can make shipment promptly; it can adjust its production to the peculiarities of its territory, and it can make adjustments with its customers more intelligently than factories which are situated at a great distance. High pressure methods of distribution do not seem tempting to such a factory. They do not tempt it for the very good reason that such a factory has no problem to which high pressure distribution offers a solution.

[1]Ralph Borsodi, *The Distribution Age* (New York and London: D. Appleton and Company, 1929), pp. 217, 228.

[2]*Ibid.*, p. 110.

[3]Quoted in *Ibid.*, pp. 160-61.

[4]*Ibid.*, p. v.

[5]*Ibid.*, p. 4.

> It is the factory which has decided to produce trade-marked, uniform, packaged, individualized, and nationally advertised products, and which has to establish itself in the national market by persuading distributors to pay a higher than normal price for its brand, which has had to turn to high pressure distribution. Such a factory has a selling problem of a very different nature from that of factories which are content to sell only where and to whom they can sell most efficiently.[1]

For those whose low overhead permits them to produce in response to consumer demand, marketing is relatively cheap. Rather than expending enormous effort to make people buy their product, they can just fill the orders that come in. When demand for the product must be created, the effort (to repeat Borsodi's metaphor) is comparable to that of making water run uphill. Mass advertising is only a small part of it. Even more costly is direct mail advertising and door-to-door canvassing by salesmen to pressure grocers in a new market to stock one's goods, and canvassing of grocers themselves by sales reps.[2] The costs of advertising, packaging, brand differentiation, etc., are all costs of overcoming sales resistance that only exist because production is divorced from demand rather than driven by it.

And this increased marginal cost of distribution for output above the natural level of demand results, in accordance with Ricardo's law of rent, in higher average price for *all* goods. This means that in the market as it exists now, the price of generic and store brand goods is not governed by production cost, as it would be if competing in a commodity market; it is governed by the bare amount it needs to be marked down to compete with brand name goods.[3]

For those who can flexibly respond to demand, also, predictability of consumer demand doesn't matter that much. Of the grocer, for example, Borsodi pointed out that the customer would always have to eat, and would continue to do so without a single penny of high pressure marketing. It was therefore a matter of indifference to the grocer whether the customer ate some particular product or brand name; he would stock whatever goods the customer preferred, as his existing stocks were used up, and change his orders in keeping with changes in customer preference. To the manufacturer, on the other hand, it is of vital importance that the customer buy (say) mayonnaise in particular—and not just mayonnaise, but his particular brand of mayonnaise.[4]

And the proliferation of brand names with loyal followings raises the cost of distribution considerably: rather than stocking generic cornflakes in bulk commodity form, and replacing the stock as it is depleted, the grocer must maintain large enough stocks of all the (almost identical) popular brands to ensure against running out, which means slower turnover and more wasted shelf space. In other words, push distribution results in the costly disruption of flow by stagnant eddies and flows, in the form of ubiquitous inventories.[5]

The advantage of brand specification, from the perspective of the producer, is that it "lifts a product out of competition":[6] "the prevalence of brand specification has all but destroyed the normal basis upon which true competitive prices can be established."[7] As Barry Stein described it, branding "convert[s] true commodities to

[1] *Ibid.*, pp. 112-113.
[2] *Ibid.*, p. 136.
[3] *Ibid.*, p. 247.
[4] *Ibid.*, pp. 83-84.
[5] *Ibid.*, p. 84.
[6] *Ibid.*, p. 162.
[7] *Ibid.* pp. 216-17.

apparent tailored goods, so as to avoid direct price competition in the marketplace."

> The distinctions introduced—elaborate packaging, exhortative advertising and promotion that asserts the presence of unmeasurable values, and irrelevant physical modification (colored toothpaste)—do not, in fact, render these competing products "different" in any substantive sense, but to the extent that consumers are convinced by these distinctions and treat them as if they were different, product loyalty is generated.[1]

Under the old regime, competition between identifiable producers of bulk goods enabled grocers to select the highest quality bulk goods, while providing them to customers at the lowest price. Brand specification, on the other hand, relieves the grocer of the responsibility for standing behind his merchandise and turns him into a mere stocker of shelves with the most-demanded brands.

The change, naturally, did not go unremarked by those profiting from it. For example, here's a bit of commentary from an advertising trade paper in 1925:

> In the statement to its stockholders issued recently by The American Sugar Refining Company, we find this statement:
>
> "Formerly, as is well known, household sugar was largely of bulk pricing. We have described the sale of package sugar and table syrup under the trade names of 'Domino' and 'Franklin' with such success that the volume of trademark packages now constitutes roundly one-half of our production that goes into households. . . . "
>
> These facts should be of vital interest to any executive who faces the problem of marketing a staple product that is hard to control because it is sold in bulk.
>
> Twenty years ago the sale of sugar in cardboard cartons under a brand name would have been unthinkable. Ten years hence this kind of history will have repeated itself in connection with many other staple commodities now sold in bulk. . . . [2]

The process went on, just as the paper predicted, until—decades later—the very idea of a return to price competition in the production of goods, instead of brand-name competition for market share, would strike manufacturers with horror. What Borsodi proposed, making "[c]ompetition . . . descend from the cloudy heights of sales appeals and braggadocio generally, to just one factor—price,"[3] is the worst nightmare of the oligopoly manufacturer and the advertising industry:

> At the annual meeting of the U.S. Association of National Advertisers in 1988, Graham H. Phillips, the U.S. Chairman of Ogilvy & Mather, berated the assembled executives for stooping to participate in a "commodity marketplace" rather than an image-based one. "I doubt that many of you would welcome a commodity marketplace in which one competed solely on price, promotion and trade deals, all of which can be easily duplicated by competition, leading to ever-decreasing profits, decay, and eventual bankruptcy." Others spoke of the importance of maintaining "conceptual value-added," which in effect means adding nothing but marketing. Stooping to compete on the basis of real value, the agencies ominously warned, would speed not just the death of the brand, but corporate death as well.[4].

[1]Stein, *Size, Efficiency, and Community Enterprise*, p. 79.

[2]*Advertising and Selling Fortnightly*, February 25, 1925, in Borsodi, *The Distribution Age*, pp. 159-60.

[3]Stuart Chase and F. J. Schlink, *The New Republic*, December 30, 1925, in *Ibid.*, p. 204.

[4]Naomi Klein, *No Logo* (New York: Picador, 1999), p. 14.

It's telling that Chandler, the apostle of the great "efficiencies" of this entire system, frankly admitted all of these things. In fact, far from regarding it as an "admission," he treated it as a feature of the system. He explicitly equated "prosperity" to the rate of flow of material through the system and the speed of production and distribution—without any regard to whether the rate of "flow" was twice as fast because people were throwing stuff in the landfills twice as fast to keep the pipelines from clogging up.

> The new middle managers did more than devise ways to coordinate the high-volume flow from suppliers of raw materials to consumers. They invented and perfected ways to expand markets and to speed up the processes of production and distribution. Those at American Tobacco, Armour, and other mass producers of low-priced packaged products perfected techniques of product differentiation through advertising and brand names that had been initially developed by mass marketers, advertising agencies, and patent medicine makers. The middle managers at Singer wee the first to systematize personal selling by means of door-to-door canvassing; those at McCormick among the first to have franchised dealers using comparable methods. Both companies innovated in installment buying and other techniques of consumer credit.[1]

In other words, the Sloanist system Chandler idealized was more "efficient" because it was better at persuading people to throw stuff away so they could buy more, and better at producing substandard shit that would *have* to be thrown away in a few years. Only a man of the mid-20th century, writing at the height of consensus capitalism, from the standpoint of an establishment liberalism was as yet utterly untainted by the thinnest veneer of greenwash, could write such a thing from the standpoint of an *enthusiast*.

Increased unit costs from idle capacity, given the high overhead of large-scale production, are the chief motive behind the push distribution model. Even so, the restrained competition of an oligopoly market limits the competitive disadvantage resulting from idle capacity—so long as the leading firms in an industry are running at roughly comparable percentages of capacity, and can pass their overhead costs onto the customer. The oligopoly mark-up included in consumer price reflects the high costs of excess capacity.

> It is difficult to estimate how large a part of the nation's production facilities are normally in use. One particularly able observer of economic tendencies, Colonel Leonard P. Ayres, uses the number of blast furnaces in operation as a barometer of business conditions. When blast furnaces are in 60 per cent. operation, conditions are normal. . . .
>
> It is obvious, if 60 per cent. represents normality, that consumers of such a basic commodity as pig iron must pay dividends upon an investment capable of producing two-thirds more pig iron than the country uses in normal times.

Borsodi also found that flour mills, steel plants, shoe factories, copper smelters, lumber mills, automobiles, and rayon manufacturers were running at similar or lower percentages of total capacity.[2] Either way, it is the consumer who pays for overaccumulation: both for the high marketing costs of distributing overproduced goods when industry runs at full capacity, and for the high overhead when the firms in an oligopoly market all run at low capacity and pass their unit costs on through administered pricing.

So cartelization and high costs from idle capacity, alongside push distribution and planned obsolescence, together constitute the twin pathologies of monopoly

[1]Chandler, *The Visible Hand*, p. 411.
[2]Borsodi, *The Distribution Age*, pp. 42-43.

capitalism. Both are expedients for dealing with the enormous capital outlays and overproduction entailed in mass-production industry, and both require that outside society be subordinated to the needs of the corporation and subjected to its control.

The worst-case scenario, from our standpoint, is that big business will attempt an end-run around the problem of excess capacity and underconsumption through measures like the abortive National Industrial Recovery Act of the New Deal era: cartelizing an industry under government auspices, so all its firms can operate at a fraction of full capacity indefinitely and use monopoly pricing to pass the cost of idle capacity on to the consumer on a cost-plus basis. Anyone tempted to see this as a solution should bear in mind that it removes all incentive to control costs or to promote efficiency. For a picture of the kind of society that would result from such an arrangement, one need only watch the movie *Brazil*.

The overall system, in short, was a "solution" in search of a problem. State subsidies and mercantilism gave rise to centralized, overcapitalized industry, which led to overproduction, which led to the need to find a way of creating demand for lots of crap that nobody wanted.

C. State Action to Absorb Surplus: Imperialism

The roots of the corporate state in the U.S., more than anything else, lie in the crisis of overproduction as perceived by corporate and state elites—especially the traumatic Depression of the 1890s—and the requirement, also as perceived by them, for state intervention to absorb surplus output or otherwise deal with the problems of overproduction, underconsumption, and overaccumulation. According to William Appleman Williams, "the Crisis of the 1890's raised in many sections of American society the specter of chaos and revolution."[1] Economic elites saw it as the result of overproduction and surplus capital, and believed it could be resolved only through access to a "new frontier." Without state-guaranteed access to foreign markets, output would fall below capacity, unit costs would go up, and unemployment would reach dangerous levels.

Accordingly, the centerpiece of American foreign policy to the present day has been what Williams called "Open Door Imperialism"[2]: securing American access to foreign markets on equal terms to the European colonial powers, and opposing attempts by those powers to divide up or close markets in their spheres of influence.

Open Door Imperialism consisted of using U.S. political power to guarantee access to foreign markets and resources on terms favorable to American corporate interests, without relying on direct political rule. Its central goal was to obtain for U.S. merchandise, in each national market, treatment equal to that afforded any other industrial nation. Most importantly, this entailed active engagement by the U.S. government in breaking down the imperial powers' existing spheres of economic influence or preference. The result, in most cases, was to treat as hostile to U.S. security interests any large-scale attempt at autarky, or any other policy whose effect was to withdraw major areas of the world from the disposal of the U.S. corporate economy. When the power attempting such policies was an equal, like the British Empire, the U.S. reaction was merely one of measured coolness.

[1]William Appleman Williams, *The Tragedy of American Diplomacy* (New York: Dell Publishing Company, 1959, 1962) 21-2.

[2]Williams, *The Contours of American History* (Cleveland and New York: The World Publishing Company, 1961).

When it was perceived as an inferior, like Japan, the U.S. resorted to more forceful measures, as events of the late 1930s indicate. And whatever the degree of equality between advanced nations in their access to Third World markets, it was clear that Third World nations were still to be subordinated to the industrialized West in a collective sense.

In the late 1930s, the American political leadership feared that Fortress Europe and the Greater East Asian Co-Prosperity sphere would deprive the American corporate economy of vitally needed raw materials, not to mention outlets for its surplus output and capital; that's what motivated FDR to maneuver the country into another world war. The State Department's internal studies at the time estimated that the American economy required, at a minimum, the resources and markets of a "Grand Area" consisting of Latin America, East Asia, and the British Empire. Japan, meanwhile, was conquering most of China (home of the original Open Door) and the tin and rubber of Indochina, and threatening to capture the oil of the Dutch East Indies as well. In Europe, the worst case scenario was the fall of Britain, followed by the German capture of some considerable portion of the Royal Navy and subsequently of the Empire. War with the Axis would have followed from these perceived threats as a matter of course, even had FDR not successfully maneuvered Japan into firing the first shot.[1]

World War II, incidentally, also went a long way toward postponing America's crises of overproduction and overaccumulation for a generation, by blowing up most of the capital in the world outside the United States and creating a permanent war economy to absorb surplus output.

The American policy that emerged from the war was to secure control over the markets and resources of the global "Grand Area" through institutions of global economic governance, as created by the postwar Bretton Woods system, and to make preventing "defection from within" by autarkic powers the centerpiece of national security policy.

The problem of access to foreign markets and resources was central to U.S. postwar planning. Given the structural imperatives of "export dependent monopoly capitalism,"[2] the threat of a postwar depression was very real. The original drive toward foreign expansion at the end of the nineteenth century reflected the fact that industry, with state capitalist encouragement, had expanded far beyond the ability of the domestic market to consume its output. Even before World War II,

[1]Laurence H. Shoup and William Minter, "Shaping a New World Order: The Council on Foreign Relations' Blueprint for World Hegemony, 1939-1945," in Holly Sklar, ed., *Trilateralism: The Trilateral Commission and Elite Planning for World Management* (Boston: South End Press, 1980), pp. 135-56

[2]"Now the price that brings the maximum monopoly profit is generally far above the price that would be fixed by fluctuating competitive costs, and the volume that can be marketed at that maximum price is generally far below the output that would be technically and economically feasible. . . . [The trust] extricates itself from this dilemma by producing the full output that is economically feasible, thus securing low costs, and offering in the protected domestic market only the quantity corresponding to the monopoly price—insofar as the tariff permits; while the rest is sold, or "dumped," abroad at a lower price. . . ."—Joseph Schumpeter, "Imperialism," in *Imperialism, Social Classes: Two Essays* by Joseph Schumpeter. Translated by Heinz Norden. Introduction by Hert Hoselitz (New York: Meridian Books, 1955) 79-80.

Joseph Stromberg, by the way, did an excellent job of integrating this thesis, generally identified with the historical revisionism of the New Left, into the theoretical framework of Mises and Rothbard, in "The Role of State Monopoly Capitalism in the American Empire" *Journal of Libertarian Studies* Volume 15, no. 3 (Summer 2001), pp. 57-93. Available online at <http://www.mises.org/journals/jls/15_3/15_3_3.pdf>.

the state capitalist economy had serious trouble operating at the level of output needed for full utilization of capacity and cost control. Military-industrial policy during the war exacerbated the problem of over-accumulation, greatly increasing the value of plant and equipment at taxpayer expense. The end of the war, if followed by the traditional pattern of demobilization, would have resulted in a drastic reduction in orders to that same overbuilt industry just as over ten million workers were being dumped back into the civilian labor force.

A central facet of postwar economic policy, as reflected in the Bretton Woods agencies, was state intervention to guarantee markets for the full output of U.S. industry and profitable outlets for surplus capital. The World Bank was designed to subsidize the export of capital to the Third World, by financing the infrastructure without which Western-owned production facilities could not be established there. According to Gabriel Kolko's 1988 estimate, almost two thirds of the World Bank's loans since its inception had gone to transportation and power infrastructure.[1] A laudatory Treasury Department report referred to such infrastructure projects (comprising some 48% of lending in FY 1980) as "externalities" to business, and spoke glowingly of the benefits of such projects in promoting the expansion of business into large market areas and the consolidation and commercialization of agriculture.[2] The Volta River power project, for example, was built with American loans (at high interest) to provide Kaiser aluminum with electricity at very low rates.[3]

D. State Action to Absorb Surplus: State Capitalism

Government also directly intervened to alleviate the problem of overproduction, by its increasing practice of directly purchasing the corporate economy's surplus output—through Keynesian fiscal policy, massive highway and civil aviation programs, the military-industrial complex, the prison-industrial complex, foreign aid, and so forth. Baran and Sweezy point to the government's rising share of GDP as "an approximate index of the extent to which government's role as a creator of effective demand and absorber of surplus has grown during the monopoly capitalist era."[4]

> If the depressive effects of growing monopoly had operated unchecked, the United States economy would have entered a period of stagnation long before the end of the nineteenth century, and it is unlikely that capitalism could have survived into the second half of the twentieth century. What, then, were the powerful external stimuli which offset these depressive effects and enabled the economy to grow fairly rapidly during the later decades of the nineteenth century and, with significant interruptions, during the first two thirds of the twentieth century? In our judgment, they are of two kinds which we classify as (1) epoch-making innovations, and (2) wars and their aftermaths.

[1]Gabriel Kolko, *Confronting the Third World: United States Foreign Policy 1945-1980* (New York: Pantheon Books, 1988), p. 120.

[2]*United States Participation in the Multilateral Development Banks in the 1980s*. Department of the Treasury (Washingon, DC: 1982), p. 9.

[3]L. S. Stavrianos, *The Promise of the Coming Dark Age* (San Francisco: W. H. Freeman and Co. 1976), p. 42.

[4]Baran and Sweezy, pp. 146-147.

By "epoch-making innovations," Baran and Sweezy refer to "those innovations which shake up the entire pattern of the economy and hence create vast investment outlets in addition to the capital which they directly absorb."[1]

As for wars, Emmanuel Goldstein described their function quite well. "Even when weapons of war are not actually destroyed, their manufacture is still a convenient way of expending labor power without producing anything that can be consumed." War is a way of "shattering to pieces, or pouring into the stratosphere, or sinking in the depths of the sea," excess output and capital.[2]

Earlier, we quoted Robin Marris on the tendency of corporate bureaucracies to emphasize, not the character of goods produced, but the skills with which their production was organized. This is paralleled at a societal level. The imperative to destroy surplus is reflected in the GDP, which measures not the utility of goods and services to the consumer but the materials consumed in producing them. The more of Bastiat's "broken windows," the more inputs consumed to produce a given output, the higher the GDP.

As we said in the last chapter, the highway-automobile complex and the civil aviation system were continuations of the process begun with the railroads and other "internal improvements" of the nineteenth century: i.e., government subsidy to market centralization and large firm size. But as we pointed out then, they also have special significance as examples of the phenomenon Paul Baran and Paul Sweezy described in *Monopoly Capitalism*: government's creation of entire new industries to soak up the surplus generated by corporate capitalism's chronic tendencies toward overinvestment and overproduction.

Of the automobile-highway complex, Baran and Sweezy wrote, "[t]his complex of private interests clustering around one product has no equal elsewhere in the economy—or in the world. And the whole complex, of course, is completely dependent on the public provision of roads and highways."[3] Not to mention the role of U.S. foreign policy in guaranteeing access to "cheap and abundant" petroleum.

> One of the major barriers to the fledgling automobile industry at the turn of the century was the poor state of the roads. One of the first highway lobbying groups was the League of American Wheelmen, which founded "good roads" associations around the country and, in 1891, began lobbying state legislatures. . . .
>
> The Federal Aid Roads Act of 1916 encouraged coast-to-coast construction of paved roads, usually financed by gasoline taxes (a symbiotic relationship if ever there was one). By 1930, the annual budget for federal road projects was $750 million. After 1939, with a push from President Franklin Roosevelt, limited-access interstates began to make rural areas accessible.[4]

It was this last, in the 1930s, that signified the most revolutionary change. From its beginning, the movement for a national superhighway network was identified, first of all, with the fascist industrial policy of Hitler, and second with the American automotive industry.

> The "most powerful pressure group in Washington" began in June, 1932, when GM President, Alfred P. Sloan, created the National Highway Users Conference,

[1]*Ibid.*, p. 219.

[2]George Orwell, *1984*. Signet Classics reprint (New York: Harcourt Brace Jovanovich, 1949, 1981), p. 157.

[3]Baran and Sweezy, pp. 173-174.

[4]Jim Motavalli, "Getting Out of Gridlock: Thanks to the Highway Lobby, Now We're Stuck in Traffic. How Do We Escape?" *E Magazine*, March/April 2002 <http://www.emagazine.com/view/?534>.

> inviting oil and rubber firms to help GM bankroll a propaganda and lobbying effort that continues to this day.[1]

One of the earliest depictions of the modern superhighway in America was the Futurama exhibit at the 1939 World's Fair in New York, sponsored by (who else?) GM.

> The exhibit . . . provided a nation emerging from its darkest decade since the Civil War a mesmerizing glimpse of the future—a future that involved lots and lots of roads. Big roads. Fourteen-lane superhighways on which cars would travel at 100 mph. Roads on which, a recorded narrator promised, Americans would eventually be able to cross the nation in a day.[2]

The Interstate's association with General Motors didn't end there, of course. Its actual construction took place under the supervision of DOD Secretary Charles Wilson, formerly the company's CEO. During his 1953 confirmation hearings, when asked whether "he could make a decision in the country's interest that was contrary to GM's interest,"

> Wilson shot back with his famous comment, "I cannot conceive of one because for years I thought what was good for our country was good for General Motors, and vice versa. The difference did not exist. Our company is too big."[3]

Wilson's role in the Interstate program was hardly that of a mere disinterested technocrat. From the time of his appointment to DOD, he "pushed relentlessly" for it. And the chief administrator of the program was "Francis DuPont, whose family owned the largest share of GM stock. . . . "[4]

Corporate propaganda, as so often in the twentieth century, played an active role in attempts to reshape the popular culture.

> Helping to keep the driving spirit alive, Dow Chemical, producer of asphalt, entered the PR campaign with a film featuring a staged testimonial from a grade school teacher standing up to her anti-highway neighbors with quiet indignation. "Can't you see this highway means a whole new way of life for the children?"[5]

Whatever the political motivation behind it, the economic effect of the Interstate system should hardly be controversial. Virtually 100% of the roadbed damage to highways is caused by heavy trucks. And despite repeated liberalization of maximum weight restrictions, far beyond the heaviest conceivable weight the Interstate roadbeds were originally designed to support,

> fuel taxes fail miserably at capturing from big-rig operators the cost of exponential pavement damage caused by higher axle loads. Only weight-distance user charges are efficient, but truckers have been successful at scrapping them in all but a few western states where the push for repeal continues.[6]

[1]Mike Ferner, "Taken for a Ride on the Interstate Highway System," MRZine (Monthly Review) June 28, 2006 <http://mrzine.monthlyreview.org/ferner280606.html>.

[2]Justin Fox, "The Great Paving How the Interstate Highway System helped create the modern economy—and reshaped the FORTUNE 500." Reprinted from *Fortune*. CNNMoney.Com, January 26, 2004 <http://money.cnn.com/magazines/fortune/fortune_archive/2004/01/26/358835/index.htm>.

[3]Edwin Black, "Hitler's Carmaker: How Will Posterity Remember General Motors' Conduct? (Part 4)" *History News Network*, May 14, 2007 <http://hnn.us/articles/38829.html>.

[4]Ferner, "Taken for a Ride."

[5]*Ibid.*

[6]Frank N. Wilner, "Give truckers an inch, they'll take a ton-mile: every liberalization has been a launching pad for further increases—trucking wants long combination vehicle

So only about half the revenue of the highway trust fund comes from fees or fuel taxes on the trucking industry, and the rest is externalized on private automobiles. Even David S. Lawyer, a skeptic on the general issue of highway subsidies, only questions whether highways receive a net subsidy from general revenues over and above total user fees on both trucks and cars; he effectively concedes the subsidy of heavy trucking by the gasoline tax.[1]

As for the civil aviation system, from the beginning it was a creature of the state. The whole physical infrastructure was built, in its early decades, with tax money.

> Since 1946, *the federal government has poured billions of dollars into airport development*. In 1992, Prof. Stephen Paul Dempsey of the University of Denver estimated that *the current replacement value of the U.S. commercial airport system*—virtually all of it developed with federal grants and tax-free municipal bonds—*at $1 trillion*.
>
> Not until 1971 did the federal government begin collecting user fees from airline passengers and freight shippers to recoup this investment. In 1988 the Congressional Budget Office found that *in spite of user fees paid into the Airport and Airways Trust Fund, the taxpayers still had to transfer $3 billion in subsidies per year to the FAA* to maintain its network of more than 400 control towers, 22 air traffic control centers, 1,000 radar-navigation aids, 250 long-range and terminal radar systems and its staff of 55,000 traffic controllers, technicians and bureaucrats.[2]

(And even aside from the inadequacy of user fees, eminent domain remains central to the building of new airports and expansion of existing ones.)

Subsidies to the airport and air traffic control infrastructure of the civil aviation system are only part of the picture. Equally important was the direct role of the state in creating the heavy aircraft industry, whose heavy cargo and passenger jets revolutionized civil aviation after WWII. The civil aviation system is, many times over, a creature of the state.

In *Harry Truman and the War Scare of 1948*, Frank Kofsky described the aircraft industry as spiraling into red ink after the end of the war, and on the verge of bankruptcy when it was rescued by Truman's new bout of Cold War spending on heavy bombers.[3] David Noble pointed out that civilian jumbo jets would never have existed without the government's heavy bomber contracts. The production runs for the civilian market alone were too small to pay for the complex and expensive machinery. The 747 is essentially a spinoff of military production.[4]

The permanent war economy associated with the Cold War prevented the U.S. from relapsing into depression after demobilization. The Cold War restored the corporate economy's heavy reliance on the state as a source of guaranteed sales. Charles Nathanson argued that "one conclusion is inescapable: major firms with huge aggregations of corporate capital owe their survival after World War II

restrictions dropped," *Railway Age*, May 1997 <http://findarticles.com/p/articles/mi_m1215/is_n5_v198/ai_19460645>.

[1]David S. Lawyer, "Are Roads and Highways Subsidized?" March 2004 <http://www.lafn.org/~dave/trans/econ/highway_subsidy.html>.

[2]James Coston, Amtrak Reform Council, 2001, in "America's long history of subsidizing transportation" <http://www.trainweb.org/moksrail/advocacy/resources/subsidies/transport.htm>.

[3]Frank Kofsky, *Harry Truman and the War Scare of 1948*, (New York: St. Martin's Press, 1993).

[4]Noble, *America by Design*, pp. 6-7.

to the Cold War. . . . "[1] According to David Noble, employment in the aircraft industry grew more than tenfold between 1939 and 1954. Whereas military aircraft amounted to only a third of industry output in 1939. By 1953, military airframe weight production was 93% of total output.[2] "The advances in aerodynamics, metallurgy, electronics, and aircraft engine design which made supersonic flight a reality by October 1947 were underwritten almost entirely by the military."[3]

As Marx pointed out in Volume Three of *Capital*, the rise of major new forms of industry could absorb surplus capital and counteract the falling direct rate of profit." Baran and Sweezy, likewise, considered "epoch-making inventions" as partial counterbalances to the ever-increasing surplus. Their chief example was the rise of the automobile industry in the 1920s, which (along with the highway program) was to define the American economy for most of the mid-20th century.[4] The high tech boom of the 1990s was a similarly revolutionary event. It is revealing to consider the extent to which both the automobile and computer industries, far more than most industries, were direct products of state capitalism.

Besides civilian jumbo jets, many other entirely new industries were also created almost entirely as a byproduct of military spending. Through the military-industrial complex, the state has socialized a major share—probably the majority—of the cost of "private" business's research and development. If anything the role of the state as purchaser of surplus economic output is eclipsed by its role as subsidizer of research cost, as Charles Nathanson pointed out. Research and development was heavily militarized by the Cold War "military-R&D complex." Military R&D often results in basic, general use technologies with broad civilian applications. Technologies originally developed for the Pentagon have often become the basis for entire categories of consumer goods.[5] The general effect has been to "substantially [eliminate] the major risk area of capitalism: the development of and experimentation with new processes of production and new products."[6]

This is the case in electronics especially, where many products originally developed by military R&D "have become the new commercial growth areas of the economy."[7] Transistors and other miniaturized circuitry were developed primarily with Pentagon research money. The federal government was the primary market for large mainframe computers in the early days of the industry; without government contracts, the industry might never have had sufficient production runs to adopt mass production and reduce unit costs low enough to enter the private market.

Overall, Nathanson estimated, industry depended on military funding for around 60% of its research and development spending; but this figure is considerably understated by the fact that a significant part of nominally civilian R&D spending is aimed at developing civilian applications for military technology.[8] It is also understated by the fact that military R&D is often used for developing produc-

[1]Charles Nathanson, "The Militarization of the American Economy," in David Horowitz, ed., *Corporations and the Cold War* (New York and London: Monthly Review Press, 1969), p. 214.

[2]David F. Noble, *Forces of Production: A Social History of American Automation* (New York: Alfred A. Knopf, 1984), pp. 5-6.

[3]*Ibid.*, p. 6.

[4]Baran and Sweezy, *Monopoly Capitalism*, p. 220.

[5]"The Militarization of the American Economy," p. 208.

[6]*Ibid.*, p. 230.

[7]*Ibid.*, p. 230.

[8]*Ibid.*, pp. 222-25.

tion technologies that become the basis for production methods throughout the civilian sector.

In particular, as described by Noble in *Forces of Production*, industrial automation, cybernetics and miniaturized electronics all emerged directly from the military-funded R&D of WWII and the early Cold War. The aircraft, electronics and machine tools industries were transformed beyond recognition by the military economy.[1]

"The modern electronics industry," Noble writes, "was largely a military creation." Before the war, the industry consisted largely of radio.[2] Miniaturized electronics and cybernetics were almost entirely the result of military R&D.

> Miniaturization of electrical circuits, the precursor of modern microelectronics, was promoted by the military for proximity fuses for bombs. . . . Perhaps the most significant innovation was the electronic digital computer, created primarily for ballistics calculations but used as well for atomic bomb analysis. After the war, the electronics industry continued to grow, stimulated primarily by military demands for aircraft and missile guidance systems, communications and control instruments, industrial control devices, high-speed electronic computers for air defense command and control networks . . . , and transistors for all of these devices. . . . In 1964, two-thirds of the research and development costs in the electrical equipment industry (e.g., those of GE, Westinghouse, RCA, Raytheon, AT&T, Philco, IBM, Sperry Rand_ were still paid for by the government.[3]

The transistor, "the outgrowth of wartime work on semi-conductors," came out of Bell Labs in 1947. Despite obstacles like high cost and reliability, and resistance resulting from path dependency in the tube-based electronic industry, the transistor won out

> through the large-scale and sustained sponsorship of the military, which needed the device for aircraft and missile control, guidance, and communications systems, and for the digital command- and-control computers that formed the core of their defense networks.[4]

In cybernetics, likewise, the electronic digital computer was developed largely in response to military needs. ENIAC, developed for the Army at the University's Moore School of Electrical Engineering, was used for ballistics calculations and for calculations in the atomic bomb project.[5] Despite the reduced cost and increased reliability of hardware, and advances in computer language software systems, "in the 1950s the main users remained government agencies and, in particular, the military. The Air Force SAGE air defense system alone, for example, employed the bulk of the country's programmers . . . "

SAGE produced, among other things, "a digital computer that was fast enough to function as part of a continuous feedback control system of enormous complexity," which could therefore "be used continuously to monitor and control a vast array of automatic equipment in 'real time'. . . . " These capabilities were key to later advances industrial automation.[6]

The same pattern prevailed in the machine tool industry, the primary focus of *Forces of Production*. The share of total machine tools in use that were under ten years old rose from 28% in 1940 to 62% in 1945. At the end of the war, three hun-

[1]Noble, *Forces of Production*, p. 5.
[2]*Ibid.*, p. 7.
[3]*Ibid.*, pp. 7-8.
[4]*Ibid.*, pp. 47-48.
[5]*Ibid.*, p. 50.
[6]*Ibid.*, p. 52.

dred thousand machine tools were declared surplus and dumped on the commercial market at fire-sale prices. Although this caused the industry to contract (and consolidate), the Cold War resulted in a revival of the machine tools industry. R&D expenditures in machine tools expanded eightfold from 1951 to 1957, thanks to military needs. In the process, the machine tool industry became dominated by the "cost plus" culture of military industry, with its guaranteed profit.[1]

The specific technologies used in automated control systems for machine tools all came out of the military economy:

> . . . [T]he effort to develop radar-directed gunfire control systems, centered at MIT's Servomechanisms Laboratory, resulted in a range of remote control devices for position measurement and precision control of motion; the drive to develop proximity fuses for mortar shells produced miniaturized transceivers, early integrated circuits, and reliable, rugged, and standardized components. Finally, by the end of the war, experimentation at the National Bureau of Standards, as well as in Germany, had produced magnetic tape, recording heads (tape readers), and tape recorders for sound movies and radio, as well as information storage and programmable machine control.[2]

In particular, World War II R&D for radar-directed gunfire control systems were the primary impetus behind the development of servomechanisms and automatic control,

> pulse generators, to convey precisely electrical information; transducers, for converting information about distance, heat, speed, and the like into electrical signals; and a whole range of associated actuating, control and sensing devices.[3]

Industrial automation was introduced in private industry by the same people who had developed the technology for the military economy. The first analog computer-controlled industrial operations were in the electrical power and petroleum refining industries in the 1950s. By 1959, Texaco's Port Arthur refinery placed production under full digital computer control, and was followed in 1960 by Monsanto's Louisiana ammonia plant and B. F. Goodrich's vinyl plant in Calvert, Kentucky. From there the revolution quickly spread to steel rolling mills, blast furnaces, and chemical processing plants. By the 1960s, computerized control evolved from open-loop to closed-loop feedback systems, with computers making adjustments automatically based on sensor feedback.[4]

Numerically controlled machine tools, in particular, were first developed with Air Force money, and first introduced (both with Air Force funding and under Air Force pressure) in the aircraft and the aircraft engines and parts industries, and in USAF contractors in the machine tool industry.[5]

So the military economy and other state-created industries were an enormous sponge for surplus capital and surplus output. The heavy industrial and high tech sectors were given a virtually guaranteed outlet, not only by U.S. military procurement, but by grants and loan guarantees for foreign military sales under the Military Assistance Program.

Although apologists for the military-industrial complex have tried to stress the relatively small fraction of total production represented by military goods, it makes more sense to compare the volume of military procurement to the amount of idle capacity. Military production runs amounting to a minor percentage of total

[1] *Ibid.*, pp. 8-9.
[2] *Ibid.*, p. 47.
[3] *Ibid.*, pp. 48-49.
[4] *Ibid.*, pp. 60-61.
[5] *Ibid.*, p. 213.

production might absorb a major part of total idle production capacity, and have a huge effect on reducing unit costs. Besides, the rate of profit on military contracts tends to be quite a bit higher, given the fact that military goods have no "standard" market price, and the fact that prices are set by political means (as periodic Pentagon budget scandals should tell us).[1] So military contracts, small though they might be as a portion of a firm's total output, might well make the difference between profit and loss.

Seymour Melman described the "permanent war economy" as a privately-owned, centrally-planned economy that included most heavy manufacturing and high tech industry. This "*state-controlled economy*" was based on the principles of "maximization of costs and of government subsidies."[2]

> It can draw on the federal budget for virtually unlimited capital. It operates in an insulated, monopoly market that makes the state-capitalist firms, singly and jointly, impervious to inflation, to poor productivity performance, to poor product design and poor production managing. The subsidy pattern has made the state-capitalist firms failure-proof. That is the state-capitalist replacement for the classic self-correcting mechanisms of the competitive, cost-minimizing, profit-maximizing firm.[3]

A great deal of what is called "progress" amounts, not to an increase in the volume of consumption per unit of labor, but to an increase in the inputs consumed per unit of consumption—namely, the increased cost and technical sophistication entailed in a given unit of output, with no real increase in efficiency.

The chief virtue of the military economy is its utter unproductivity. That is, it does not compete with private industry to supply any good for which there is consumer demand. But military production is not the only such area of unproductive government spending. Neo-Marxist Paul Mattick elaborated on the theme in a 1956 article. The overbuilt corporate economy, he wrote, ran up against the problem that "[p]rivate capital formation . . . finds its limitation in diminishing market-demand." The State had to absorb part of the surplus output; but it had to do so without competing with corporations in the private market. Instead, "[g]overnment-induced production is channeled into non-market fields—the production of non-competitive public-works, armaments, superfluities and waste.[4] As a necessary result of this state of affairs,

> so long as the principle of competitive capital production prevails, steadily growing production will in increasing measure be a "production for the sake of production," benefiting neither private capital nor the population at large.
>
> This process is somewhat obscured, it is true, by the apparent profitability of capital and the lack of large-scale unemployment. Like the state of prosperity, profitability, too, is now largely government manipulated. Government spending and taxation are managed so as to strengthen big business at the expense of the economy as a whole. . . .
>
> In order to increase the scale of production and to accummulate [sic] capital, government creates "demand" by ordering the production of non-marketable goods, financed by government borrowings. This means that the government avails itself of productive resources belonging to private capital which would otherwise be idle.[5]

[1]Nathanson, "The Militarization of the American Economy," p. 208.

[2]Seymour Melman, *The Permanent War Economy: American Capitalism in Decline* (New York: Simon and Schuster, 1974), p. 11.

[3]*Ibid.*, p. 21.

[4]Paul Mattick, "The Economics of War and Peace," *Dissent* (Fall 1956), p. 377.

[5]*Ibid.*, pp. 378-379.

Such consumption of output, while not always directly profitable to private industry, serves a function analogous to foreign "dumping" below cost, in enabling industry to operate at full capacity despite the insufficiency of private demand to absorb the entire product at the cost of production.

It's interesting to consider how many segments of the economy have a guaranteed market for their output, or a "conscript clientele" in place of willing consumers. The "military-industrial complex" is well known. But how about the state's education and penal systems? How about the automobile-trucking-highway complex, or the civil aviation complex? Foreign surplus disposal ("export dependant monopoly capitalism") and domestic surplus disposal (government purchases) are different forms of the same phenomenon.

E. Mene, Mene, Tekel, Upharsin (A Critique of Sloanism's Defenders)

Although Galbraith and Chandler commonly justified the corporation's power over the market in terms of its social benefits, they had things exactly backward. The "technostructure" can survive because it is enabled to be *less* responsive to consumer demand. An oligopoly firm in a cartelized industry, in which massive, inefficient bureaucratic corporations share the same bureaucratic culture, is protected from competition. The "innovations" Chandler so prized are made by a leadership completely out of touch with reality. These "innovations" succeed because they are determined by the organization for its own purposes, and the organization has the power to impose top-down "change" on a cartelized market, with little regard to consumer preferences, instead of responding flexibly to them. "Innovative strategies" are based, not on finding out what people want and providing it, but on inventing ever-bigger hammers and then forcing us to be nails. The large corporate organization is not more efficient at accomplishing goals received from outside; it is more efficient at accomplishing goals it sets for itself for its own purposes, and then using its power to adapt the rest of society to those goals.

So to turn to our original point, the apostles of mass production have all, at least tacitly, identified the superior efficiency of the large corporation with its control over the external environment. Sloanist mass production subordinates the consumer, and the rest of outside society, to the institutional needs of the corporation.

Chandler himself admitted as much, in discussing what he called a strategy of "productive expansion." Big business added new outlets that permitted it to make "more complete use" of its "centralized services and facilities."[1] In other words, "efficiency" is defined by the existence of "centralized facilities," as such; efficiency is then promoted by finding ways to make people buy the stuff the centralized facilities can produce running at full capacity.

The authoritarianism implicit in such thinking is borne out by Chandler disciple William Lazonick's circular understanding of "organizational success," as he discusses it in his survey of "innovative organizations" in Part III of *Business Organization and the Myth of the Market Economy*.[2] The centralized, managerialist technostructure is the best vehicle for "organizational success"—defined as what best suits the interests of the centralized, managerialist technostructure. And of

[1]Chandler, *The Visible Hand*, p. 487.

[2]William Lazonick, *Business Organization and the Myth of the Market Economy* (Cambridge, 1991).

course, such "organizational success" has little or nothing to do with what society outside that organization might decide, on its own initiative, that it wants. Indeed (as Galbraith argued), "organizational success" requires institutional mechanisms to prevent outside society from doing what it wants, in order to provide the levels of stability and predictable demand that the technostructure needs for its long planning horizons. These theories amount, in practice, to a circular argument that oligopoly capitalism is "successful" because it is most efficient at achieving the ends of oligopoly capitalism.

Lazonick's model of "*successful* capitalist development" raises the question "successful" for whom? His "innovative organization" is no doubt "successful" for the people who make money off it—but not for those at whose expense they make money. It is only "success" if one posits the goals and values of the organization as those of society, and acquiesces in whatever organizational supports are necessary to impose those values on the rest of society.

His use of the expression "value-creating capabilities" seems to have very little to do with the ordinary understanding of the word "value" as finding out what people want and then producing it more efficiently than anyone else. According to his (and Chandler's and Galbraith's) version of value, rather, the organization decides what it wants to produce based on the interests of its hierarchy, and then uses its organizational power to secure the stability and control it needs to carry out its self-determined goals without interference from the people who actually buy the stuff.

This parallels Chandler's view of "organizational capabilities," which he seemed to identify with an organization's power over the external environment. A telling example, as we saw in Chapter One, is Chandler's book on the tech industry.[1] For Chandler, "organizational capabilities" in the consumer electronics industry amounted to the artificial property rights by which the firm was able to exercise ownership rights over technology and over the skill and situational knowledge of its employees, and to prevent the transfer of technology and skill across corporate boundaries. Thus, his chapter on the history of the consumer electronics industry through the mid-20th century is largely an account of what patents were held by which companies, and who subsequently bought them.

The "innovation" Chandler and Lazonick lionize means, in practice, 1) developing processes so capital-intensive and high-tech that, if all costs were fully internalized in the price of the goods produced, consumers would prefer simpler and cheaper models; or 2) developing products so complex and prone to breakdown that, if cartelized industry weren't able to protect its shared culture from outside competition, the consumer would prefer a more durable and user-friendly model. Cartelized, over-built industry deals with overproduction through planned obsolescence, and through engineering a mass-consumer culture, and succeeds because cartelization restricts the range of consumer choice.

The "innovative products" that emerge from Chandler's industrial model, all too often, are what engineers call "gold-plated turds": horribly designed products with proliferating features piled one atop another with no regard to the user's needs, ease of use, dependability or reparability. For a good example, compare the acceptable Word 2003 to the utterly godawful Word 2007.[2]

[1]Alfred D. Chandler, Jr., *Inventing the Electronic Century* (New York: The Free Press, 2001), pp. 13-49.

[2]Alan Cooper's *The Inmates are Running the Asylum: Why High-Tech Products Drive Us Crazy and How to Restore the Sanity* (Indianapolis: Sams, 1999) is an excellent survey of

Chandler's version of "successful development" is a roaring success indeed, if we start with the assumption that society should be reengineered to desire what the technostructure wants to produce.

Robin Marris described this approach quite well. The bureaucratic culture of the corporation, he wrote,

> is likely to divert emphasis from the character of the goods and services produced to the skill with which these activities are organized. . . . The concept of consumer need disappears, and the only question of interest . . . is whether a sufficient number of consumers, irrespective of their "real need" can be persuaded to buy [a proposed new product]."[1]

As the satirist John Gall put it, the large organization tends to redefine the consumption of inputs as outputs.

> A giant program to conquer cancer is begun. At the end of five years, cancer has not been conquered, but one thousand research papers have been published. In addition, one million copies of a pamphlet entitled "You and the War Against Cancer" have been distributed. These publications will absolutely be regarded as Output rather than Input.A giant program to conquer cancer is begun. At the end of five years, cancer has not been conquered, but one thousand research papers have been published. In addition, one million copies of a pamphlet entitled "You and the War Against Cancer" have been distributed. These publications will absolutely be regarded as Output rather than Input.[2]

The marketing "innovations" Chandler trumpeted in *Scale and Scope*—in foods, for example, the techniques for "refining, distilling, milling, and processing"[3]—were actually expedients for ameliorating the inefficiencies imposed by large-scale production and long-distance distribution: refined white flour, inferior in taste and nutrition to fresh-milled local flour, but which would keep for long-term storage; gas-ripened rubber tomatoes and other vegetables grown for transportability rather than taste; etc. The standard American diet of refined white flour, hydrogenated oils, and high fructose corn syrup is in large part a tribute to Chandler.

F. The Pathologies of Sloanism

Not only are the large and capital-intensive manufacturing corporations themselves characterized by high overhead and bureaucratic style; their organizational culture contaminates the entire system, becoming a hegemonic norm copied even by small organizations, labor-intensive firms, cooperatives and nonprofits. In virtually every field of endeavor, as Goodman put it, there is a "need for amounts of capital out of proportion to the nature of the enterprise." Every aspect of social life becomes dominated by the high overhead organization.

Goodman classifies organizations into a schema. Categories A and B, respectively, are "enterprises extrinsically motivated and interlocked with the other centralized systems," and "enterprises intrinsically motivated and tailored to the con-

the tendency of American industry to produce gold-plated turds without regard to the user.

[1]Quoted in Stein, *Size, Efficiency, and Community Enterprise*, p. 55.

[2]John Gall, *Systemantics: How Systems Work and Especially How They Fail* (New York: Pocket Books, 1975), p. 74.

[3]Alfred Chandler, *Scale and Scope: The Dynamics of Industrial Capitalism* (Cambridge and London: The Belknap Press of Harvard University Press, 1990), p. 262.

crete product or service." The two categories are each subdivided, roughly, into profit and nonprofit classes.

The interesting thing is that the large institutional nonprofits (Red Cross, Peace Corps, public schools, universities, etc.) are not counterweights to for-profit culture. Rather, they share the *same* institutional culture: "status salaries and expense accounts are equally prevalent, excessive administration and overhead are often more prevalent, and there is less pressure to trim costs."

Rather than the state and large nonprofits acting as a "countervailing power" on large for-profit enterprise, in Galbraith's schema, what happens more often is a coalition of the large for-profit and large nonprofit:

> . . . the military-industrial complex, the alliance of promoters, contractors, and government in Urban Renewal; the alliance of universities, corporations, and government in research and development. This is the great domain of cost-plus.[1]

Goodman contrasts the bureaucratic organization with the small, libertarian organization. "What swell the costs in enterprises carried on in the interlocking centralized systems of society, whether commercial, official, or non-profit institutional,"

> are all the factors of organization, procedure, and motivation that are not directly determined to the function and to the desire to perform it. These are patents and rents, fixed prices, union scales, featherbedding, fringe benefits, status salaries, expense accounts, proliferating administration, paper work, permanent overhead, public relations and promotion, waste of time and sill by departmentalizing task-roles, bureaucratic thinking that is penny-wise and pound-foolish, inflexible procedure and tight scheduling that exaggerate contingencies and overtime.
>
> But when enterprises can be carried on autonomously by professionals, artists, and workmen intrinsically committed to the job, there are economies all along the line. People make do on means. They spend on value, not convention. They flexibly improvise procedures as opportunity presents and they step in in emergencies. They do not watch the clock. The available skills of each person are put to use. They eschew status and in a pinch accept subsistence wages. Administration and overhead are *ad hoc*. The task is likely to be seen in its essence rather than abstractly.

Instead of expensive capital outlays, the ad hoc organization uses spare capacity of existing small-scale capital goods its members already own, along with recycled or vernacular building materials. The staff of a small self-managed organization are free to use their own judgment and ingenuity in formulating solutions to unforeseen problems, cutting costs, and so forth. And because the staff is often the source of the capital investments, they are likely to be quite creative in finding ways to save money.

A couple of things come to mind here. First, Friedrich Hayek's treatment of distributed knowledge: those directly engaged in a task are usually the best source of ideas for improving its efficiency. And second, Milton Friedman's ranking of the relative efficiencies achieved by 1) people spending other people's money on other people; 2) people spending other people's money on themselves; 3) people spending their own money on other people; and 4) people spending their own money on themselves.

The staff of a small, self-directed undertaking can afford to throw themselves into maximizing their effectiveness, because they know the efficiency gains they produce won't be appropriated by absentee owners or senior management who

[1]Paul Goodman, *People or Personnel*, pp. 114-115.

simply use the higher productivity to skim more profit off the top or to lay off some of the staff. Most of the features of Weberian bureaucracy and hierarchical systems of control—job descriptions, tracking forms and controls, standard procedures, and the like—result from the fact that the workforce has absolutely no rational interest in expending effort or working effectively, beyond the bare minimum required to keep the employer in business and to avoid getting fired.

Goodman's chapter on "Comparative Costs" in *People or Personnel* is a long series of case studies contrasting the cost of bureaucratic to ad hoc organizations.[1] He refers, for example, to the practices at a large corporate TV station ("the usual featherbedding of stagehands to provide two chairs," or paying technicians "twice $45 to work the needle on a phonograph")—jobs that would be done by the small permanent staff at a nonprofit station run out of City College of New York.[2] The American Friends' Voluntary International Service Assignments carried almost no administrative costs, compared to the Peace Corps' enormous cost of thousands of dollars per volunteer.[3]

The Housing Board's conventional Urban Renewal proposal in Greenwich Village would have bulldozed a neighborhood containing many useful villages, to be replaced by "the usual bureaucratically designed tall buildings," at a cost of $30 million and a net increase of 300 dwelling units. The neighborhood offered a counter-proposal that ruled out demolishing anything salvageable or relocating anyone against their wishes; it would have provided a net increase of 475 new units at a cost of $8.5 million. Guess which one was chosen?[4]

Most of the per pupil cost of conventional urban public schools, as opposed to alternative or experimental schools, results from administrative overhead and the immense cost of buildings and other materials built to a special set of specifications at some central location on some of the most expensive real estate in town. His hypothetical cooperative prep school cost about a third as much per pupil as the typical high school.[5] This is a thought experiment I'd repeatedly conducted for myself long before ever reading Goodman: figuring the cost for twenty or so parents to set up their own schooling cooperative, renting a house for classroom space and hiring a few part-time instructors, and then trying to imagine how one could *possibly* waste enough money to come up with the $8,000 or more per-pupil that the public schools typically spend.

In the nearby town of Siloam Springs, Ark., not long after voters rejected a millage increase for the schools, the administration announced the cancellation of its planned purchase of new computers and its decision instead to upgrade existing ones. The cost of adding RAM, it was said, would be a small fraction of replacement—and yet it would result in nearly the same performance improvement. But it's a safe guess the administration would never have considered such a thing if it hadn't been forced to.

Another similar case is Goodman's contrast of the tuition costs of the typical large, institutional college, to those of an "alternative" school like Black Mountain College (run by the faculty, on the same "scholars' guild" model as the medieval universities). Much of the physical plant of the latter was the work of faculty and staff, and indeed for its first eight years (1933-1941) the "campus" consisted of build-

[1]*Ibid.*, pp. 94-122.
[2]*Ibid.*, pp. 102-104.
[3]*Ibid.*, pp. 107-110.
[4]*Ibid.*, pp. 110-111.
[5]*Ibid.*, p. 105.

ings rented from a YMCA. Without any endowment or contributions, the tuition was still far lower than that of a conventional college.[1]

A more contemporary example might be the enormous cost of conventional Web 2.0 firms compared to that of their free culture counterparts. The Pirate Bay's file-sharing operations, for example, cost only $3,000 a month—compared to estimated *daily* operating costs for YouTube ranging from $130,000 to a million![2]

The contrasting styles of the ad hoc, self-managed organization and the bureaucratic, institutional organization were brought home to me in my personal experience with two libraries.

At the University of Arkansas (Fayetteville), until a few years ago, non-students were discouraged from applying for library cards by an application form that asked whether their needs could not be met instead by, among other things, relying on Interlibrary Loan services. Then the policy changed so that a library card (with $40 annual fee) was required to use Interlibrary Loan. Never mind that a library official professed unawareness (while hardly bothering to conceal her disbelief), in her best "Oceania has always at war with Eastasia" manner, that the library had ever promoted Interlibrary Loan as an *alternative* to a library card. The interesting thing was that she justified the new card purchase requirement on grounds of equity: it cost, she claimed, some $25 to process every Interlibrary Loan request. I was utterly dumbfounded. If this were true, you'd think the ILL bureaucracy would be *ashamed* to admit it. How does Amazon.Com or AbeBooks manage to stay in business when *buying* a used book and shipping it cross-country usually costs me less than that—shipping and handling included? The only answer must be that the library bureaucracy has far higher levels of bureaucratic overhead than even a large bureaucratic corporation, for performing an analogous function.

At the Springdale, Ark. public library, I submitted a written complaint to their Technology Coordinator regarding the abysmally poor performance of their new desktop software after the recent "upgrade," compared to what they had had before.

> Comment: Please don't automatically upgrade the desktops to the latest version of Windows and other MS accessories.
>
> In general, if you already have something from Microsoft that works in a minimally acceptable manner, you should quit while you're ahead; if Bill Gates offers you something "new" and "better," run in the opposite direction as fast as you can.
>
> Since you "upgraded" the computers, if you can call it that, usability has suffered a nosedive. I used to have no problem emailing myself attachments and opening them up here to work on. Now if I want to print something out, I have to open it as a Google Document and paste it into a new Word file. What's more, I can't edit the file here and save it to the desktop so I can email it to myself again. Any time I attempt to save a textfile on your computers I'm blocked from doing so.
>
> In addition, if you compare Word 2007 to the Word 2003 you previously had on the desktop menu, the former is a classic example of what engineers call a "gold-plated turd." It's got so many proliferating "features" that the editing dashboard has to be tabbed to fit them all in.
>
> To summarize: your computers worked just fine for all my purposes before the so-called "upgrade," and now they're godawful. Please save yourselves money

[1] *Ibid.*, p. 106; "Black Mountain College," *Wikipedia* <http://en.wikipedia.org/wiki/Black_Mountain_College> (captured March 30, 2009).

[2] Janko Roettgers, "The Pirate Bay: Distributing the World's Entertainment for $3,000 a Month," *NewTeeVee.Com*, July 19, 2009 <http://newteevee.com/2009/07/19/the-pirate-bay-distributing-the-worlds-entertainment-for-3000-a-month/>.

> in future and stick with what works instead of being taken in by Microsoft's latest poorly designed crap.

The Coordinator, C.M., replied (rather lamely in my opinion) that "the recent upgrade to MicroSoft Office 2007 on both the Library's public and staff computers is in line with what other libraries and companies across the country currently offer/use as office productivity software." And the refusal to save files to desktop, which the previous software had done without a problem, was "a standard security feature."

Now, this would be perfectly understandable from a grandma, who uses the computer mainly to read email from her grandkids, and buys her granddaughter a PC with Vista and Word 2007 installed because "I heard it's the latest thing." But this was an IT officer—someone who's supposed to be at least vaguely aware of what's going on.

So I told her the software was a piece of crap that didn't work, and Ms. C. M. (although I'm sure it wasn't her intention) told me *why* it was a piece of crap that didn't work: Springdale's library adopted it because it was what all the other libraries and corporations use. I replied, probably a little too testily:

> . . . I'm afraid the fact that an upgrade "in line with what other libraries and companies across the country currently offer/use" actually made things worse reflects unflatteringly on the institutional culture that predominates in organizations across the country, and in my opinion suggests the folly of being governed by the institutional culture of an industry rather than bottom-up feedback from one's own community of users.
>
> I've worked in more than one job where company policy reflected the common institutional culture of the industry, and whatever "best practice" du jour the other CEOs solemnly assured our CEO was working like gangbusters. Had there been less communication between the people at the tops of the pyramids, and more communication between the top of each pyramid with those below, the people in direct contact with the situation might have cut through the . . . official happy talk and told them what a total clusterf*** their policies had resulted in.

For some reason, I never heard back.

The state and its affiliated corporate system, by mandating minimum levels of overhead for supplying all human wants, creates what Ivan Illich called "radical monopolies."

> I speak about radical monopoly when one industrial production process exercises an exclusive control over the satisfaction of a pressing need, and excludes nonindustrial activities from competition. . . .
>
> Radical monopoly exists where a major tool rules out natural competence. Radical monopoly imposes compulsory consumption and thereby restricts personal autonomy. It constitutes a special kind of social control because it is enforced by means of the imposed consumption of a standard product that only large institutions can provide.[1]
>
> Radical monopoly is first established by a rearrangement of society for the benefit of those who have access to the larger quanta; then it is enforced by compelling all to consume the minimum quantum in which the output is currently produced. . . .[2]

[1]Ivan Illich, *Tools for Conviviality* (New York, Evanston, San Francisco, London: Harper & Row, 1973), pp. 52-53.

[2]Illich, *Energy and Equity* (1973), Chapter Six (online edition courtesy of Ira Woodhead and Frank Keller) <http://www.cogsci.ed.ac.uk/~ira/illich/texts/energy_and_equity/energy_and_equity.html>.

The goods supplied by a radical monopoly can only be obtained at comparably high expense, requiring the sale of wage labor to pay for them, rather than direct use of one's own labor to supply one's own needs. The effect of radical monopoly is that capital-, credential- and tech-intensive ways of doing things crowd out cheaper and more user-friendly, more libertarian and decentralist, technologies. The individual becomes increasingly dependent on credentialed professionals, and on unnecessarily complex and expensive gadgets, for all the needs of daily life. He experiences an increased cost of subsistence, owing to the barriers that mandatory credentialing erects against transforming one's labor directly into use-value (Illich's "convivial" production), and the increasing tolls levied by the licensing cartels and other gatekeeper groups.

> People have a native capacity for healing, consoling, moving, learning, building their houses, and burying their dead. Each of these capacities meets a need. The means for the satisfaction of these needs are abundant so long as they depend on what people can do for themselves, with only marginal dependence on commodities. . . .
>
> These basic satisfactions become scarce when the social environment is transformed in such a manner that basic needs can no longer be met by abundant competence. The establishment of a radical monopoly happens when people give up their native ability to do what they can do for themselves and each other, in exchange for something "better" that can be done for them only by a major tool. Radical monopoly reflects the industrial institutionalization of values. . . . It introduces new classes of scarcity and a new device to classify people according to the level of their consumption. This redefinition raises the unit cost of valuable services, differentially rations privileges, restricts access to resources, and makes people dependent.[1]

The overall process is characterized by "the replacement of general competence and satisfying subsistence activities by the use and consumption of commodities;"

> the monopoly of wage-labor over all kinds of work; redefinition of needs in terms of goods and services mass-produced according to expert design; finally, the arrangement of the environment . . . [to] favor production and consumption while they degrade or paralyze use-value oriented activities that satisfy needs directly.[2]

Leopold Kohr observed that "what has actually risen under the impact of the enormously increased production of our time is not so much the standard of living as the level of subsistence."[3] Or as Paul Goodman put it, "decent poverty is almost impossible."[4]

For example: subsidized fuel, freeways, and automobiles generate distance between things, so that "[a] city built around wheels becomes inappropriate for feet."[5] The car becomes an expensive necessity; feet and bicycle are rendered virtually useless, and the working poor are forced to earn the additional wages to own and maintain a car just to be *able* to work at all.

[1]Illich, *Tools for Conviviality*, p. 54.

[2]Illich, *Vernacular Values* (1980), "Part One: The Three Dimensions of Social Choice," online edition courtesy of The Preservation Institute <http://www.preservenet.com/theory/Illich/Vernacular.html>.

[3]Leopold Kohr, *The Overdeveloped Nations: The Diseconomies of Scale* (New York: Schocken Books, 1978, 1979), pp. 27-28.

[4]Goodman, *Compulsory Miseducation*, in *Compulsory Miseducation* and *The Community of Scholars* (New York: Vintage books, 1964, 1966), p. 108.

[5]Illich, *Disabling Professions* (New York and London: Marion Boyars, 1977), p. 28.

Radical monopoly has a built-in tendency to perpetuate itself and expand. First of all, those running large hierarchical organizations tend to solve the problems of bureaucracy by adding more of it. In the hospital where I work, this means that problems resulting from understaffing are "solved" by new tracking forms that further reduce nurses' available time for patient care—when routine care already frequently goes undone, and nurses stay over two or three hours past the end of a twelve-hour shift to finish paperwork.

They solve problems, in general, with a "more of the same" approach. In Illich's excellent phrase, it's an attempt to "solve a crisis by escalation."[1] It's what Einstein referred to as trying to solve problems "at the same level of thinking we were at when we created them." Or as E. F. Schumacher says of intellectuals, technocrats "always tend to try and cure a disease by intensifying its causes."[2]

The way the process works, in Paul Goodman's words, is that "[a] system destroys its competitors by pre-empting the means and channels and then proves that it is the only conceivable mode of operating."[3]

The effect is to make subsistence goods available only through institutional providers, in return for money earned by wages, at enormous markup. As Goodman put it, it makes decent poverty impossible. To take the neoliberals' statistical gushing over increased GDP and stand it on its head, "[p]eople who were poor and had food now cannot subsist on ten or fifty times the income."[4] "Everywhere one turns . . . there seems to be a markup of 300 and 400 percent, to do anything or make anything."[5] And paradoxically, the more "efficiently" an organization is run, "the more expensive it is per unit of net value, if we take into account the total social labor involved, both the overt and the covert overhead."[6]

Goodman points to countries where the official GDP is one fourth that of the U.S., and yet "these unaffluent people do not seem four times 'worse off' than we, or hardly worse off at all."[7] The cause lies in the increasing portion of GDP that goes to support and overhead, rather than direct consumption. Most of the costs do not follow from the technical requirements of producing direct consumption goods themselves, but from the mandated institutional structures for producing and consuming them.

> It is important to notice how much the various expensive products and services of corporations and government make people subject to repairmen, fees, commuting, queues, unnecessary work, dressing just for the job; and these things often prevent satisfaction altogether.[8]

A related phenomenon is what Kenneth Boulding called the "non-proportional change" principle of structural development: the larger an institution grows, the larger the proportion of resources that must be devoted to secondary, infrastructure and support functions rather than the actual primary function of the institution. "As any structure grows, the proportions of the parts and of its significant variables *cannot* remain constant. . . . This is because a uniform increase in

[1]Illich, *Tools for Conviviality*, p. 9.

[2]E. F. Schumacher, *Small is Beautiful: Economics as if People Mattered* (New York, Hagerstown, San Francisco, London: Harper & Row, Publishers, 1973), p. 38.

[3]Goodman, *People or Personnel*, p. 70.

[4]*Ibid.*, p. 70.

[5]*Ibid.*, p. 120.

[6]Goodman, *The Community of Scholars*, in *Compulsory Miseducation* and *The Community of* Scholars, p. 241.

[7]Goodman, *People or Personnel*, p. 120.

[8]*Ibid.*, p. 117.

the linear dimensions of a structure will increase all its areas as the square, and its volume as the cube, of the increase in the linear dimension. . . . "[1]

Leopold Kohr gave the example of a skyscraper: the taller the building, the larger the percentage of floorspace that must taken up with elevator shafts and stairwells, heating and cooling ducts, and so forth. Eventually, the building reaches the point where the space on the last floor added will be cancelled out by the increased space required for support structures. This is hardly theoretical: Kohr gave the example in the 1960s of a $25 billion increase in GNP, $18 billion (or 72%) of which went to administrative and support costs of various sorts.[2]

G. Mandatory High Overhead

As a pathology, this phenomenon deserves a separate section of its own. It is a pathology not only of the Sloanist mass-production economy, but also of local economies under the distorting effects of zoning, licensing, "safety" and "health" codes, and other regulations whose primary effect is to put a floor under overhead costs. Social regulations and commercial prohibitions, as Thomas Hodgskin said, "compel us to employ more labour than is necessary to obtain the prohibited commodity," or "to give a greater quantity of labour to obtain it than nature requires," and put the difference into the pockets of privileged classes.[3]

Such artificial property rights enable the privileged to appropriate productivity gains for themselves, rather than allowing their benefits to be socialized through market competition.

But they do more than that: they make it possible to collect tribute for the "service" of *not* obstructing production. As John R. Commons observed, the alleged "service" performed by the holder of artificial property rights, in "contributing" some "factor" to production, is defined entirely by his ability to obstruct access to it. As I wrote in *Studies in Mutualist Political Economy*, marginalist economics

> treated the existing structure of property rights over "factors" as a given, and proceeded to show how the product would be distributed among these "factors" according to their marginal contribution. By this method, if slavery were still extant, a marginalist might with a straight face write of the marginal contribution of the slave to the product (imputed, of course, to the slave-owner), and of the "opportunity cost" involved in committing the slave to one or another use.[4]

Such privileges, Maurice Dobb argued, were analogous to a state grant of authority to collect tolls, (much like the medieval robber barons who obstructed commerce between their petty principalities):

> Suppose that toll-gates were a general institution, rooted in custom or ancient legal right. Could it reasonably be denied that there would be an important sense in which the income of the toll-owning class represented "an appropriation of goods produced by others" and not payment for an "activity directed to the production or transformation of economic goods?" Yet toll-charges would be fixed in competition with alternative roadways, and hence would, presumably, represent prices fixed "in an open market. . . . " Would not the opening and shutting of toll-gates become an essential factor of production, according to most current

[1] Kenneth Boulding, *Beyond Economics* (Ann Arbor: University of Michigan Press, 1968), p. 75.

[2] Kohr, *The Overdeveloped Nations*, pp. 36-37.

[3] Thomas Hodgskin, *Popular Political Economy: Four Lectures Delivered at the London Mechanics' Institution* (London: Printed for Charles and William Tait, Edinburgh, 1827), pp. 33-34.

[4] Kevin Carson, *Studies in Mutualist Political Economy* (Blitzprint, 2004), p. 79.

> definitions of a factor of production, with as much reason at any rate as many of the functions of the capitalist entrepreneur are so classed to-day? This factor, like others, could then be said to have a "marginal productivity" and its price be regarded as the measure and equivalent of the service it rendered. At any rate, where is a logical line to be drawn between toll-gates and property-rights over scarce resources in general?[1]

Thorstein Veblen made a similar distinction between property as capitalized serviceability, versus capitalized disserviceability. The latter consisted of power advantages over rivals and the public which enabled owners to obstruct production.[2]

At the level of the national corporate economy, a centtral function of government is to artificially inflate the levels of capital outlay and overhead needed to undertake production.

The single biggest barrier to modular design for common platforms is probably "intellectual property." If it were abolished, there would be no legal barrier against many small companies producing competing modular components or accessories for the same platform, or even big companies producing modular components designed for interoperability with other companies products.

What's more, with the barrier to such competition removed, there would be a great deal of competitive advantage from designing one's product so as to be conducive to production of modular components by other companies. In a market where the consumer preferred the highest possible degree of interoperability and cross-compatibility, to maximize his own freedom to mix 'n' match components, or to maximize his options for extending the lifetime of the product, a product that was designed with such consumer behavior in mind would have a leg up on competing products designed to be incompatible with other companies' accessories and modules. In other words, products designed to be easily used with other people's stuff would sell better. Imagine if

- Ford could produce engine blocks that were compatible with GM chasses, and vice versa;
- if a whole range of small manufacturers could produce competing spare parts and modular accessories for Ford or GM vehicles;
- such small companies, individually or in networks, could produce entire competing car designs around the GM or Ford engine block;
- or many small assembly plants sprang up to put together automobiles from engine blocks ordered from Ford or GM, combined with other components produced by themselves or a wide variety of other small companies on the Emilia-Romagna networked model.

Under those circumstances, there would be no legal barrier to other companies producing entire, modularization-friendly design platforms for use around Ford or GM products, and Ford and GM would find it to their competitive advantage to facilitate compatibility with such designs.

In keeping with Sloanism's emphasis on planned obsolescence to generate artificially high levels of product turnover, products are deliberately designed to discourage or impede repair by the user.

> . . . [A]n engineering culture has developed in recent years in which the object is to "hide the works," rendering the artifacts we use unintelligible to direct inspec-

[1]Maurice Dobb, *Political Economy and Capitalism: Some Essays in Economic Tradition*, 2nd rev. ed. (London: Routledge & Kegan Paul Ltd, 1940, 1960), p. 66

[2]Thorstein Veblen, *The Place of Science in Modern Civilization and other Essays*, p. 352, in John R. Commons, *Institutional Economics* (New York: Macmillan, 1934), p. 664.

> tion. . . . This creeping concealedness takes various forms. The fasteners holding small appliances together now often require esoteric screwdrivers not commonly available, apparently to prevent the curious or the angry from interrogating the innards. By way of contrast, older readers will recall that until recent decades, Sears catalogues included blown-up parts diagrams and conceptual schematics for all appliances and many other mechanical goods. It was simply taken for granted that such information would be demanded by the consumer.[1]

Julian Sanchez gives the specific example of Apple's iPhone. The scenario, as he describes it, starts when

> 1) Some minor physical problem afflicts my portable device—the kind of thing that just happens sooner or later when you're carting around something meant to be used on the go. In this case, the top button on my iPhone had gotten jammed in, rendering it nonfunctional and making the phone refuse to boot normally unless plugged in.
>
> 2) I make a pro forma trip to the putative "Genius Bar" at an Apple Store out in Virginia. Naturally, they inform me that since this doesn't appear to be the result of an internal defect, it's not covered. But they'll be only too happy to service/replace it for something like $250, at which price I might as well just buy a new one. . . .
>
> 3) I ask the guy if he has any tips if I'm going to do it myself—any advice on opening it, that sort of thing. He's got no idea. . . .
>
> 4) Pulling out a couple of tiny screwdrivers, I start in on the satanic puzzle-box casing Apple locks around all its hardware. I futz with it for at least 15 minutes before cracking the top enough to get at the inner works.
>
> 5) Once this is done, it takes approximately five seconds to execute the necessary repair by unwedging the jammed button.
>
> I have two main problems with this. First, you've got what's obviously a simple physical problem that can very probably be repaired in all of a minute flat with the right set of tools. But instead of letting their vaunted support guys give this a shot, they're encouraging customers—many of whom presumably don't know any better—to shell out a ludicrous amount of money to replace it and send the old one in. I appreciate that it's not always obvious that a problem can be this easily remedied on site, but in the instance, it really seems like a case of exploiting consumer ignorance.
>
> Second, the iPhone itself is pointlessly designed to deter self service. Sure, the large majority of users are never going to want to crack their phone open. Then again, most users probably don't want to crack their desktops or laptops open, but we don't expect manufacturers to go out of their way to make it difficult to do.[2]

The iPhone is a textbook example of a "blobject," the product of industrial design geared toward the cheap injection-molding of streamlined plastic artifacts. Eric Hunting writes:

> Blobjects are also often deliberately irreparable and un-upgradeable -sometimes to the point where they are engineered to be unopenable without being destroyed in the process. This further facilitates planned obsolescence while also imposing limits on the consumer's own use of a product as a way to protect market share and technology propriety. Generally, repairability of consumer goods is now impractical as labor costs have made repair frequently more expensive than replacement, where it isn't already impossible by design. In the 90s car companies actually toyed with the notion of welding the hoods of new cars shut on the premise that the engineering of components had reached the state where noth-

[1]Matthew B. Crawford, "Shop Class as Soulcraft," *The New Atlantis*, Number 13, Summer 2006, pp. 7-24 <http://www.thenewatlantis.com/publications/shop-class-as-soulcraft>.

[2]Julian Sanchez, "Dammit, Apple," *Notes from the Lounge*, June 2, 2008 <http://www.juliansanchez.com/2008/06/02/dammit-apple/>.

ing in the engine compartment needed to be serviceable over a presumed 'typical' lifetime for a car. (a couple of years) This, of course, would have vastly increased the whole replacement rate for cars and allowed companies to hide a lot of dirty little secrets under that welded hood.[1]

"Intellectual property" in onboard computer software and diagnostic equipment has essentially the same effect.

> As cars become vastly more complicated than models made just a few years ago, [independent mechanic David] Baur is often turning down jobs and referring customers to auto dealer shops. Like many other independent mechanics, he does not have the thousands of dollars to purchase the online manuals and specialized tools needed to fix the computer-controlled machines. . . .
>
> Access to repair information is at the heart of a debate over a congressional bill called the Right to Repair Act. Supporters of the proposal say automakers are trying to monopolize the parts and repair industry by only sharing crucial tools and data with their dealership shops. The bill, which has been sent to the House Committee on Energy and Commerce, would require automakers to provide all information to diagnose and service vehicles.
>
> Automakers say they spend millions in research and development and aren't willing to give away their intellectual property. They say the auto parts and repair industry wants the bill passed so it can get patented information to make its own parts and sell them for less. . . .
>
> Many new vehicles come equipped with multiple computers controlling everything from the brakes to steering wheel, and automakers hold the key to diagnosing a vehicle's problem. In many instances, replacing a part requires reprogramming the computers —— a difficult task without the software codes or diagrams of the vehicle's electrical wires. . . .
>
> Dealership shops may be reaping profits from the technological advancements. A study released in March by the Automotive Aftermarket Industry Association found vehicle repairs cost an average of 34 percent more at new car dealerships than at independent repair shops, resulting in $11.7 billion in additional costs for consumers annually.
>
> The association, whose members include Autozone, Jiffy Lube and other companies that provide replacement parts and accessories, contend automakers want the bill rejected so they can continue charging consumers more money.
>
> "You pay all this money for your car, you should be able to decide where to get it repaired," said Aaron Lowe, the association's vice president of government affairs.
>
> Opponents of the bill counter that the information and tools to repair the vehicles are available to those willing to buy them.[2]

As Mike Masnick sums it up:

> Basically, as cars become more sophisticated and computerized, automakers are locking up access to those computers, and claiming that access is protected by copyrights. Mechanics are told they can only access the necessary diagnostics if they pay huge sums—meaning that many mechanics simply can't repair certain

[1]Eric Hunting, "On Defining a Post-Industrial Style (1): from Industrial blobjects to post-industrial spimes," *P2P Foundation Blog*, November 2, 2009 <http://blog.p2pfoundation.net/on-defining-a-post-industrial-style-1-from-industrial-blobjects-to-post-industrial-spimes/2009/11/02>.

[2]Daisy Nguyen, "High tech vehicles pose trouble for some mechanics," *North County Times*, December 26, 2009 <http://nctimes.com/news/state-and-regional/article_4ea03fd6-090d-5c2e-bd91-dfb5508495ef.html>.

cars, and car owners are forced to go to dealers, who charge significantly higher fees.[1]

One of Masnick's readers at *Techdirt* pointed out that a primary effect of "intellectual property" law in this case is to give manufacturers "an incentive to build crappy cars." If automakers have "an exclusive right to fix their own products," they will turn repair operations into a "cash cow." (Of course, that's exactly the same business model currently followed by companies that sell cheap platforms and make money off proprietary accessories and spare parts.) "Suddenly, the money made from repairing automobiles would outweigh the cost of selling them."

In a free market, of course, it wouldn't be necessary to pay for the information, or to pay proprietary prices for the tools, because software hacks and generic versions of the tools would be freely available without any legal impediment. That Congress is considering legislation to mandate the sharing of information protected by "intellectual property" law is a typical example of government's Rube Goldberg nature: all that's really needed is to eliminate the "intellectual property" in the fist place.

One effect of the shift in importance from tangible to intangible assets is that a growing portion of product prices consists of embedded rents on "intellectual property" and other artificial property rights rather than the material costs of production. Tom Peters cited former 3M strategic planner George Hegg on the increasing portion of product "value" made up of "intellect" (i.e., the amount of final price consisting of tribute to the owners of "intellectual property"): "We are trying to sell more and more intellect and less and less materials." Peters produces a long string of such examples:

> . . . My new Minolta 9xi is a lumpy object, but I suspect I paid about $10 for its plastic casing, another $50 for the fine-ground optical glass, and the rest, about $640, for its intellect . . . [2]
>
> It is a soft world. . . . Nike contracts for the production of its spiffy footwear in factories around the globe, but it creates the enormous stock value via superb design and, above all, marketing skills. Tom Silverman, founder of upstart Tommy Boy Records, says Nike was the first company to understand that it was in the lifestyle business. . . . Shoes? Lumps? Forget it! Lifestyle. Image. Speed. Value via intellect and pizazz.[3]
>
> "Microsoft's only factory asset is the human imagination," observed The New York Times Magazine writer Fred Moody. In seminars I've used the slide on which those words appear at least a hundred times, yet every time that simple sentence comes into view on the screen I feel the hairs on the back of my neck bristle.[4]
>
> A few years back, Philip Morris purchased Kraft for $12.9 billion, a fair price in view of its subsequent performance. When the accountants finished their work, it turned out that Philip Morris had bought $1.3 billion worth of "stuff" (tangible assets) and $11.6 billion of "Other." What's the other, the 116/129?
>
> . . . . Call it intangibles, good-will (the U.S. accountants' term), brand equity, or the ideas in the heads of thousands of Kraft employees around the world.[5]

[1]Mike Masnick, "How Automakers Abuse Intellectual Property Laws to Force You to Pay More For Repairs," *Techdirt*, December 29, 2009 <http://techdirt.com/articles/20091228/0345127515.shtml>.

[2]Tom Peters, *The Tom Peters Seminar: Crazy Times Call for Crazy Organizations* (New York: Vantage Books, 1999), p. 10.

[3]*Ibid.*, pp. 10-11.

[4]*Ibid.*, p. 11.

[5]*Ibid.* p. 12.

Regarding Peters' Minolta example, as Benkler points out the marginal cost of reproducing "its intellect" is virtually zero. So about 90% of the price of that new Minolta comes from tolls to corporate gatekeepers, who have been granted control of that "intellect."

The same goes for Nike's sneakers. I suspect the amortization cost of the physical capital used to manufacture the shoes in those Asian sweatshops, plus the cost of the sweatshop labor, is less than 10% of the price of the shoes. The wages of the workers could be tripled or quadrupled with negligible impact on the retail price.

In an economy where software and product design were the product of peer networks, unrestricted by the "intellectual property" of old corporate dinosaurs, 90% of the product's price would evaporate overnight. To quote Michael Perelman,

> the so-called weightless economy has more to do with the legislated powers of intellectual property that the government granted to powerful corporations. For example, companies such as Nike, Microsoft, and Pfizer sell stuff that has high value relative to its weight only because their intellectual property rights insulate them from competition.[1]

"Intellectual property" plays exactly the same protectionist role for global corporations that tariffs did for the old national industrial economies. Patents and copyrights are barriers, not to the movement of physical goods, but to the diffusion of technique and technology. The one, as much as the other, constitutes a monopoly of productive capability. "Intellectual property" enables the transnational corporation to benefit from the moral equivalent of tariff barriers, regardless of where it is situated. In so doing, it breaks the old link between geography and protectionism. "Intellectual property," exactly like tariffs, serves the primary function of legally restricting who can produce a given thing for a given market. With an American tariff on a particular kind of good, the corporations producing that good have a monopoly on it only within the American market. With the "tariff" provided by a patent on the industrial technique for producing that good, the same corporations have an identical monopoly in every single country in the world that adheres to the international patent regime.

How many extra hours does the average person work each week to pay tribute to the owners of the "human imagination"?

The Consumer Product Safety Improvement Act (CPSIA) is a good illustration of how regulations put a floor under overhead. To put it in perspective, first consider how the small apparel manufacturer operates. According to Eric Husman, an engineer who blogs on lean manufacturing and whose wife is in the apparel industry, a small apparel manufacturer comes up with a lot of designs, and then produces whatever designs sell, switching back and forth between products as the orders come in. Now consider the effect the CPSIA has on this model. Its most onerous provision is its mandate of third party testing and certification, not of materials, but of every component of each separate product.

> The testing and certification requires that finished products be tested, not materials, and that every component of every item must be tested separately. A price quote from a CPSIA-authorized testing facility says that testing Learning Resources' product Let's Tackle Kindergarten, a tackle box filled with learning tools—flash cards, shapes, counters and letters—will cost $6,144.

[1]Michael Perelman, "The Political Economy of Intellectual Property," *Monthly Review*, January 2003 <http://www.monthlyreview.org/0103perelman.htm>.

> Items made from materials known not to contain lead, or items tested to other comparable standards, must still be tested. A certified organic cotton baby blanket appliquéd with four fabrics must be tested for lead at $75 per component material. Award-winning German toy company Selecta Spielzeug—whose sustainably harvested wood toys are colored with nontoxic paints, sealed with beeswax, and compliant with European testing standards—pulled out of the United States market at the end of 2008, stating that complying with the CPSIA would require them to increase their retail prices by at least 50 percent. Other European companies are expected to follow suit.

The total cost of testing can range from $100 to thousands of dollars per product. With this level of mandated overhead per product, obviously, the only way to amortize such an enormous capital outlay is large batch production. So producing on a just-in-time basis, with low overhead, using small-scale capital goods, is for all intents and purposes criminalized.[1]

The Design Piracy Prohibition Act, which Sen. Charles Schumer recently introduced for the fourth time, would have a similar effect on fashion. Essentially a DMCA for the fashion industry, it would require thousands of dollars in legal fees to secure CYA documentation of the originality of each design. Not only would it impose such fees on apparel producers of any scale, no matter how small, who produce their own designs, but—because it fails to indemnify apparel manufacturers or retailers—it would deter small producers and retailers from producing or selling the designs of small independent designers who had not paid for such a legal investigation.[2]

NAIS, which requires small family farms to ID chip their livestock at their own expense, operates on the same principle.

At the local level, one of the central functions of so-called "health" and "safety" codes, and occupational licensing is to prevent people from using idle capacity (or "spare cycles") of what they already own anyway, and thereby transforming them into capital goods for productive use. Such regulations mandate minimum levels of overhead (for example, by outlawing a restaurant run out of one's own home, and requiring the use of industrial-sized ovens, refrigerators, dishwashers, etc.), so that the only way to service the overhead and remain in business is to engage in large batch production.

You can't do just a few thousand dollars worth of business a year, because the state mandates capital equipment on the scale required for a large-scale business if you engage in the business at all. Consider all the overhead costs imposed on this chef, who wanted to open a restaurant on the first floor of a hotel:

> That's when the fun began.
>
> I sketched some plans and had them drawn up by an architect ($1000).
>
> I submitted them for review to the County building Dept. ($300).
>
> Everything was OK, except for the bathrooms. They were not ADA compliant. Newly built bathrooms must have a 5' radius turning space for a wheelchair. No problem. I tried every configuration I could think of to accomodate the larger bathroom space without losing seating which would mean losing revenue. No luck. I would have to eat into my storage space and replace it with a separate exterior walk-in cooler($5,000). I would also have to reduce the dining room space slightly so I had to plan on banquettes along the exterior wall to retain the same

[1]Kathryn Geurin, "Toybox Outlaws," *Metroland Online*, January 29, 2009 <http://www.metroland.net/back_issues/vol32_no05/features.html>.

[2]Kathleen Fasanella, "IP Update: DPPA & Fashion Law Blog," *Fashion Incubator*, March 10, 2010 <http://www.fashion-incubator.com/archive/ip-update-dppa-fashion-law-blog/>.

> number of seats (banquettes vs. separate stand alone tables ($5,000) Revised plans ($150). Re-review ($100)
>
> Next came the Utility Dept. It seems the water main was insufficient even for the current use, a 24 suite hotel, and would need to be replaced ($10,000).
>
> Along comes the Historical Preservation Society, a purely advisory group of starched collar, pince nez wearing fuddy-duddies (well, not literally) to offer their "better take it or else" advice, or maybe lose the Historic Status tax break for the hotel.
>
> It seems that the mushroom for the kitchen exhaust fan would be visible from the street, so could I please relocate it to the rear of the building? Pretty please? Extra ducting and more powerful fan ($5,000).
>
> Hello Fire Dept! My plans showed a 40 seat dining room, 2 restrooms , a microscopic office, and a kitchen. My full staffing during tourist season was 4 servers, 1 dishwasher and 1 seasonal cook-total occupancy 47, myself included.
>
> The Fire Inspector said the space could accomodate 59. "But I only have 40 seats. I want luxurious space around the tables." I pleaded. "No. It goes by square footage. 48 seats, 4 servers, 3 cooks, one dishwasher, 1 person in the office and 2 people in the restrooms." "Why would I need 4 cooks for 40 seats when I am capable of doing that alone? And if the cooks are cooking, the servers are serving, the officer is officing, the diners are dining, then who the H#$% is in the bathrooms?"
>
> "Square footage. Code!" And therefore it went from Class B to Class A, requiring a sprinkler system for the dining room and a third exit ($10,000) in addition to the existing front door and the back kitchen door. It would have to be punched through the side wall and have a lit EXIT sign.
>
> Could it be behind the screen shielding the patrons from viewing the inside of the bathrooms every time the door opened? Oh, no! It might not be visible. The door would have to be located where 4 guests at the banquette plus their opposite companions were seated-loss of 20% of seating unless I squeezed them into smaller tables destroying the whole planned luxurious ambience.
>
> Pro Forma:
> $250K sales.
> $75K Food and Beverage purchases
> $75K Labor cost
> $75K Expenses
> $25K net before taxes.
> Result of above experience=Fugget Aboud It!!!
> Loss to community-$100K income plus tips +$20K Sales tax.
>
> Another "Gifte Shoppe" went into the space and closed a month after the end of tourist season. When we left town 2 years later to go sailing the Caribbean, the space was still vacant.

I might add that I had advice in all this from a retired executive who volunteered his time (small donation to Toys 4 Tots gratefully accepted) through a group that connected us. He said that in his opinion that my project budgeted at $200K would cost upward of $1 million in NYC and perhaps SF due to higher permits and fees.[1]

At the smaller end of the spectrum, consider restrictions on informal, unlicensed daycare centers operated out of people's homes.

> MIDDLEVILLE, Mich. (WZZM)—A West Michigan woman says the state is threatening her with fines and possibly jail time for babysitting her neighbors' children.

[1]Quoted by Charles Hugh Smith, in "The Travails of Small Business Doom the U.S. Economy," *Of Two Minds*, August 17, 2009 <http://charleshughsmith.blogspot.com/2009/08/he-travails-of-small-business-doom-us.html>.

> Lisa Snyder of Middleville says her neighborhood school bus stop is right in front of her home. It arrives after her neighbors need to be at work, so she watches three of their children for 15-40 minutes until the bus comes.
>
> The Department of Human Services received a complaint that Snyder was operating an illegal child care home. DHS contacted Snyder and told her to get licensed, stop watching her neighbors' kids, or face the consequences.
>
> "It's ridiculous." says Snyder. "We are friends helping friends!" She added that she accepts no money for babysitting.
>
> Mindy Rose, who leaves her 5-year-old with Snyder, agrees. "She's a friend . . . I trust her."
>
> State Representative Brian Calley is drafting legislation that would exempt people who agree to care for non-dependent children from daycare rules as long as they're not engaged in a business.
>
> "We have babysitting police running around this state violating people, threatening to put them in jail or fine them $1,000 for helping their neighbor (that) is truly outrageous" says Rep. Calley.
>
> A DHS spokesperson would not comment on the specifics of the case but says they have no choice but to comply with state law, which is designed to protect Michigan children.[1]

Another good example is the medallion system of licensing taxicabs, where a license to operate a cab costs into the hundreds of thousands of dollars. The effect of the medallion system is to criminalize the countless operators of gypsy cab services. For the unemployed person or unskilled laborer, driving carless retirees around on their errands for an hourly fee seems like an ideal way to transform one's labor directly into a source of income without doing obesiance to the functionaries of some corporate Human Resources department.

The primary purpose of the medallion system is not to ensure safety. That could be accomplished just as easily by mandating an annual vehicle safety inspection, a criminal background check, and a driving record check (probably all the licensed taxi firms do anyway, and with questionable results based on my casual observation of both vehicles and drivers). And it would probably cost under a hundred bucks rather than three hundred thousand. No, the primary purpose of the medallion system is to allow the owners of licenses to screw both the consumer and the driver.

Local building codes amount to a near-as-dammit lock-in of conventional techniques, regulating the pace of innovation in building techniques in accordance with the preferences of the consensus of contracting firms. As a result, building contractors are protected against vigorous competition from cheap, vernacular local materials, and from modular or prefab designs that are amenable to self-building.

In the case of occupational licensing, a good example is the entry barriers to employment as a surveyor today, as compared to George Washington's day. As Vin Suprynowicz points out, Washington had no formal schooling until he was eleven, only two years of it thereafter, and still was able to learn enough geometry, trigonometry and surveying to get a job paying $100,000 annually in today's terms.

> How much government-run schooling would a youth of today be told he needs before he could contemplate making $100,000 a year as a surveyor—a job which has not changed except to get substantially easier, what with hand-held comput-

[1]Jeff Quackenbush, Jessica Puchala , "Middleville woman threatened with fines for watching neighbors' kids," WZZM13.Com, September 24, 2009 <http://www.wzzm13.com/news/news_story.aspx?storyid=114016&catid=14#>.

ers, GPS scanners and laser range-finders? Sixteen years, at least—18, more likely.[1]

The licensing of retailers protects conventional retail establishments against competition from buying clubs and other low-overhead establishments run out of people's homes, by restricting their ability to sell to the general public. For example, a family-run food-buying co-op in LaGrange, Ohio, whose purpose was to put local farmers into direct contact with local consumers, was raided by sheriff's deputies for allegedly operating as an unlicensed retail establishment.

> A spokeswoman at the Department of Agriculture said its officers were at the scene in an advisory role. A spokeswoman at the county health agency refused to comment except to explain it was a "licensing" issue regarding the family's Manna Storehouse.[2]

Never mind the illegitimacy of the legal distinction between a private bulk food-buying club and a public retail establishment, or the licensing requirement for selling to the general public. The raid was a textbook entrapment operation, in which and undercover agent had persistently badgered the family to sell him eggs. Apparently the family had gotten on the bad side of local authorities by responding in an inadequately deferential manner to peremptory accusations that they were running a store.

> The confrontation began developing several years ago when local health officials demanded the family hold a retail food license in order to run their co-op. Thompson said the family wrote a letter questioning that requirement and asking for evidence that would suggest they were operating a food store and how their private co-op was similar to a WalMart.
>
> The Stowers family members simply "take orders from (co-op) members . . . then divide up the food," Thompson explained.
>
> "The health inspector didn't like the tone of the letter," Thompson said, and the result was that law enforcement officials planned, staged and carried out the Dec. 1 SWAT-style raid on the family's home.
>
> Thompson said he discussed the developments of the case with the health inspector personally.
>
> "He didn't think the tone of that letter was appropriate," Thompson said. " I've seen the letter. There's not anything there that's belligerent."
>
> Thompson explained the genesis of the raid was a series of visits to the family by an undercover agent for the state agriculture agency.
>
> "He showed up (at the Stowers' residence) unannounced one day," Thompson explained, and "pretended" to be interested in purchasing food.
>
> The family explained the co-op was private and they couldn't provide service to the stranger.
>
> The agent then returned another day, stayed for two hours, and explained how he thought his sick mother would be helped by eggs from range-fed chickens to which the Stowers had access.
>
> The family responded that they didn't sell food and couldn't help. When he refused to leave, the family gave him a dozen eggs to hasten his departure, Thompson explained.
>
> Despite protests from the family, the agent left some money on a counter and departed.
>
> On the basis of that transaction, the Stowers were accused of engaging in the retail sale of food, Thompson said. . . .

[1]Vin Suprynowicz, "Schools guarantee there can be no new Washingtons," *Review Journal*, February 10, 2008 <http://www.lvrj.com/opinion/15490456.html>.

[2]Bob Unruh, "Food co-op hit by SWAT raid fights back," *WorldNetDaily*, December 24, 2008 <http://www.wnd.com/index.php?fa=PAGE.view&pageId=84445>.

> He said the state agency came from "nowhere" and then worked to get the family involved "in something that might require a license." . . .
>
> Pete Kennedy of the Farm-to-Consumer Legal Defense Fund said the case was government "overreaching" and was designed more to intimidate and "frighten people into believing that they cannot provide food for themselves."
>
> "This is an example where, once again, the government is trying to deny people their inalienable, fundamental right to produce and consume the foods of their choice," said Gary Cox, general counsel for the FTCLDF. "The purpose of our complaint is to correct that wrong."[1]

As much as I love the local brew pub I visit on a weekly basis, I was taken aback by the manager's complaint about street hot dog vendors being allowed to operate during street festivals. It was unfair for the city to allow it, he said, because an established indoor business with all its associated overhead costs couldn't compete.

The system is effectively rigged to ensure that nobody can start a small business without being rich. Everyone else can get by on wage labor and *like* it (and of course that works out pretty well for the people trying to hire wage labor on the most advantageous terms, don't you think?). Roderick Long asks,

> In the absence of licensure, zoning, and other regulations, how many people would start a restaurant *today* if all they needed was their living room and their kitchen? How many people would start a beauty salon *today* if all they needed was a chair and some scissors, combs, gels, and so on? How many people would start a taxi service *today* if all they needed was a car and a cell phone? How many people would start a day care service *today* if a bunch of working parents could simply get together and pool their resources to pay a few of their number to take care of the children of the rest? These are not the sorts of small businesses that receive SBIR awards; they are the sorts of small businesses that get hammered down by the full strength of the state whenever they dare to make an appearance without threading the lengthy and costly maze of the state's permission process.[2]

[1]Bob Unruh, "SWAT raid on food co-op called 'entrapment'," *WorldNetDaily*, December 26, 2008 <http://www.wnd.com/index.php?fa=PAGE.view&pageId=84594>. See also Andrea Zippay, "Organic food co-op raid sparks case against health department, ODA," FarmAndDairy.Com, December 19, 2008 <http://www.farmanddairy.com/news/organic-food-co-op-raid-sparks-court-case-against-health-department-oda/10752.html>.

[2]Roderick Long, "Free Market Firms: Smaller, Flatter, and More Crowded," *Cato Unbound*, Nov. 25, 2008 <http://www.cato-unbound.org/2008/11/25/roderick-long/free-market-firms-smaller-flatter-and-more-crowded>.

3

Babylon is Fallen

Introduction

If you watch the mainstream cable news networks and Sunday morning interview shows, you've no doubt seen, many times, talking head commentators rolling their eyes at any proposal for reform that's too radically different from the existing institutional structure of society. That much of a departure would be completely unrealistic, they imply, because it is an imposition on all of the common sense people who prefer things the way they are, and because "the way things are" is a natural state of affairs that came about by being recognized, through a sort of tacit referendum of society at large, as self-evidently the most efficient way of doing things.

But in fact the present system is, itself, radical. The corporate economy was created in a few short decades as a radical departure from what prevailed before. And it did not come about by natural evolutionary means, or "just happen"; it's not just "the way things are." It was imposed from above (as we saw in Chapter One) by a conscious, deliberate, *radical* social engineering effort, with virtually no meaningful democratic input from below. The state-imposed corporatization of the economy in the late nineteenth century could be compared in scope and severity, without much exaggeration, to Stalin's collectivization of agriculture and the first Five Year Plan. Although the Period is sometimes called the Gilded Age or the Great Barbecue, John Curl prefers to call it the Great Betrayal.[1] In the Tilden-Hayes dispute, Republicans ended military Reconstruction and handed the southern states back over to the planter class and segregation, in return for a free hand in imposing corporate rule at the national level.

All social systems include social reproduction apparatuses, whose purpose is to produce a populace schooled to accept "the way things are" as the only possible world, and the only natural and inevitable way of doing things. So the present system, once established, included a cultural, ideological and educational apparatus (lower and higher education, the media, etc.) run by people with exactly the same ideology and the same managerial class background as those running the large corporations and government agencies.

All proposals for "reform" within the present system are designed to be implemented within existing institutional structures, by the sorts of people currently running the dominant institutions. Anything that fundamentally weakened or altered the present pattern of corporate-state domination, or required eliminating the power of the elites running the dominant institutions, would be—by definition—"too radical."

[1]John Curl, *For All the People: Uncovering the Hidden History of Cooperation, Cooperative Movements, and Communalism in America* (Oakland, CA: PM Press, 2009).

The system of power, consequently, can only be undermined by forces beyond its control. Fortunately, it faces a mutually reinforcing and snowballing series of terminal crises which render it unsustainable.

The present system's enculturation apparatus functions automatically to present it as inevitable, and to suppress any consciousness that "other worlds are possible." But not only are other worlds possible; under the conditions of Sloanist mass production described in Chapter Two, the terminal crises of the present system mean that *this* world, increasingly, is becoming *impossible*.

A. Resumption of the Crisis of Overaccumulation

State capitalism, with industry organized along mass-production lines, has a chronic tendency to overaccumulation: in other words, its overbuilt plant and equipment are unable to dispose of their full output when running at capacity, and the system tends to generate a surplus that only worsens the crisis over time.

Paul Baran and Paul Sweezy, founders of the neo-Marxist *Monthly Review*, described the Great Depression as "the normal outcome of the workings of the American economic system." It was the culmination of the "stagnationist tendencies inherent in monopoly capitalism," and far from being a deviation from economic normality was "the realization in practice of the theoretical norm toward which the system is always tending."[1]

Fortunately for corporate capitalism, World War Two postponed the crises for a generation or so, by blowing up most of the plant and equipment in the world outside the United States. William Waddell and Norman Bodek, in *The Rebirth of American Industry*, describe the wide-open field left for the American mass-production model:

> General Motors, Ford, General Electric and the rest converted to war production and were kept busy, if not prosperous, for the next four years. When the war ended, they had vast, fully functional factories filled with machine tools. They also had plenty of cash, or at least a pocket full of government IOUs. More important, they also had the entire world market to themselves. The other emerging automobile makers, electric product innovators, consumer product companies, and machine tool builders of Europe and Asia were in ruins.[2]

Harry Magdoff and Paul Sweezy of the *Monthly Review* group described it, in similar terms, as a virtual rebirth of American capitalism.

> The Great Depression was ended, not by a spontaneous resurgence of the accumulation process but by the Second World War. And . . . the war itself brought about vast changes in almost every aspect of the world capitalist system. Much capital was destroyed; the diversion of production to wartime needs left a huge backlog of unfilled consumer demand; both producers and consumers were able to pay off debts and build up unprecedented reserves of cash and borrowing power; important new industries (e.g., jet planes) grew from military technologies; drastically changed power relations between and among victorious and defeated nations gave rise to new patterns of trade and capital flows. In a real

[1]Paul Baran and Paul Sweezy, *Monopoly Capital: An Essay in the American Economic and Social Order* (New York: Monthly Review Press, 1966) p. 240.

[2]William Waddell and Norman Bodek, *Rebirth of American Industry: A Study of Lean Management* (Vancouver, WA: PCS Press, 2005) p. 94.

> sense, world capitalism was reborn on new foundations and entered a period in important respects similar to that of its early childhood.[1]

Even so, the normal tendency was toward stagnation even during the early postwar "Golden Age." In the period after WWII, "actual GNP has equaled or exceeded potential" in only ten years. And eight of those were during the Korean and Vietnam conflicts. The only two peacetime years in which the economy reached its potential, 1956 and 1973, had notably worse levels of employment than 1929.[2]

The tendency postwar, as before it, was for the productive capacity of the economy to far outstrip the ability of normal consumption to absorb. The difference:

> Whereas in the earlier period this tendency worked itself out in a catastrophic collapse of production—during the 1930s as a whole, unemployment and utilization of productive capacity averaged 18 percent and 63 percent respectively—in the postwar period economic energies, instead of lying dormant, have increasingly been channelled into a variety of wasteful, parasitic, and generally unproductive uses. . . . [T]he point to be emphasized here is that far from having eliminated the stagnationist tendencies inherent in today's mature monopoly capitalist economy, this process has forced these tendencies to take on new forms and disguises.[3]

The destruction of capital in World War II postponed the crisis of overaccumulation until around 1970, when the industrial capacity of Europe and Japan had been rebuilt. By that time, according to Piore and Sabel, American domestic markets for industrial goods had become saturated.[4]

This saturation was simply a resumption of the normal process described by Marx in the third volume of *Capital*, which World War II had only temporarily set back.

Leaving aside more recent issues of technological development tunneling through the cost floor and reducing the capital outlays needed for manufacturing by one or more orders of magnitude (about which more below), it is still natural for investment opportunities to decline in mature capitalism. According to Magdoff and Sweezy, domestic opportunities for the extensive expansion of capitalist investment were increasingly scarce as the domestic noncapitalist environment shrunk in relative size and the service sectors were increasingly industrialized. And quantitative needs for investment in producer goods decline steadily as industrialization proceeds:

> . . . [T]he demand for investment capital to build up Department I, a factor that bulked large in the later nineteenth and early twentieth centuries, is of relatively minor importance today in the advanced capitalist countries. They all have highly developed capital-goods industries which, even in prosperous times, normally operate with a comfortable margin of excess capacity. The upkeep and modernization of these industries—and also of course of existing industries in

[1]Harry Magdoff and Paul M. Sweezy, "Capitalism and the Distribution of Income and Wealth," Magdoff and Sweezy, *The Irreversible Crisis: Five Essays by Harry Magdoff and Paul M. Sweezy* (New York: Monthly Review Press, 1988), p. 38

[2]John F. Walker and Harold G. Vattner, "Stagnation—Performance and Policy: A Comparison of the Depression Decade with 1973-1974," *Journal of Post Keynesian Economics*, Summer 1986, in Magdoff and Sweezy, "Stagnation and the Financial Explosion," Magdoff and Sweezy, *The Irreversible Crisis*, pp. 12-13.

[3]Magdoff and Sweezy, "Stagnation and the Financial Explosion," p. 13.

[4]Piore and Sabel, *The Second Industrial Divide*, p. 184.

> Department II (consumer goods)—is provided for by depreciation reserves and generates no new net demand for investment capital.[1]
>
> . . . [T]he need for new investment, relative to the size of the system as a whole, had steadily declined and has now reached an historic low. The reproduction of the system is largely self-financing (through depreciation reserves), and existing industries are for the most part operating at low levels of capacity utilization. New industries, on the other hand, are not of the heavy capital-using type and generate a relatively minor demand for additional capital investment.[2]

"Upkeep and modernization" of existing industry is funded almost entirely by retained earnings, and those retained earnings are in fact often far in excess of investment needs. Corporate management generally finances capital expansion as much as possible through retained earnings, and resorts to bond issues or new stock only as a last resort. And as Martin Hellwig points out, this does not by any means necessarily operate as a constraint on management resources, or force management to ration investment. If anything, the glut of retained earnings is more likely to leave management at a loss as to what to spend it all on.[3]

And as we saw in Chapter Two, the traditional investment model, in oligopoly industry, is tacit collusion between cartelized firms in spooning out investment in new capital assets only as fast as the old ones wear out . Schumpeter's "creative destruction," in a free market, would lead to the constant scrapping and replacement of functional capital assets. But cartelized firms are freed from competitive pressure to scrap obsolete machinery and replace it before it wears out. What's more, as we shall see in the next chapter, in the economically uncertain conditions of the past thirty years, established industry has increasingly shifted new investment from expensive product-specific machinery in the mass-production core to far less expensive general-purpose craft machinery in flexible manufacturing supplier networks.

If anything, Magdoff's and Sweezy's remarks on the reduced capital outlays required by new industries were radically understated, given developments of the subsequent twenty years. Newly emerging forms of manufacturing, as we shall see in Chapter Five, require *far* less capital to undertake production. The desktop revolution has reduced the capital outlays required for music, publishing and software by two orders of magnitude; and the newest open-source designs for computerized machine tools are being produced by hardware hackers for a few hundred dollars.

The result, according to Magdoff and Sweezy, is that "a developed capitalist system such as that of the United States today has the capacity to meet the needs of reproduction and consumption with little or no net investment."[4] From the early days of the industrial revolution, when "the demand for investment capital seemed virtually unlimited, [and] the supply was narrowly restricted," mature capitalism has evolved to the point where the opposite is true: the overabundant supply of investment capital is confronted by a dearth of investment opportunities.[5]

Marx, in the third volume of *Capital*, outlined a series of tendencies that might absorb surplus investment capital and thereby offset the general trend toward a falling direct rate of profit in mature capitalism. And these offsetting ten-

[1]Magdoff and Sweezy, "Capitalism and the Distribution of Income and Wealth," p. 31.

[2]*Ibid.*, p. 39.

[3]Martin Hellwig, "On the Economics and Politics of Corporate Finance and Corporate Control," in Xavier Vives, ed., *Corporate Governance: Theoretical and Empirical Perspectives* (Cambridge: Cambridge University Press, 2000), pp. 114-115.

[4]Magdoff and Sweezy, "Capitalism and the Distribution of Income and Wealth," p. 32.

[5]*Ibid.*, p. 33.

dencies theorized by Marx coincide to a large extent with the expedients actually adopted under developed capitalism. According to Walden Bello the capitalist state, after the resumed crisis of the 1970s, attempted to address the resumed crisis of overproduction with a long series of expedients—including a combination of neoliberal restructuring, globalization, the creation of the tech sector, the housing bubble and intensified suburbanization, and the expansion of the FIRE economy (finance, insurance and real estate)—as successive attempts to soak up surplus capital.[1]

Unfortunately for the state capitalists, the neoliberal model based on offshoring capital has reached its limit; China itself has become saturated with industrial capital.[2] The export-oriented industrialization model in Asia is hitting the walls of both Peak Oil and capital saturation.

The choice of export-oriented industrialization reflected a deliberate calculation by Asian governments, based on the realization that

> import substitution industrialization could continue only if domestic purchasing power were increased via significant redistribution of income and wealth, and this was simply out of the question for the region's elites. Export markets, especially the relatively open US market, appeared to be a painless substitute.

Today, however, as "goods pile up in wharves from Bangkok to Shanghai, and workers are laid off in record numbers, people in East Asia are beginning to realize they aren't only experiencing an economic downturn but living through the end of an era." The clear lesson is that the export-oriented industrial model is extremely vulnerable to both increased shipping costs and decreases in Western purchasing power—a lesson that has "banished all talk of decoupling" a growing Asian economy from the stagnating West. Asia's manufacturing sector is "linked to debt-financed, middle-class spending in the United States, which has collapsed."[3] The Asian export economy, as a result, has fallen through the floor.

> Worldwide, industrial production has ground to a halt. Goods are stacking up, but nobody's buying; the *Washington Post* reports that "the world is suddenly awash in almost everything: flat-panel televisions, bulldozers, Barbie dolls, strip malls, Burberry stores." A Hong Kong-based shipping broker told *The Telegraph* that his firm had "seen trade activity fall off a cliff. Asia-Europe is an unmitigated disaster." *The Economist* noted that one can now ship a container from China to Europe for free—you only need to pick up the fuel and handling costs—but half-empty freighters are the norm along the world's busiest shipping routes. Global airfreight dropped by almost a quarter in December alone; Giovanni Bisignani, who heads a shipping industry trade group, called the "free fall" in global cargo "unprecedented and shocking."[4]

If genuine decoupling is to take place, it will require a reversal of the strategic assessments and policy decisions which led to the choice of export-oriented industrialization over import substitution in the first place. It will require, in particular, rethinking the unthinkable: putting the issues of local income distribution and purchasing power back on the table. That means, in concrete terms, that Asian

[1]Walden Bello, "A Primer on Wall Street Meltdown," *MR Zine*, October 3, 2008 <http://mrzine.monthlyreview.org/bello031008.html>.

[2]*Ibid.*

[3]Walden Bello, "Asia: The Coming Fury," *Asia Times Online*, February 11, 2009 <http://www.atimes.com/atimes/Asian_Economy/KB11Dk01.html>.

[4]Joshua Holland, "The Spectacular, Sudden Crash of the Global Economy," *Alternet*, February 24, 2009 <http://www.alternet.org/module/printversion/128412/the_spectacular%2C_sudden_crash_of_the_global_economy/>.

manufacturers currently engaged in the Nike ("outsource everything") model of distributed manufacturing must treat the Western corporate headquarters as nodes to be bypassed, repudiate their branding and other "intellectual property," and reorient production to the domestic market with prices that reflect something like the actual cost of production without brand-name markup. It also requires that Asian governments cease their modern-day reenactment of the "primitive accumulation" of eighteenth century Britain, restore genuine village control of communal lands, and otherwise end their obsessive focus on attracting foreign investment through policies that suppress the bargaining power of labor and drive people into the factories like wild beasts. In other words, those Nike sneakers piling up on the wharves need to be marketed to the local population minus the swoosh, at an 80% markdown. At the same time, agriculture needs to shift from cash crop production for the urban and export market to a primary focus on subsistence production and production for the domestic market.

Bello points out that 75% of China's manufacturers were already complaining of excess capacity and demand stagnation, even before the bubble of debt-fueled demand collapsed. Interestingly, he also notes that the Chinese government is trying to bolster rural demand as an alternative to collapsing demand in the export market, although he's quite skeptical of the policy's prospects for success. The efforts to promote rural purchasing power, he argues, are too little and too late—merely chipping at the edges of a 25-year policy of promoting export-oriented industrialization "on the back of the peasant." China's initial steps toward market liberalization in the 1970s were centered on the prosperity of peasant smallholders. In the '80s, the policy shifted toward subsidizing industry for the export market, with a large increase in the rural tax burden and as many as three hundred million peasants evicted from their land in favor of industrial use. But any hope at all for China's industrial economy depends on restoring the prosperity of the agricultural sector as a domestic source of demand.[1]

Suburbanization, thanks to Peak Oil and the collapse of the housing bubble, has also ceased to be a viable outlet for surplus capital.

The stagnation of the economy from the 1970s on—every decade since the postwar peak of economic growth in the 1960s has seen lower average rates of annual growth in real GDP compared to the previous decade, right up to the flat growth of the present decade—was associated with a long-term trend in which demand was stimulated mainly by asset bubbles.[2] In 1988, a year after the 1987 stock market crash and on the eve of the penultimate asset bubble (the dotcom bubble of the '90s), Sweezy and Magdoff summed up the previous course of financialization in language that actually seems understated in light of subsequent asset bubbles.

> Among the forces counteracting the tendency to stagnation, none has been more important or less understood by economic analysts than the growth, beginning in the 1960s and rapidly gaining momentum after the severe recession of the mid-1970s, of the country's debt structure . . . at a pace far exceeding the sluggish expansion of the underlying "real" economy. The result has been the emergence of an unprecedentedly huge and fragile financial superstructure subject to

[1]Walden Bello, "Can China Save the World from Depression?" *Counterpunch*, May 27, 2009 <http://www.counterpunch.org/bello05272009.html>.

[2]John Bellamy Foster and Fred Magdoff, "Financial Implosion and Stagnation: Back to the Real Economy," *Monthly Review*, December 2008 <http://www.monthlyreview.org/081201foster-magdoff.php>.

> stresses and strains that increasingly threaten the stability of the economy as a whole.

Between the 1960s and 1987, the debt-to-GNP ratio rose from 1.5 to 2.25.[1]

But it was only after the collapse of the tech bubble that financialization—the use of derivatives and securitization of debt as surplus capital sponges to soak up investment capital for which no outlet existed in productive industry—really came into its own. As Joshua Holland noted, in most recessions the financial sector contracted along with the rest of the economy; but after the 2000 tech bust it just kept growing, ballooning up to ten percent of the economy.[2] We're seeing now how that worked out.

Financialization was a way of dealing with a surplus of productive capacity, whose output the population lacked sufficient purchasing power to absorb—a problem exacerbated by the fact that almost all increases in productivity had gone to increasing the wealth of the upper class. Financialization enabled the upper class to lend its increased wealth to the rest of the population, at interest, so they could buy the surplus output.

Conventional analysts and editorialists frequently suggest, to the point of cliche, that the shift from productive investment to speculation in the finance sector is the main cause of our economic ills. But as Magdoff and Sweezy point out, it's the other way around. The expansion of investment capital against the backdrop of a sluggish economy led to a shift in investment to financial assets, given the lack of demand for further investment in productive capital assets.

> It should be obvious that capitalists will not invest in additional capacity when their factories and mines are already able to produce more than the market can absorb. Excess capacity emerged in one industry after another long before the extraordinary surge of speculation and finance in the 1970s, and this was true not only in the United States but throughout the advanced capitalist world. The shift in emphasis from industrial to pecuniary pursuits is equally international in scope.[3]

In any case the housing bubble collapsed, government is unable to reinflate housing and other asset values even with trillion-dollar taxpayer bailouts, and an alarming portion of the population is no longer able to service the debts accumulated in "good times." Not only are there no inflated asset values to borrow against to fuel demand, but many former participants in the Ditech spending spree are now becoming unemployed or homeless in the Great Deleveraging.[4]

Besides, the problem with debt-inflated consumer demand was that there was barely enough demand to keep the wheels running and absorb the full product of overbuilt industry even when everyone maxed out their credit cards and tapped into their home equity to replace everything they owned every five years. And we'll never see that kind of demand again. So there's no getting around the fact that a major portion of existing plant and equipment will be rust in a few years.

State capitalism seems to be running out of safety valves. Barry Eichengreen and Kevin O'Rourke suggest that, given the scale of the decline in industrial out-

[1] Magdoff and Sweezy, "Stagnation and the Financial Explosion," pp. 13-14.

[2] Joshua Holland, "Let the Banks Fail: Why a Few of the Financial Giants Should Crash," *Alternet*, December 15, 2008 <http://www.alternet.org/workplace/112166/let_the_banks_fail%3A_why_a_few_of_the_financial_giants_should_crash_/>.

[3] Magdoff and Sweezy, "Stagnation and the Financial Explosion," p. 23.

[4] Charles Hugh Smith, "Globalization and China: Neoliberal Capitalism's Last 'Fix'," *Of Two Minds*, June 29, 2009 <http://www.oftwominds.com/blogjune09/globalization06-09.html>.

put and global trade, the term "Great Recession" may well be over-optimistic. Graphing the rate of collapse in global industrial output and trade from spring 2008 to spring 2009, they found the current rate of decline has actually been steeper than that of 1929-1930. From appearances in early 2009, it was "a Depression-sized event," with the world "currently undergoing an economic shock every bit as big as the Great Depression shock of 1929-30."[1]

Left-Keynesian Paul Krugman speculated that the economy narrowly escaped another Great Depression in early 2009.

> A few months ago the possibility of falling into the abyss seemed all too real. The financial panic of late 2008 was as severe, in some ways, as the banking panic of the early 1930s, and for a while key economic indicators—world trade, world industrial production, even stock prices—were falling as fast as or faster than they did in 1929-30.
>
> But in the 1930s the trend lines just kept heading down. This time, the plunge appears to be ending after just one terrible year.
>
> So what saved us from a full replay of the Great Depression? The answer, almost surely, lies in the very different role played by government.
>
> Probably the most important aspect of the government's role in this crisis isn't what it has done, but what it hasn't done: unlike the private sector, the federal government hasn't slashed spending as its income has fallen.[2]

This is not to suggest that the Keynesian state is a desirable model. Rather, it is made necessary by state capitalism. But make no mistake: so long as we have state capitalism, with state promotion of overaccumulation and the maldistribution of purchasing power that results from privilege, state intervention to manage aggregate demand is necessary to avert depression. Given state capitalism, we have only two alternatives: 1) eliminate the privileges and subsidies to overaccumulation that result in chronic crisis tendencies; or 2) resort to Keynesian stabilizing measures. Frankly, I can't work up much enthusiasm for the mobs of teabaggers demanding an end to the Keynesian stabilizing measures, when those mobs reflect an astroturf organizing effort funded by the very people who benefit from the privileges and subsidies that contribute to chronic crisis tendencies.

And we should bear in mind that it's far from clear the worst has, in fact, been averted. Karl Denninger argues that the main reason GDP fell only 1% in the second quarter of 2009, as opposed to 6% in the first, was increased government spending. As he points out, the fall of investment slowed in the second quarter; but given that it was already cut almost in half, there wasn't much further it *could* fall. Exports fell "only" 7% and imports 15.1%; but considering they had already fallen 29.9% and 36.4%, respectively, in the first quarter, this simply means that exports and imports have "collapsed." Consumer spending fell in the second quarter more than in the first, with a second quarter increase in the rate of "savings" (or rather, of paying down debt). If the rate of collapse is slowing, it's because there's so much less distance to fall. Denninger's take: "The recession is not 'easing', it is DEEPENING."[3]

The reduction in global trade is especially severe, considering that the very modest uptick in summer 2009 still left the shortfall from baseline levels far lower in the Great Recession than it was at a comparable point in the Great Depression.

[1]Barry Eichengreen and Kevin H. O'Rourke, "A Tale of Two Depressions," *VoxEU.Org*, June 4, 2009 <http://www.voxeu.org/index.php?q=node/3421>.

[2]Paul Krugman, "Averting the Worst," *New York Times*, August 9, 2009 <http://www.nytimes.com/2009/08/10/opinion/10krugman.html>.

[3]Karl Denninger, "GDP: Uuuuggghhhh—UPDATED," *The Market Ticker*, July 31, 2009 <http://market-ticker.denninger.net/archives/1276-GDP-Uuuuggghhhh.html>.

As of late summer 2009, world trade was some 20% below the pre-recession baseline, compared to only 8% the same number of months into the Depression. Bear in mind that the collapse of world trade in the Depression is widely regarded as the catastrophic result of the Smoot-Hawley tariff, and to have been a major exacerbating factor in the continuing progression of the economic decline in the early '30s. The current reduction in volume of world trade, far greater than that of the Great Depression, has occurred *without* Smoot-Hawley![1]

Stoneleigh, a former writer for *The Oil Drum Canada*, argues that the asset deflation has barely begun:

> Banks hold extremely large amounts of illiquid 'assets' which are currently marked-to-make-believe. So long as large-scale price discovery events can be avoided, this fiction can continue. Unfortunately, a large-scale loss of confidence is exactly the kind of circumstance that is likely to result in a fire-sale of distressed assets. . . .
>
> A large-scale mark-to-market event of banks illiquid 'assets' would reprice entire asset classes across the board, probably at pennies on the dollar. This would amount to a very rapid destruction of staggering amounts of putative value. This is the essence of deflation. . . .

The currently celebrated "green shoots," which she calls "gangrenous," are comparable to the suckers' stock market rally of 1930.[2]

In any case, if Keynesianism is *necessary* for the survival of state capitalism, we're reaching a point at which it is no longer *sufficient*. If pessimists like Denninger are wrong, and Keynesian policies have indeed turned the free fall into a slow motion collapse, the fact remains that they are insufficient to restore "normalcy"—because normalcy is no longer an option. Keynesianism was sufficient during the postwar "Consensus Capitalism" period only because of the worldwide destruction of plant and equipment in WWII, which postponed the crisis of overaccumulation for a generation or so.

Bello makes the very good point that Keynesianism is not a long-term solution to the present economic difficulties because it ceased to be a solution the *first* time around.

> The Keynesian-inspired activist capitalist state that emerged in the post-World War II period seemed, for a time, to surmount the crisis of overproduction with its regime of relatively high wages and technocratic management of capital-labor relations. However, with the addition of massive new capacity from Japan, Germany, and the newly industrializing countries in the 1960s and 1970s, its ability to do this began to falter. The resulting stagflation—the coincidence of stagnation and inflation—swept throughout the industrialized world in the late 1970s.[3]

Conventional left-Keynesian economists are at a loss to imagine some basis on which a post-bubble economy can ever be reestablished with anything like current levels of output and employment. This is especially unfortunate, given the focus of both the Bush and Obama administrations' banking policies on restoring asset prices to something approaching their pre-collapse value, and the focus of their

[1]Cassander, "It's Hard Being a Bear (Part Three): Good Economic History," *Steve Keen's Debtwatch*, September 5, 2009 <http://www.debtdeflation.com/blogs/2009/09/05/it%E2%80%99s-hard-being-a-bear-part-three-good-economic-history/>.

[2]"October 30 2009: An interview with Stoneleigh—The case for deflation," *The Automatic Earth* <http://theautomaticearth.blogspot.com/2009/10/october-30-2009-interview-with.html>.

[3]Walden Bello, "Keynes: A Man for This Season?" *Share the World's Resources*, July 9, 2009 <http://www.stwr.org/globalization/keynes-a-man-for-this-season.html>.

economic policies on at least partially reinflating the bubble economy as a source of purchasing power, so that—as James Kunstler so eloquently puts it—

> the US public could resume a revolving credit way-of-life within an economy dedicated to building more suburban houses and selling all the needed accessories from supersized "family" cars to cappuccino machines. This would keep everyone employed at the jobs they were qualified for—finish carpenters, realtors, pool installers, mortgage brokers, advertising account executives, Williams-Sonoma product demonstrators, showroom sales agents, doctors of liposuction, and so on.[1]

Both the Paulson and Geithner TARP plans involve the same kind of Hamiltonian skullduggery: borrowing money, to be repaid by taxpayers with interest, to purchase bad assets from banks at something much closer to face value than current market value in order to increase the liquidity of banks to the point that they might lend money back to the public—should they deign to do so—at interest. Or as Michael Hudson put it, TARP "aims at putting in place enough new bank-lending capacity to start inflating prices on credit all over again."[2]

Charles Hugh Smith describes the parallel between Japan's "Lost Decade" and the current economic crisis:

> Ushinawareta junen is the Japanese phrase for "Lost Decade." The term describes the 1991-2000 no-growth decade in which Japan attempted to defeat debt-liquidation deflationary forces with massive government borrowing and spending, and a concurrent bailout of "zombie" (insolvent) banks with government funds.
>
> The central bank's reflation failed. By any measure, the Lost Decade is now the Lost Decades. Japan's economy enjoyed a brief spurt from America's real estate bubble and China's need for Japanese factory equipment and machine tools. But now that those two sources of demand have ebbed, Japan is returning to its deflationary malaise. . . .
>
> . . . It seems the key parallel is this: an asset bubble inflated with highly leveraged debt pops and the value of real estate and stocks declines. But the high levels of debt taken on to speculate in stocks and housing remain.
>
> Rather than let the private-sector which accepted the high risks and took the enormous profits take staggering losses and writedowns, the government and central bank shift the losses from the private sector to the public balance sheet via bailouts and outright purchases of toxic/impaired private debt.[3]

The problem is that pre-collapse levels of output can only be absorbed by debt-financed and bubble-inflated purchasing power, and that another bubble on the scale of the tech and real estate booms just ain't happening.

Keynesianism might be viable as a long-term strategy if deficit stimulus spending were merely a way of bridging the demand shortfall until consumer spending could be restored to normal levels, after which it would use tax revenues in good times to pay down the public debt. But if normal levels of consumer spending *won't* come back, it amounts to the U.S. government borrowing $2 tril-

[1]James Kunstler, "Note: Hope = Truth," *Clusterfuck Nation*, April 20, 2009 <http://jameshowardkunstler.typepad.com/clusterfuck_nation/2009/04/note-hope-truth.html>.

[2]Michael Hudson, "What Wall Street Wants," *Counterpunch*, February 11, 2009 <http://www.counterpunch.org/hudson02112009.html> (see also expanded version, "Obama's Awful Financial Recovery Plan," *Counterpunch*, February 12, 2009 <http://www.counterpunch.org/hudson02122009.html>).

[3]Charles Hugh Smith, "Welcome to America's Lost Decade(s)," *Of Two Minds*, September 18, 2009 <http://charleshughsmith.blogspot.com/2009/09/welcome-to-americas-lost-decades.html>.

lion this year to shore up consumer spending *for this year*—with consumer spending falling back to Depression levels next year if *another* $2 trillion isn't spent.

> We estimate that absent all the forms of government stimulus in the second quarter, real GDP would have contracted at a decidedly brown-shooty 6% annual rate as opposed to the posted 1% decline. And, while consensus forecasts are centered around 3.0-3.5% for current quarter growth, again the pace of economic activity would be flat-to-negative absent Cash-for-Clunkers, government auto purchases, and first-time homebuyer subsidies, not to mention the FHA's best efforts to recreate the housing and credit bubble. . . . [1]

So capitalism might be sustainable, in terms of the demand shortfall taken in isolation—if the state is prepared to run a deficit of $1 or $2 trillion a year, every single year, indefinitely. But there will never again be a tax base capable of paying for these outlays, because the implosion of production costs from digital production and small-scale manufacturing technology is destroying the tax base. What we call "normal" levels of demand are a thing of the past. As Paul Krugman points out, as of late fall 2009 stimulus spending is starting to run its course, with no sign of sufficient self-sustaining demand to support increased industrial production; the increasingly likely result is a double dip recession with Part Two in late 2010 or 2011.[2]

So the crisis of overaccumulation exacerbates the fiscal crisis of the state (about which more below).

It might be possible to sustain such spending on a permanent basis via something like the "Social Credit" proposals of Major Douglas some eighty years ago (simply creating the money out of thin air instead of borrowing it or funding it with taxes, and depositing so much additional purchasing power in every citizen's checking account each month). But that would undermine the basic logic of capitalism, removing the incentive to accept wage labor on the terms offered, and freeing millions of people to retire on a subsistence income from the state while participating in the non-monetized gift or peer economy. Even worse, it would create the economic basis for continuing subsidized waste and planned obsolescence until the ecosystem reached a breaking point—a state of affairs analogous to the possibility, contemplated with horror by theologians, that Adam and Eve in their fallen state might have attained immortality from the Tree of Life.

Those who combine some degree of "green" sympathy with their Keynesianism have a hard time reconciling the fundamental contradiction involved in the two sides of modern "Progressivism." You can't have all the good Michael Moore stuff about full employment and lifetime job security, without the bad stuff about planned obsolescence and vulgar consumerism. Krugman is a good case in point:

> I'm fairly optimistic about 2010.
>
> But what comes after that? Right now everyone is talking about, say, two years of economic stimulus—which makes sense as a planning horizon. Too much of the economic commentary I've been reading seems to assume, however, that that's really all we'll need—that once a burst of deficit spending turns the economy around we can quickly go back to business as usual.
>
> In fact, however, things can't just go back to the way they were before the current crisis. And I hope the Obama people understand that.
>
> The prosperity of a few years ago, such as it was—profits were terrific, wages not so much—depended on a huge bubble in housing, which replaced an earlier

[1]David Rosenberg, *Lunch with Dave*, September 4, 2009 <http://www.scribd.com/doc/19430778/Lunch-With-Dave-090409>.

[2]Paul Krugman, "Double dip warning," Paul Krugman Blog, *New York Times*, Dec. 1, 2009 <http://krugman.blogs.nytimes.com/2009/12/01/double-dip-warning/>.

> huge bubble in stocks. And since the housing bubble isn't coming back, the spending that sustained the economy in the pre-crisis years isn't coming back either.
>
> To be more specific: the severe housing slump we're experiencing now will end eventually, but the immense Bush-era housing boom won't be repeated. Consumers will eventually regain some of their confidence, but they won't spend the way they did in 2005-2007, when many people were using their houses as ATMs, and the savings rate dropped nearly to zero.
>
> So what will support the economy if cautious consumers and humbled homebuilders aren't up to the job?[1]

(I would add that, whatever new standard of post-bubble "normalcy" prevails, in the age of Peak Oil and absent previous pathological levels of consumer credit, it's unlikely the U.S. will ever see a return to automobile sales of 18 million a year. If anything, the current output of ca. ten million cars is probably enormously inflated.)[2]

And Krugman himself, it seems, is not entirely immune to the delusion that a sufficient Keynesian stimulus will restore the levels of consumer demand associated with something like "normalcy."

Krugman first compares the longer duration and greater severity of depressions without countercyclical government policy to those with, and then cites Keynes as an authority in estimating the length of the current Great Recession without countercyclical stimulus spending: "a recession would have to go on until 'the shortage of capital through use, decay and obsolescence causes a sufficiently obvious scarcity to increase the marginal efficiency.'"[3]

But as he himself suggested in his earlier column, the post-stimulus economy may have much lower "normal" levels of demand than the pre-recession economy, in which case the only effect of the stimulus will be to pump up artificial levels of demand so long as the money is still being spent. In that case, as John Robb argues, the economy will eventually have to settle into a new equilibrium with levels of demand set at much lower levels.

> The assumption is that new homes will eventually need to be built to accommodate population growth and new cars will be sold to replace old stock. However, what if there is a surge in multi-generational housing (there is) or people start to drive much less (they are) or keep their cars until they drop (most people I know are planning this). If that occurs, you have to revise the replacement level assumption to a far lower level than before the start of the downturn. What's that level? I suspect it is well below current sales levels, which means that there is much more downside movement possible.[4]

The truth of the matter is, the present economic crisis is not cyclical, but structural. There is excess industrial capacity that will be rust in a few years because we are entering a period of *permanently* low consumer demand and frugality. As Peter Kirwan at *Wired* puts it, the mainstream talking heads are mistaking

[1]Paul Krugman, "Life Without Bubbles," *New York Times*, January 6, 2009 <http://www.nytimes.com/2008/12/22/opinion/22krugman.html?ref=opinion>.

[2]Despite exuberance in the press over Cash for Clunkers, auto sales went flat—in fact reaching a low for the year—as soon as the program ended. Associated Press, "Retail sales fall after Cash for Clunkers ends," MSNBC, October 14, 2009 <http://www.msnbc.msn.com/id/33306465/ns/business-retail/>.

[3]Paul Krugman, "Use, Delay, and Obsolescence," *The Conscience of a Liberal*, February 13, 2009 <http://krugman.blogs.nytimes.com/2009/02/13/use-delay-and-obsolescence/>.

[4]John Robb, "Below Replacement Level," *Global Guerrillas*, February 20, 2009 <http://globalguerrillas.typepad.com/johnrobb/2009/02/below-replacement-level.html>.

for a cyclical downturn what is really "permanent structural change" and "industrial collapse."[1]

Both the bailout and stimulus policies, under the late Bush and Obama administrations, have amounted to standing in the path of these permanent structural changes and yelling "Stop!" The goal of U.S. economic policy is to prevent the deflation of asset bubbles, and restore sufficient demand to utilize the idle capacity of mass-production industry. But this only delays the inevitable structural changes that must take place, as Richard Florida points out:

> The bailouts and stimulus, while they may help at the margins, also pose an enormous opportunity costs [sic]. On the one hand, they impede necessary and long-deferred economic adjustments. The auto and auto-related industries suffer from massive over-capacity and must shrink. The housing bubble not only helped spur the financial crisis, it also produced an enormous mis-allocation of resources. Housing prices must come a lot further down before we can reset the economy—and consumer demand—for a new round of growth. The financial and banking sector grew massively bloated—in terms of employment, share of GDP and wages, as the detailed research of NYU's Thomas Phillipon has shown—and likewise have to come back to earth.[2]

The new frugality, to the extent that it entails more common-sense consumer behavior, threatens the prevailing Nike model of outsourcing production and charging a price consisting almost entirely of brand-name markup. A *Wall Street Journal* article cites a Ms. Ball: "After years of spending $17 on bottles of Matrix shampoo and conditioner, 28-year-old Ms. Ball recently bought $5 Pantene instead. . . . 'I don't know that you can even tell the difference.'" Procter & Gamble has been forced to scale back its prices considerably, and offer cheaper and less elaborate versions of many of its products. William Waddell comments:

> Guess what P&G—Ms. Ball and millions like her will not come back to your hollow brands once the economy comes back now that she knows the $5 stuff is exactly the same as the $17 stuff.[3]

A permanent, mass shift from brand-name goods to almost identical generic and store brand goods would destroy the basis of push-distribution capitalism. We already saw, in the previous chapter, quotes from advertising industry representatives stating in the most alarmist terms what would happen if their name brand goods had to engage in direct price competition like commodities. The min-revolt against brand-name goods during the downturn of the early '90s—the so-called "Marlboro Friday"—was a dress rehearsal for just such an eventuality.

> On April 2, 1993, advertising itself was called into question by the very brands the industry had been building, in some cases, for over two centuries. That day is known in marketing circles as "Marlboro Friday," and it refers to a sudden announcement from Philip Morris that it would slash the price of Marlboro cigarettes by 20 percent in an attempt to compete with bargain brands that

[1]Peter Kirwan, "Bad News: What if the money's not coming back?" *Wired.Co.Uk*, August 7, 2009 <http://www.wired.co.uk/news/archive/2009-08/07/bad-news-what-if-the-money%27s-not-coming-back.aspx>.

[2]Richard Florida, "Are Bailouts Saving the U.S. from a New Great Depression," *Creative Class*, March 18, 2009 <http://www.creativeclass.com/creative_class/2009/03/18/are-the-bailouts-saving-us-from-a-new-great-depression/>.

[3]Ellen Byron, "Tide Turns 'Basic' for P&G in Slump," *WSJ* online, August 6, 2009 <http://online.wsj.com/article/SB124946926161107433.html>; in William Waddell, "But You Can't Fool All the People All the Time," *Evolving Excellence*, August 25, 2009 <http://www.evolvingexcellence.com/blog/2009/08/but-you-cant-foool-all-the-people-all-the-time.html>.

> were eating into its market. The pundits went nuts, announcing in frenzied unison that not only was Marlboro dead, all brand names were dead. The reasoning was that if a "prestige" brand like Marlboro, whose image had been carefully groomed, preened and enhanced with more than a billion advertising dollars, was desperate enough to compete with no-names, then clearly the whole concept of branding had lost its currency. The public had seen the advertising, and the public didn't care. . . . The implication that Americans were suddenly thinking for themselves en masse reverberated through Wall Street. The same day Philip Morris announced its price cut, stock prices nose-dived for all the household brands: Heinz, Quaker Oats, Coca-Cola, PepsiCo, Procter and Gamble and RJR Nabisco. Philip Morris's own stock took the worst beating.
>
> Bob Stanojev, national director of consumer products marketing for Ernst and Young, explained the logic behind Wall Street's panic: "If one or two powerhouse consumer products companies start to cut prices for good, there's going to be an avalanche. Welcome to the value generation.

As Klein went on to write, the Marlboro Man eventually recovered from his setback, and brand names didn't exactly become obsolete in the ensuing age of Nike and The Gap. But even if the panic was an "overstated instant consensus," it was nevertheless "not entirely without cause."

> The panic of Marlboro Friday was not a reaction to a single incident. Rather, it was the culmination of years of escalating anxiety in the face of some rather dramatic shifts in consumer habits that were seen to be eroding the market share of household-name brands, from Tide to Kraft. Bargain-conscious shoppers, hit hard by the recession, were starting to pay more attention to price than to the prestige bestowed on their products by the yuppie ad campaigns of the 1980s. The public was suffering from a bad case of what is known in the industry as "brand blindness."
>
> Study after study showed that baby boomers, blind to the alluring images of advertising and deaf to the empty promises of celebrity spokespersons, were breaking their lifelong brand loyalties and choosing to feed their families with private-label brands from the supermarket—claiming, heretically, that they couldn't tell the difference . . . It appeared to be a return to the proverbial shopkeeper dishing out generic goods from the barrel in a prebranded era.
>
> The bargain craze of the early nineties shook the name brands to their core. Suddenly it seemed smarter to put resources into price reductions and other incentives than into fabulously expensive ad campaigns. This ambivalence began to be reflected in the amounts companies were willing to pay for so-called brand-enhanced advertising. Then, in 1991, it happened: overall advertising spending actually went down by 5.5 percent for the top 100 brands. It was the first interruption in the steady increase of U.S. ad expenditures since a tiny dip of 0.6 percent in 1970, and the largest drop in four decades.
>
> It's not that top corporations weren't flogging their products, it's just that to attract those suddenly fickle customers, many decided to put their money into promotions such as giveaways, contests, in-store displays and (like Marlboro) price reductions. In 1983, American brands spent 70 percent of their total marketing budgets on advertising, and 30 percent on these other forms of promotion. By 1993, the ratio had flipped: only 25 percent went to ads, with the remaining 75 percent going to promotions.[1]

And Ms. Ball, mentioned above, may prefigure a more permanent shift to the same sort of behavior in the longer and deeper Great Recession of the 21st century.

While Krugman lamely fiddles around with things like a reduction of the U.S. trade deficit as a possible solution to the demand shortfall, liberal blogger Mat-

[1]Naomi Klein, *No Logo* (New York: Picador, 2000, 2002), pp. 12-14.

thew Yglesias has a more realistic idea of what a sustainable post-bubble economy might actually entail.

> I would say that part of the answer may well involve taking a larger share of our productivity gains as increased leisure rather than increased production and incomes. . . . A structural shift to less-work, less-output dynamic could be catastrophic if that means a structural shift to a very high rate of unemployment. But if it means a structural shift toward six-week vacations and fewer 60 hour weeks then that could be a good thing.[1]

Exactly. But a better way of stating it would be "a structural shift toward a less-work, less-output, less-planned-obsolescence, and less-embedded-rents-on-IP-and-ephemera dynamic, with no reduction in material standard of living. A structural dynamic toward working fewer hours to produce less stuff because it lasts longer instead of going to the landfill after a brief detour in our living rooms, would indeed be a good thing.

Michel Bauwens ventures a somewhat parallel analysis from a different perspective, that of Kondratiev's long-wave theory and neo-Marxist theories of the social structure of accumulation (particularly the idea of a new social structure of accumulation as necessary to resolve the crises of the previous structure[2]). According to Bauwens, 1929 was the sudden systemic shock of the last system, and from it emerged the present system, based on Fordist mass-production and the New Deal/organized labor social contract, the automobile, cheap fossil fuels—you know the drill. The system's golden age lasted from WWII to the early 1970s, when its own series of systemic shocks began: the oil embargo, the saturation of world industrial capital, and all the other systemic crises we're considering in this chapter. According to Bauwens, each long wave is characterized by a new energy source, a handful of technological innovations (what the neo-Marxists would call "epoch-making industries"), a new mode of financial system, and a new social contract. Especially interesting, each long wave presents "a new 'hyperproductive' way to 'exploit the territory,'" which parallels his analysis (which we will examine in later chapters) of the manorial economy as a path of intensive development when the slave economy reached its limits of expansion, and of netarchical capitalism as a way to extract value intensively when extensive addition of capital inputs is no longer feasible.

According to Bauwens, the emerging long wave will be characterized by renewable energy and green technology, crowdsourced credit and microlending, relocalized networked manufacturing, a version of small-scale organic agriculture that applies the latest findings of biological science, and a mode of economic organization centered on civil society and peer networks.[3]

However, to the extent that the capture of value through "intellectual property" is no longer feasible (see below), it seems unlikely that any such new paradigm can function on anything resembling the current corporate capitalist model.

It's a fairly safe bet we're in for a period of prolonged economic stagnation and decline, measured in conventional terms. The imploding capital outlays required

[1]Matthew Yglesias, "The Elusive Post-Bubble Economy," *Yglesias/ThinkProgress.Org*, December 22, 2008 <http://yglesias.thinkprogress.org/archives/2008/12/the_elusive_post_bubble_economy.php>.

[2]David Gordon, "Stages of Accumulation and Long Economic Cycles," in Terence K. Hopkins and Immanuel Wallerstein, eds., *Processes of the World-System* (Beverly Hills, Calif.: Sage, 1980), pp. 9-45.

[3]Michel Bauwens, "Conditions for the Next Long Wave," *P2P Foundation Blog*, May 28, 2009 <http://blog.p2pfoundation.net/conditions-for-the-next-long-wave/2009/05/28>.

for manufacturing, thanks to current technological developments, mean that the need for investment capital falls short of available investment funds by at least an order of magnitude. The increasing unenforcability of "intellectual property" means that attempts to put a floor under either mandated capital outlays, overhead, or commodity price, as solutions to the crisis, will fail. Established industry will essentially cut off all net new investment in capital equipment and begin a prolonged process of decay, with employment levels suffering accordingly.

Those who see this as leading to a sudden, catastrophic increase in technological unemployment are probably exaggerating the rate of progression of the crisis. What we're more likely to see is what Alan Greenspan called a Great Malaise, gradually intensifying over the next couple of decades. Given the toolkit of anti-deflationary measures available to the central bankers, he argued in 1980, the collapse of asset bubbles would never again be allowed to follow its natural course—a "cascading set of bankruptcies" leading to a chain reaction of debt deflation. The central banks, he continued, would "flood the world's economies with paper claims at the first sign of a problem," so that a "full-fledged credit deflation" on the pattern of the early 1930s could not happen. And indeed, Sweezy and Magdoff argue, had the government not intervened following the stock market crash of 1987, it's quite likely the aftermath would have been a deflationary collapse like that of the Depression.

Greenspan's successor Ben "Helicopter" Bernanke, whose nickname comes from his stated willingness to airdrop cash to maintain liquidity, made good on such guarantees in the financial crisis of Fall 2008. The federal government also moved far more quickly than in the 1930s, as we saw above, to use deficit spending to make up a significant part of the demand shortfall.

The upshot of this is that the crisis of overaccumulation and underconsumption is likely to be reflected, not in a sudden deflationary catastrophe, but—in Greenspan's words—a Great Malaise.

> Thus in today's political and institutional environment, a replay of the Great Depression is the Great Malaise. It would not be a period of falling prices and double-digit unemployment, but rather an economy racked with inflation, excessive unemployment (8 to 9 percent), falling productivity, and little hope for a more benevolent future.[1]

That kind of stagnation is essentially what happened in the late '30s, after FDR succeeded in pulling the economy back from the cliff of full-scale Depression, but failed to restore anywhere near normal levels of output. From 1936 or so until the beginning of WWII, the economy seemed destined for long-term stagnation with unemployment fluctuating around 15%. In today's Great Malaise, likewise, we can expect long-term unemployment from 10% to 15%, and utilization of industrial capacity in the 60% range, with a simultaneous upward creeping of part-time work and underemployment, and the concealment of real unemployment levels as more people stop looking for work and drop from the unemployment rolls.

Joshua Cooper Ramo notes that employment has fallen much more rapidly in the Great Recession than Okun's Law (which states the normal ratio of GDP decline to job losses) would have predicted. Instead of the 8.5% unemployment predicted by Okun's Law, we're at almost 10%.

> Something new and possibly strange seems to be happening in this recession. Something unpredicted by the experts. "I don't think," Summers told the

[1]Greenspan remarks from 1980, quoted by Magdoff and Sweezy, "The Great Malaise," in Magdoff and Sweezy, *The Irreversible Crisis*, pp. 58-60.

> Peterson Institute crowd—deviating again from his text—"that anyone fully understands this phenomenon." And that raises some worrying questions. Will creating jobs be that much slower too? Will double-digit unemployment persist even after we emerge from this recession? Has the idea of full employment rather suddenly become antiquated? . . .
>
> When compiling the "worst case" for stress-testing American banks last winter, policymakers figured the most chilling scenario for unemployment in 2009 was 8.9%—a figure we breezed past in May. From December 2007 to August 2009, the economy jettisoned nearly 7 million jobs, according to the Bureau of Labor Statistics. That's a 5% decrease in the total number of jobs, a drop that hasn't occurred since the end of World War II. The number of long-term unemployed, people who have been out of work for more than 27 weeks, was the highest since the BLS began recording the number in 1948. . . .
>
> America now faces the direst employment landscape since the Depression. It's troubling not simply for its sheer scale but also because the labor market, shaped by globalization and technology and financial meltdown, may be fundamentally different from anything we've seen before. And if the result is that we're stuck with persistent 9%-to-11% unemployment for a while . . . we may be looking at a problem that will define the first term of Barack Obama's presidency. . . . The total number of nonfarm jobs in the U.S. economy is about the same now—roughly 131 million—as it was in 1999. And the Federal Reserve is predicting moderate growth at best. That means more than a decade without real employment expansion.[1]

To put things in perspective, the employment-to-population ratio—since its peak of 64.7% in 2000—has fallen to 58.8%.[2] That means the total share of the population which is employed has fallen by about a tenth over the past nine years. And the employment-to-population ratio is a statistic that's a lot harder to bullshit than the commonly used official unemployment figures. The severity of the latter is generally concealed by discouraged job-seekers dropping off the unemployment rolls; the official unemployment figure is consistently understated because of shrinkage of the job market, and counts only those who are still bothering to look for work. The reason unemployment only rose rose to 9.8% in September 2009, instead of 10%, is that 571,000 discouraged workers dropped out of the job market that month. Another statistic, the hours-worked index, has also displayed a record decline (8.6% from the prerecession peak, compared to only 5.8% in the 1980-82 recession).[3]

A much larger portion of total unemployed in this recession are long-term unemployed. 53% (or eight million) of the unemployed in August were not on temporary layoff, and of those five million had sought work unsuccessfully for six months or more—both record highs.[4] Although total unemployment levels as of November 2009 have yet to equal their previous postwar peak in 1983, the percentage of the population who have been seeking jobs for six months or more is now

[1] Joshua Cooper Ramo, "Jobless in America: Is Double-Digit Unemployment Here to Stay?" *Time*, September 11, 2009 <http://www.time.com/time/printout/0,8816,1921439,00.html>.

[2] Brad DeLong, "Another Bad Employment Report (I-Wish-We-Had-a-Ripcord-to-Pull Department)," *Grasping Reality with All Eight Tentacles*, October 2, 2009 <http://delong.typepad.com/sdj/2009/10/another-bad-employment-report-i-wish-we-had-a-ripcord-to-pull-department.html>.

[3] *Ibid.*

[4] "U.S. Suffering Permanent Destruction of Jobs," *Washington's Blog*, October 5, 2009 <http://www.washingtonsblog.com/2009/10/us-suffering-permanent-destruction-of.html>

2.3%—compared to only 1.6% in 1983.[1] The Bureau of Labor Statistics announced in January 2010 that the rate of long-term unemployment was the highest since 1948, when it began measuring it; those who had been out of work for six months or longer comprised 40% of all unemployed.[2]

And we face the likely prospect that the economy will continue to shed jobs even after the resumption of growth in GDP; in other words not just a "jobless recovery," but a recovery with job losses.[3] As J. Bradford DeLong points out, the economy is shedding jobs *despite* an increase in demand for domestically manufactured goods.

> Real spending on American-made products is rising at a rate of about 3.5% per year right now and has been since May.
>
> The point is that even though spending on American products is rising, employment in America is still falling.[4]

Three quarters after recovery began in the 1981 recession, employment was up 1.5%. Three quarters into this recovery, it's down 0.6%. The recent surge in employment, despite enthusiastic celebration in the press, is hardly enough to keep pace with population growth and prevent unemployment from worsening.[5] And according to Neil Irwin, the massive debt deleveraging which is yet to come means there will be insufficient demand to put the unemployed back to work.

> American households are trying to reduce debt to stabilize finances. But they are doing so slowly, with total household debt at 94 percent of gross domestic product in the fourth quarter down just slightly from 96 percent when the recession began in late 2007.
>
> By contrast, that ratio of household debt to economic output was 70 percent in 2000. To get back to that level, Americans would need to pay down $3.4 trillion in debt—and if they do, that money wouldn't be available to spend on goods and services.[6]

In such a period of stagnation, capital goods investment is likely to lag far behind even the demand for consumer goods; investment in plant and equipment, generally, tends to fall much lower than capacity utilization of consumer goods industry in economic downturns, and to be much slower rebounding in the recovery. In the 1930s, investment in plant and equipment was cut by 70% to 80%. Machine tool builders shut down production for prolonged periods, and depreciated industrial capital stock was not replaced for years. In 1939, despite consumer demand 12% over its peak in the 1920s, investment in plant and equipment was at

[1]"Long-Term Unemployment," *Economist's View*, November 9, 2009 <http://economistsview.typepad.com/economistsview/2009/10/longterm-unemployment.html>.

[2]Ron Scherer, "Number of long-term unemployed hits highest rate since 1948," *Christian Science Monitor*, January 8, 2010 <http://www.csmonitor.com/USA/2010/0108/Number-of-long-term-unemployed-hits-highest-rate-since-1948>.

[3]Quiddity, "Job-loss recovery," *uggabugga*, October 25, 2009 <http://uggabugga.blogspot.com/2009/10/job-loss-recovery-experts-see.html#comments>.

[4]DeLong, "Jobless Recovery: Quiddity Misses the Point," *J. Bradford DeLong's Grasping Reality with All Eight Tentacles*, October 25, 2009 <http://delong.typepad.com/sdj/2009/10/jobless-recovery-quiddity-misses-the-point.html>.

[5]Ezra Klein, "A Fast Recovery? Or a Slow One?" Washington Post, April 14, 2010 <http://voices.washingtonpost.com/ezra-klein/2010/04/a_fast_recovery_or_a_slow_one.html>.

[6]Neil Irwin, "Economic data don't point to boom times just yet," *Washington Post*, April 13, 2010 <http://www.washingtonpost.com/wp-dyn/content/article/2010/04/12/AR2010041204236.html>.

less than 60% of the 1929 level.[1] Investment in plant and equipment only began to come back with heavy government Lend-Lease spending (the machinery industry expanded output 30% in 1940).[2] In the coming period, as we shall see below, we can expect a virtual freeze of investment in the old mass-production industrial core.

Charles Hugh Smith expects "a decades-long period of structural unemployment in which there will not be enough jobs for tens of millions of citizens": the employment rolls will gradually shrink from their present level of 137 million to 100 million or so, and then stagnate at that level indefinitely.[3] Economist Mark Zandi of Moody's Economy.com predicts "the unemployment rate will be permanently higher, or at least for the foreseeable future."[4] Of course, it's quite plausible that the harm will be mitigated to some extent by a greater shift to job-sharing, part-time work by all but one member of a household, or even a reduction of the standard work week to 32 hours.

The hope—*my* hope—is that these increasing levels of underemployment and unemployment will be offset by increased ease of meeting subsistence needs outside the official economy, by the imploding cost of goods manufactured in the informal sector, and by the rise of barter networks as the means of providing an increasing share of consumption needs by direct production for exchange between producers in the informal sector. As larger and larger shares of total production disappear as sources of conventional wage employment, and cease to show up in the GDP figures, the number of hours it's necessary to work to meet needs outside the informal sector will also steadily decline, and the remaining levels of part-time employment for a majority of the population will be sufficient to maintain a positive real material standard of living.

B. Resource Crises (Peak Oil)

In recent decades, the centerpiece of both the energy policy and a major part of the national security policy of the U.S. government has been to guarantee "cheap, safe and abundant energy" to the corporate economy. It was perhaps exemplified most forcefully in the Carter Doctrine of 1980: "An attempt by any outside force to gain control of the Persian Gulf region will be regarded as an assault on the vital interests of the United States of America, and such an assault will be repelled by any means necessary, including military force."[5]

This is no longer possible: the basic idea of Peak Oil is that the rate of extraction of petroleum has peaked, or is about to peak. On the downside of the peak, the supply of oil will gradually contract year by year. Although the total amount of oil reserves in the ground may be roughly comparable to those extracted to date, they will be poorer in quality, and more expensive in both dollar terms and energy to extract.

[1]Harry Magdoff and Paul Sweezy, *The End of Prosperity: The American Economy in the 1970s* (New York and London: Monthly Review Press, 1977), pp. 95, 120-121.

[2]*Ibid.*, p. 96.

[3]Smith, "Unemployment: The Gathering Storm," *Of Two Minds*, September 26, 2009 <http://charleshughsmith.blogspot.com/2009/09/unemployment-gathering-storm.html>.

[4]"Uh, oh, higher jobless rates could be the new normal," *New York Daily News*, October 23, 2009 <http://www.nydailynews.com/money/work_career/2009/10/19/2009-10-19_uh_oh_higher_jobless_rates_could_be_the_new_normal.html>.

[5]"Carter Doctrine," *Wikipedia*, accessed December 23, 2009 <http://en.wikipedia.org/wiki/Carter_Doctrine>.

All the panaceas commonly put forth for Peak Oil—oil shale, tar sands, offshore drilling, algae—turn out to be pipe dreams. The issue isn't the absolute amount of oil in offshore reserves or tar sands, but the *cost* of extracting them and the maximum feasible *rate* of extraction. In terms of the net energy surplus left over after the energy cost of extraction (Energy Return on Energy Investment, or EROEI), all the "drill baby drill" gimmicks are far more costly—cost far more BTUs per net BTU of energy produced—than did petroleum in the "good old days." The maximum rate of extraction from all the newly discovered offshore oil bonanzas the press reports, and from unconventional sources like tar sands, doesn't begin to compensate for the daily output of old wells in places like the Persian Gulf that will go offline in the next few years. And the oil from such sources is far more costly to extract, with much less net energy surplus.[1]

The list of false panaceas includes coal, by the way. It's sometimes argued that Peak Coal is some time away, and that increased coal output (e.g. China's much-vaunted policy of building another coal-fired generator every week) will compensate for decreased oil output in the intermediate term. But estimates of coal reserves have been revised radically downward in the last two decades—by some 55%, as a matter of fact. In virtually every country where coal reserves have been reestimated since the 1990s, such a downward revision has recurred. Poland, the largest coal producer in the EU, had its reserve estimates downgraded by 50%, and Germany by 90%. UK reserve estimates were revised from 45 billion tons to 0.22 billion tons. And interestingly, the countries with some of the highest estimated coal reserves (e.g. China) are also the countries whose estimates are the oldest and most out of date. The most recent figures for China, for an estimated 55 years' reserves, date back all the way to 1992—and Chinese production since then has amounted to some 20% of those total reserves.

> The Energy Watch Group report gives projected production profiles showing that China is likely to experience peak coal production in the next 10-15 years, followed by a steep decline. It should also be noted that these production profiles do not take into account uncontrolled coal fires which—according to satellite based estimates—add around 5-10% to regular consumption. Since China's production dwarfs that of any other country (being almost double that of the second largest producer, the USA) the global coal production peak will be heavily influenced by China's production profile.[2]

The Energy Watch Group's estimate for peak coal energy is 2025.[3] And even assuming increased coal output for another decade or more, Richard Heinberg forecasts total fossil fuel energy production peaking around 2010 or so.[4]

Peak Oil skeptics frequently argue that a price spike like the one in 2008 is caused, not by Peak Oil, but "instead" by some special circumstance like a specific supply disruption or speculative bubble. But that misses the point.

The very fact that supply has reached its peak, and that price is entirely determined by the amount of demand bidding for a fixed supply, means that the price of oil is governed by the same speculative boom-bust cycle Henry George observed in land. Given the prospect of a fixed supply of land or oil, the rational in-

[1]Rob Hopkins, *The Transition Handbook: From Oil Dependency to Local Resilience* (Totnes: Green Books, 2008), p. 23.

[2]Chris Vernon, "Peak Coal—Coming Soon?" *The Oil Drum: Europe*, April 5, 2007 <http://europe.theoildrum.com/node/2396>.

[3]*Ibid.*

[4]Richard Heinberg, *Peak Everything: Waking Up to the Century of Declines* (Gabriola Island, B.C.: New Society Publishers, 2007), p. 12.

terest of the oil industry, like that of real estate speculators, will lead them to hold greater or lesser quantities off the market, or dump them on the market, based on their estimate of the future movement of price. Hence the inconvenient fact, during the "drill here drill now" fever of the McCain-Palin campaign, that the oil companies were already sitting on large offshore oil reserves that they were failing to develop in anticipation of higher prices.

> The oil companies already have access to some 34 billion barrels of offshore oil they haven't even developed yet, but ending the federal moratorium on offshore drilling would probably add only another 8 billion barrels (assuming California still blocks drilling off its coast). Who thinks adding under 100,000 barrels a day in supply sometime after 2020—some one-thousandth of total supply—would be more than the proverbial drop in the ocean? Remember the Saudis couldn't stop prices from rising now by announcing that they will add 500,000 barrels of oil a day by the end of this year!
>
> Here is the key data from EIA:
>
> Look closely. As of 2003, oil companies had available for leasing and development 40.92 billion barrels of offshore oil in the Gulf of Mexico. I asked the EIA analyst how much of that (estimated) available oil had been discovered in the last five years. She went to her computer and said "about 7 billion barrels have been found." That leaves about 34 billion still to find and develop.
>
> The federal moratorium only blocks another 18 billion barrels of oil from being developed.[1]

And given the prospect of fixed supplies of oil, the greater the anticipated future scarcity value of oil, the greater will be the rational incentive for terrorists to leverage their power by disrupting supply. The infrastructure for extracting and distributing oil is unprecedentedly fragile, precisely because of a decline in productive capacity. Between 1985 and 2001, OPEC's excess production capacity fell from 25% of global demand to 2%. In 2003, the International Energy Agency estimated available excess capacity was at its lowest level in thirty years.[2]

According to Jeff Vail, speculative hoarding of petroleum and terrorist actions against oil pipelines are not *alternative* explanations *in place of* Peak Oil, but the results of a positive feedback process created by Peak Oil itself.

> It is quite common to hear "experts" explain that the current tight oil markets are due to "above-ground factors," and not a result of a global peaking in oil production. It seems more likely that it is geological peaking that is driving the geopolitical events that constitute the most significant "above-ground factors" such as the chaos in Iraq and Nigeria, the nationalization in Venezuela and Bolivia, etc. Geological peaking spawns positive feedback loops within the geopolitical system. Critically, these loops are not separable from the geological events—they are part of the broader "system" of Peak Oil.
>
> Existing peaking models are based on the logistics curves demonstrated by past peaking in individual fields or oil producing regions. Global peaking is an entirely different phenomenon—the geology behind the logistics curves is the same, but global peaking will create far greater geopolitical side-effects, even in regions with stable or rising oil production. As a result, these geopolitical side-effects of peaking global production will accelerate the rate of production decline, as well as increase the impact of that production decline by simultaneously increasing marginal demand pressures. The result: the right side of the global oil production curve will not look like the left . . . whatever logistics curve is fit to

[1]Joseph Romm, "McCain's Cruel Offshore Drilling Hoax," *CommonDreams.Org*, July 11, 2008 <http://www.commondreams.org/archive/2008/07/11/10301>.

[2]Richard Heinberg, *Powerdown* (Gabriola Island, British Columbia: New Society Publishers, 2004), pp. 27-28.

> the left side of the curve (where historical production increased), actual declines in the future will be sharper than that curve would predict.
>
> Here are five geopolitical processes, each a positive-feedback loop, and each an accelerant of declining oil production:
>
> 1. Return on Investment: Increased scarcity of energy, as well as increased prices, increase the return on investment for attacks that target energy infrastructure. . . .
>
> 2. Mercantilism: To avoid the dawning "bidding cycles" between crude oil price increases and demand destruction, Nation-States are increasingly returning to a mercantilist paradigm on energy. This is the attitude of "there isn't enough of it to go around, and we can't afford to pay the market price, so we need to lock up our own supply. . . .
>
> 3. "Export-Land" Model: Jeffrey Brown, a commentator at The Oil Drum, has proposed a geopolitical feedback loop that he calls the "export-land" model. In a regime of high or rising prices, a state's existing oil exports brings in great revenues, which trickles into the state's economy, and leads to increasing domestic oil consumption. This is exactly what is happening in most oil exporting states. The result, however, is that growth in domestic consumption reduces oil available for export. . . .
>
> 4. Nationalism: Because our Westphalian system is fundamentally broken, the territories of nations and states are rarely contiguous. As a result, it is often the case that a nation is cut out of the benefits from its host state's oil exports. . . . As a result, nations or sectarian groups within states will increasingly agitate for a larger share of the pie. . . . This process will develop local variants on the tactics of infrastructure disruption, as well as desensitize energy firms to ever greater rents for the security of their facilities and personnel—both of which will drive the next loop. . . .
>
> 5. Privateering: Nationalist insurgencies and economies ruined by the downslide of the "export-land" effect will leave huge populations with no conventional economic prospects. High oil prices, and the willingness to make high protection payments, will drive those people to become energy privateers. We are seeing exactly this effect in Nigeria, where a substantial portion of the infrastructure disruption is no longer carried out by politically-motivated insurgents, but by profit-motivated gangs. . . .[1]

Mercantilism, in particular, probably goes a long way toward explaining America's invasion of Iraq and the Russian-American "Great Game" in Central Asia in recent years. The United States' post-9/11 drive for basing rights in the former Central Asian republics of the old USSR, and the rise of the Shanghai Cooperation Organization as a counterweight to American power, are clearly more meaningful in the light of the Caspian Sea basin oil reserves.

And the evidence is clear that price really is governed entirely by the fluctuation of demand, and that supply—at least on the upward side—is extremely inelastic. Just consider the movement of oil supplies after the price shock of the late '70s and early eighties to that of the past few years. As "transition town" movement founder Rob Hopkins points out, the supply of oil has increased little if any since 2005—fluctuating between 84 and 87 mbd—despite record price levels.[2]

Peak Oil is likely to throw a monkey-wrench into the gears of the Chinese model of state-sponsored capitalism. China heavily subsidizes energy and transportation inputs, pricing them at artificially low levels to domestic industrial consumers, just as did the USSR. This accounting gimmick won't work externally—the Saudis want cash on the barrel head, at the price they set for crude petroleum—

[1]Jeff Vail, "Five Geopolitical Feedback-Loops in Peak Oil," *JeffVail.Net*, April 23, 2007 <http://www.jeffvail.net/2007/04/five-geopolitical-feedback-loops-in.html>.

[2]Hopkins, *The Transition Handbook*, p. 22.

and the increased demand for subsidized energy inputs by wasteful domestic Chinese producers will just cause China to bankrupt itself buying oil abroad.

Overall, the effect of Peak Oil is likely to be a radical shortening of corporate supply and distribution chains, a resurrection of small-scale local manufacturing in the United States, and a reorientation of existing manufacturing facilities in China and other offshore havens toward production for their own domestic markets.

The same is true of relocalized agriculture. The lion's share of in-season produce is apt to shift back to local sourcing, and out of season produce to become an expensive luxury. As Jeff Rubin describes it,

> As soaring transport costs take New Zealand lamb and California blueberries off Toronto menus and grocery-store shelves, the price of locally grown lamb and blueberries will rise. The higher they rise, the more they will encourage people to raise sheep and grow blueberries. Ultimately, the price will rise so high that now unsaleable real estate in the outer suburbs will be converted back into farmland. That new farmland will then help stock the grocery shelves in my supermarket, just like it did thirty or forty years ago.[1]

This was a common theme during the oil shocks of the 1970s, and has been revived in the past few years. In the late '70s Warren Johnson, in *Muddling Toward Frugality*, predicted that rising energy prices would lead to a radical shortening of industrial supply chains, and the relocalization of manufacturing and agriculture.[2] Although he jumped the gun by thirty years, his analysis is essentially sound in the context of today's Peak Oil concerns. The most pessimistic (not to say catastrophic) Peak Oil scenario is that of James Kunstler, outlined not only in *The Long Emergency* but fictionally in *World Made by Hand*.[3] Kunstler's depiction of a world of candles and horse-drawn wagons, in my opinion, greatly underestimates the resilience of market economies in adjusting to energy shocks. Brian Kaller's "return to Mayberry scenario" is much less alarmist.

> In fact, peak oil will probably not be a crash, a moment when everything falls apart, but a series of small breakdowns, price hikes, and local crises. . . .
>
> Take one of the more pessimistic projections of the future, from the Association for the Study of Peak Oil, and assume that by 2030 the world will have only two-thirds as much energy per person. Little breakdowns can feed on each other, so crudely double that estimate. Say that, for some reason, solar power, wind turbines, nuclear plants, tidal power, hydroelectric dams, bio-fuels, and new technologies never take off. Say that Americans make only a third as much money, or their money is worth only a third as much, and there is only a third as much driving. Assume that extended families have to move in together to conserve resources and that we must cut our flying by 98 percent.
>
> Many would consider that a fairly clear picture of collapse. But we have been there before, and recently. Those are the statistics of the 1950s—not remembered as a big time for cannibalism.[4]

[1]Jeff Rubin, *Why Your World is About to Get a Whole Lot Smaller: Oil and the End of Globalization* (Random House, 2009), p. 220.

[2]Warren Johnson, *Muddling Toward Frugality* (San Francisco: Sierra Club Books, 1978).

[3]James Howard Kunstler, *The Long Emergency: Surviving the End of Oil, Climate Change, and Other Converging Catastrophes of the Twenty-First Century* (Grove Press, 2006); Kunstler, *World Made by Hand* (Grove Press, 2009).

[4]Brian Kaller, "Future Perfect: the future is Mayberry, not Mad Max," *Energy Bulletin*, February 27, 2009 (from *The American Conservative*, August 2008) <http://www.energybulletin.net/node/48209>.

Like Kaller, Jeff Rubin presents the world after Peak Oil as largely "a return to the past . . . in terms of the re-emergence of local economies."[1]

But despite the differences in relative optimism or pessimism among these various Peak Oil thinkers, their analyses all have a common thread running through them: the radical shortening of industrial supply and distribution chains, and an end to globalization based on the export of industry to low-wage sweatshop havens like China.

To quote a Rubin article from May 2008, two months before oil prices peaked, rising transportation costs had more than offset the Chinese wage differential. The cost of shipping a standard 40-ft container, he wrote, had tripled since 2000, and could be expected to double again as oil prices approached $200/barrel.[2] What's more, "the explosion in global transport costs has effectively offset all the trade liberalization efforts of the last three decades." A rise in oil prices from $20 to $150/barrel has the same effect on international trade as an increase in tariffs from 3% to 11%—i.e., to their average level in the 1970s.[3] According to Richard Milne,

> Manufacturers are abandoning global supply chains for regional ones in a big shift brought about by the financial crisis and climate change concerns, according to executives and analysts.
>
> Companies are increasingly looking closer to home for their components, meaning that for their US or European operations they are more likely to use Mexico and eastern Europe than China, as previously.[4]

Domestically, sustained oil prices at or above mid-2008 levels will cause a radical contraction in the trucking and airline industries. Estimates were widespread in the summer of 2008 that airlines would shut down 20% of their routes in the near-term of oil prices of $140/barrel or more persisted, and long-haul truckers were under comparable pressure. Joseph Romm, an energy analyst, argues that the airline industry is "barely viable" at $150/barrel. Sustained oil prices of $200/barrel will cause air travel to become a luxury good (as in the days when those who could afford it were referred to as the "jet set").[5]

C. Fiscal Crisis of the State

The origins of corporate capitalism and the mass-production economy are associated with massive government subsidies; since then the tendency of corporate capital to socialize its operating costs has never abated. As a matter of basic economics, whenever you subsidize something and make it available to the user for less than its real cost, demand for it will increase. American capitalism, as a result, has followed a pattern of expansion skewed toward extensive additions of subsi-

[1]David Parkinson, "A coming world that's 'a whole lot smaller,'" *The Globe and Mail*, May 19, 2009 <http://docs.google.com/Doc?id=dg5dgmrv_79hjb66vc3>.

[2]Jeffrey Rubin, "The New Inflation," *StrategEcon* (CIBC World Markets), May 27, 2008 <http://research.cibcwm.com/economic_public/download/smay08pdf>.

[3]Jeffrey Rubin and Benjamin Tal, "Will Soaring Transport Costs Reverse Globalization?" *StrategEcon*, May 27, 2008, p. 4.

[4]Richard Milne, "Crisis and climate force supply chain shift," *Financial Times*, August 9, 2009 <http://www.ft.com/cms/s/0/65a709ec-850b-11de-9a64-00144feabdc0.html>. See also Fred Curtis, "Peak Globalization: Climate change, oil depletion and global trade," *Ecological Economics* Volume 69, Issue 2 (December 15, 2009).

[5]Sam Kornell, "Will PeakOil Turn Flying into Something Only Rich People Can Afford?" *Alternet*, May 7, 2010 <http://www.alternet.org/economy/146769/will_peak_oil_turn_flying_into_something_only_rich_people_can_afford>.

dized inputs, rather than more intensive use of existing ones. As James O'Connor describes the process,

> Transportation costs and hence the fiscal burden on the state are not only high but also continuously rising. It has become a standard complaint that the expansion of road transport facilities intensifies traffic congestion. The basic reason is that motor vehicle use is subsidized and thus the growth of the freeway and highway systems leads to an increase in the demand for their use.[1]
>
> There is another reason to expect transportation needs (and budgets) to expand. The development of rapid transport and the modernization of the railroads, together with the extension of the railroad systems, will push the suburbs out even further from urban centers, putting still more distance between places of work, residence, and recreation. Far from contributing to an environment that will free suburbanites from congestion and pollution, rapid transit will, no doubt, extend the traffic jams and air pollution to the present perimeters of the suburbs, thus requiring still more freeway construction, which will boost automobile sales.[2]

And the tendency of monopoly capitalism to generate surplus capital and output also increases the amount of money that the state must spend to absorb the surplus.

Monopoly capitalism, according to O'Connor, is therefore plagued by a "fiscal crisis of the state." " . . . [T]he socialization of the costs of social investment and social consumption capital increases over time and increasingly is needed for profitable accumulation by monopoly capital."[3]

> . . . [A]lthough the state has socialized more and more capital costs, the social surplus (including profits) continues to be appropriated privately. . . . The socialization of costs and the private appropriation of profits creates a fiscal crisis, or "structural gap," between state expenditures and state revenues. The result is a tendency for state expenditures to increase more rapidly than the means of financing them.[4]

In short, the state is bankrupting itself providing subsidized inputs to big business, while big business's demand for those subsidized inputs increases faster than the state can provide them. As Ivan Illich put it,

> queues will sooner or later stop the operation of any system that produces needs faster than the corresponding commodity. . . .[5]
>
> . . . [I]nstitutions create needs faster than they can create satisfaction, and in the process of trying to meet the needs they generate, they consume the Earth.[6]

The distortion of the price system, which in a free market would tie quantity demanded to quantity supplied, leads to ever-increasing demands on state services. Normally price functions as a form of feedback, a homeostatic mechanism much like a thermostat. Putting a candle under a thermostat will result in an ice-cold house When certain hormonal feedback loops are distorted in an organism, you get gigantism; the victim dies crushed by his own weight. Likewise, when the consumption of some factor is subsidized by the state, the consumer is protected from the real cost of providing it, and unable to make a rational decision about

[1]James O'Connor, *The Fiscal Crisis of the State* (New York: St. Martin's Press, 1973), p. 106.

[2]*Ibid.*, pp. 109-110.

[3]*Ibid.*, p. 8.

[4]*Ibid.*, p. 9.

[5]Illich, *Disabling Professions* (New York and London: Marion Boyars, 1977), p. 30.

[6]Illich, *Deschooling Society* (New York, Evanston, San Francisco, London: Harper & Row, 1973).

how much to use. So the state capitalist sector tends to add factor inputs extensively, rather than intensively; that is, it uses the factors in larger amounts, rather than using existing amounts more efficiently. The state capitalist system generates demands for new inputs from the state geometrically, while the state's ability to provide new inputs increases only arithmetically. The result is a process of snowballing irrationality, in which the state's interventions further destabilize the system, requiring yet further state intervention, until the system's requirements for stabilizing inputs finally exceed the state's resources. At that point, the state capitalist system reaches a breaking point.

Eventually, therefore, state capitalism hits a wall at which the state is no longer able to increase the supply of subsidized inputs. States approach the condition described by John Robb's term "hollow state":

> The hollow state has the trappings of a modern nation-state ("leaders", membership in international organizations, regulations, laws, and a bureaucracy) but it lacks any of the legitimacy, services, and control of its historical counter-part. It is merely a shell that has some influence over the spoils of the economy.[1]
>
> . . . A hollow state is different from a failed state in that it continues to exist on the international stage. It has all the standard edifices of governance although most are heavily corrupted and in thrall to global corporate/monied elites. It continues to deliver political goods (albeit to a vastly diminished group, usually around the capital) and maintains a military. Further, in sections of the country, there is an appearance of normal life.
>
> However, despite this facade, the hollow state has abdicated (either explicitly as in Lebanon's case or de facto as in Mexico's) vast sections of its territory to networked tribes (global guerrillas). Often, these groups maintain a semblance of order, as in rules of Sao Paulo's militias or the Taliban's application of sharia. Despite the fact that these group [sic] control/manipulate explicit economic activity and dominate the use/application of violence at the local level, these groups often grow the local economy. How? By directly connecting it to global supply chains of illegal goods—from people smuggling to drugs to arms to copy-theft to money laundering.
>
> The longer this state of affairs persists, the more difficult it is to eradicate. The slate of alternative political goods delivered by these non-state groups, in contrast to the ineffectiveness of the central government, sets the stage for a shift in legitimacy. Loyalties shift. Either explicitly through membership in tribal networks, or acknowledgement of the primacy of these networks over daily life.[2]

The entente between American and Iraqi government military forces, on the one hand, and the Sunni militias in Al Anbar province, on the other, is a recent example of a hollowed state coming to terms with "Fourth Generation Warfare" networks as de facto local governments. An early example was the Roman imperial state of the fifth century, delegating de facto territorial control to German tribal entities in return for de jure fealty to Rome.

And of course, in Robb's preferred scenario—as we will see in Chapter Six—loyalties shift from the state to resilient communities.

If the state does not become completely hollowed out by Robb's criteria, it nevertheless is forced to retreat from an ever increasing share of its former functions owing to its shrinking resources: a collapse of the value of official currency,

[1]John Robb, "Onward to a Hollow State," *Global Guerrillas*, September 22, 2009 <http://globalguerrillas.typepad.com/globalguerrillas/2008/09/onward-to-a-hol.html>.

[2]Robb, "HOLLOW STATES vs. FAILED STATES," *Global Guerrillas*, March 24, 2009 <http://globalguerrillas.typepad.com/globalguerrillas/2009/03/hollow-states-vs-failed-states.html>.

combined with a catastrophic decline in tax revenues. The state delegates more and more functions to private entities nominally operating pursuant to state policy but primarily in the interest of self-aggrandizement, becomes prey to kleptocrats, leaves unenforced more and more laws that are technically on the books, and abandons ever increasing portions of its territory to the black market and organized criminal gangs.

In many ways, this is a positive development. Local sheriffs may decide that evicting mortgage defaulters and squatters, enforcing regulatory codes against household microenterprises, and busting drug users fall very low on their list of priorities, compared to dealing with murder and robbery. Governments may find themselves without the means of financing corporate welfare.

Something like this happened in Poland in the 1980s, with Gen. Jaruzelski—in a classic example of joining 'em when you can't beat 'em—finally deciding to legalize banned groups and hold open elections because Poland had become "ungovernable." Solidarity activist Wiktor Kulerski, in what should be an extremely suggestive passage for those of us who dream of an unenforceable regime of patent and copyright, zoning and licensing laws, wrote of his vision for a hollow state in Poland:

> This movement should create a situation in which authorities will control empty stores, but not the market; the employment of workers, but not their livelihood; the official media, but not the circulation of information; printing plants, but not the publishing movement; the mail and telephones, but not communications; and the school system, but not education.[1]

But to the extent that the current economic structure is heavily dependent on government activity, and adjustment to the withdrawal of subsidized infrastructure and services may take time, an abrupt retreat of state activity may result in a catastrophic period of adjustment.

The fiscal crisis dovetails with Peak Oil and other resource crises, in a mutually reinforcing manner. The imperative of securing strategic access to foreign oil reserves, and keeping the sea lanes open, results in costly wars. The increased cost of asphalt intensifies the already existing tendency, of demand for subsidized transportation infrastructure to outstrip the state's ability to supply it. As the gap expands, the period between deterioration of roads and the appropriation of money to repair them lengthens. The number of miles of high-volume highway the state is able to keep in a reasonable state of repair falls from one year to the next, and the state is continually forced to retreat and regroup and relegate an ever-larger share of highways to second-tier status. As James Kunstler points out, a highway is either kept in repair, or it quickly deteriorates.

> Another consequence of the debt problem is that we won't be able to maintain the network of gold-plated highways and lesser roads that was as necessary as the cars themselves to make the motoring system work. The trouble is you have to keep gold-plating it, year after year. Traffic engineers refer to this as "level-of-service." They've learned that if the level-of-service is less than immaculate, the highways quickly enter a spiral of disintegration. In fact, the American Society of Civil Engineers reported several years ago that the condition of many highway bridges and tunnels was at the "D-minus" level, so we had already fallen far be-

[1]Lawrence W. Reed, "A Tribute to the Polish People," *The Freeman: Ideas on Liberty*, October 2009 <http://www.thefreemanonline.org/columns/ideas-and-consequences/a-tribute-to-the-polish-people/>.

> hind on a highway system that had simply grown too large to fix even when we thought we were wealthy enough to keep up.[1]

It doesn't take many years of neglect before deterioration and axle-breaking potholes render a highway unusable to heavy trucks, so that a growing share of the highway network will for all intents and purposes be abandoned.[2]

So each input crisis feeds the other, and we have a perfect storm of terminal crises. As described by Illich,

> The total collapse of the industrial monopoly on production will be the result of synergy in the failure of multiple systems that fed its expansion. This expansion is maintained by the illusion that careful systems engineering can stabilize and harmonize present growth, while in fact it pushes all institutions simultaneously toward their second watershed.[3]

D. Decay of the Cultural Pseudomorph

What Mumford called the "cultural pseudomorph," as we saw it described in Chapter One, was actually only the first stage. It has since decayed into a second, much weaker stage, unforeseen by Mumford, and shows signs of its final downfall. In the first stage, as Mumford observed, neotechnic methods (i.e., electrically powered machinery) were integrated into a mass-production framework fundamentally opposed to the technology's real potential. But this stage reached its limit by the 1970s.

In the second stage, mass production on the Sloan model is being replaced by flexible, networked production with general-purpose machinery, with the production process organized along lines much closer to the original neotechnic ideal.

Piore and Sabel describe the "lean" revolution of recent decades as the discovery, after a long interlude of mass production, of the proper way of organizing an industrial economy. "[T]he mass-production paradigm had unforeseen consequences: it took almost a century (from about 1870 to 1960) to discover how to organize an economy to reap the benefits of the new technology."[4]

According to those authors, the shift to lean production in America from the 1980s on was in large part a response to the increasing environment of macroeconomic uncertainty that prevailed after the resumption of the crisis of overaccumulation, and the oil shocks of the '70s. Mass-production industry is extremely brittle—i.e., it "does not adjust easily to major changes in its environment." The question is not just how industry will react to resource depletion, but how it will react to wildly fluctuating prices and erratic supplies.[5] Economic volatility and uncertainty means mass production industry will be hesitant to invest in specialized production machinery that may be unpredictably rendered superfluous by "changes in raw materials prices, interest rates, and so on."[6] As we saw in Chapter Two, long-term capital investment in costly technologies requires predictability;

[1]James Howard Kunstler, "Lagging Recognition," *Clusterfuck Nation*, June 8, 2009 <http://kunstler.com/blog/2009/06/lagging-recognition.html>

[2]Kunstler, *The Long Emergency*, pp. 264-265.

[3]Illich, *Tools for Conviviality* (New York, Evanston, San Francisco, London: Harper & Row, 1973), p. 103.

[4]Piore and Sabel, *Second Industrial Divide*, p. 48.

[5]*Ibid.*, p. 192.

[6]Piore and Sabel, "Italian Small Business Development: Lessons for U.S. Industrial Policy," in John Zysman and Laura Tyson, eds., *American Industry in International Competition: Government Policies and Corporate Strategies* (Ithaca and London: Cornell University Press, 1983), p. 397.

and the environment associated with Peak Oil and other input and cyclical crises is just about the opposite of what conduces to the stability of mass-production industry.

Conversely, though, the system prevailing in industrial districts like Emilia-Romagna is called "flexible manufacturing" for a reason. It is able to reallocate dedicated capital goods and shift contractual relationships, and do so quite rapidly, in response to sudden changes in the environment.

Although craft production has always tended to expand relative to mass-production industry during economic downturns, it was only in the prolonged stagnation of the 1970s and '80s that it began permanently to break out of its peripheral status.

> From the second industrial revolution at the end of the nineteenth century to the present, economic downturns have periodically enlarged the craft periphery with respect to the mass-production core—but without altering their relationship. Slowdowns in growth cast doubt on subsequent expansion; in an uncertain environment, firms either defer mass-production investments or else switch to craft-production techniques, which allow rapid entry into whatever markets open up. The most straightforward example is the drift toward an industrial-subsistence, or -repair, economy: as markets stagnate, the interval between replacements of sold goods lengthens. This lengthened interval increases the demand for spare parts and maintenance services, which are supplied only by flexibly organized firms, using general-purpose equipment. The 1930s craftsman with a tool kit going door to door in search of odd jobs symbolizes the decreased division of labor that accompanies economic retrocession: the return to craft methods.
>
> But what is distinctive about the current crisis is that the shift toward greater flexibility is provoking technological sophistication—rather than regression to simple techniques. As firms have faced the need to redesign products and methods to address rising costs and growing competition, they have found new ways to cut the costs of customized production. . . . In short, craft has challenged mass production as the paradigm.[1]

In the case of small Japanese metalworking firms, American minimills and the Pratese textile industry, the same pattern prevailed. Small subcontractors of larger manufacturing firms "felt the increasing volatility of their clients' markets; in response, they adopted techniques that reduced the time and money involved in shifting from product to product, and that also increased the sophistication and quality of the output."[2]

In the Third Italy in particular, large mass-production firms outsourced an increasing share of components to networks of small, flexible manufacturers. The small firms, initially, were heavily dependent on the large ones as outlets. But new techniques and machine designs made production increasingly efficient in the small firms.

> In some cases . . . the larger equipment is miniaturized. In other cases, however, artisan-like techniques of smelting, enameling, weaving, cutting, or casting metal are designed into new machines, some of which are controlled by sophisticated microprocessors.

At the same time, small firms which previously limited themselves to supplying components to a large manufacturer's blueprints instead began marketing products of their own.[3]

[1]Piore and Sabel, *Second Industrial Divide*, p. 207.
[2]*Ibid.*, p. 218.
[3]Piore and Sabel, "Italian Small Business Development," pp. 397-398.

While small manufacturers in the late 1960s were still dependent on a few or even one large client, there was a wholesale shift in the 1970s.

> To understand how this dependence was broken in the course of the 1970s, and a new system of production created, imagine a small factory producing transmissions for a large manufacturer of tractors. Ambition, the joy of invention, or fear that he and his clients will be devastated by an economic downturn lead the artisan who owns the shop to modify the design of the tractor transmission to suit the need of a small manufacturer of high-quality seeders. . . . But once the new transmission is designed, he discovers that to make it he needs precision parts not easily available on the market. If he cannot modify his own machines to make these parts, he turns to a friend with a special lathe, who like himself fears being too closely tied to a few large manufacturers of a single product. Soon more and more artisans with different machines and skills are collaborating to make more and more diverse products.[1]

So a shift has taken place, with the work formerly done by vertically integrated firms being outsourced to flexible manufacturing networks, and with a smaller and smaller share of essential functions that can only be performed by the core mass-production firm. As Eric Hunting observed:

> In the year 2000 our civilization reached an important but largely unnoticed milestone. For the first time the volume of consumer goods produced in 'job shop' facilities—mostly in Asia—exceeded the volume produce in traditional Industrial Age factories. This marks a long emerging trend of demassification of production capability driven by the trends in machine-tool evolution (smaller, smarter, cheaper) that is producing a corresponding demassification of capital and a homogenization of labor values around the globe. Globalization has generally sought profit through geographical spot-market value differences in resources and labor. But now those differences are disappearing faster the more they're exploited and capital has to travel ever faster and farther in search of shrinking margins.[2]

The organization of physical production, in both the Toyota Production System and in the Emilia-Romagna model of local manufacturing networks, is beginning—after a long mass-production interlude—to resemble the original neotechnic promise of integrating power machinery into craft production.

But the neotechnic, even though it has finally begun to emerge as the basis of a new, coherent production model governed by its own laws, is still distorted by the pseudomorph in a weaker form: the new form of production still takes place within a persistent corporate framework of marketing, finance and "intellectual property."

Andy Robinson, a member of the P2P Research email list, argued that "given recent studies showing equal productivity in factories in North and South,"

> the central mechanism of core-periphery exploitation has moved from technological inequality (high vs low value added) to rent extraction on IP. Since the loss of IP would make large companies irrelevant, they fight tooth and nail to preserve it, even beyond strict competitiveness, and behave in otherwise quite "irrational" ways to prevent their own irrelevance (e.g. the MPAA and RIAA's alienating of customers).[3]

[1]Piore and Sabel, "Italy's High-Technology Cottage Industry," *Transatlantic Perspectives* 7 (December 1982), p. 7.

[2]Eric Hunting, private email, August 4, 2008.

[3]Andy Robinson, "[p2p research] CAD files at the Pirate Bay? (Follow up," October 28, 2009 <http://listcultures.org/pipermail/p2presearch_listcultures.org/2009-October/005403.html>.

And despite the admitted control of distributed manufacturing within a corporate framework, based on corporate ownership of "intellectual property," Robinson suggests that the growing difficulty of enforcing IP will cause that framework to erode in the near future:

> . . . [I]t may be more productive to look at the continuing applicability or enforceability of IP, rather than whether businesses will continue to use it. While this is very visible in the virtual and informational sphere ("pirating" and free duplication of games, software, console systems, music, film, TV, news, books, etc), it is also increasingly the case in terms of technological hardware. Growing Southern economies—China being especially notorious—tend to have either limited IP regimes or lax enforcement, meaning that everything that a MNC produces there, will also be copied or counterfeited at the same quality for the local market, and in some cases traded internationally. I have my suspicions that Southern regimes are very aware of the centrality of IP to core-periphery exploitation and their laxity is quite deliberate. But, in part it also reflects the limits of the Southern state in terms of capacity to dominate society, and the growing sophistication of transnational networks (e.g. organised crime networks), which can evade, penetrate and fight the state very effectively.[1]

Elsewhere, Robinson brilliantly drew the parallels between the decay of the pseudomorph in the industrial and political realms:

> I think part of the crisis of the 70s has to do with networks and hierarchies. The "old" system was highly hierarchical, but was suffering problems from certain kinds of structural weaknesses in relation to networks—the American defeat in Vietnam being especially important. . . . And ever since the 70s the system has been trying to find hybrids of network and hierarchy which will harness and capture the power of networks without leading to "chaos" or system-breakdown. We see this across a range of fields: just-in-time production, outsourcing and downsizing, use of local subsidiaries, contracting-out, Revolution in Military Affairs, full spectrum dominance, indirect rule through multinational agencies, the Nixon Doctrine, joined-up governance, the growing importance of groups such as the G8 and G20, business networks, lifelong learning, global cities, and of course the development of new technologies such as the Internet. . . .
>
> In the medium term, the loss of power to networks is probably irreversible, and capital and the state will either go down fighting or create more-or-less stable intermediary forms which allow them to persist for a time. We are already seeing the beginnings of the latter, but the former is more predominant. The way I see the crisis deepening is that large areas will drift outside state and capitalist control, integrated marginally or not at all (this is already happening at sites such as Afghanistan, NWFP, the Andes, Somalia, etc., and in a local way in shanty-towns and autonomous centres). I also expect the deterritorialised areas to spread, as a result of the concentration of resources in global cities, the ecological effects of extraction, the neoliberal closing of mediations which formerly integrated, and the growing stratum of people excluded either because of the small number of jobs available or the growing set of requirements for conformity. Eventually these marginal spaces will become sites of a proliferation of new forms of living, and a pole of attraction compared to the homogeneous, commandist, coercive core.[2]

So long as the state successfully manages to prop up the centralized corporate economic order, libertarian and decentralist technologies and organizational forms will be incorporated into the old centralized, hierarchical framework. As the system approaches its limits of sustainability, those elements become increasingly

[1]*Ibid.*

[2]Andy Robinson, "[p2p research] Berardi essay," P2P Research email list, May 25, 2009 <http://listcultures.org/pipermail/p2presearch_listcultures.org/2009-May/003079.html>.

destabilizing forces within the present system, and prefigure the successor system. When the system finally reaches those limits, those elements will (to paraphrase Marx) break out of their state capitalist integument and become the building blocks of a fundamentally different society. We are, in short, building the foundations of the new society within the shell of the old.

And the second stage of the pseudomorph is weakening. For example, although the Nike model of "outsourcing everything" and retaining corporate control of an archipelago of small manufacturing shops still prevails to a considerable extent among U.S.-based firms, small subcontractors elsewhere have increasingly rebelled against the hegemony of their large corporate clients. In Italy and Japan the subcontractors have federated among themselves to create flexible manufacturing networks and reduce their dependence on any one outlet for their products.[1] The result is that the corporate headquarters, increasingly, is becoming a redundant node in a network—a redundant node that can be bypassed.

Indeed, the Nike model is itself extremely vulnerable to such bypassing. As David Pollard observes:

> In their famous treatise explaining the Internet phenomenon, Doc Searls, Dave Weinberger et al said that what made the Internet so powerful and so resilient was that it had no control 'centre' and no hierarchy: All the value was added, by millions of people, at the 'ends'. And if someone tried to disrupt it, these millions of users would simply work around the disruption. There is growing evidence that the same phenomenon is happening in businesses, which have long suffered from diseconomies of scale and bureaucracy that stifle innovation and responsiveness. Think of this as a kind of 'outsourcing of everything'. . . . Already companies like Levi Strauss make nothing at all—they simply add their label to stuff made by other companies, and distribute it (largely through independent companies they don't own either).[2]

If the people actually producing and distributing the stuff ever decide they have the right to market an identical product, Levi Strauss's ownership of the label notwithstanding, Levi's is screwed.

As a general phenomenon, the shift from physical to human capital as the primary source of productive capacity in so many industries, along with the imploding price and widespread dispersion of ownership of capital equipment in so many industries, means that corporate employers are increasingly hollowed out and only maintain control over the physical production process through legal fictions. When so much of actual physical production is outsourced to the small sweatshop or the home shop, the corporation becomes a redundant "node" that can be bypassed; the worker can simply switch to independent production, cut out the middleman, and deal directly with suppliers and outlets.

A good example of the weakness of the second stage of the pseudomorph is the relationship of the big automakers with parts suppliers today, compared to when Galbraith wrote forty years ago. As portrayed in *The New Industrial State*, the relationship between large manufacturers and their suppliers was one of unilateral market control. Today, Toyota's American factories share about two-thirds of their auto parts suppliers with the Detroit Three.[3] According to Don Tapscott and Anthony Williams, more than half of a vehicle's value already consists of electrical

[1]Piore and Sabel, *Second Industrial Divide*, pp. 226-227.

[2]David Pollard, "Ten Important Business Trends," *How to Save the World*, May 12, 2009 <http://blogs.salon.com/0002007/2009/05/12.html#a2377>.

[3]Dan Strumpf, "Exec Says Toyota Prepared for GM Bankruptcy," Associated Press, April 8, 2009 <http://abcnews.go.com/Business/wireStory?id=7288650>.

systems, electronics and software rather than the products of mechanical engineering, and by 2015 suppliers will conduct most R&D and production.[1]

Taking into account only the technical capabilities of the suppliers, it's quite feasible for parts suppliers to produce generic replacement parts in competition with the auto giants, to produce competing modular components designed for a GM or Toyota platform, or even to network to produce entirely new car designs piggybacked on a GM or Toyota chassis and engine block. The only thing stopping them is trademark and patent law.

And in fact supplier networks are beginning to carry out design functions among themselves, albeit on contract to large corporate patrons. For example, Boeing's designers used to do all the work of developing detailed specs for each separate part, with suppliers just filling the order to the letter; Boeing then assembled the parts in its own plant. But now, according to Don Tapscott and Anthony Williams, "suppliers codesign airplanes from scratch and deliver complete subassemblies to Boeing's factories. . . . " Rather than retaining control of all R&D in-house, Boeing is now "handing significant responsibility for innovation over to suppliers. . . . "[2]

An early indication that things may be reaching a tipping point is China's quasi-underground "shanzhai" enterprises which, despite being commonly dismissed as mere producers of knockoffs, are in fact extremely innovative not only in technical design but in supply chain efficiency and the speed of their reactions to change. The shanzhai economy resembles the flexible manufacturing networks of the Third Italy. Significantly, supplier networks for transnational corporations have begun to operate underground to supply components for shanzhai enterprises.

> Tapping into the supply chains of big brands is easy, producers say. "It's really common for factories to do a night shift for other companies," says Zhang Haizhen, who recently ran a shanzhai company here. "No one will refuse an order if it is over 5,000 mobile phones."[3]

The Chinese motorcycle industry is a good illustration of these trends. Many of its major designs are reverse-engineered from Japanese products, and the industry's R&D model is based on networked collaborative design efforts between many small, independent actors. And the reverse-engineered bikes are not simple copies of the original Japanese designs in all their major details; they build on the original designs that are in many ways superior to it. "Rather than copy Japanese models precisely, suppliers take advantage of the loosely defined specifications to amend and improve the performance of their components, often in collaboration with other suppliers."[4]

And recently, according to Bunnie Huang, there have been indications that native Chinese auto firms have been producing an unauthorized version of the Corolla. Huang spotted what appeared to be a Toyota Corolla bearing the logo of the Chinese BYD auto company.

> So when I saw this, I wasn't sure if it was a stock Corolla to which a local enthusiast attached a BYD badge, or if it was a BYD copycat of our familiar brand-name Toyota car. Or, by some bizarre twist, perhaps Toyota is now using BYD to

[1]Don Tapscott and Anthony D. Williams, *Wikinomics: How Mass Collaboration Changes Everything* (New York: Portfolio, 2006), p. 231.

[2]Tapscott and Williams, pp. 217-218.

[3]David Barboza, "In China, Knockoff Cellphones are a Hit," *New York Times*, April 27, 2009 <http://www.nytimes.com/2009/04/28/technology/28cell.html>.

[4]Tapscott and Williams, pp. 221-222.

> OEM their cars in China through a legitimized business relationship. I don't know which is true, but according to the rumors I heard from people who saw this photo, this is actually a copycat Toyota made using plans purchased on the black market that were stolen from Toyota. Allegedly, someone in China who studies the automobile industry has taken one of these apart and noted that the welds are done by hand. In the original design, the welds were intended to be done by machine. Since the hand-welds are less consistent and of lower quality than the robotic welds, the car no longer has adequate crash safety. There are also other deviations, such as the use of cheap plastic lenses for the headlights. But, I could see that making a copycat Corolla is probably an effective exercise for giving local engineers a crash-course in world-class car manufacture.[1]

Generally speaking, the corporate headquarters' control over the supplier is growing increasingly tenuous. As long ago as a decade ago, Naomi Klein pointed out that the "competing labels . . . are often produced side by side in the same factories, glued by the very same workers, stitched and soldered on the very same machines."[2]

E. Failure to Counteract Limits to Capture of Value by Enclosure of the Digital Commons

As Michel Bauwens describes it, it is becoming increasingly impossible to capture value from the ownership of ideas, designs, and technique—all the "ephemera" and "intellect" that Tom Peters writes about as a component of the price of manufactured goods—leading to a crisis of sustainability for capitalism. "Cognitive capitalism" is capital's attempt to adjust to the shift from physical to human capital, and to capture value from the immaterial realm. Bauwens cites McKenzie Wark's theory that a new "vectoralist" class "has arisen which controls the vectors of information, i.e. the means through which information and creative products have to pass, for them to realize their exchange value." This describes "the processes of the last 40 years, say the post-1968 period, which saw a furious competition through knowledge-based competition and for the acquisition of knowledge assets, which led to the extraordinary weakening of the scientific and technical commons."[3]

Cognitive capitalism arose as a solution to the unsustainability of the older pattern of capitalist growth, based on extensive addition of physical inputs and expansion into new geographical areas. Bauwens uses the analogy of the ancient slave economy, which became untenable when avenues of extensive development (i.e. expansion into new territory, and acquisition of new slaves) were closed off. When the slave system reached its limits of external expansion, it turned to intensive development via the feudal manor system, transforming the slave into a peasant who had an incentive to work the land more efficiently.

> The alternative to extensive development is intensive development, as happened in the transition from slavery to feudalism. But notice that to do this, the system had to change, the core logic was no longer the same. The dream of our current economy is therefore one of intensive development, to grow in the immaterial

[1]Bunnie Huang, "Copycat Corolla?" *bunnie's blog*, December 13, 2009 <http://www.bunniestudios.com/blog/?p=749>.

[2]Klein, *No Logo*, p. 203.

[3]Michel Bauwens, *P2P and Human Evolution*. Draft 1.994 (Foundation for P2P Alternatives, June 15, 2005) <http://integralvisioning.org/article.php?story=p2ptheory1>; Although I've read Wark, his abstruse postmodern style generally obfuscates what Bauwens summarizes with great clarity clarifty.

> field, and this is basically what the experience economy means. The hope that it expresses is that business can simply continue to grow in the immaterial field of experience.[1]

And the state, as enforcer of the total surveillance society and copyright lockdown, is central to this business model. Johann Soderberg relates the crisis of realization under state capitalism to capital's growing dependence on the state to capture value from social production and redistribute it to private corporate owners. This takes the form both of "intellectual property" law, as well as direct subsidies from the taxpayer to the corporate economy. He compares, specifically, the way photocopiers were monitored in the old USSR to protect the power of elites in that country, to the way the means of digital reproduction are monitored in this country to protect corporate power.[2] The situation is especially ironic, Cory Doctorow notes, when you consider the pressure the U.S. has put on the post-Soviet regime to enforce the global digital copyright regime: "post-Soviet Russia forgoes its hard-won freedom of the press to protect Disney and Universal!"[3] That's doubly ironic, considering the use of the term "Samizdat pirate" under the Soviet regime.

James O'Connor's theme, of the ever-expanding portion of the operating expenses of capital which come from the state, is also relevant here, considering the extent to which the technical prerequisites of the digital revolution were developed with state financing.

The ability to capture value from efficiency increases, through artificial scarcity and artificial property rights, is central to the New Growth Theory of Paul Romer. Consider his remarks in an interview with *Reason*'s Ron Bailey:

> **reason:** Yet there is a mechanism in the market called patents and copyright, for quasi-property rights in ideas.
>
> **Romer:** That's central to the theory. To the extent that you're using the market system to refine and bring ideas into practical application, we have to create some kind of control over the idea. That could be through patents. It could be through copyright. It might even be through secrecy. A firm can keep secret a lot of what it knows how to do. . . . So for relying on the market—and we do have to rely on the market to develop a lot of ideas—you have to have some mechanisms of control and some opportunities for people to make a profit developing those ideas.
>
> . . .
>
> **Romer:** There was an old, simplistic notion that monopoly was always bad. It was based on the realm of objects—if you only have objects and you see somebody whose cost is significantly lower than their price, it would be a good idea to break up the monopoly and get competition to reign freely. So in the realm of

[1]Michel Bauwens, "Can the experience economy be capitalist?" *P2P Foundation Blog*, September 27, 2007 <http://blog.p2pfoundation.net/can-the-experience-economy-be-capitalist/2007/09/27>. Joseph Tainter's thesis, that the collapse of complex societies results from the declining marginal productivity of increases in complexity or expansion, is relevant here; *The Collapse of Complex Societies* (Cambridge, New York, New Rochelle, Melbourne, Sydney: Cambridge University Press, 1988). In particular, he echoes Bauwens' thesis that classical civilization failed as a result of the inability to continue extensive addition of inputs through territorial expansion. As we will see shortly below, it is the inability to capture sufficient marginal returns on new increments of capital investment and innovation, in an era of "Free," that is destroying the existing economic system.

[2]Soderberg, *Hacking Capitalism*, pp. 144-145.

[3]Cory Doctorow, "Happy Meal Toys versus Copyright: How America Chose Hollywood and Wal-Mart, and Why It's Doomed Us, and How We Might Survive Anyway," in Doctorow, *Content: Selected Essays on Technology, Creativity, Copyright, and the Future of the Future* (San Francisco: Tachyon Publications, 2008), p. 39.

> things, of physical objects, there is a theoretical justification for why you should never tolerate monopoly. But in the realm of ideas, you have to have some degree of monopoly power. There are some very important benefits from monopoly, and there are some potential costs as well. What you have to do is weigh the costs against the benefits.
>
> Unfortunately, that kind of balancing test is sensitive to the specifics, so we don't have general rules. Compare the costs and benefits of copyrighting books versus the costs and benefits of patenting the human genome. They're just very different, so we have to create institutions that can respond differentially in those cases.

Although Romer contrasts the realm of "science" with the realm of "the market," and argues that there should be some happy medium between their respective open and proprietary cultures, it's interesting that he identifies "intellectual property" as an institution of "the market."

And Romer makes it clear that what he means by "growth" is economic growth, in the sense of monetized exchange value:

> **Romer:**. . . . Now, what do I mean when I say growth can continue? I don't mean growth in the number of people. I don't even mean growth in the number of physical objects, because you clearly can't get exponential growth in the amount of mass that each person controls. We've got the same mass here on Earth that we had 100,000 years ago and we're never going to get any more of it. What I mean is growth in value, and the way you create value is by taking that fixed quantity of mass and rearranging it from a form that isn't worth very much into a form that's worth much more.[1]

Romer's thought is another version of Daniel Bell's post-industrialism thesis. As summarized by Manuel Castells, that thesis held that:

> (1) The source of productivity and growth lies in the generation of knowledge, extended to all realms of economic activity through information processing.
>
> (2) Economic activity would shift from goods production to services delivery. . . .
>
> (3) The new economy would increase the importance of occupations with a high informational and knowledge content in their activity. Managerial, professional, and technical occupations would grow faster than any other occupational position and would constitute the core of the new social structure.[2]

The problem is that post-industrialism is self-liquidating: technological progress destroys the conditions necessary for capturing value from technological progress.

By their nature technological innovation and increased efficiency *destroy* growth. Anything that lowers the cost of inputs to produce a given output, in a free market with competition unfettered by entry barriers, will result in the reduction of exchange value (i.e. price). And since GDP is an accounting mechanism that measures the total value of inputs consumed, increased efficiency will *reduce* the size of "the economy."

Romer's model is essentially Schumpeterian. Recouping outlays for innovation requires prices that reflect average cost rather than marginal cost. Hence Romer's Schumpeterian schema precludes price-taking behavior in a competitive market;

[1]Ronald Bailey, "Post-Scarcity Prophet: Economist Paul Romer on growth, technological change, and an unlimited human future," *Reason*, December 2001 <http://reason.com/archives/2001/12/01/post-scarcity-prophet/>.

[2]Manuel Castells, *The Rise of the Network Society* (Blackwell Publishers, 1996), pp. 203-204

rather, it presupposes some form of market power ("monopolistic competition") by which firms can set prices to cover costs. Romer argues that his model of economic growth based on innovation is incompatible with price-taking behavior. A firm that invested significant sums in innovation, but sold only at marginal cost, could not survive as a price-taker. It is necessary, therefore, that the benefits of innovation—even though non-rival by their nature—be at least partially excludable through "intellectual property" law.[1]

Some right-wing libertarians mock big government liberals for a focus on "jobs" as an end in themselves, rather than as a means to an end. But Romer's focus on "growth" and "increased income," rather than on the amount of labor required to obtain a consumption good, is an example of the very same fallacy (and Bailey cheers him on, of course).

Jeff Jarvis sparked a long chain of discussions by arguing that innovation, by increasing efficiency, results in "shrinkage" rather than growth. The money left in customers' pockets, to the extent that it is reinvested in more productive venues, may affect the small business sector and not even show up in econometric statistics.[2]

Anton Steinpilz, riffing off Jarvis, suggested that the reduced capital expenditures might not reappear as increased spending *anywhere*, but might (essentially a two-sided coin) be pocketed by the consumer in the form of increased leisure and/or forced on the worker in the form of technological unemployment.[3] And Eric Reasons, writing about the same time, argued that innovation was being passed on to consumers, resulting in "massive deflation" and "less money involved" overall.[4]

Reasons built on this idea, massive deflation resulting from increased efficiency, in a subsequent blog post. The problem, Reasons argued, was that while the deflation of prices in the old proprietary content industries benefited consumers by leaving dollars in their pockets, many of those consumers were employees of industries made obsolete by the new business models.

> Effectively, the restrictions that held supply in check for IP are slowly falling away. As effective supply rises, price plummets. Don't believe me? You probably spend less money now on music than you did 15 years ago, and your collection is larger and more varied than ever. You probably spend less time watching TV news, and less money on newspapers than you did 10 years ago, and are better informed.
>
> I won't go so far as to say that the knowledge economy is going to be no economy at all, but it is a shrinking one in terms of money, both in terms of cost to the consumer, and in terms of the jobs produced in it.[5]

[1]Paul M. Romer, "Endogenous Technological Change" (December 1989). NBER Working Paper No. W3210.

[2]Jeff Jarvis, "When innovation yields efficiency," BuzzMachine, June 12, 2009 <http://www.buzzmachine.com/ 2009/06/ 12/when-innovation-yields-efficiency/>.

[3]Anton Steinpilz, "Destructive Creation: BuzzMachine's Jeff Jarvis on Internet Disintermediation and the Rise of Efficiency," *Generation Bubble*, June 12, 2009 <http://generationbubble.com/2009/06/12/destructive-creation-buzzmachines-jeff-jarvis-on-internet-disintermediation-and-the-rise-of-efficiency/>.

[4]Eric Reasons, "Does Intellectual Property Law Foster Innovation?" *The Tinker's Mind*, June 14, 2009 <http://blog.ericreasons.com/2009/06/does-intellectual-property-law-foster.html>.

[5]Reasons, "Intellectual Property and Deflation of the Knowledge Economy," *The Tinker's Mind*, June 21, 2009 <http://blog.ericreasons.com/2009/06/intellectual-property-and-deflation-of.html>.

And the issue is clearly shrinkage, not just a shift of superfluous capital and purchasing power to new objects. Craigslist employs fewer people than the industries it destroyed, for example. The ideal, Reasons argued, is for unproductive activity to be eliminated, but for falling work hours to be offset by lower prices, so that workers experience the deflation as a reduction in the ratio of effort to consumption:

> Given the amount of current consumption of intellectual property (copyrighted material like music, software, and newsprint; patented goods like just about everything else), couldn't we take advantage of this deflation to help cushion the blow of falling wages? How much of our income is dedicated to intellectual property, and its derived products? If wages decrease at the same time as cost-of-living decreases, are we really that bad off? Deflation moves in both directions, as it were. . . .
>
> Every bit of economic policy coming out of Washington is based on trying to maintain a status quo that can not be maintained in a global marketplace. This can temporarily inflate some sectors of our economy, but ultimately will leave us with nothing but companies that make the wrong things, and people who perform the wrong jobs. You know what they say: "As GM goes, so goes the country."[1]

Contrary to "Free" optimists like Chris Anderson and Kevin Kelley, Reasons suspects that reduced rents on proprietary content cannot be replaced by monetization in other areas. The shrinkage of proprietary content industries will not be replaced by growth elsewhere, or the reduced prices offset by a shift of demand elsewhere, on a one-to-one basis.[2]

Mike Masnick, of *Techdirt*, praised Reasons' analysis, but suggested—from a fairly conventional standpoint—that it was incomplete:

> So this is a great way to think about the *threat* side of things. Unfortunately, I don't think Eric takes it all the way to the next side (the opportunity side), which we tried to highlight in that first link up top, here. Eric claims that this "deflation" makes the sector shrink, but I don't believe that's right. It makes companies who rely on business models of artificial scarcity to shrink, but it doesn't make the overall sector shrink if you define the market properly. Economic efficiency may make certain segments of the market shrink (or disappear), but it expands the overall market.
>
> Why? Because efficiency gives you more output for the same input (bigger market!). The tricky part is that it may move around where that output occurs. And, when you're dealing with what I've been calling "infinite goods" you can have a multiplicative impact on the market. That's because a large part of the "output" is now infinitely reproduceable at no cost. For those who stop thinking of these as "goods that are being copied against our will" and start realizing that they're "*inputs* into a wider market where we don't have to pay for any of the distribution or promotion!" there are much greater opportunities. It's just that they don't come from artificial scarcity any more. They come from abundance.[3]

Reasons responded, in a comment below Masnick's post (aptly titled "The glass is twice the size it needs to be . . . "), that "this efficiency will make the economic markets they affect "shrink" in terms of economy and capital. It doesn't

[1]Reasons, "The Economic Reset Button," *The Tinker's Mind*, July 2, 2009 <http://blog.ericreasons.com/2009/07/economic-reset-button.html>.

[2]Reasons, "Innovative Deflation," *The Tinker's Mind*, July 5, 2009 <http://blog.ericreasons.com/2009/07/innovative-deflation.html>.

[3]Mike Masnick, "Artificial Scarcity is Subject to Massive Deflation," *Techdirt*, <http://techdirt.com/articles/ 20090624/ 0253385345.shtml>.

mean that the number of variation of the products available will shrink, just the capital involved."[1]

He stated this assessment in even sharper terms in a comment under Michel Bauwens's blog post on the exchange:

> While I certainly wouldn't want to go toe-to-toe with Mike Masnick on the subject, I did try to clarify in comments that it isn't that I don't see the opportunity in the "knowledge economy", but simply that value can be created where capital can't be captured from it. The trick is to reap that value, and distinguish where capital can and where it cannot add value. Of course there's money to be made in the knowledge economy—ask Google or Craigslist—but by introducing such profound efficiencies, they deflate the markets they touch at a rate far faster than the human capital can redeploy itself in other markets. Since so much capital is dependent upon consumerism generated by that idled human capital, deflation follows.[2]

Neoclassical economists would no doubt dismiss Reasons' argument, and other theories of technological unemployment, as variations on the "lump of labor fallacy." But their dismissal of it, under that trite label, itself makes an implicit assumption that's hardly self-evident: that demand is infinitely, upwardly elastic.

That assumption is stated, in the most vulgar of terms, from an Austrian standpoint by a writer at LewRockwell.com:

> You know, properly speaking, the "correct" level of unemployment is zero. Theoretically, the demand for goods and services is infinite. My own desire for goods and services has no limit, and neither does anyone else's. So even if everyone worked 24/7, they could never satisfy all the potential demand. It's just a matter of allowing people to work at wages that others are willing and able to pay.[3]

Aside from the fact that this implicitly contradicts Austrian arguments that increased labor productivity from capital investment are responsible for reduced working hours (see, e.g., George Reisman, quoted elsewhere in this chapter), this is almost cartoonish nonsense. If the demand for goods and services is unconstrained by the disutility of labor, then it follows that absent a minimum wage people would be working at least every possible waking hour—even if not "24/7." On the other hand if there *is* a tradeoff between infinite demand and the disutility of labor, then demand is *not* infinitely upwardly elastic. Some productivity increases will be lost through "leakages" in the form of increased leisure, rather than consumption of increased output of goods. That means that the demand for labor, even if somewhat elastic, will not grow as quickly as labor productivity.

Tom Walker (aka Sandwichman), an economist who has devoted most of his career to unmasking the "lump of labor" caricature as a crude strawman, confesses a degree of puzzlement as to why orthodox economists are so strident on the issue. After all, what they denounce as the "lump of labor fallacy" is based on what, "[w]hen economists do it, . . . is arcane and learned *ceteris paribus* hokus pokus."[4]

[1]Reasons comment under *Ibid.*, "The glass is twice the size it needs to be" <http://techdirt.com/article.php?sid=20090624/0253385345#c257>.

[2]Comment under Michel Bauwens, "The great internet/p2p deflation," *P2P Foundation Blog*, November 11, 2009 <http://blog.p2pfoundation.net/the-great-internetp2p-deflation/2009/11/11>.

[3]"Doug Casey on Unemployment," *LewRockwell.Com*, January 22, 2010. Interviewed by Louis James, editor, *International Speculator* <http://www.lewrockwell.com/casey/ casey38.1.html>.

[4]Tom Walker, "The Doppelganger Effect," *EconoSpeak*, January 2, 2010 <http://econospeak.blogspot.com/ 2010/01/ doppelg-effect.html>.

Given existing levels of demand for consumer goods, any increase in labor productivity will result in a reduction in total work hours available.

Of course the orthodox economist will argue that *ceteris* is never *paribus*. But that demand freed up by reduced wage expenditures in one sector will automatically translate, on a one-to-one basis, into increased demand (and hence employment) in another sector is itself by no means self-evident. And an assumption that such will occur, so strong that one feels sufficiently confident to invent a new "fallacy" for those who argue otherwise, strikes me as a belief that belongs more in the realm of theology than of economics.

P. M. Lawrence, in a discussion sparked by Casey's argument, expressed similar views in a private email:

> I always thought that "lump" reasoning was perfectly sound in any area in analysing instantaneous responses, as there's a lag before it changes while supply and demand respond—which means, it's important for matters of survival until those longer runs, and also you can use it in mathematically or verbally modelling how the lump does in fact change over time. . . . [1]

These shortcomings of Romer's New Growth apply, more particularly, to the "progressive" and "green" strands of cognitive capitalism. Bill Gates and Richard

[1]P. M. Lawrence, private email, January 25, 2010. Lawrence subsequently requested I add the following explanatory material:

> . . . people might not understand just how you can use the idea of a "fixed" value in intermediate calculations on the way to getting a better description of how it really does vary.
>
> So you should probably refer people to more detail in the footnote, particularly on these areas:-
>
> - Successive relaxation; see http://en.wikipedia.org/wiki/Gauss%E2%80%93Seidel_method. Related topics include "accelerated convergence" (see http://en.wikipedia.org/wiki/Series_acceleration), which can be combined directly with that in successive over-relaxation (see http://en.wikipedia.org/wiki/Successive_over-relaxation).
> - The method of perturbations; see http://en.wikipedia.org/wiki/Perturbation_theory, which states "This general procedure is a widely used mathematical tool in advanced sciences and engineering: start with a simplified problem and gradually add corrections that make the formula that the corrected problem matches closer and closer to the formula that represents reality". (Successive relaxation is applying that general approach in one particular area.) The part of my email you cut read "oversimplifying the technique just a little, as an engineering approximation you assume it's fixed, then you run it through the figures in a circular way to get a new contradictory value—and that's the value it changes to, after a corresponding time step; repeat indefinitely for a numerical model, or work out the time dependent equations that match that and solve them analytically". Your footnote should edit this and connect it to the same general approach, bringing out the idea that the first simplification is to pretend that the value is constant (as in a "lump of labour", say), and saying that since the whole point is to use an incorrect description to get to a better description, "incorrect" doesn't mean "invalid"—and, over a short enough term, even that first simplification of being fixed can be useful and meaningful as people really do have to get through those very short terms.
> - Simultaneous differential equations, rigidly coupled and otherwise. . . .
>
> I brought some of these issues out in an unpublished letter to the Australian Financial Review, written 6.7.98, available at
>
> http://users.beagle.com.au/peterl/publicns.html#AFRLET3.

Florida are typical of this tendency. Florida specifically refers to Romer's New Growth Theory, "which assigns a central role to creativity or idea generation." But he never directly addresses the question of just how such "idea generation" can be the source of economic growth, unless it is capitalized as the source of rents through artificial property rights. He quotes, without seeming to grasp its real significance, this remark of Romer's: "We are not used to thinking of ideas as economic goods, but they are surely the most significant ones that we produce." "Economic goods" are goods with exchange value; and ideas can only have exchange value when they are subject to monopoly. Florida continues to elaborate on Romer's theory, arguing that an idea can be used over and over again, "an in fact grows in value the more it is used. It offers not diminishing returns, but *increasing returns*." This displays a failure to grasp the distinction between use-value and exchange value. An idea can, indeed, result in exponential increases in our standard of living the more they are used, by reducing the labor and material inputs required to produce a unit of consumption. But in so doing, it *reduces* exchange value and causes marginal returns to fall to zero. Innovation causes economic value to implode.[1]

Florida himself, for all his celebration of networks and free agency, assumes a great deal of continuity with the existing corporate economy.

> In tracing economic shifts, I often say that our economy is moving from an older corporate-centered system defined by large companies to a more people-driven one. This view should not be confused with the unfounded and silly notion that big companies are dying off. Nor do I buy the fantasy of an economy organized around small enterprises and independent "free agents." Companies, including very big ones, obviously still exist, are still influential and probably always will be.[2]
>
> A related myth is that the age of large corporations is over—that they have outlived their usefulness, their power has been broken, and they will eventually fade away along with other big organizational forms. The classic metaphor is the lumbering dinosaur made obsolete and susurped by small, nimble mammals—the usurpers in this case being small, nimble startup companies. . . .
>
> But big companies are by no means going away. Microsoft and Intel continue to control much of the so-called information economy, along with Oracle, Cisco, IBM and AOL Time Warner. Big industrial concerns, from General Motors to General Electric, General Dynamics and General Foods, still turn out most of the nation's goods. Our money is managed not by upstarts but by large financial institutions. The resources that power our economy are similarly managed and controlled by giant corporations. . . .
>
> The economy, like nature, is a dynamic system. New companies form and help us to propel it forward, with some dying out while others carry on to grow quite large themselves, like Microsoft and Intel. An economy composed only of small, short-lived entities would be no more sustainable than an ecosystem composed only of insects.[3]

Florida fails to explain just *why* large organizations are necessary. Large, hierarchical organizations originally came into existence as a result of the enormous capital outlays required for production, and the need to manage and control those capital assets. When physical capital outlays collapse by one or two orders of magnitude for most kinds of production, what further need is there for the large organizations? The large size of Microsoft and Intel results, in most cases (aside from the enormous capital outlay required for a microchip foundry, of course), from

[1]Richard Florida, *The Rise of the Creative Class* (New York: Basic Books, 2002), p.36.
[2]*Ibid.* p. 6.
[3]*Ibid.*, pp. 26-27.

patents on hardware, software copyrights, and the like, that artificially increase required capital outlays, otherwise raise entry barriers, and thereby lock them into an artificial position of control.

And the purported instabilities of an economy of small firms, over which Florida raises so much alarm, are a strawman. Networked industrial ecologies of small firms achieve stability and permanence, as we shall see in Chapter Six, from modular design for common platforms. The individual producers may come and go, but the common specifications and protocols live on.

Florida's focus on individual career paths based on free agency, and on internal corporate cultures of "creativity," at the expense of genuine changes in institutional structure and size, remind me of Charles Reich's approach in *The Greening of America*. The great transformation Reich envisioned amounted to little more than leaving the giant corporations and central government agencies in place, but staffing them entirely with people in beads and bell-bottoms who, you know, had their heads in the right place, man.

But this approach is now failing in the face of the increasing inability to capture value from the immaterial realm. The strategy of shifting the burden of realization onto the state is untenable. Strong encryption, coupled with the proliferation of bittorrent and episodes like the DeCSS uprising (see later in this chapter), have shown that "intellectual property" is ultimately unenforceable. J. A. Pouwelse and his coauthors estimate that the continuing exponential advance of file-sharing technology will make copyright "impossible to enforce by 2010."[1] In particular, they mention

> anonymous downloading, uploading, and injection of content using a darknet. A darknet inhibits both Internet censorship and enforcement of copyright law. The freenetproject.org has in 2000 already produced a darknet, but it was slow, difficult to use, and offered little content. Darknets struggle with the second cardinal feature of P2P platforms. Full anonymity costs both extra bandwidth and is difficult to combine with enforcement of resource contributions. By 2010 darknets should be able to offer the same performance as traditional P2P software by exploiting social networking. No effective legal or technological method currently exits [sic] to stop darknets, with the exception of banning general-purpose computing. Technologies such as secure computing and DRM are convincingly argued to be unable to stop darknets.[2]

And in fact, as reported by *Ars Technica* back in 2007, attempts by university administrators to ban P2P at the RIAA's behest have caused students to migrate to darknets in droves.[3]

The rapid development of circumvention technology intersects—powerfully so—with the cultural attitudes of a generation for which industry "anti-songlifting" propaganda is as gut-bustingly hilariously as *Reefer Madness*. Girlintraining, commenting under a Slashdot post, had this to say of such propaganda:

> I used to read stuff like this and get upset. But then I realized that my entire generation knows it's baloney. They can't explain it intellectually. They have no real understanding of the subtleties of the law, or arguments about artists' rights

[1]J.A. Pouwelse, P. Garbacki, D.H.J. Epema, and H.J. Sips, "Pirates and Samaritans: a Decade of Measurements on Peer Production and their Implications for Net Neutrality and Copyright" (The Netherlands: Delft University of Technology, 2008) <http://www.tribler.org/trac/wiki/PiratesSamaritans>., p. 20.

[2]*Ibid.*, p. 15.

[3]Ken Fisher, "Darknets live on after P2P ban at Ohio U," *Ars Technica*, Mqy 9, 2007 <http://arstechnica.com/tech-policy/news/2007/05/darknets-live-on-after-p2p-ban-at-ohio-u.ars>.

> or any of that. All they really understand is there is are large corporations charging private citizens tens, if not hundreds of thousands of dollars, for downloading a few songs here and there. And it's intuitively obvious that it can't possibly be worth that.
>
> An entire generation has disregarded copyright law. It doesn't matter whether copyright is useful or not anymore. They could release attack dogs and black helicopters and it wouldn't really change people's attitudes. It won't matter how many websites they shut down or how many lives they ruin, they've already lost the culture war because they pushed too hard and alienated people wholesale. The only thing these corporations can do now is shift the costs to the government and other corporations under color of law in a desperate bid for relevance. And that's exactly what they're doing.
>
> What does this mean for the average person? It means that we google and float around to an ever-changing landscape of sites. We communicate by word of mouth via e-mail, instant messaging, and social networking sites where the latest fix of free movies, music, and games are. If you don't make enough money to participate in the artificial marketplace of entertainment goods—you don't exclude yourself from it, you go to the grey market instead. All the technological, legal, and philosophical barriers in the world amount to nothing. There is a small core of people that understand the implications of what these interests are doing and continually search for ways to liberate their goods and services for "sale" on the grey market. It is (economically and politically) identical to the Prohibition except that instead of smuggling liquor we are smuggling digital files.
>
> Billions have been spent combating a singularily simple idea that was spawned thirty years ago by a bunch of socially-inept disaffected teenagers working out of their garages: Information wants to be free. Except information has no wants—it's the people who want to be free. And while we can change attitudes about smoking with aggressive media campaigns, or convince them to cast their votes for a certain candidate, selling people on goods and services they don't really need, what we cannot change is the foundations upon which a generation has built a new society out of.[1]

Cory Doctorow, not overly fond of the more ideologically driven wing of the open-source movement (or as he calls them, "patchouli-scented info-hippies"), says it isn't about whether "information wants to be free." Rather, the simple fact of the matter is "that computers are machines for copying bits and that once you . . . turn something into bits, they will get copied. . . . [I]f your business model is based on bits not getting copied you are screwed."[2]

> Raise your hand if you're thinking something like, "But DRM doesn't have to be proof against smart attackers, only average individuals! . . . "
>
> . . . I don't have to be a cracker to break your DRM. I only need to know how to search Google, or Kazaa, or any of the other general-purpose search tools for the cleartext that someone smarter than me has extracted.[3]
>
> It used to be that copy-prevention companies' strategies went like this: "We'll make it easier to buy a copy of this data than to make an unauthorized copy of it. That way, only the *uber*-nerds and the cash-poor/time rich classes will bother to copy instead of buy." But every time a PC is connected to the Internet and its owner is taught to use search tools like Google (or The Pirate Bay), a third option appears: you can just download a copy from the Internet. . . .
>
> As I write this, I am sitting in a hotel room in Shanghai, behind the Great Firewall of China. Theoretically, I can't access blogging services that carry nega-

[1]Girlintraining comment under Soulskill, "Your Rights Online," *Slashdot*, January 9, 2010 <http://yro.slashdot.org/story/10/01/09/0341208/Politicians-Worldwide-Asking-Questions-About-ACTA>.

[2]Bascha Harris, "A very long talk with Cory Doctorow, part 1," redhat.com, January 2006 <http://www.redhat.com/magazine/015jan06/features/doctorow/>.

[3]Doctorow, "Microsoft DRM Research Talk," in *Content*, pp. 7-8.

> tive accounts of Beijing's doings, like WordPress, Blogger, and LiveJournal, nor the image-sharing site Flickr, nor Wikipedia. The (theoretically) omnipotent bureaucrats at the local Minitrue have deployed their finest engineering talent to stop me. Well, these cats may be able to order political prisoners executed and their organs harvested for Party members, but they've totally failed to keep Chinese people . . . off the world's Internet. The WTO is rattling its sabers at China today, demanding that they figure out how to stop Chinese people from looking at Bruce Willis movies without permission—but the Chinese government can't even figure out how to stop Chinese people from looking at seditious revolutionary tracts online.[1]

File-sharing networks spring up faster than they can be shut down. As soon as Napster was shut down, the public migrated to Kazaa and Gnutella. When Kazaa was shut down, its founders went on to create Skype and Joost. Other file-sharing services also sprang up in Kazaa's niche, like the Russian AllofMP3, which reappeared under a new name as soon as the WTO killed it.[2]

The proliferation of peer production and the open-source model, and the growing unenforceability of the "intellectual property" rules on which the capture of value depends, is creating "a vast new information commons . . . , which is increasingly out of the control of cognitive capitalism."[3] Capital, as a result, is incapable of realizing returns on ownership in the cognitive realm. As Bauwens explains it:

> 1) The creation of non-monetary value is exponential
> 2) The monetization of such value is linear
>
> In other words, we have a growing discrepancy between the direct creation of use value through social relationships and collective intelligence . . . , [and the fact that] only a fraction of that value can actually be captured by business and money. Innovation is becoming . . . an emergent property of the networks rather than an internal R & D affair within corporations; capital is becoming an a posteriori intervention in the realization of innovation, rather than a condition for its occurrence. . . .
>
> What this announces is a crisis of value . . . , but also essentially a crisis of accumulation of capital. Furthermore, we lack a mechanism for the existing institutional world to re-fund what it receives from the social world. So on top of all of that, we have a crisis of social reproduction. . . .

Thus, while markets and private ownership of physical capital will persist, "the core logic of the emerging experience economy, operating as it does in the world of non-rival exchange, is unlikely to have capitalism as its core logic."[4]

A good example is the way in which digital culture, according to Douglas Rushkoff, destroyed California's economy:

> The fact is, most Internet businesses don't require venture capital. The beauty of these technologies is that they decentralize value creation. Anyone with a PC and bandwidth can program the next Twitter or Facebook plug-in, the next iPhone app, or even the next social network. While a few thousand dollars might be nice, the hundreds of millions that venture capitalists want to—need to—invest, simply aren't required. . . .
>
> The banking crisis began with the dot.com industry, because here was a business sector that did not require massive investments of capital in order to grow. (I spent an entire night on the phone with one young entrepreneur who

[1]Doctorow, "It's the Information Economy, Stupid," *Ibid.*, p. 60.

[2]Doctorow, "Why is Hollywood Making a Sequel to the Napster Wars?" in *Content*, p. 47.

[3]Bauwens, *P2P and Human Evolution.*

[4]Bauwens, "Can the experience economy be capitalist?"

secured $20 million of capital from a venture firm, trying to figure out how to possibly spend it. We could only come up with $2 million of possible expenditures.) What's a bank to do when its money is no longer needed? . . .

So they fail, the tax base decreases, companies based more on their debt structures than their production fail along with them, and we get an economic crisis. Yes, the Internet did all this.

But that's also why the current crisis should be seen as a cause for celebration as well: the Internet actually did what it was supposed to by decentralizing our ability to create and exchange value.

This was the real dream, after all. Not simply to pass messages back and forth, but to dis-intermediate our exchanges. To cut out the middleman, and let people engage and transact directly.

This is, quite simply, cheaper to do. There's less money in it. Not necessarily less money for us, the people doing the exchanging, but less money for the institutions that have traditionally extracted value from our activity. If I can create an application or even a Web site like this one without borrowing a ton of cash from the bank, then I am also undermining America's biggest industry—finance.

While we rightly mourn the collapse of a state's economy, as well as the many that are to follow, we must—at the very least—acknowledge the real culprit. For digital technology not only killed the speculative economy, but stands ready to build us a real one.[1]

The actual physical capital outlays required for digital creation are simply unable to absorb anything like the amounts of surplus capital in search of a profitable investment outlet—unless artificial property rights and artificial scarcity can be used to exclude independent production by all but the corporate owners of "intellectual property," and mandate outlays totally unrelated to the actual physical capital requirements for production. Since such artificial property rights are, in fact, becoming increasingly unenforceable, corporate capital is unable either to combat the growing superfluity of its investment capital in the face of low-overhead production, or to capture value through artificial scarcity by suppressing low-cost competition.

If we view the transition from the perspective of innovators rather than venture capitalists, of course, it's a much more positive development. Michel Bauwens described the collapse of the dot-com bubble and the rise of Web 2.0 as the decoupling of innovation and entrepreneurship from capital, and the shift of innovation to networked communities.

As an internet entrepreneur, I personally experienced both the manic phase, and the downturn, and the experience was life changing because of the important discovery I and others made at that time. All the pundits where predicting, then as now, that without capital, innovation would stop, and that the era of high internet growth was over for a foreseeable time. In actual fact, the reality was the very opposite, and something apparently very strange happened. In fact, almost everything we know, the Web 2.0, the emergence of social and participatory media, was born in the crucible of that downturn. In other words, innovation did not slow down, but actually increased during the downturn in investment. This showed the following new tendency at work: capitalism is increasingly being divorced from entrepreneurship, and entrepreneurship becomes a networked activity taking place through open platforms of collaboration.

The reason is that internet technology fundamentally changes the relationship between innovation and capital. Before the internet, in the Schumpeterian world, innovators need capital for their research, that research is then protected

[1]Douglas Rushkoff, "How the Tech Boom Terminated California's Economy," *Fast Company*, July 10, 2009 <http://www.fastcompany.com/article/how-tech-boom-terminated-californias-economy?page=0%2C1>.

> through copyright and patents, and further funds create the necessary factories. In the post-schumpeterian world, creative souls congregate through the internet, create new software, or any kind of knowledge, create collaboration platforms on the cheap, and paradoxically, only need capital when they are successful, and the servers risk crashing from overload. As an example, think about Bittorrent, the most important software for exchanging multimedia content over the internet, which was created by a single programmer, surviving through a creative use of some credit cards, with zero funding. But the internet is not just for creative individual souls, but enables large communities to cooperate over platforms. Very importantly, it is not limited to knowledge and software, but to everything that knowledge and software enables, which includes manufacturing. Anything that needs to be physically produced, needs to be 'virtually designed' in the first place.
>
> This phenomena is called social innovation or social production, and is increasingly responsible for most innovation.[1]

As we will see in Chapter Five, initial capital outlay requirements for physical production are imploding in exactly the same way, which means that venture capital will lose most of its outlets in manufacturing as well.

For this reason the Austrian dogma of von Mises, that the only way to raise real wages is to increase the amount of capital invested, is shown to rely on a false assumption: the assumption that there is some necessary link between productivity and the sheer quantity of capital invested. George Reisman displays this tendency at its crudest.

> The truth, which real economists, from Adam Smith to Mises, have elaborated, is that in a market economy, the wealth of the rich—of the capitalists—is overwhelmingly invested in means of production, that is, in factories, machinery and equipment, farms, mines, stores, and the like. This wealth, this capital, produces the goods which the average person buys, and as more of it is accumulated and raises the productivity of labor higher and higher, brings about a progressively larger and ever more improved supply of goods for the average person to buy.[2]

But it has been at the heart of most twentieth century assumptions about economy of scale, and an unquestioned assumption behind the work of liberal managerialists like Chandler and Galbraith.

For the same reason that the Austrian fixation on the quantity of capital investment as a source of productivity is obsolete, Marxist theories of the "social structure of accumulation" as an engine of growth are likewise obsolete. Technical innovation, in such theories, provides the basis for a new long-wave of investment to soak up surplus capital. The creation of some sort of new infrastructure is both a long-term sink for capital, and the foundation for new levels of productivity.

Gopal Balakrishnan, in *New Left Review*, correctly observes capitalism's inability, this time around, to gain a new lease on life through a new Kondratieff long-wave cycle: i.e., "a new socio-technical infrastructure, to supersede the existing fixed-capital grid." But he mistakenly sees it as the result either of an inability to bear the expense (as if productivity growth required an enormous capital outlay), or of technological stagnation. His claim of "technological stagnation," frankly, is utterly astonishing. He equates the outsourced production in job-shops, on the flexible manufacturing model that prevails in various forms in Shenzhen, Emilia-

[1]Michel Bauwens, "Asia needs a Social Innovation Stimulus plan," *P2P Foundation Blog*, March 23, 2009 <http://blog.p2pfoundation.net/asia-needs-a-social-innovation-stimulus-plan/2009/03/23>.

[2]George Reisman, "Answer to Paul Krugman on Economic Inequality," *The Webzine*, March 3, 2006 <http://thewebzine.com/articles/030306ReismanAnswer.html>.

Romagna, and assorted corporate supplier networks, with a lower level of technological advancement.[1] But the shift of production from the old expensive, capital-intensive, product-specific infrastructure of mass-production industry to job-shops is in fact the result of an amazing level of technological advance: namely, the rise of cheap CNC machine tools scaled to small shops that are more productive than the old mass-production machinery. By technological stagnation, apparently, Balakrishnan simply means that less money is being invested in new generations of capital; but the crisis of capitalism results precisely from the fact that new forms of technology permit unprecedented levels of productivity with physical capital costs an order of magnitude lower. Both the Austrians and the neo-Marxists, in their equation of progress and productivity with the mass of capital invested, are stuck in the paleotechnic age.

This shows why the "cognitive capitalism" model of Gates, Romer, etc. is untenable. The natural tendency of technical innovation is not to add to GDP, but to destroy it. GDP measures, not the utility of production outputs to the consumer, but the value of inputs consumed in production.[2] So anything that reduces the total labor and material inputs required to produce a given unit of output should reduce GDP, unless artificial scarcity puts a floor under commodity price and prevents prices from falling to the new cost of production.

This is essentially what we saw Eric Reasons point out above. As Chris Anderson argues in *Free*, Microsoft's launch of Encarta on CD-Rom in the 1990s resulted in $100 billion in sales for Encarta—while destroying some $600 billion in sales for the traditional encyclopedia industry. And Wikipedia, in turn, destroyed still more sales for both traditional encyclopedias and Encarta.[3]

As Niall Cook describes it, enterprise software vendors are experiencing similar deflationary pressure.

> 'The design of business applications is more important than ever, says Joe Kraus, CEO of JobSpot. 'If I'm a buyer at a manufacturing company and I'm using Google Earth to look at the plants of my competition, and the Siebel sales rep asks me to spend $2 million on glorified database software, that causes a real disconnect.'

[1]Gopal Balakrishnan, "Speculations on the Stationary State," *New Left Review*, September-October 2009 <http://www.newleftreview.org/A2799>.

[2]Balakrishnan, in *Ibid.*, points to an interesting parallel between national accounting in the Soviet bloc and the neoliberal West:

. . . During the heyday of Reaganism, official Western opinion had rallied to the view that the bureaucratic administration of things was doomed to stagnation and decline because it lacked the ratio of market forces, coordinating transactions through the discipline of competition. Yet it was not too long after the final years of what was once called socialism that an increasingly debt- and speculation-driven capitalism began to go down the path of accounting and allocating wealth in reckless disregard of any notionally objective measure of value. The balance sheets of the world's greatest banks are an imposing testimony to the breakdown of standards by which the wealth of nations was once judged.

In their own ways, both bureaucratic socialism and its vastly more affluent neo-liberal conqueror concealed their failures with increasingly arbitrary tableaux économiques. By the 80s the gdr's reported national income was revealed to be a statistical artifact that grossly inflated its cramped standards of living. But in the same decade, an emerging circuit of global imbalances was beginning to generate considerable problems for the measurement of capitalist wealth. The coming depression may reveal that the national economic statistics of the period of bubble economics were fictions, not wholly unlike those operative in the old Soviet system.

[3]Chris Anderson, *Free: The Future of a Radical Price* (New York: Hyperion, 2009), pp. 129-130.

> In the 1990s some enterprise software vendors were busy telling customers that even the simplest problems needed large, complex systems to solve them. Following the dot-com crash at the start of the millennium few of these vendors survived, usurped by cheap—if not free—alternatives. This trend continues unabated in the form of social software. As Peter Merholz . . . , president and founder of user expereince firm Adaptive Path, put it, 'enterprise software is being eaten away from below'.[1]

The usual suspects proclaim that demand is upwardly elastic, and endlessly so, so that a reduction of costs in one industry will simply free up demand for increased output elsewhere. But it's unlikely, as Reasons pointed out, that there will be a one-to-one transfer of the demand freed up by lower prices from falling production costs to new forms of consumer goods, for the same reason that there's a backward-bending supply curve for labor. What economists mean by this latter wonkish-sounding term is that labor doesn't follow the upward sloping supply curve as most normal commodities, with higher wages resulting in willingness to work longer hours. Rather, part of the increase in income from higher wages is likely to be used to reduce work hours; rather than workers increasing demand for new products to absorb the total increase, it's more likely that total demand will grow less than the wage increase, and it will take fewer hours to earn the desired level of consumption. The reason is that the expenditure of labor carries disutility. For the same reason, rather than reduced production costs and prices in one industry simply freeing up demand for an equal value in new products elsewhere, it's likely that total GDP, i.e. total expenditure of labor and material inputs, will decline.

Rushkoff's reference to the collapsing tax base is especially interesting. As we have already seen, in an economy of subsidized inputs, the demand for such inputs grows exponentially, faster than the state can meet them. The state capitalist system will soon reach a point at which, thanks to the collapse of the portion of value comprised of rents on artificial property, the base of taxable value is imploding at the very time big business most needs subsidies to stay afloat. In the words of Charles Hughes Smith,

> what if the "end of paying work" will bring down the entire credit/consumption-dependent economy and the Federal government which depends on tax revenues from all that financial churn? . . .
>
> What if the Web, which is busily (creatively) destroying print media, the music industry, the movie business, Microsoft and many other rentier-type enterprises, ends up destroying income and profit-based tax revenues? How can the government support a status quo which requires $2 trillion in new borrowing every year just to keep from collapsing? What if that debt load is unsustainable?[2]

So the fiscal crisis of the state is accelerated not only by Peak Oil, but by the collapse of proprietary information as a source of value.

The growing importance of human capital relative to physical capital, another effect of the implosion of material outlays and overhead for production, is also creating governability problems for the standard absentee-owned, hierarchical corporate enterprise. At the same time, there is a growing inability to enforce corporate boundaries on human capital because of the unenforceability of "intellectual property." Fifty years ago, enormous outlays on physical capital were the main struc-

[1]Niall Cook, *Enterprise 2.0: How Social Software Will Change the Future of Work* (Burlington, Vt.: Gower, 2008), p. 24.

[2]Charles Hugh Smith, "What if the (Debt Based) Economy Never Comes Back?" *Of Two Minds*, July 2, 2009 <http://www.oftwominds.com/blogjuly09/what-if07-09.html>.

tural basis for the corporation as a locus of control over physical assets. Today, for a growing number of industries, the physical capital requirements for entering the market have imploded, and "intellectual property" is the main structural support to corporate boundaries.

In this environment, the only thing standing between the old information and media dinosaurs and their total collapse is their so-called "intellectual property" rights—at least to the extent they're still enforceable. Ownership of "intellectual property" becomes the new basis for the power of institutional hierarchies, and the primary structural bulwark for corporate boundaries. Without them, in any industry where the basic production equipment is affordable to all, and bottom-up networking renders management obsolete, it is likely that self-managed, cooperative production will replace the old managerial hierarchies. The network revolution, if its full potential is realized,

> will lead to substantial redistribution of power and money from the twentieth century industrial producers of information, culture, and communications—like Hollywood, the recording industry, and perhaps the broadcasters and some of the telecommunications giants—to a combination of widely diffuse populations around the globe, and the market actors that will build the tools that make this population better able to produce its own information environment rather than buying it ready-made."[1]

The same thing is true in the physical realm, of course. As we shall see in Chapter Five, the revolution in cheap CNC machine tools (including homebrew 3-D printers, cutting/routing tables, etc., that cost a few hundred dollars in parts) is having almost as radical an effect on the capital outlays required for physical production as the desktop revolution had on the immaterial production. And the approach of the old corporate dinosaurs—trying to maintain artificial scarcity and avoid having to compete with falling production costs—is exactly the same in the physical as in the immaterial realm.

F. Networked Resistance, Netwar, and Asymmetric Warfare Against Corporate Management

We already mentioned the corporate governance issues caused by the growing importance of human relative to physical capital, and the untenability of "intellectual property" as a legal support for corporate boundaries. Closely related is the vulnerability of corporate hierarchies to asymmetric warfare by networked communities of consumers and their own employees. Centralized, hierarchical institutions are increasingly vulnerable to open-source warfare.

In the early 1970s, in the aftermath of a vast upheaval in American political culture, Samuel Huntington wrote of a "crisis of democracy"; the American people, he feared, were becoming ungovernable. In *The Crisis of Democracy*, he argued that the system was collapsing from demand overload, because of an excess of democracy. Huntington's analysis is illustrative of elite thinking behind the neoliberal policy agenda of the past thirty years.

For Huntington, America's role as "hegemonic power in a system of world order" depended on a domestic system of power; this system of power, variously referred to in this work as corporate liberalism, Cold War liberalism, and the wel-

[1]James C. Bennett, "The End of Capitalism and the Triumph of the Market Economy," from *Network Commonwealth: The Future of Nations in the Internet Era* (1998, 1999) <http://www.pattern.com/bennettj-endcap.html>.

fare-warfare state, assumed a general public willingness to stay out of government affairs.[1] And this was only possible because of a domestic structure of political authority in which the country "was governed by the president acting with the support and cooperation of key individuals and groups in the Executive office, the federal bureaucracy, Congress, and the more important businesses, banks, law firms, foundations, and media, which constitute the private establishment."[2]

America's position as defender of global capitalism required that its government have the ability "to mobilize its citizens for the achievement of social and political goals and to impose discipline and sacrifice upon its citizens in order to achieve these goals."[3] Most importantly, this ability required that democracy be largely nominal, and that citizens be willing to leave major substantive decisions about the nature of American society to qualified authorities. It required, in other words, "some measure of apathy and non-involvement on the part of some individuals and groups."[4]

Unfortunately, these requirements were being gravely undermined by "a breakdown of traditional means of social control, a delegitimation of political and other means of authority, and an overload of demands on government, exceeding its capacity to respond."[5]

The overload of demands that caused Huntington to recoil in horror in the early 1970s must have seemed positively tame by the late 1990s. The potential for networked resistance created by the Internet exacerbated Huntington's crisis of democracy beyond his wildest imagining.

Networked resistance is based on a principle known as stigmergy. "Stigmergy" is a term coined by biologist Pierre-Paul Grasse in the 1950s to describe the process by which termites coordinated their activity. Social insects like termites and ants coordinate their efforts through the independent responses of individuals to environmental triggers like chemical trails, without any need for a central coordinating authority.[6]

Applied by way of analogy to human society, it refers primarily to the kinds of networked organization associated with wikis, group blogs, and "leaderless" organizations organized along the lines of networked cells.

Matthew Elliott contrasts stigmergic coordination with social negotiation. Social negotiation is the traditional method of organizing collaborative group efforts, through agreements and compromise mediated by discussions between individuals. The exponential growth in the number of communications with the size of the group, obviously, imposes constraints on the feasible size of a collaborative group, before coordination must be achieved by hierarchy and top-down authority. Stigmergy, on the other hand, permits collaboration on an unlimited scale by individuals acting independently. This distinction between social negotiation and stigmergy is illustrated, in particular, by the contrast between traditional models

[1]Samuel P. Huntington, Michael J. Crozier, Joji Watanuki, *The Crisis of Democracy*. Report on the Governability of Democracies to the Trilateral Commission: Triangle Paper 8 (New York: New York University Press, 1975), pp. 105-6.

[2]*Ibid.*, p. 92.

[3]*Ibid.*, pp. 7-8.

[4]*Ibid.*, pp. 113-5.

[5]*Ibid.*, pp. 7-8.

[6]Mark Elliott, "Stigmergic Collaboration: The Evolution of Group Work," *M/C Journal*, May 2006 <http://journal.media-culture.org.au/0605/03-elliott.php>.

of co-authoring and collaboration in a wiki.[1] Individuals communicate indirectly, "via the stigmergic medium."[2]

The distinction between social negotiation and stigmergic coordination parallels Elliott's distinction, elsewhere, between "discursive collaboration" and "stigmergic collaboration." The "discursive elaboration of shared representations (ideas)" is replaced by "the annotation of material and digital artefacts as embodiments of these representations. "Additionally, when stigmergic collaboration is extended by computing and digital networks, a considerable augmentation of processing capacity takes place which allows for the bridging of the spatial and temporal limitations of discursive collaboration, while subtly shifting points of negotiation and interaction away from the social and towards the cultural."[3]

There is a wide body of literature on the emergence of networked modes of resistance in the 1990s, beginning with the Rand studies on netwar by David Ronfeldt, John Arquilla and other writers. In their 1996 paper "The Advent of Netwar," Arquilla and Ronfeldt wrote that technological evolution was working to the advantage of networks and the detriment of hierarchies. Although their focus was on the military aspect (what has since been called "Fourth Generation Warfare"), they also mentioned governability concerns in civil society much like those Huntington raised earlier. "Intellectual property pirates," "militant single-issue groups" and "transnational social activists," in particular, were "developing netwar-like attributes."

> Now . . . the new information technologies and related organizational innovations increasingly enable civil-society actors to reduce their isolation, build far-flung networks within and across national boundaries, and connect and coordinate for collective action as never before. As this trend deepens and spreads, it will strengthen the power of civil-society actors relative to state and market actors around the globe. . . .
>
> For years, a cutting edge of this trend could be found among left-leaning activist NGOs concerned with human-rights, environmental, peace, and other social issues at local, national, and global levels. Many of these rely on APC affiliates for communications and aim to construct a "global civil society" strong enough to counter the roles of state and market actors. In addition, the trend is spreading across the political spectrum. Activists on the right—from moderately conservative religious groups, to militant antiabortion groups—are also building national and transnational networks based in part on the use of new communications systems.[4]

In "Tribes, Institutions, Markets, Networks" (1996) Ronfeldt focused on the special significance of the network for networked global civil society.

> . . . [A]ctors in the realm of civil society are likely to be the main beneficiaries. The trend is increasingly significant in this realm, where issue-oriented multiorganizational networks of NGOs—or, as some are called, nonprofit organizations (NPOs), private voluntary organizations (PVOs), and grassroots organizations (GROs)—continue to multiply among activists and interest groups who identify with civil society. Over the long run, this realm seems likely to be

[1] *Ibid.*

[2] Mark Elliott, "Some General Off-the-Cuff Reflections on Stigmergy," *Stigmergic Collaboration*, May 21, 2006 <http://stigmergiccollaboration.blogspot.com/2006/05/some-general-off-cuff-reflections-on.html>.

[3] Mark Elliott, *Stigmergic Collaboration: A Theoretical Framework for Mass Collaboration*. Doctoral Dissertation, Centre for Ideas, Victorian College of the Arts, University of Melbourne (October 2007) , pp. 9-10

[4] John Arquilla and David Ronfeldt, *The Advent of Netwar* MR-789 (Santa Monica, CA: RAND, 1996) <http://www.rand.org/pubs/monograph_reports/MR789/>.

> strengthened more than any other realm, in relative if not also absolute terms. While examples exist across the political spectrum, the most evolved are found among progressive political advocacy and social activist NGOs—e.g., in regard to environmental, human-rights, and other prominent issues—that depend on using new information technologies like faxes, electronic mail (e-mail), and on-line conferencing systems to consult and coordinate. This nascent, yet rapidly growing phenomenon is spreading across the political spectrum into new corners and issue areas in all countries.
>
> The rise of these networks implies profound changes for the realm of civil society. In the eighteenth and nineteenth centuries, when most social theorists focused on state and market systems, liberal democracy fostered, indeed required, the emergence of this third realm of activity. Philosophers such as Adam Ferguson, Alexis de Tocqueville, and G. W. F. Hegel viewed civil society as an essential realm composed of all kinds of independent nongovernmental interest groups and associations that acted sometimes on their own, sometimes in coalitions, to mediate between state and society at large. However, civil society was also considered to be a weaker realm than the state or the market. And while theorists treated the state and the market as systems, this was generally not the case with civil society. It was not seen as having a unique form of organization equivalent to the hierarchical institution or the competitive market, although some twentieth century theorists gave such rank to the interest group.
>
> Now, the innovative NGO-based networks are setting in motion new dynamics that promise to reshape civil society and its relations with other realms at local through global levels. Civil society appears to be the home realm for the network form, the realm that will be strengthened more than any other—either that, or a new, yet-to-be-named realm will emerge from it. And while classic definitions of civil society often encompassed state- and market-related actors (e.g., political parties, businesses and labor unions), this is less the case with new and emerging definitions—the separation of "civil society" from "state" and "market" realms may be deepening.
>
> The network form seems particularly well suited to strengthening civil-society actors whose purpose is to address social issues. At its best, this form may thus result in vast collaborative networks of NGOs geared to addressing and helping resolve social equity and accountability issues that traditional tribal, state, and market actors have tended to ignore or are now unsuited to addressing well.
>
> The network form offers its best advantages where the members, as often occurs in civil society, aim to preserve their autonomy and to avoid hierarchical controls, yet have agendas that are interdependent and benefit from consultation and coordination.[1]

In *The Zapatista "Social Netwar" in Mexico*,[2] Arquilla, Ronfeldt et al expressed grave concern over the possibilities of decentralized "netwar" techniques for destabilizing the political and economic order. They saw ominous signs of such a movement in the global political support network for the Zapatistas. Loose, ad hoc coalitions of affinity groups, organizing through the Internet, could throw together large demonstrations at short notice, and "swarm" the government and mainstream media with phone calls, letters, and emails far beyond their capacity to cope. Ronfeldt and Arquilla noted a parallel between such techniques and the "leaderless resistance" ideas advocated by right-wing white supremacist Louis Beam, circulating in some Constitutionalist/militia circles.

[1]David F. Ronfeldt, *Tribes, Institutions, Markets, Networks* P-7967 (Santa Monica: RAND, 1996) <http://www.rand.org/pubs/papers/P7967/>.

[2]John Arquilla, David Ronfeldt, Graham Fuller, and Melissa Fuller. *The Zapatista "Social Netwar" in Mexico* MR-994-A (Santa Monica: Rand, 1998) <http://www.rand.org/pubs/monograph_reports/MR994/index.html>.

The interesting thing about the Zapatista netwar, according to Ronfeldt and Arquilla, is that to all appearances it started out as a run-of-the-mill Third World army's suppression of a run-of-the-mill local insurgency. Right up until Mexican troops entered Chiapas, there was every indication the uprising would be suppressed quickly, and that the world outside Mexico would "little note nor long remember" it. It looked that way until Subcommandante Marcos and the Zapatistas made their appeal to global civil society and became the center of a networked movement that stirred activists the world over. The Mexican government was blindsided by the global reaction.[1]

Similarly, global corporations have been caught off guard when what once would have been isolated and easily managed conflicts become global political causes.

> Natural-resource companies had grown accustomed to dealing with activists who could not escape the confines of their nationhood: a pipeline or mine could spark a peasants' revolt in the Philippines or the Congo, but it would remain contained, reported only by the local media and known only to people in the area. But today, every time Shell sneezes, a report goes out on the hyperactive "shell-nigeria-action" listserve, bouncing into the in-boxes of all the far-flung organizers involved in the campaign, from Nigerian leaders living in exile to student activists around the world. And when a group of activists occupied part of Shell's U.K. Headquarters in January 1999, they made sure to bring a digital camera with a cellular linkup, allowing them to broadcast their sit-in on the Web, even after Shell officials turned off the electricity and phones. . . .
>
> The Internet played a similar role during the McLibel Trial, catapulting London's grassroots anti-McDonald's movement into an arena as global as the one in which its multinational opponent operates. "We had so much information about McDonald's, we thought we should start a library," Dave Morris explains, and with this in mind, a group of Internet activists launched the McSpotlight Web site. The site not only has the controversial pamphlet online, it contains the complete 20,000-page transcript of the trial, and offers a debating room where McDonald's workers can exchange horror stories about McWork under the Golden Arches. The site, one of the most popular destinations on the Web, has been accessed approximately sixty-five million times.
>
> . . . [This medium is] less vulnerable to libel suits than more traditional media. [McSpotlight programmer] Ben explains that while McSpotlight's server is located in the Netherlands, it has "mirror sites" in Finland, the U.S. New Zealand and Australia. That means that if a server in one country is targeted by McDonald's lawyers, the site will still be available around the world from the other mirrors.[2]

In "Swarming & the Future of Conflict," Ronfeldt and Arquilla focused on swarming, in particular, as a technique that served the entire spectrum of networked conflict—including "civic-oriented actions."[3] Despite the primary concern with swarming as a military phenomenon, they also gave some attention to networked global civil society—and the Zapatista support network in particular—as examples of peaceful swarming with which states were ill-equipped to deal:

> A recent example of swarming can be found in Mexico, at the level of what we call activist "social netwar" (see Ronfeldt et al., 1998). Briefly, we see the Za-

[1]David Ronfeldt and Armando Martinez, "A Comment on the Zapatista Netwar," in Ronfeldt and Arquilla, *In Athena's Camp: Preparing for Conflict in th Information Age* (Santa Monica: Rand, 1997), pp. 369-371.

[2]Klein, *No Logo*, pp. 393-395.

[3]Arquilla and Ronfeldt, *Swarming & the Future of Conflict* DB-311 (Santa Monica, CA: RAND, 2000), iii <http://www.rand.org/pubs/documented_briefings/DB311/>.

patista movement, begun in January 1994 and continuing today, as an effort to mobilize global civil society to exert pressure on the government of Mexico to accede to the demands of the Zapatista guerrilla army (EZLN) for land reform and more equitable treatment under the law. The EZLN has been successful in engaging the interest of hundreds of NGOs, who have repeatedly swarmed their media-oriented "fire" (i.e., sharp messages of reproach) against the government. The NGOs also swarmed in force—at least initially—by sending hundreds of activists into Chiapas to provide presence and additional pressure. The government was able to mount only a minimal counterswarming "fire" of its own, in terms of counterpropaganda. However, it did eventually succeed in curbing the movement of activists into Chiapas, and the Mexican military has engaged in the same kind of "blanketing" of force that U.S. troops employed in Haiti—with similar success.[1]

At present, our best understanding of swarming—as an optimal way for myriad, small, dispersed, autonomous but internetted maneuver units to coordinate and conduct repeated pulsing attacks, by fire or force—is best exemplified in practice by the latest generation of activist NGOs, which assemble into transnational networks and use information operations to assail government actors over policy issues. These NGOs work comfortably within a context of autonomy from each other; they also take advantage of their high connectivity to interact in the fluid, flexible ways called for by swarm theory.

The growing number of cases in which activists have used swarming include, in the security area, the Zapatista movement in Mexico and the International Campaign to Ban Landmines (ICBL). The former is a seminal case of "social netwar," in which transnationally networked NGOs helped deter the Mexican government and army from attacking the Zapatistas militarily. In the latter case, a netwar-like movement, after getting most nations to sign an international antilandmine treaty, won a Nobel Peace Prize. Swarming tactics have also been used, to a lesser degree, by pro-democracy movements aiming to put a dictatorship on the defensive and/or to alter U.S. trade and other relations with that dictatorship. Burma is an example of this.

Social swarming is especially on the rise among activists that oppose global trade and investment policies. Internet-based protests helped to prevent approval of the Multilateral Agreement on Investment (MAI) in Europe in 1998. Then, on July 18, 1999—a day that came to be known as J18—furious anticapitalist demonstrations took place in London, as tens of thousands of activists converged on the city, while other activists mounted parallel demonstrations in other countries. J18 was largely organized over the Internet, with no central direction or leadership. Most recently, with J18 as a partial blueprint, several tens of thousands of activists, most of them Americans but many also from Canada and Europe, swarmed into Seattle to shut down a major meeting of the World Trade Organization (WTO) on opening day, November 30, 1999—in an operation known to militant activists and anarchists as N30, whose planning began right after J18. The vigor of these three movements and the effectiveness of the activists' obstructionism came as a surprise to the authorities.

The violent street demonstrations in Seattle manifested all the conflict formations discussed earlier—the melee, massing, maneuver, and swarming. Moreover, the demonstrations showed that information-age networks (the NGOs) can prevail against hierarchies (the WTO and the Seattle police), at least for a while. The persistence of this "Seattle swarming" model in the April 16, 2000, demonstrations (known as A16) against the International Monetary Fund and the World Bank in Washington, D.C., suggests that it has proven effective enough to continue to be used.

From the standpoints of both theory and practice, some of the most interesting swarming was conducted by black-masked anarchists who referred to themselves collectively as the N30 Black Bloc, which consisted of anarchists from

[1] *Ibid.*, p. 39.

> various affinity groups around the United States. After months of planning, they took to the field individually and in small groups, dispersed but internetted by two-way radios and other communications measures, with a concept of collective organization that was fluid and dynamic, but nonetheless tight. They knew exactly what corporate offices and shops they intended to damage—they had specific target lists. And by using spotters and staying constantly in motion, they largely avoided contact with the police (instead, they sometimes clashed with "peace keepers" among the protesters). While their tactics wrought physical destruction, they saw their larger philosophical and strategic goals in disruptive informational terms, as amounting to breaking the "spell" of private property, corporate hegemony, and capitalism over society.
>
> In these social netwars—from the Zapatistas in 1994, through the N30 activists and anarchists in 1999—swarming appears not only in real-life actions but also through measures in cyberspace. Swarms of email sent to government figures are an example. But some "hacktivists" aim to be more disruptive—pursuing "electronic civil disobedience." One notable recent effort associated with a collectivity called the Electronic Disturbance Theater is actually named SWARM. It seeks to move "digital Zapatismo" beyond the initial emphasis of its creators on their "FloodNet" computer system, which has been used to mount massive "ping" attacks on government and corporate web sites, including as part of J18. The aim of its proponents is to come up with new kinds of "electronic pulse systems" for supporting militant activism. This is clearly meant to enable swarming in cyberspace by myriad people against government, military, and corporate targets. [1]

Swarming—in particular the swarming of public pressure through letters, phone calls, emails, and public demonstrations, and the paralysis of communications networks by such swarms—is the direct descendant of the "overload of demands" Huntington wrote of in the 1970s.

Netwar, Ronfeldt and Arquilla wrote elsewhere, is characterized by "the networked organizational structure of its practitioners—with many groups actually being leaderless —and the suppleness in their ability to come together quickly in swarming attacks."[2]

Jeff Vail discusses netwar techniques, in his *A Theory of Power* blog, using a term of his own: "Rhizome." Vail predicts that the political struggles of the 21st century will be defined by the structural conflict between rhizome and hierarchy.

> Rhizome structures, media and asymmetric politics will not be a means to support or improve a centralized, hierarchical democracy—they will be an alternative to it.
>
> Many groups that seek change have yet to identify hierarchy itself as the root cause of their problem . . . , but are already beginning to realize that rhizome is the solution.[3]

Many open-source thinkers, going back to Eric Raymond in *The Cathedral and the Bazaar*, have pointed out the nature of open-source methods and network culture as force-multipliers.[4] Open-source design communities pick up the innovations of individual members and quickly distribute them wherever they are needed, with maximum economy. By way of analogy, recall the argument from

[1] *Ibid.*, pp. 50-52.

[2] John Arquilla and David Ronfeldt, "Introduction," in Arquilla and Ronfeldt, eds., "Networks and Netwars: The Future of Terror, Crime, and Militancy" MR-1382-OSD (Santa Monica: Rand, 2001) <http://www.rand.org/pubs/monograph_reports/MR1382/> ix.

[3] Jeff Vail, *A Theory of Power* (iUniverse, 2004) <http://www.jeffvail.net/atheoryofpower.pdf>.

[4] Eric S. Raymond, *The Cathedral and the Bazaar* <http://catb.org/~esr/writings/homesteading>.

Cory Doctorow we saw above: proprietary content owners—who still don't "get" network culture—think if they only make DRM too difficult for the average consumer to circumvent, the losses to hard-core geeks who have the time and skills to get around it will be insignificant (" . . . DRM doesn't have to be proof against smart attackers, only average individuals!"). But network culture makes it unnecessary to figure out a way to route around DRM obstructions more than once; as soon as the first person does it, it becomes part of the common pool of intelligence, available to anyone who can search The Pirate Bay (or whatever TPB successor exists at any given time).

Australia, in fact, was recently the location of a *literal* "geeks helping grandmas" story, as geeks at The Pirate Party provided technical expertise to seniors wishing to circumvent government blockage of right-to-die websites:

> Exit International is an assisted suicide education group in Australia, whose average member is over 70 years old. The Exit International website ~~has been~~ will likely be blocked by the Great Firewall of Australia, so Exit International has turned to Australia's Pirate Party and asked for help in producing a slideshow explaining firewall circumvention for seniors. It's a pretty informative slideshow—teachers could just as readily use it for schoolkids in class in a teaching unit on getting access to legit educational materials that's mistakenly blocked by school censorware.[1]

Open-source insurgency follows a similar development model, with each individual innovation quickly becoming part of a common pool of intelligence. John Robb writes:

> The decentralized, and seemingly chaotic guerrilla war in Iraq demonstrates a pattern that will likely serve as a model for next generation terrorists. This pattern shows a level of learning, activity, and success similar to what we see in the open source software community. I call this pattern the bazaar. The bazaar solves the problem: how do small, potentially antagonistic networks combine to conduct war? Lessons from Eric Raymond's "The Cathedral and the Bazaar" provides a starting point for further analysis. Here are the factors that apply (from the perspective of the guerrillas):
>
> - Release early and often. Try new forms of attacks against different types of targets early and often. Don't wait for a perfect plan.
> - Given a large enough pool of co-developers, any difficult problem will be seen as obvious by someone, and solved. Eventually some participant of the bazaar will find a way to disrupt a particularly difficult target. All you need to do is copy the process they used.
> - Your co-developers (beta-testers) are your most valuable resource. The other guerrilla networks in the bazaar are your most valuable allies. They will innovate on your plans, swarm on weaknesses you identify, and protect you by creating system noise.[2]

Tom Knapp provides a good practical example of the Bazaar in operation—the G-20 protests in Philadelphia:

> During the G-20 summit in the Pittsburgh area last week, police arrested two activists. These particular activists weren't breaking windows. They weren't

[1]Cory Doctorow, "Australian seniors ask Pirate Party for help in accessing right-to-die sites," *Boing Boing*, April 9, 2010 <http://www.boingboing.net/2010/04/09/australian-seniors-a.html >.

[2]John Robb, "THE BAZAAR'S OPEN SOURCE PLATFORM," *Global Guerrillas*, September 24, 2004 <http://globalguerrillas.typepad.com/globalguerrillas/2004/09/bazaar_dynamics.html>.

> setting cars on fire. They weren't even parading around brandishing giant puppets and chanting anti-capitalist slogans.
>
> In fact, they were in a hotel room in Kennedy, Pennsylvania, miles away from "unsanctioned" protests in Lawrenceville . . . listening to the radio and availing themselves of the hotel's Wi-Fi connection. Now they stand accused of "hindering apprehension, criminal use of a communication facility and possessing instruments of crime."
>
> The radio they were listening to was (allegedly) a police scanner. They were (allegedly) using their Internet access to broadcast bulletins about police movements in Lawrenceville to activists at the protests, using Twitter. . . .
>
> Government as we know it is engaged in a battle for its very survival, and that battle, as I've mentioned before, looks in key respects a lot like the Recording Industry Association of America's fight with peer-to-peer "file-sharing" networks. The RIAA can—and is—cracking down as hard as it can, in every way it can think of, but it is losing the fight and there's simply no plausible scenario under which it can expect to emerge victorious. The recording industry as we know it will change its business model, or it will go under.
>
> The Pittsburgh Two are wonderfully analogous to the P2P folks. Their arrest boils down, for all intents and purposes, to a public debugging session. Pittsburgh Two 2.0 will set their monitoring stations further from the action (across jurisdictional lines), use a relay system to get the information to those stations in a timely manner, then retransmit that information using offshore and anonymizing proxies. The cops won't get within 50 miles of finding Pittsburgh Two 2.0, and anything they do to counter its efficacy will be countered in subsequent versions.[1]

Two more recent examples are the use of Twitter in Maricopa County to alert the Latino community to raids by Sherrif Joe Arpaio, and to alert drivers to sobriety checkpoints.[2]

One especially encouraging development is the stigmergic sharing of innovations in the technologies of resistance between movements around the world, aiding each other across national lines and bringing combined force to bear against common targets. The Falun Gong has played a central role in this effort:

> When these dissident Iranians chatted with each other and the outside world, they likely had no idea that many of their missives were being guided and guarded by 50 Falun Gong programmers spread out across the United States. These programmers, who almost all have day jobs, have created programs called Freegate and Ultrasurf that allow users to fake out Internet censors. Freegate disguises the browsing of its users, rerouting traffic using proxy servers. To prevent the Iranian authorities from cracking their system, the programmers must constantly switch the servers, a painstaking process.
>
> The Falun Gong has proselytized its software with more fervor than its spiritual practices. It distributes its programs for free through an organization called the Global Internet Freedom Consortium (GIFC), sending a downloadable version of the software in millions of e-mails and instant messages. In July 2008, it introduced a Farsi version of its circumvention tool.
>
> While it is hardly the only group to offer such devices, the Falun Gong's program is particularly popular thanks to its simplicity and relative speed. . . .

[1]Thomas L. Knapp, "The Revolution Will Not Be Tweeted," *Center for a Stateless Society*, October 5, 2009 <http://c4ss.org/content/1179>.

[2]Katherine Mangu-Ward, "The Sheriff is Coming! The Sheriff is Coming!" *Reason Hit & Run*, January 6, 2010 <http://reason.com/blog/2010/01/06/the-sheriff-is-coming-the-sher>; Brad Branan, "Police: Twitter used to avoid DUI checkpoints," *Seattle Times*, December 28, 2009 <http://seattletimes.nwsource.com/html/nationworld/2010618380_twitterdui29.html>.

For all their cleverness, [Falun Gong] members found themselves constantly outmaneuvered. They would devise a strategy that would break past China's filtering tools, only to find their new sites quickly hacked or stymied. In 2002, though, they had their Freegate breakthrough. According to David Tian, a programmer with the GIFC and a research scientist at nasa, Freegate was unique because it not only disguised the ISP addresses, or Web destinations, but also cloaked the traffic signatures, or the ways in which the Chinese filters determined whether a Web user was sending an e-mail, navigating a website, sending an instant message, or using Skype. "In the beginning, Freegate was rudimentary, then the communists analyzed the software, they tried to figure out how we beat them. They started to block Freegate. But then, we started hiding the traffic signature," says Mr. Tian. "They have not been able to stop it since.". . . .

The Falun Gong was hardly alone in developing this kind of software. In fact, there's a Coke-Pepsi rivalry between Freegate and the other main program for skirting the censors: The Onion Router, or TOR. Although TOR was developed by the U.S. Navy—to protect Internet communication among its vessels—it has become a darling of the libertarian left. The TOR project was originally bankrolled, in part, by the Electronic Frontier Foundation (EFF), the group that first sued the U.S. government for warrantless wiretapping. Many libertarians are drawn to TOR because they see it as a way for citizens to shield themselves from the prying eyes of government.

TOR uses an algorithm to route traffic randomly across three different proxy servers. This makes it slow but extremely secure—so secure that both the FBI and international criminal gangs have been known to use it. Unlike the Falun Gong, the TOR programmers have a fetish for making their code available to anyone.

There's an irony in the EFF's embrace of TOR, since the project also receives significant funding from the government. The Voice of America has contributed money so that its broadcasts can be heard via the Internet in countries that have blocked their site, a point of envy for the GIFC. For the past four years, the Falun Gong has also been urging the U.S. government to back Freegate financially, going so far as to enlist activists such as Michael Horowitz, a Reagan administration veteran, and Mark Palmer, a former ambassador to Hungary, to press Congress. (Neither was paid for his work.) But, when the two finally persuaded Congress to spend $15 million on anti-censorship software last year, the money was redirected to a program for training journalists. Both Palmer and Horowitz concluded that the State Department despised the idea of funding the Falun Gong.

That's a reasonable conclusion. The Chinese government views the Falun Gong almost the way the United States views Al Qaeda. As Richard Bush, a China expert at the Brookings Institution, puts it, "An effort to use U.S. government resources in support of a Falun Gong project would be read in the worst possible way by the Chinese government."

Still, there will no doubt be renewed pressure to direct money to the likes of the GIFC and TOR. In the wake of the Iran demonstrations, three bills to fund anti-censorship software are rocketing through Congress, with wide support. Tom Malinowski, the Washington director for Human Rights Watch, argues that such software "is to human rights work today what smuggling mimeograph machines was back in the 1970s, except it reaches millions more people."[1]

The last three paragraphs are suggestive concerning the internal contradictions of state capitalism and its IP regime. The desire of would-be hegemons to aid each other's internal resistance often leads to the creation of virally replicable technologies of benefit to their own internal resistance; on the other hand, this danger sometimes sparks a sense of honor among thieves in which competing hegemons refrain from supporting each other's resistance. But overall, global inter-

[1]Eli Lake, "Hacking the Regime," *The New Republic*, September 3, 2009 <http://www.tnr.com/article/politics/hacking-the-regime>.

state conflict is a source of technologies that can be exploited by non-state actors for internal resistance against the state.

Of course the conflict continues—but the resistance seems to be capable of developing counter-countermeasures before the state's counter-measures are even implemented.

> And, while the Falun Gong has managed to win the upper hand in its battle with the Chinese government, it has reason to be less sanguine about the future. The Chinese have returned to the cyber-nanny model that U.S. libraries have deployed. This notorious project is called the Green Dam, or, more precisely, the Green Dam Youth Escort. Under the Green Dam, every new Chinese computer is required to come with a stringent filter pre-installed and, therefore, nearly impossible to remove. As the filter collects data on users, it relies on a government database to block sites. If anything, the Green Dam is too comprehensive. In its initial run, the software gummed up computers, crashing browsers and prohibiting virtually every Web search. In August, Beijing announced that it would delay the project indefinitely. Still, China had revealed a model that could, in theory, defeat nearly every Web-circumvention tool.
>
> When I asked David Tian, the GIFC programmer, about Green Dam, he spoke about it with a mix of pride and horror. The pride comes from the fact that the GIFC's successes have placed the Chinese on the defensive. "One of the reasons they started this Green Dam business and moved the filter to the computer is because they cannot stop our products with the current filters," he said. But he conceded that Green Dam will render Freegate useless.
>
> In the world of product development—and freedom fighting—you innovate or die. The Falun Gong is determined not to go the way of the Commodore 64 into technological irrelevance. It has released a beta version of a new piece of software to overcome the Green Dam. Without a real chance to test it, it's hard to tell whether it will work. But it has overcome the first hurdle of product development. It has marketed its product with a name that captures the swagger of the enterprise. It is called Green Tsunami.[1]

We will examine the general principles of the Bazaar and network culture, as they relate to the superior agility and resilience of the alternative economy as a whole, in Chapter Seven.

The concept of networked resistance is especially interesting, from our standpoint, as it relates to two things: the kind of anti-corporate "culture jamming" Naomi Klein describes in *No Logo*, and to labor struggle as a form of asymmetric warfare.

In both cases, governments and corporations, hierarchies of all kinds, are learning to their dismay that, in a networked age, it's impossible to suppress negative publicity. As Cory Doctorow put it, "Paris Hilton, the Church of Scientology, and the King of Thailand have discovered . . . [that] taking a piece of information off the Internet is like getting food coloring out of a swimming pool. Good luck with that."[2]

It's sometimes called the Streisand effect, in honor of Barbra Streisand (whose role in its discovery—about which more below—was analogous to Sir Isaac Newton's getting hit on the head by an apple).

One of the earliest examples of the phenomenon was the McLibel case in Britain, in which McDonald's attempt to suppress a couple of embarrassing pamphleteers with a SLAPP lawsuit wound up bringing them far worse publicity as a direct result. The pamphleteers were indigent and represented themselves in court much of the time, and repeatedly lost appeals in the British court system throughout the

[1]*Ibid.*

[2]Doctorow, "It's the Information Economy, Stupid," p. 60.

nineties (eventually they won an appeal in the European Court of Human Rights). But widespread coverage of the case on the Internet, coupled with the defendants' deliberate use of the courtroom as a bully pulpit to examine the factual issues, caused McDonald's one of the worst embarrassments in its history.[1] (Naomi Klein called it "the corporate equivalent of a colonoscopy.")[2]

Two important examples in 2004, the Sinclair Media boycott and the Diebold corporate emails, both decisively demonstrated the impossibility of suppressing online information in an age of mirror sites. A number of left-wing websites and liberal bloggers organized a boycott of Sinclair Media after its stations aired an anti-Kerry documentary by the Swift Boat campaign.

> In the ensuing boycott campaign, advertisers were deluged with more mail and phone calls than they could handle. By October 13, some sponsors were threatening litigation, viewing unsolicited boycott emails as illegal SPAM. Nick Davis, creator of one of the boycott sites, posted legal information explaining that anti-SPAM legislation applied only to commercial messages, and directed threatening sponsors to that information. At the same time, some Sinclair affiliates threatened litigation against sponsors who withdrew support in response to the boycott. Davis organized a legal support effort for those sponsors. By October 15, sponsors were pulling ads in droves. The price of Sinclair stock crashed, recovering only after Sinclair reversed its decision to air the documentary.[3]

Diebold, similarly, attempted to shut down websites which hosted leaked corporate emails questioning the security of the company's electronic voting machines. But the data was widely distributed among student and other activist databases, and the hosting sites were mirrored in jurisdictions all over the world.

> In August, someone provided a cache of thousands of Diebold internal emails to *Wired* magazine and to Bev Harris. Harris posted the emails on her site. Diebold threatened litigation, demanding that Harris, her ISP, and other sites reproducing the emails take them down. Although the threatened parties complied, the emails had been so widely replicated and stored in so many varied settings that Diebold was unable to suppress them. Among others, university students at numerous campuses around the U.S. stored the emails and scrutinized them for evidence. Threatened by Diebold with provisions of the DMCA that required Web-hosting companies to remove infringing materials, the universities ordered the students to remove the materials from their sites. The students responded with a campaign of civil disobedience, moving files between students' machines, duplicating them on FreeNet (an "anti-censorship peer-to-peer publication network") and other peer-to-peer file-sharing systems. . . . They remained publicly available at all times.[4]

An attempt to suppress information on the Wikileaks hosting site, in 2007, resulted in a similar disaster.

> Associated Press (via the first amendment center) reports that "an effort at (online) damage control has snowballed into a public relations disaster for a Swiss bank seeking to crack down on Wikileaks for posting classified information about some of its wealthy clients. While Bank Julius Baer claimed it just wanted stolen and forged documents removed from the site (rather than close it down), **instead of the information disappearing, it rocketed through cy-**

[1]"McDonald's Restaurants v Morris & Steele," *Wikipedia* <http://en.wikipedia.org/wiki/McLibel_case> (accessed December 26, 2009).

[2]Klein, *No Logo*, p. 330.

[3]Yochai Benkler, *The Wealth of Networks: How Social Production Transforms Markets and Freedom* (New Haven and London: Yale University Press, 2006), pp. 220-223.

[4]*Ibid.*, pp. 227-231.

> **berspace,** landing on other Web sites and Wikileaks' own "mirror" sites outside the U.S. . . .
>
> The **digerati call the online phenomenon of a censorship attempt backfiring into more unwanted publicity the "Streisand effect."** Techdirt Inc. chief executive Mike Masnick coined the term on his popular technology blog after the actress Barbra Streisand's 2003 lawsuit seeking to remove satellite photos of her Malibu house. Those photos are now easily accessible, just like the bank documents. "It's a perfect example of the Streisand effect," Masnick said. "This was a really small thing that no one heard about and now it's everywhere and everyone's talking about it."[1]

The so-called DeCSS uprising, in which corporate attempts to suppress publication of a code for cracking the DRM on DVDs failed in the face of widespread defiance, is one of the most inspiring episodes in the history of the free culture movement.

> Journalist Eric Corley—better known as Emmanuel Goldstein, a nom de plume borrowed from Orwell's *1984*—posted the code for DeCSS (so called because it decrypts the Content Scrambling System that encrypts DVDs) as a part of a story he wrote in November for the well-known hacker journal 2600. The Motion Picture Association of America (MPAA) claims that Corley defied anticircumvention provisions of the Digital Millennium Copyright Act (DMCA) by posting the offending code. . . .
>
> The whole affair began when teenager Jon Johansen wrote DeCSS in order to view DVDs on a Linux machine. The MPAA has since brought suit against him in his native Norway as well. Johansen testified on Thursday that he announced the successful reverse engineering of a DVD on the mailing list of the Linux Video and DVD Project (LiViD), a user resource center for video- and DVD-related work for Linux. . . .
>
> The judge in the case, the honorable Lewis Kaplan of the US District Court in southern New York, issued a preliminary injunction against posting DeCSS. Corley duly took down the code, but did not help his defense by defiantly linking to myriad sites which post DeCSS. . . .

True to their hacker beliefs, Corley supporters came to the trial wearing the DeCSS code on t-shirts. There are also over 300 Websites that still link to the decryption code, many beyond the reach of the MPAA.[2]

In the Usmanov case of the same year, attempts to suppress embarrassing information led to similar Internet-wide resistance.

> The Register, UK: Political websites have lined up in defence of a former diplomat whose blog was deleted by hosting firm Fasthosts after threats from lawyers acting for billionaire Arsenal investor Alisher Usmanov.
>
> Four days after Fasthosts pulled the plug on the website run by former UK ambassador to Uzbekistan Craig Murray it remains offline. Several other political and freedom of speech blogs in the UK and abroad have picked up the gauntlet however, and reposted the article that originally drew the takedown demand.
>
> The complaints against Murray's site arose after a series of allegations he made against Usmanov. . . .
>
> After being released from prison, and pardoned, Usmanov became one of a small group of oligarchs to make hay in the former USSR's post-communist asset carve-up. . . .
>
> On his behalf, libel law firm Schillings has moved against a number of Arsenal fan sites and political bloggers repeating the allegations. . . .[1]

[1]"PR disaster, Wikileaks and the Streisand Effect" PRdisasters.com, March 3, 2007 <http://prdisasters.com/pr-disaster-via-wikileaks-and-the-streisand-effect/>.

[2]Deborah Durham-Vichr. "Focus on the DeCSS trial," CNN.Com, July 27, 2000 <http://archives.cnn.com/2000/TECH/computing/07/27/decss.trial.p1.idg/index.html>.

That reference to "[s]everal other political and freedom of speech blogs," by the way, is like saying the ocean is "a bit wet." An article at *Chicken Yoghurt* blog provides a list of all the venues that have republished Murray's original allegations, recovered from Google's caches of the sites or from the Internet Archive. It is a very, very long list[2]—so long, in fact, that *Chicken Yoghurt* helpfully provides the html code with URLs already embedded in the text, so it can be easily cut and pasted into a blog post. In addition, *Chicken Yoghurt* provided the IP addresses of Usmanov's lawyers as a heads-up to all bloggers who might have been visited by those august personages.

A badly edited photo of a waif in a Ralph Lauren ad, which made the model appear not just emaciated but deformed, was highlighted on the Photoshop Disasters website. Lauren sent the site legal notices of DMCA infringement, and got the site's ISP to take it down. In the process, though, the photo—and story—got circulated all over the Internet. Doctorow issued his defiance at *BoingBoing*:

> So, instead of responding to their legal threat by suppressing our criticism of their marketing images, we're gonna mock them. Hence this post. . . .
>
> . . . And every time you threaten to sue us over stuff like this, we will:
>
> a) Reproduce the original criticism, making damned sure that all our readers get a good, long look at it, and;
>
> b) Publish your spurious legal threat along with copious mockery, so that it becomes highly ranked in search engines where other people you threaten can find it and take heart; and
>
> c) Offer nourishing soup and sandwiches to your models.[3]

The Trafigura case probably represents a new speed record, in terms of the duration between initial thuggish attempts to silence criticism and the company lawyers' final decision to cave. The Trafigura corporation actually secured a court injunction against *The Guardian*, prohibiting it from reporting a question by an MP on the floor of Parliament about the company's alleged dumping of toxic waste in Africa. Without specifically naming either Trafigura or the MP, reporter Alan Rusbridger was able to comply with the terms of the injunction and still include enough hints in his cryptic story for readers to scour the Parliamentary reports and figure it out for themselves. By the time he finished work that day, "Trafigura" was already the most-searched-for term on Twitter; by the next morning Trafigura's criminal acts—plus their attempt at suppressing the story—had become front-page news, and by noon the lawyers had thrown in the towel.[4]

John Robb describes the technical potential for information warfare against a corporation, swarming customers, employees, and management with propaganda and disinformation (or the most potent weapon of all, I might add—the truth), and in the process demoralizing management.

> As we move forward in this epochal many to many global conflict, and given many early examples from wide variety of hacking attacks and conflicts, we are

[1]Chris Williams, "Blogosphere shouts 'I'm Spartacus' in Usmanov-Murray case: Uzbek billionaire prompts Blog solidarity," *The Register*, September 24, 2007 <http://www.theregister.co.uk/2007/09/24/usmanov_vs_the_internet/>.

[2]"Public Service Announcement—Craig Murray, Tim Ireland, Boris Johnson, Bob Piper and Alisher Usmanov . . . " *Chicken Yoghurt*, September 20, 2007 <http://www.chickyog .net/2007/09/20/public-service-announcement/>.

[3]Doctorow, "The criticism that Ralph Lauren doesn't want you to see!" *BoingBoing*, October 6, 2009 <http://www.boingboing.net/2009/10/06/the-criticism-that-r.html>.

[4]Alan Rusbridge, "First Read: The Mutualized Future is Bright," *Columbia Journalism Review*, October 19, 2009 <http://www.cjr.org/reconstruction/the_mutualized_future_is_brigh.php>.

likely to see global guerrillas come to routinely use information warfare against corporations. These information offensives will use network leverage to isolate corporations morally, mentally, and physically. . . . Network leverage comes in three forms:

- Highly accurate lists of targets from hacking "black" marketplaces. These lists include all corporate employee e-mail addresses and phone numbers—both at work and at home. ~<$0.25 a dossier (for accurate lists).
- Low cost e-mail spam. Messages can be range from informational to phishing attacks. <$0.1 a message.
- Low cost phone spam. Use the same voice-text messaging systems and call centers that can blanket target lists with perpetual calls. Pennies a call. . . .

In short, the same mechanisms that make spamming/direct marketing so easy and inexpensive to accomplish, can be used to bring the conflict directly to the employees of a target corporation or its partner companies (in the supply chain). Executives and employees that are typically divorced/removed from the full range of their corporation's activities would find themselves immediately enmeshed in the conflict. The objective of this infowar would be to increase . . . :

- Uncertainty. An inability to be certain about future outcomes. If they can do this, what's next? For example: a false/troll e-mail or phone campaign from the CEO that informs employees at work and at home that it will divest from the target area or admits to heinous crimes.
- Menace. An increase [sic] personal/familial risk. The very act of connecting to directly to employee [sic] generates menace. The questions it should evoke: should I stay employed here given the potential threat?
- Mistrust. A mistrust of the corporations moral and legal status. For example: The dissemination of information on a corporations actions, particularly if they are morally egregious or criminal in nature, through a NGO charity fund raising drive.

With an increase in uncertainty, menace, and mistrust within the target corporation's ranks and across the supply chain partner companies, the target's connectivity (moral, physical, and mental) is likely to suffer a precipitous fall. This reduction in connectivity has the potential to create non-cooperative centers of gravity within the targets as cohesion fails. Some of these centers of gravity would opt to leave the problem (quit or annul contractual relationships) and some would fight internally to divest themselves of this problem.[1]

More generally, hierarchical institutions are finding that the traditional means of suppressing communication, that worked as recently as twenty years ago, are useless. Take something as simple as suppressing a school newspaper whose content violates the administrators' sensibilities. An increasingly common response is to set up an informal student newspaper online, and if necessary to tweak the hosting arrangements to thwart attempts at further suppression.[2]

Corporations are immensely vulnerable to informational warfare, both by consumers and by workers. The last section of Naomi Klein's *No Logo* discusses in depth the vulnerability of large corporations and brand name images to netwar campaigns.[3] She pays special attention to "culture jamming," which involves riffing off of corporate logos and thereby "tapping into the vast resources spent to make

[1]John Robb, "INFOWAR vs. CORPORATIONS," *Global Guerrillas*, October 1, 2009 <http://globalguerrillas.typepad.com/globalguerrillas/2009/10/infowar-vs-corporations.html>.

[2]Mike Masnick, "Yet Another High School Newspaper Goes Online to Avoid District Censorship," *Techdirt*, January 15, 200 <http://www.techdirt.com/articles/20090112/1334043381.shtml>.

[3]Klein, *No Logo*, pp. 279-437.

[a] logo meaningful."[1] A good example is the anti-sweatshop campaign by the National Labor Committee, headed by Charles Kernaghan.

> Kernaghan's formula is simple enough. First, select America's most cartoonish icons, from literal ones like Mickey Mouse to virtual ones like Kathie Lee Gifford. Next, create head-on collisions between image and reality. "They live by their image," Kernaghan says of his corporate adversaries. "That gives you a certain power over them . . . these companies are sitting ducks."[2]

At the time she wrote, technological developments were creating unprecedented potential for culture jamming. Digital design and photo editing technology made it possible to make incredibly sophisticated parodies of corporate logos and advertisements.[3] Interestingly, a lot of corporate targets shied away from taking culture jammers to court for fear a public might side with the jammers against the corporate plaintiffs. The more intelligent corporate bosses understand that "legal battles . . . will clearly be fought less on legal than on political grounds." In the words of one advertising executive, "No one wants to be in the limelight because they are the target of community protests or boycotts."[4]

Klein riffed off of Saul Alinsky's term "political jujitsu" to describe "using one part of the power structure against another part." Culture jamming is a form of political jujitsu that uses the power of corporate symbols—symbols deliberately developed to tap into subconscious drives and channel them in directions desired by the corporation—against their corporate owners.[5]

> Anticorporate activism enjoys the priceless benefits of borrowed hipness and celebrity—borrowed, ironically enough, from the brands themselves. Logos that have been burned into our brains by the finest image campaigns money can buy, . . . are bathed in a glow. . . .
>
> . . . Like a good ad bust, anticorporate campaigns draw energy from the power and mass appeal of marketing, at the same time as they hurl that energy right back at the brands that have so successfully colonized our everyday lives.
>
> You can see this jujitsu strategy in action in what has become a staple of many anticorporate campaigns: inviting a worker from a Third World country to come visit a First World superstore—with plenty of cameras rolling. Few newscasts can resist the made-for-TV moment when an Indonesian Nike worker gasps as she learns that the sneakers she churned out for $2 a day sell for $120 at San Francisco Nike Town.[6]

The effect of "sully[ing] some of the most polished logos on the brandscape," as Klein characterized Kernaghan's efforts,[7] is much like that of "Piss Christ." He plays on the appeal of the dogs in *101 Dalmatians* by comparing the living conditions of the animals on the set to those of the human sweatshop workers who produce the tie-in products. He shows up for public appearances with "his signature shopping bag brimming with Disney clothes, Kathie Lee Gifford pants and other logo gear," along with pay slips and price tags used as props to illustrate the discrepancy between worker pay and retail price. In El Salvador, he pulls items out of the bag with price tags attached to show workers what their products fetch in the U.S. After a similar demonstration of Disney products in Haiti, "workers screamed with shock, disbelief, anger, and a mixture of pain and sadness, as their eyes fixed

[1]*Ibid.*, p. 281.
[2]*Ibid.*, p. 351.
[3]*Ibid.* p. 285.
[4]*Ibid.*, p. 288.
[5]*Ibid.*, p. 281.
[6]*Ibid.*, pp. 349-350.
[7]*Ibid.*, p. 351.

on the Pocahontas shirt"—a reaction captured in the film *Mickey Mouse Goes to Haiti*.[1]

Culture jamming is also an illustration of the effects of network culture. Although corporate imagery is still created by people thinking in terms of one-way broadcast communication, the culture jammers have grown up in an age where audiences can talk back to the advertisement or mock it to one another. The content of advertising becomes just another bit of raw material for mashups, as products once transmitted on a one-way conveyor belt from giant factory to giant retailer to consumer have now become raw material for hacking and reverse-engineering.[2]

The Wobbly idea of "direct action on the job" was a classic example of asymmetric warfare. And modern forms of networked resistance are ideally suited to labor struggle. In particular, network technology creates previously unimaginable possibilities for the Wobbly tactic of "open-mouth sabotage." As described in "How to Fire Your Boss":

> Sometimes simply telling people the truth about what goes on at work can put a lot of pressure on the boss. Consumer industries like restaurants and packing plants are the most vulnerable. And again, as in the case of the Good Work Strike, you'll be gaining the support of the public, whose patronage can make or break a business.
>
> Whistle Blowing can be as simple as a face-to-face conversation with a customer, or it can be as dramatic as the P.G.&E. engineer who revealed that the blueprints to the Diablo Canyon nuclear reactor had been reversed. . . .
>
> Waiters can tell their restaurant clients about the various shortcuts and substitutions that go into creating the faux-haute cuisine being served to them. Just as Work to Rule puts an end to the usual relaxation of standards, Whistle Blowing reveals it for all to know.[3]

The authors of *The Cluetrain Manifesto* are quite expansive on the potential for frank, unmediated conversations between employees and customers as a way of building customer relationships and circumventing the consumer's ingrained habit of blocking out canned corporate messages.[4] They characterize the typical corporate voice as "sterile happytalk that insults the intelligence," "the soothing, humorless monotone of the mission statement, marketing brochure, and your-call-is-important-to-us busy signal."[5]

When employees engage customers frankly about the problems they experience with the company's product, and offer useful information, customers usually respond positively.

What the *Cluetrain* authors *don't* mention is the potential for disaster, from the company's perspective, when disgruntled workers see the customer as a potential ally against a common enemy. What would happen if employees decided, not that they wanted to help their company by rescuing it from the tyranny of PR and the official line and winning over customers with a little straight talk—but that

[1]*Ibid.*, p. 353.

[2]*Ibid.*, p. 294.

[3]"How to Fire Your Boss: A Worker's Guide to Direct Action" <http://www.iww.org/organize/strategy/strikes.shtml> (originally a Wobbly Pamphlet, it is reproduced in all its essentials at the I.W.W. Website under the heading of "Effective Strikes and Economic Actions"—although the Wobblies no longer endorse it in its entirety).

[4]"Markets are Conversations," in Rick Levine, Christopher Locke, Doc Searls and David Weinberger, *The Cluetrain Manifesto: The End of Business as Usual* (Perseus Books Group, 2001) <http://www.cluetrain.com/book/index.html>.

[5]"95 theses," in *Ibid.*

they hated the company and that its management was evil? What if, rather than simply responding to a specific problem with what the customer had needed to know, they'd aired all the dirty laundry about management's asset stripping, gutting of human capital, hollowing out of long-term productive capability, gaming of its own bonuses and stock options, self-dealing on the job, and logrolling with directors?

Corporate America, for the most part, still views the Internet as "just an extension of preceding mass media, primarily television." Corporate websites are designed on the same model as the old broadcast media: a one-to-many, one-directional communications flow, in which the audience couldn't talk back. But now the audience *can* talk back.

> Imagine for a moment: millions of people sitting in their shuttered homes at night, bathed in that ghostly blue television aura. They're passive, yeah, but more than that: they're isolated from each other.
>
> Now imagine another magic wire strung from house to house, hooking all these poor bastards up. They're still watching the same old crap. Then, during the touching love scene, some joker lobs an off-color aside—and everybody hears it. Whoa! What was that? . . . The audience is suddenly connected to itself.
>
> What was once The Show, the hypnotic focus and tee-vee advertising carrier wave, becomes . . . an excuse to get together. . . . Think of Joel and the 'bots on Mystery Science Theater 3000. The point is not to watch the film, but to outdo each other making fun of it.
>
> And for such radically realigned purposes, some bloated corporate Web site can serve as a target every bit as well as Godzilla, King of the Monsters. . . .
>
> So here's a little story problem for ya, class. If the Internet has 50 million people on it, and they're not all as dumb as they look, but the corporations trying to make a fast buck off their asses are as dumb as they look, how long before Joe is laughing as hard as everyone else?
>
> The correct answer of course: not long at all. And as soon as he starts laughing, he's not Joe Six-Pack anymore. He's no longer part of some passive couch-potato target demographic. Because the Net connects people to each other, and impassions and empowers through those connections, the media dream of the Web as another acquiescent mass-consumer market is a figment and a fantasy.
>
> The Internet is inherently seditious. It undermines unthinking respect for centralized authority, whether that "authority" is the neatly homogenized voice of broadcast advertising or the smarmy rhetoric of the corporate annual report.[1]
>
> . . . . Look at how this already works in today's Web conversation. You want to buy a new camera. You go to the sites of the three camera makers you're considering. You hastily click through the brochureware the vendors paid thousands to have designed, and you finally find a page that actually gives straightforward factual information. Now you go to a Usenet discussion group, or you find an e-mail list on the topic. You read what real customers have to say. You see what questions are being asked and you're impressed with how well other buyers—strangers from around the world—have answered them. . . .
>
> Compare that to the feeble sputtering of an ad. "SuperDooper Glue—Holds Anything!" says your ad. "Unless you flick it sideways—as I found out with the handle of my favorite cup," says a little voice in the market. "BigDisk Hard Drives—Lifetime Guarantee!" says the ad. "As long as you can prove you oiled it three times a week," says another little voice in the market. What these little voices used to say to a single friend is now accessible to the world. No number of ads will undo the words of the market. How long does it take until the market conversation punctures the exaggerations made in an ad? An hour? A day? The speed of word of mouth is now limited only by how fast people can type. . . .[2]

[1]"Chapter One. Internet Apocalypso," in *Ibid.*
[2]"Chapter Four. Markets Are Conversations," in *Ibid.*

> . . . Marketing has been training its practitioners for decades in the art of impersonating sincerity and warmth. But marketing can no longer keep up appearances. People talk.[1]

Even more important for our purposes, employees talk. It's just as feasible for the corporation's workers to talk directly to its customers, and for workers and customers together to engage in joint mockery of the company.

In an age when unions have virtually disappeared from the private sector workforce, and downsizings and speedups have become a normal expectation of working life, the vulnerability of employer's public image may be the one bit of real leverage the worker has over him—and it's a doozy. If they go after that image relentlessly and systematically, they've got the boss by the short hairs.

Web 2.0, the "writeable web," is fundamentally different from the 1990s vision of an "information superhighway" (one-way, of course), a more complex version of the old unidirectional hub-and-spoke architecture of the broadcast era—or as Tapscott and Williams put it, "one big content-delivery mechanism—a conveyor belt for prepackaged, pay-per-use content" in which "publishers . . . exert control through various digital rights management systems that prevent users from repurposing or redistributing content."[2] Most large corporations still see their websites as sales brochures, and Internet users as a passive audience. But under the Web 2.0 model, the Internet is a platform in which users are the active party.

Given the ease of setting up anonymous blogs and websites (just think of any company and then look up the URL employernamesucks.com), the potential for using comment threads and message boards, the possibility of anonymous saturation emailing of the company's major suppliers and customers and advocacy groups concerned with that industry. . . . well, let's just say the potential for "swarming" and "netwar" is corporate management's worst nightmare.

It's already become apparent that corporations are quite vulnerable to bad publicity from dissident shareholders and consumers. For example, Luigi Zingales writes,

> shareholders' activist Robert Monks succeeded [in 1995] in initiating some major changes at Sears, not by means of the norms of the corporate code (his proxy fight failed miserably) but through the pressure of public opinion. He paid for a full-page announcement in the *Wall Street Journal* where he exposed the identities of Sears' directors, labeling them the "non-performing assets" of Sears. . . . The embarrassment for the directors was so great that they implemented all the changes proposed by Monks.[3]

There's no reason to doubt that management would be equally vulnerable to embarrassment by such tactics from disgruntled production workers, in today's networked world.

For example, although Wal-Mart workers are not represented by NLRB-certified unions, in any bargaining unit in the United States, the "associates" have been quite successful at organized open-mouth sabotage through Wake Up Wal-Mart and similar activist organizations.

Consider the public relations battle over Wal-Mart "open availability" policy. Corporate headquarters in Bentonville quickly moved, in the face of organized public criticism, to overturn the harsher local policy announced by management in Nitro, West Virginia.

[1]*Ibid.*

[2]Tapscott and Williams, p. 271.

[3]Luigi Zingales, "In Search of New Foundations," *The Journal of Finance*, vol. lv, no. 4 (August 2000), pp. 1627-1628.

> A corporate spokesperson says the company reversed the store's decision because Wal-Mart has no policy that calls for the termination of employees who are unable to work certain shifts, the Gazette reports.
>
> "It is unfortunate that our store manager incorrectly communicated a message that was not only inaccurate but also disruptive to our associates at the store," Dan Fogleman tells the Gazette. "We do not have any policy that mandates termination."[1]

The Wal-Mart Workers' Association acts as an unofficial union, and has repeatedly obtained concessions from store management teams in several publicity campaigns designed to embarrass and pressure the company.[2] As Ezra Klein noted,

> This is, of course, entirely a function of the pressure unions have exerted on Wal-Mart—pressure exerted despite the unions having almost no hope of actually unionizing Wal-Mart. Organized Labor has expended tens of millions of dollars over the past few years on this campaign, and while it hasn't increased union density one iota, it has given a hundred thousand Wal-Mart workers health insurance, spurred Wal-Mart to launch an effort to drive down prescription drug prices, drove them into the "Divided We Fail" health reform coalition, and contributed to the company's focus on greening their stores (they needed good press to counteract all the bad).[3]

Another example is the IWW-affiliated Starbucks union, which publicly embarrassed Starbucks Chairman Howard Schultz. It organized a mass email campaign, notifying the Co-op Board of a co-op apartment he was seeking to buy into of his union-busting activities.[4]

Charles Johnson points to the Coalition of Imolakee Workers as an example of an organizing campaign outside the Wagner framework, relying heavily on the open mouth:

> They are mostly immigrants from Mexico, Central America, and the Caribbean; many of them have no legal immigration papers; they are pretty near all mestizo, Indian, or Black; they have to speak at least four different languages amongst themselves; they are often heavily in debt to coyotes or labor sharks for the cost of their travel to the U.S.; they get no benefits and no overtime; they have no fixed place of employment and get work from day to day only at the pleasure of the growers; they work at many different sites spread out anywhere from 10–100 miles from their homes; they often have to move to follow work over the course of the year; and they are extremely poor (most tomato pickers live on about $7,500–$10,000 per year, and spend months with little or no work when the harvesting season ends). But in the face of all that, and across lines of race, culture, nationality, and language, the C.I.W. have organized themselves anyway, through efforts that are nothing short of heroic, and *they have done it as a wildcat union with no recognition from the federal labor bureaucracy and little outside help from the organized labor establishment.* By using creative nonviolent tactics that would be completely illegal if they were subject to the bureaucratic discipline of the Taft-Hartley Act, the C.I.W. has won major victories on wages and conditions over the past two years. They have bypassed the approved chan-

[1]"Wal-Mart Nixes 'Open Availability' Policy," *Business & Labor Reports* (Human Resources section), June 16, 2005 <http://hr.blr.com/news.aspx?id=15666>.

[2]Nick Robinson, "Even Without a Union, Florida Wal-Mart Workers Use Collective Action to Enforce Rights," *Labor Notes*, January 2006. Reproduced at Infoshop, January 3, 2006 <http://www.infoshop.org/inews/article.php?story=20060103065054461>.

[3]Ezra Klein, "Why Labor Matters," *The American Prospect*, November 14, 2007 <http://www.prospect.org/csnc/blogs/ezraklein_archive?month=11&year=2007&base_name=why_labor_matters>.

[4]"Say No to Schultz Mansion Purchase," Starbucks Union <http://www.starbucksunion.org/node/1903>.

nels of collective bargaining between select union reps and the boss, and gone up the supply chain to pressure the tomato buyers, because they realized that they can exercise a lot more leverage against highly visible corporations with brands to protect than they can in dealing with a cartel of government-subsidized vegetable growers that most people outside of southern Florida wouldn't know from Adam.

The C.I.W.'s creative use of moral suasion and secondary boycott tactics have already won them agreements with Taco Bell (in 2005) and then McDonald's (this past spring), which almost doubled the effective piece rate for tomatoes picked for these restaurants. They established a system for pass-through payments, under which participating restaurants agreed to pay a bonus of an additional penny per pound of tomatoes bought, which an independent accountant distributed to the pickers at the farm that the restaurant bought from. Each individual agreement makes a significant but relatively small increase in the worker's effective wages . . . [,] but each victory won means a concrete increase in wages, and an easier road to getting the pass-through system adopted industry-wide, which would in the end nearly *double* tomato-pickers' annual income.

Burger King held out for a while after this, following Taco Bell's earlier successive strategies of ignoring, stonewalling, slick PR, slander (denouncing farm workers as "richer than most minimum-wage workers," consumer boycotts as extortion, and C.I.W. as scam artists), and finally even an attempt at federal prosecution for racketeering.[1]

As Johnson predicted, the dirty tricks were of no avail. He followed up on this story in May 2008, when Burger King caved in. Especially entertaining, after the smear campaign and other dirty tricks carried out by the Burger King management team, was this public statement by BK CEO John Chidsey:

We are pleased to now be working together with the CIW to further the common goal of improving Florida tomato farmworkers' wages, working conditions and lives. The CIW has been at the forefront of efforts to improve farm labor conditions, exposing abuses and driving socially responsible purchasing and work practices in the Florida tomato fields. We apologize for any negative statements about the CIW or its motives previously attributed to BKC or its employees and now realize that those statements were wrong.[2]

Of course corporations are not entirely oblivious to these threats. The corporate world is beginning to perceive the danger of open-mouth sabotage, as well. For example, one Pinkerton thug almost directly equates sabotage to the open mouth, to the near exclusion of all other forms of direct action. According to Darren Donovan, a vice president of Pinkerton's eastern consulting and investigations division,

[w]ith sabotage, there's definitely an attempt to undermine or disrupt the operation in some way or slander the company. . . . There's a special nature to sabotage because of the overtness of it—and it can be violent. . . . Companies can replace windows and equipment, but it's harder to replace their reputation. . . . I

[1]Charles Johnson, "Coalition of Imolakee Workers marches in Miami," *Rad Geek People's Daily*, November 30, 2007 <http://radgeek.com/gt/2007/11/30/coalition_of/>.

[2]Coalition of Immokalee Workers. "Burger King Corp. and Coalition of Immokalee Workers to Work Together," May 23, 2008 <http://www.ciw-online.org/BK_CIW_joint_release.html>. Charles Johnson, "¡Sí, Se Puede! Victory for the Coalition of Imolakee Workers in the Burger King penny-per-pound campaign," *Rad Geek People's Daily*, May 23, 2008 <http://radgeek.com/gt/2008/05/23/si_se/>.

> think that's what HR execs need to be aware of because it is a crime, but it can be different from stealing or fraud.[1]

As suggested by both the interest of a Pinkerton thug and his references to "crime," there is a major focus in the corporate world on identifying whistleblowers and leakers through surveillance technology, and on the criminalization of free speech to combat negative publicity.

And if Birmingham Wragge is any indication, there's a market for corporations that seek to do a Big Brother on anonymous detractors.

> Birmingham's largest law firm has launched a new team to track down people who make anonymous comments about companies online.
>
> The Cyber Tracing team at Wragge & Co was set up to deal with what the law firm said was a rising problem with people making anonymous statements that defamed companies, and people sharing confidential information online.
>
> And Wragge boasted the new team would ensure there was "nowhere to hide in cyberspace".
>
> The four-strong team at the Colmore Row firm is a combination of IT litigation and employment law specialists.
>
> One of the members of the team said redundancies and other reorganisations caused by the recession meant the numbers of disgruntled employees looking to get their own back on employers or former employers was also on the rise.
>
> Adam Fisher said: "Organisations are suffering quite a lot from rogue employees at the moment, partly because of redundancies or general troubles.
>
> "We have had a number of problematic cases where people have chosen to put things online or have shared information on their company email access."
>
> He said much of the job involved trying to get Internet Service Providers to give out details of customers who had made comments online. . . .
>
> A spokeswoman for Wragge said: "Courts can compel Internet Service Providers or telephone service providers to make information available regarding registered names, email addresses and other key account holder information.[2]

But if corporate managers think this will actually work, they're even stupider than I thought they were. Firms like Birmingham Wragge, and policies like RIAA lawsuits and "three strikes" cutoff of ISPs, will have only one significant effect: the rapid mainstreaming of proxy servers and encryption.

In late 2004 and 2005, the phenomenon of "Doocing" (the firing of bloggers for negative commentary on their workplace, or for the expression of other non-approved opinions on their blogs) began to attract mainstream media attention, and exemplified a specialized case of the Streisand Effect. Employers, who fired disgruntled workers out of fear for the bad publicity their blogs might attract, were blindsided by the far worse publicity–far, far worse–that resulted from news of the firing (the term "Doocing" itself comes from Dooce, the name of a blog whose owner was fired). Rather than an insular blog audience of a few hundred reading that "it sucks to work at Employer X," or "Employer X gets away with treating its customers like shit," it became a case of tens of millions of readers of the major newspapers of record and wire services reading that "Employer X fires blogger for revealing how bad it sucks to work at Employer X." Again, the bosses are learning that, for the first time since the rise of the giant corporation and the broadcast culture, workers and consumers can talk back–and not only is there absolutely no way

[1]Jennifer Kock, "Employee Sabotage: Don't Be a Target!" <http://www.workforce.com/archive/features/22/20/88/mdex-printer.php>.

[2]Tom Scotney, "Birmingham Wragge team to focus on online comment defamation," *Birmingham Post*, October 28, 2009 <http://www.birminghampost.net/birmingham-business/birmingham-business-news/legal-business/2009/10/28/birmingham-wragge-team-to-focus-on-online-comment-defamation-65233-25030203/>.

to shut us up, but we actually just keep making more and more noise the more they try to do so.[1]

There's a direct analogy between the Zapatista netwar and assymetrical warfare by labor and other anti-corporate activists. The Zapatistas turned an obscure and low-level military confrontation within an isolated province into a global political struggle. They waged their netwar with the Mexican government mostly outside Chiapas, isolating the authorities and pitting them against the force of world opinion. Similarly, networked labor activists turn labor disputes within a corporation into society-wide economic, political and media struggle, isolating corporate management and exposing it to swarming from an unlimited number of directions. Netwarriors choose their own battlefield.

The problem with authoritarianism like that of the Pinkertons and Birmingham Wragge, from the standpoint of the bosses and their state, is that before you can waterboard open-mouth saboteurs at Gitmo you've got to *catch them* first. If the litigation over Diebold's corporate files and emails teaches anything, it's that court injunctions and similar expedients are virtually useless against guerrilla netwar. The era of the SLAPP lawsuit is over, except for those cases where the offender is considerate enough to volunteer his home address to the target. Even in the early days of the Internet, the McLibel case turned into "the most expensive and most disastrous public-relations exercise ever mounted by a multinational company."[2] As we already noted, the easy availability of web anonymity, the "writeable web" in its various forms, the feasibility of mirroring shut-down websites, and the ability to replicate, transfer, and store huge volumes of digital information at zero marginal cost, means that it is simply impossible to shut people up. The would-be corporate information police will just wear themselves out playing whack-a-mole. They will be exhausted and destroyed in exactly the same way that the most technically advanced army in the world was defeated by a guerrilla force in black pajamas.

Whether it be disgruntled consumers, disgruntled workers, or networked public advocacy organizations, the basic principles are the same. Jon Husband, of *Wirearchy* blog, writes of the potential threat network culture and the free flow of information pose to traditional hierarchies.

> Smart, interested, engaged and articulate people exchange information with each other via the Web, using hyperlinks and web services. Often this information . . . is about something that someone in a position of power would prefer that other people (citizens, constituents, clients, colleagues) not know. . . .
>
> The exchanged-via-hyperlinks-and-web-services information is retrievable, re-usable and when combined with other information (let's play connect-the-dots here) often shows the person in a position of power to be a liar or a spinner, or irresponsible in ways that are not appropriate. This is the basic notion of transparency (which describes a key facet of the growing awareness of the power of the Web). . . .
>
> Hyperlinks, the digital infrastructure of the Web, the lasting retrievability of the information posted to the Web, and the pervasive use of the Web to publish, distribute and transport information combine to suggest that there are large shifts in power ahead of us. We have already seen some of that . we will see much more unless the powers that be manage to find ways to control the toings-and-froings on the Web.

[1]Todd Wallack, "Beware if your blog is related to work," *San Francisco Chronicle*, January 25, 2005 <http://www.sfgate.com/cgi-bin.article.cgi?f=/c/a/2005/01/24/ BIGCEAT1l01.DTL>.

[2]"270-day libel case goes on and on . . . ," *Daily* Telegraph, June 28, 1996 <http://www.mcspotlight.org/media/thisweek/jul3.html>.

> .... [T]he hoarding and protection of sensitive information by hierarchical institutions and powerful people in those institutions is under siege. ...[1]

Chris Dillow, of *Stumbling and Mumbling* blog, argues we're now at the stage where the leadership of large, hierarchical organizations has achieved "negative credibility." The public, in response to a public statement by Gordon Brown, seemingly acted on the assumption that the truth was the direct opposite.

> Could it be that the ruling class now has negative credibility? Maybe people are now taking seriously the old *Yes, Minister* joke—that one should never believe anything until it's officially denied.
>
> If so, doesn't this have serious implications? It means not merely that the managerial class has lost one of the weapons it can use to control us, but that the weapon, when used, actually fires upon its user.[2]

Thanks to network culture, the cost of "manufacturing consent" is rising at an astronomical rate. The communications system is no longer the one described by Edward Herman, with the state and its corporate media allies controlling a handful of expensive centralized hubs and talking to us via one-way broadcast links. We can all talk directly to each other now, and virally circulate evidence that calls the state's propaganda into doubt. For an outlay of well under $1000, you can do what only the White House Press Secretary or a CBS news anchor could do forty years ago. The forces of freedom will be able to contest the corporate state's domination over public consciousness, for the first time in many decades, on even terms.

We have probably already passed a "singularity," a point of no return, in the use of networked information warfare. It took some time for employers to reach a consensus that the old corporate liberal labor regime no longer served their interests, and to take note of and fully exploit the union-busting potential of Taft-Hartley. But once they began to do so, the implosion of Wagner-style unionism was preordained. Likewise, it will take time for the realization to dawn on workers that things are only getting worse, that there's no hope in traditional unionism, and that in a networked world they have the power to bring the employer to his knees by their own direct action. But when they do, the outcome is also probably preordained. The twentieth century was the era of the giant organization. By the end of the twenty-first, there probably won't be enough of them left to bury.

[1]Jon Husband, "How Hard is This to Understand?" *Wirearchy*, June 22, 2007 <http://blog.wirearchy.com/blog/ _archives/2007/6/22/3040833.html>.

[2]Chris Dillow, "Negative Credibility," *Stumbling and Mumbling*, October 12, 2007 <http://stumblingandmumbling.typepad.com/stumbling_and_mumbling/2007/10/negative-credib.html>.

Appendix
Three Works on Abundance and Technological Unemployment

William M. Dugger and James T. Peach. *Economic Abundance: An Introduction* (Armonk, New York and London, England: M.E. Sharpe, 2009).

Adam Arvidsson. "The Makers—again: or the need for keynesian management of abundance," *P2P Foundation Blog*, February 25, 2010.[1]

Martin Ford. *The Lights in the Tunnel: Automation, Accelerating Technology and the Economy of the Future* (Acculant Publishing, 2009).

Introduction

I've grouped these three authors together because their focus overlaps in one particular: their approach to abundance, to the imploding requirements for labor and/or capital to produce a growing share of the things we consume, is in some way to guarantee full employment of the idle labor and capital.

They all share, in some sense, a "demand-side" focus on the problem of abundance: assuming that the prices of goods and services either will or should be propped up despite the imploding cost of production, and then looking for ways to provide the population with sufficient purchasing power to buy those goods. My approach, which will gradually be developed below, is just the opposite—a "supply-side" approach. That means, in practical terms, flushing artificial scarcity rents of all kinds out of the system so that people will no longer need as many hours of wage labor to pay for stuff. . . .

I

I get the impression that Dugger and Peach are influenced by Veblen's *The Engineers and the Price System*, which likewise focused on the social and institutional barriers to running industry at the technical limits of its output capacity and then distributing the entire output. The most important task from their standpoint is to solve the problem of inadequate demand, in order to eliminate idle industrial capacity and unemployment. They accept as normal, for the most part, the mass-production industrial model of the mid-twentieth century, and seek only to remove barriers to disposing of its full product.

For Dugger and Peach, scarcity is a problem of either the incomplete employment of all available production inputs, or the unequal distribution of purchasing power for production outputs. Their goal is to achieve "universal employment."

> Instead of the natural rate of unemployment or full employment, we propose driving the unemployment rate down closer and closer to absolute zero. Provide universal employment and the increased production will provide the wherewithal to put abundance within our grasp.

That's the kind of vision I'd identify more with Michael Moore than, say, Chris Anderson: a society in which virtually everyone works a forty hour week, the wheels of industry run at full capacity churning out endless amounts of stuff, and people earn enough money to keep buying all that stuff.

[1]<http://blog.p2pfoundation.net/the-makers-again-or-the-need-for-keynesian-management-of-abundance/2010/02/25>

But in our existing economy, the volume of stuff produced is mainly a response to the problem of overaccumulation: the need to find new ways to keep people throwing stuff away and replacing it so that our overbuilt industry can keep running at capacity. If goods were not designed to become obsolete, and it took much smaller industrial capacity to produce what we consume, some people might view it as silly to think up all sorts of new things to consume just so they could continue working forty hours a week and keep industry running at full capacity. They might prefer to liquidate a major portion of industrial capacity and work fewer hours, rather than churning out more and more products to earn the money to buy more and more products to keep themselves employed producing more and more products so they could keep consuming more and more, ad nauseam.

In failing to distinguish between natural and artificial scarcity, Dugger and Peach conflate the solutions to two different problems.

When scarcity is natural—i.e. where it costs money or effort to produce a good—then the main form of economic injustice is the broken link between effort and consumption. Privilege enables some people to consume at others' expense. The peasant must work harder to feed a landlord in addition to himself, and the factory worker must produce a surplus consumed by the idle rentier. The problem of privilege, and the zero-sum relationship that results from it, is genuine. And it is almost entirely the focus of Dugger's and Peach's analysis. What's more, their focus on the distribution of claims to the product as a solution is entirely appropriate in the case of natural scarcity. But natural scarcity and the unjust distribution of scarce goods are nothing new; they're problems that have existed, in what amounts to its present form, from the beginning of class society. Their analysis, which treats inequitable distribution of naturally scarce goods as the whole of scarcity, is completely irrelevant to the problem of artificial scarcity—i.e., artificially inflated input costs or prices that embody rents on artificial property rights. The solution to this latter problem is not to find ways to keep everyone on the treadmill forty hours a week, but to eliminate the artificial scarcity component of price so that people can work less.

The real problem, in short, is not to achieve full employment, but to reduce the amount of employment it takes to purchase our present standard of living.

II

In the first installment of this review essay, I dealt with *Economic Abundance* by William Dugger and James Peach. I found it only tangentially related, at best, to the post-scarcity tradition we're familiar with.

Adam Arvidsson and Martin Ford both write from something much closer to that tradition.

Arvidsson, following up on his initial review of *Makers* by Cory Doctorow[1], set out to explain the difference between his views and mine.

In my review of *Makers*[2], I argued that the central cause of the economic crisis was (first) the excess capacity of mass-production industry, and (second) the superfluous investment capital which lacked any profitable outlet thanks to the imploding cost of micromanufacturing technology. Arvidsson responded:

[1]Adam Arvidsson, "Review: Cory Doctorow, *The Makers*," *P2P Foundation Blog*, February 24, 2010 <http://blog.p2pfoundation.net/review-cory-doctorow-the-makers/2010/02/ 24>.

[2]Kevin Carson, "Cory Doctorow. Makers," *P2P Foundation Blog*, October 25, 2009 <http://blog.p2pfoundation.net/cory-doctorow-makers/2009/10/25>.

> However an oversupply of capital is only that in relation to an insufficient demand. The reason why hundreds of thousands or even millions of ventures can not prosper is that there is insufficient demand for their products. This suggests that an economy of abundance (also a relative concept- the old industrial economy was surely an economy of abundance in relation to the old artisanal economy) needs a Keynesian regime of regulation. That is, the state or some other state-like actor must install a mechanism for the redistribution of value that guarantees a sustained demand for new products. To accomplish this entails two things. First, to redistribute the new value that is generated away from the restricted flows of corporate and financial rent that circulate among Kettlewell and his investors and to larger swats of the population (thus activating the multiplier effect!). Since the Maker boom builds on highly socialized, or even ubiquitous productivity, it seems logical that such a redistribution takes the form of some kind of guaranteed minimum income. Second, the state (or state-like actor) must guarantee a direction of market expansion that is sustainable in the future. In our present situation that would probably mean to offer incentives to channel the productivity of a new maker culture into providing solutions to the problem of transitioning to sustainability within energy, transport and food production systems. This would, no doubt open up new sources of demand that would be able to sustain the new economy of abundance for a long time, and after that we can go into space ! Without such a Keynesian governance, a future economy of abundance is doomed to collapse, just like the industrial economy of abundance collapsed in 1929.

This might have been true of the excess industrial capacity of the 1930s, when the primary problem was overinvestment and the maldistribution of purchasing power rather than a rapid decline in the money price of capital goods. Under those circumstances, with the technical means themselves changing in a fairly gradual manner, the size of the gap between existing demand and demand on a scale necessary to run at full capacity might well be small enough to solve with a guaranteed income, or social credit, or some similar expedient.

But the problem in *Makers* is entirely different. It's not simply excess industrial capacity in an environment of gradual and stable technological advance. It takes place in an environment in which the cost of capital goods required for industrial production has fallen a hundredfold. In that environment, the only way to avoid superfluous investment capital with no profitable outlet would be if demand increased a hundredfold in material terms. If a given consumption good produced in a million dollar factory can now be produced in a $10,000 garage shop, that would mean I'd have to buy a hundred of that good where I'd bought only one before, in order to cause a hundred times as many garage shops to be built and soak up the excess capital. Either that, or I'd have to think of a hundred times as many material goods to create sufficient demand to expand industrial capacity a hundredfold. I don't think demand is anywhere near that upwardly elastic. The oversupply of capital in Makers is mainly in relation to the cost of producer goods.

So the solution, in my opinion, is—again—to approach the problem from the supply side. Allow the embedded scarcity rents in the prices of our goods to evaporate, and the bubble-inflated values of real estate and other assets along with them, so that it takes less money and fewer hours of work to obtain the things we need.

III

Of the three works considered in this series of review essays, Ford's pays by far the most attention to the issue of technological unemployment. It's the central theme of his book.

Members of the P2P Research and Open Manufacturing lists are probably familiar with the worst-case scenarios for technological unemployment frequently outlined in the posts of member Paul Fernhout. Coupled with draconian social controls and strong IP enforcement, it's the scenario of Marshall Brain's *Manna*. Still others are surely familiar with similar projections in Jeremy Rifkin's *The End of Work*.

Ford writes very much in the same tradition.

But there are significant mitigating features to technological unemployment which Ford fails to address—features which I've also raised on-list in debates with Fernhout. Most important is the imploding price of means of production.

Most discussions of technological unemployment by people like Rifkin and Ford implicitly assume a capital-intensive mass production model, using expensive, product-specific machines: conventional factories, in other words, in just about every particular except the radically reduced need for people to work in them. They seem to be talking about something like a GM factory, with microcontrollers and servomotors in place of workers, like the Ithaca works in Vonnegut's Player Piano. If such expensive, capital-intensive, mass-production methods constituted the entire world of manufacturing employment, as they were in 1960, then the Rifkin/Ford scenario would indeed be terrifying.

But the mass-production model of manufacturing in large factories has drastically shrunk in significance over the past thirty years, as described by Michel Piore and Charles Sabel in *The Second Industrial Divide*. Manufacturing corporations have always deferred investments in plant and equipment in economic downturns, because—as John Kenneth Galbraith pointed out in *The New Industrial State*—the kinds of expensive product-specific machinery used in Sloanist mass production require full utilization to amortize fixed costs, which in turn requires a high degree of confidence in the stability of demand before companies will invest in them. During recessions, therefore, manufacturing corporations tend to expand production when necessary by contracting out to the craft periphery. But the economic crisis of the 1970s was the beginning of a prolonged period of economic stagnation, with each decade's economic growth slower than the previous and anemic levels of employment and demand. And it was also the beginning of a long-term structural trend toward shifting production capacity from the mass-production core to the craft periphery. Around the turn of the century, the total share of industrial production carried out in job-shops using general purpose machinery surpassed the amount still carried out in conventional mass-production industry.

On pp. 76 and 92, Ford argues that some jobs, like auto mechanic or plumber, are probably safe from automation for the time being because of the nature of the work: a combination of craft skills and general-purpose machinery. But manufacturing work, to the extent that it has shifted to small shops like those in Emilia-Romagna and Shenzhen, using general-purpose machinery for short production runs, has taken on the same character in many instances. If manufacturing continues to be organized primarily on a conventional assembly-line model using automated, highly specialized machines, but with the *additional* step of automating all handing off of goods from one step to the next, then the threat of 100% automation will be credible. But if most manufacturing shifts to the small shop, with a craftsman setting up general purpose machines and supplying feed stock by hand, then Ford's auto mechanic/housekeeper model is much more relevant.

Indeed, the shift toward lean production methods like the Toyota Production System have been associated with the conscious choice of general-purpose machinery and skilled labor *in deliberate preference* to automated mass-production machinery. The kinds of product-specific machinery that are most conducive to automation are directly at odds with the entire lean philosophy, because they require subordinating the organization of production and marketing to the need to keep the expensive machines running at full capacity. Conventional Sloanist mass-production optimized the efficiency of each separate stage in the production process by maximizing throughput to cut down the unit costs on each expensive product-specific machine; but it did so at the cost of pessimizing the production process as a whole (huge piles of in-process inventory piled up between machines, waiting for somebody downstream to actually need it, and warehouses piled full of finished goods awaiting orders). Lean production achieves sharp reduction in overall costs by using "less efficient," more generalized machinery at each stage in the production process, in order to site production as close as possible to the market, scale the overall flow of production to orders, and scale the machinery to the flow of production.

Ford himself concedes that the high capital outlays for automating conventional mass-production industry may delay the process in the medium term (p. 215). And indeed, the pathological behaviors (like optimizing the efficiency of each stage at the expense of pessimizing the overall production flow we saw immediately above) that result from the high cost of automated product-specific machinery, are precisely what Toyota pursued a different production model to avoid. Large-scale, automated, product-specific machinery creates fixed costs that inevitably require batch production, large inventories and push distribution.

What's more, Ford's scenario of the motivation of the business owner in adopting automation technology to cut costs implicitly assumes a model of production and ownership that may not be warranted. As the costs of machinery fall, the conventional distinctions between worker and owner and between machinery and tools are eroding, and the idea of the firm as a large agglomeration of absentee-owned capital hiring wage workers will become less and less representative of the real world. Accordingly, scenarios in which the "business owner" is the primary actor deciding whether to buy automated machinery or hire workers are apt to be less relevant. The more affordable and smaller in scale production tools become, the more frequently the relevant decisionmakers in the capital vs. labor tradeoff will be people working for themselves.

Besides the shift that's already taken place under the Toyota Production System and flexible manufacturing networks like Emilia-Romagna, the shift toward small scale, low cost, general purpose machinery is continuing with the ongoing micromanufacturing revolution as it's currently being worked out in such venues as Factor e Farm, hackerspaces, Fab Labs, tech shops, Ponoko, and 100kGarages.

Technological unemployment, as described in the various scenarios of Rifkin, Brain and Ford, is meaningful mainly because of the divorce of capital from labor which resulted from the high price of producer goods during the mass production era. Indeed, the very concept of "employment" and "jobs," as the predominant source of livelihood, was a historical anomaly brought about by the enormous cost of industrial machinery (machinery which only the rich, or enterprises with large aggregations of rich people's capital, could afford). Before the industrial revolution, the predominant producer goods were general-purpose tools affordable to individual laborers or small shops. The industrial revolution, with the shift from affordable tools to expensive machinery, was associated with a shift from an economy based primarily on self-employed farmers and artisans, and subsistence production for direct use in the household sector, to an economy where most people were hired as wage laborers by the owners of the expensive machinery and purchased most consumption goods with their wages.

But the threat of technological unemployment becomes less meaningful if the means of production fall in price, and there is a retrograde shift from expensive machinery to affordable tools as the predominant form of producer good. And we're in the middle of just such a shift, as a few thousand dollars can buy general-purpose CNC machine tools with the capabilities once possessed only by a factory costing hundreds of thousands of dollars. The same forces making more and more jobs superfluous are simultaneously reducing barriers to the direct ownership of production tools by labor.

So rather than Ford's scenario of the conventional factory owner deciding whether to invest in automated machinery or hire workers, we're likely to see an increasing shift to a scenario in which the typical actor is a group of workers deciding to spend a few thousand workers to set up a garage factory to supply their neighborhood with manufactured goods in exchange for credit in the barter network, and in turn purchasing the output of other micromanufacturing shops or the fruit, vegetables, bread, cheese, eggs, beer, clothing, haircare services, unlicensed cab service, etc., available within the same network. Unlike Ford, as we will see in the next section, I see our primary task as eliminating the barriers to this state of affairs.

I do agree with Ford that we've been experiencing a long-term trend toward longer jobless recoveries and lower levels of employment (p. 134). Total employment has declined 10% since it peaked in 2000, for example. And despite all the Republican crowing over Obama's projection that unemployment would reach only 8.5% in 2009, that's exactly the level of unemployment that Okun's law would have predicted with the decline in GDP that we actually experienced. Our conventional econometric rules of thumb for predicting job losses with a given scale of economic downturn have become worthless because of the long-term structural reduction in demand for labor, and long-term unemployment is at the highest level since the Great Depression.

But while some of this is probably due to technological change that reduces the labor inputs required for a given unit of output, I think the lion's share of it is explained by the overaccumulation thesis of neo-Marxists like Paul Sweezy, Harry Magdoff, and other members of the *Monthly Review* group. The main reason for rising unemployment is corporate capitalism's same chronic tendenices to overinvestment and underconsumption that caused the Great Depression. Cartelized state capitalist industry accumulates excessive surpluses and invests them in so much plant and equipment that it can't dispose of its entire output running at capacity. This crisis was postponed by WWII, which destroyed most plant and equipment in the world outside the U.S., and created a permanent warfare state to absorb a portion of surplus production. But even so, by 1970 Japan and Europe had rebuilt their industrial economies and global capital markets were saturated. Since 1970, one expedient after another has been adopted to absorb surplus capital in an era when consumer demand is insufficient for even existing plant and equipment to operate profitably.

Ford is also correct that rising oil (and hence shipping) costs will provide a strong economic incentive to distributed manufacturing with factories located as close as possible to consumers, which—intersecting with trends to automation—will lead to "much smaller and more flexible factories located in direct proximity to markets . . . " (p. 126) But I think he underestimates the extent to which the shift in economies of scale he describes has already taken place. The flexible manufacturing trend has been toward small job-shops like those in Shenzhen described by Tom Igoe, with ever cheaper general purpose machinery. And the model of automation for such small-scale CNC machinery is most conducive to craft production using general-purpose tools. Coupled with the cutting-edge trend to even cheaper CNC machinery affordable by individuals, a major part of the relocalization of industry in the U.S. is likely to be associated with self-employed artisan producers or small cooperative shops churning out manufactured goods for neighborhood market areas of a few thousand people. Of those cheap tools, Tom Igoe writes:

> **Cheap tools.** Laser cutters, lathes, and milling machines that are affordable by an individual or a group. This is increasingly coming true. The number of colleagues I know who have laser cutters and mills in their living rooms is increasing. . . . There are some notable holes in the open hardware world that exist partially because the tools aren't there. Cheap injection molding doesn't exist yet, but injection molding services do, and they're accessible via the net. But when they're next door (as in Shenzen), you've got a competitive advantage: your neighbor.

Ford also equates automation to increasing capital-intensiveness (pp. 131-132). The traditional model presupposes that "capital-intensive" methods are more costly because capital equipment is expensive, and the most capital-intensive forms of production use the most expensive, product-specific forms of machinery. Production is "capital-intensive" in the sense that expenditures are shifted from labor compensation to machinery, and "high-tech" necessarily means "high-cost." But in fact the current trajectory of technical project in manufacturing hardware is toward drastically reduced cost, bringing new forms of micromanufacturing machinery affordable to average workers. This means that the term "capital-intensive," as conventionally understood, becomes meaningless.

He goes on to argue that manufacturing will become too capital-intensive to maintain existing levels of employment.

> Beyond this threshold or tipping point, the industries that make up our economy will no longer be forced to hire enough new workers to make up for the job losses resulting from automation; they will instead be able to meet any increase in demand primarily by investing in more technology. (p. 133)

But again, this presupposes that capital equipment is expensive, and that access to it is controlled by employers rich enough to afford it. And as the cost of machines fall to the point where they become affordable tools for workers, the "job" becomes meaningless for a growing share of our consumption needs.

Even before the rise of micromanufacturing, there was already a wide range of consumption goods whose production was within the competence of low-cost tools in the informal and household sector. As Ralph Borsodi showed as far back as the 1920s and 1930s, small electrically powered machinery scaled to household production could make a wide range of consumer goods at far lower unit cost than the factories. Although the unit cost of production was somewhat lower for factory goods, this was more than offset by drastic reductions in distribution cost when production was at or near the point of consumption, and by the elimination of supply-push marketing costs when production was directly driven by the consumer. Vegetables grown and canned at home, clothing produced on a home sewing machine from fabric woven on an efficiently designed power loom, bread baked in a kitchen oven from flour grown in a kitchen mill, all required significantly less labor to produce than the labor required to earn the wages to buy them at a store. What's more, directly transforming one's own labor into consumption goods with one's own household tools was not subject to disruption by the loss of wage employment.

If anything, Borsodi underestimated the efficiency advantage. He assumed that the household subsistence economy would be autarkic, with each household having not only its own basic food production, but weaving and sewing, wood shop, etc. He opposed the production of a surplus for external sale, because the terms of commercial sale would be so disadvantageous that it would be more efficient to devote the same time to labor in the wage economy to earn "foreign exchange" to purchase things beyond the production capacity of the household. So for Borsodi, all consumption goods were either produced by the household for itself, or factory made and purchased with wages. He completely neglected the possibility of a division of labor within the informal economy. When such a division is taken into account, efficiencies increase enormously. Instead of each house having its own set of underutilized capital equipment for all forms of small-scale production, a single piece of capital equipment can serve the neighborhood barter network and be fully utilized. Instead of the high transaction costs and learning curve from each household learning how to do everything well, like Odysseus, a skilled seamstress can concentrate on producing clothing for the neighbors and a skilled baker can concentrate on bread—but achieve these efficiencies while still keeping their respective labors in the household economy, without the need either for a separate piece of commercial real estate or for expensive capital goods beyond those scaled to the ordinary household.

Most technological unemployment scenarios assume the automation of conventional, mass-production industry, in a world where manufacturing machinery remains extremely expensive. But when the cost barriers to owning manufacturing machinery are lowered, the threat becomes a lot less terrifying.

By way of analogy: If a Star Trek-style matter replicator can replace human labor for producing most goods, but it costs so much that only a large corporation can own it, then the threat of technological unemployment is real. But if anyone can own such a replicator for a few hundred dollars, then the way we supply a major part of our needs will simply shift from selling labor for wages to producing them for ourselves on a cheap replicator.

In a world where most production is with affordable tools, employers will no longer be able to restrict our access to the means of production. It will become feasible to produce a growing share of our total consumption needs either directly for ourselves, or for exchange with other household producers, without the intermediation of the corporate money economy.

Paul Fernhout's emails (which you probably read regularly if you're on the P2P Research or Open Manufacturing email list) include a quote in the sig line about today's problems resulting from an attempt to deal with abundance in a scarcity framework. Dugger and Peach, as we saw above, failed to recognize the nature of abundance at all, and despite their use of the term worked from an ideological framework entirely adapted to scarcity. Ford, on the other hand, is halfway there. He recognizes the new situation created by abundance of consumer goods and the falling need for labor to produce them. But his solution is still adapted to a framework in which, while consumer goods are abundant, means of production remain scarce and expensive.

When means of production are cheap and readily available, the "need" for labor becomes irrelevant. The need for labor is only relevant when the amount needed is determined by someone other than the worker who controls access to the means of production. By way of analogy, when a subsistence farmer figured out a way to cut in half the labor required to perform some task on his own farm, he didn't lament the loss of "work." He didn't try to do things in a way that required twice the effort in order to keep himself "employed" or achieve "job security." He celebrated it because, being in a position to fully appropriate the benefits of his own productivity, everything came down to the ratio between his personal effort and his personal consumption. In your own home, you don't deliberately store the dishes in a cupboard as far as possible from the sink in order to guarantee yourself "sufficient work." Likewise, when the worker himself can obtain the means of production as cheap, scalable tools, and the cost of producing subsistence needs directly for oneself in the informal economy (or for exchange with other such producers), the question of the amount of labor "needed" for a unit of output is as meaningless as it would have been for the farmer.

Ford also raises the question of how the increasingly plausible prospect of stagnating employment will destabilize long-term consumer behavior. As people come to share a consensus that jobs will be fewer and harder to get in the future, and pay less, their propensity to spend will decrease. The same consumer pessimism that leads to the typical recessionary downward wage-demand spiral, thanks to technological unemployment, will become a permanent structural trend. (p. 109)

But this neglects the possibility that these trends will spur underemployed workers to meet more of their consumption needs through free alternatives in the informal economy. Even as technological change reduces the need for wage labor, it is simultaneously causing an increasing share of consumption goods to shift into the realm of things either available for free, or by direct production in the informal-household sector using low-cost tools. As a result, an increasing portion of what we consume is available independently of wage labor.

Ford argues that "free market forces" and automation, absent some government intervention to redistribute purchasing power, will lead to greater and greater concentration of incomes and consequently a constantly worsening crisis of underconsumption. The ultimate outcome of skyrocketing productivity, coupled with massive technological unemployment, is a society in which 95% of the population are impoverished and live on a subsistence level, while most income goes to the remaining 5% (p. 181). But this state of affairs could never come about in a genuine free market. The enormous wealth and incomes of the plutocracy result from rents on artificial scarcity; they are only able to become super-rich from technological innovation when artificial property rights like patents enable them to capitalize the increased productivity as a source of rents, rather than allowing the competitive market to "socialize" it in the form of lower prices to consumers.

Indeed Ford himself goes on, in the passage immediately following, to admit "the reality" that this level of income polarization would never come about, because the economic decline from insufficient purchasing power would cause asset values to collapse. Exactly! But my proposal (in the next section) is precisely to allow such collapse of asset values, and allow the collapse of the price of goods from the imploding marginal cost of production, so that it takes less wage income to buy them.

The collapse of exchange value is a good thing, from the perspective of the underemployed worker, who experiences the situation Bruce Sterling wrote of (I suspect about three-quarters facetiously, although it's hard to tell with him):

> *Waiting for the day of realization that Internet knowledge-richness actively MAKES people economically poor. "Gosh, Craigslist has such access to ultra-cheap everything now . . . hey wait a second, where did my job go?"

> *Someday the Internet will offer free food and shelter. At that point, hordes simply walk away. They abandon capitalism the way a real-estate bustee abandons an underwater building.

Ford draws a parallel between the mechanization of agriculture in the 20th century, and the ongoing automation of manufacturing and service industries (pp. 124-125). But the parallel works against him, in a sense.

The mechanization of agriculture may, to a considerable extent, have resulted in "a massive and irreversible elimination of jobs." That is, it has eliminated agriculture for many people as a way to earn money by working and then to spend that money buying food. But it has not, by any means, eliminated the possibility of using our own labor to feed ourselves by growing food. Likewise, developments in manufacturing technology, at the same time as they eliminate jobs in manufacturing as a source of income to buy stuff, are making tools for direct production more affordable.

In the particular case of agriculture, as Ralph Borsodi showed eighty years ago, the total labor required to feed ourselves growing and canning our own food at home is considerably less than that required to earn the money to buy it at the store. And nobody can "fire" you from the "job" of feeding yourself with your own labor.

What's more, the allegedly superior efficiencies of mechanized large-scale agriculture are to a large extent a myth perpetuated in the propaganda of corporate agribusiness and the USDA. The efficiencies of mechanization are legitimate for cereal crops, although economies of scale still top out on a family farm large enough to fully utilize one complete set of farming machinery. But cereal crops occupy a disproportionate share of the total food production spectrum precisely because of government subsidies to cereal crop production at the expense of fruits and vegetables.

In the case of most fruits and vegetables, the economies of mechanization are largely spurious, and reflect (again) an agitrop campaign to legitimize government subsidies to corporate agribusiness. Even small-scale conventional farming is more efficient in terms of output per acre, if not in terms of output per man-hour—to say nothing of soil-intensive forms of raised-bed horticulture like that developed John Jeavons (biointensive horticulture can feed one person on a tenth of an acre). And while large scale production may be more efficient in terms of labor inputs at the point of production, it is probably less efficient in labor terms when the wages required to pay the embedded costs of supply-push marketing and distribution are included. Although it may take more labor for me to grow a tomato than it takes a factory farm to grow it, it probably takes less labor for me to grow it myself than to pay for the costs of shipping and marketing it in addition to factory farming it. So absent government subsidies and preferences to large-scale agribusiness, the most efficient method for producing a considerable portion of our food is probably something like Ford's housekeeping or auto repair labor model.

Likewise, it's quite plausible that it would cost a decent home seamstress more in total labor time to earn the money to buy clothing even from a totally automated textile mill, when the costs of high inventories and supply-push distribution are taken into account, than to make them herself.

Besides, if I'm unemployed or working a twenty hour week, labor is something I have plenty of, and (again) I can't be "fired" from using my own labor to feed and clothe myself. The more forms of production that can be carried out in the informal sector, using our own labor with individually affordable tools, the less of what we consume depends on a boss's whim. And the higher the levels of unemployment, the stronger the incentives will be to adopt such methods. Just as economic downturns are associated with a shift of production from the mass-production core to the craft periphery, they're also (as James O'Connor described in *Accumulation Crisis*) associated with a shift of production from wage labor to the informal sector.

This is not meant, by any means, to gloss over or minimize the dislocations will occur in the meantime. Plummeting average housing prices don't mean that many won't be left homeless, or live precarious existences as squatters in their own foreclosed homes or in shantytowns. The falling price of subsistence relative to an hour's wage doesn't mean many won't lack sufficient income to scrape by.Getting from here to there will involve many human tragedies, and how to minimize the pain transition is a very real and open question. My only purpose here is to describe the trends in play, and the end-state they're pointing toward—not to deny the difficulty of the transition.

So while Ford argues that "consumption, rather than production, will eventually have to become the primary economic contribution made by the bulk of average people" (p. 105), I believe just the opposite: the shrinking scale and cost, and increasing productivity, of tools for production will turn the bulk of average people into genuine producers—as opposed to extensions of machines mindlessly obeying the orders of bosses—for the first time in over a century.

This whole discussion parallels a similar one I've had with Marxists like Christian Siefkes. Competitive markets, he argues, have winners and losers, so how do you keep the losers from being unemployed, bankrupt and homeless while the winners buy out their facilities and concentrate production in fewer and fewer hands? My answer, in that case as in the one raised by Ford, is that,with falling prices of producer goods and the rise of networked models of production, the distinction between "winners" and "losers" becomes less and less meaningful. There's no reason to have any permanent losers at all. First of all, the overhead costs are so low that it's possible to ride out a slow period indefinitely. Second, in low-overhead flexible production, in which the basic machinery for production is widely affordable and can be easily reallocated to new products, there's really no such thing as a "business" to go out of. The lower the capitalization required for entering the market, and the lower the overhead to be borne in periods of slow business, the more the labor market takes on a networked, project-oriented character—like, e.g., peer production of software. In free software, and in any other industry where the average producer owns a full set of tools and production centers mainly on self-managed projects, the situation is likely to be characterized not so much by the entrance and exit of discrete "firms" as by a constantly shifting balance of projects, merging and forking, and with free agents constantly shifting from one to another.

Education has a special place in Ford's vision of the abundant society (p. 173). As it is, he is dismayed by the prospect that technological unemployment may lead to large-scale abandonment of higher education, as knowledge work is downsized and the skilled trades offer the best hopes for stable employment.

On the other hand, education is one of the centerpieces of Ford's post-scarcity agenda (about which more below) for dealing with the destabilizing effects of abundance. As part of his larger agenda of making an increasing portion of purchasing power independent of wage labor, he proposes paying people to learn (p. 174).

But for me one of the up-sides of post-scarcity is that the same technological trends are decoupling the love of learning from careerism, dismantling the entire educational-HR complex as a conveyor belt for human raw material, and ending "education" as a professionalized process shaping people for meritocratic "advancement" or transforming them into more useful tools.

The overhead costs of the network model of education are falling, and education is becoming a free good like music or open-source software. MIT's Open Courseware project, which puts complete course syllabuses online for the university's entire catalog of courses, is only the most notable offering of its kind. Projects like Google Books, Project Gutenberg, specialized ventures like the Anarchist Archives and Marxist.Org (which has digitized most of Marx's and Engels' *Collected Works* and the major works of many other Marxist thinkers from Lenin and Trotsky to CLR James), not to mention a whole host of "unauthorized" scanning projects, make entire libraries of scholarly literature available for free. Academically oriented email discussion lists offer unprecedented opportunities for the self-educated to exchange ideas with established academicians. It's never been easier to contact a scholar with some special question or problem, by using Google to track down their departmental email.

In short, there have never been greater opportunities for independent and amateur scholars to pursue knowledge for its own sake, or to participate in freely accessible communities of scholars outside brick-and-mortar universities. The Internet is creating, in the real world, something like the autonomous and self-governing learning networks Ivan Illich described in *Deschooling Society*. But instead of the local mainframe computer at the community center pairing lists of would-be learners with expert volunteers, or renting out tape-recorded lectures, the technical possibilities of today's open education initiatives taking advantage of communications technology beyond Illich's imagining at the time he wrote.

Likewise, it's becoming increasingly feasible to pursue a technical education by the same means, in order to develop one's own capabilities as a producer in the informal economy. Someone might, say, use the engineering curriculum in something like MIT's Open Courseware in combination with mentoring by peers in a hackerspace, and running questions past the membership of a list like Open Manufacturing. Open hardware projects are typically populated by people teaching themselves programming languages or tinkering with hardware on the Edison model, who are at best tangentially connected to the "official" educational establishment.

Phaedrus' idea of the Church of Reason in *Zen and the Art of Motorcycle Maintenance* is relevant. He describes the typical unmotivated drifter who currently predominates in higher education, when deprived of the grades and meritocratic incentives for getting a career or "good job," finally dropping out for lack of interest or motivation.

> The student's biggest problem was a slave mentality which had been built into him by years of carrot-and-whip grading, a mule mentality which said, "If you don't whip me, I won't work." He didn't get whipped. He didn't work. And the cart of civilization, which he supposedly was being trained to pull, was just going to have to creak along a little slower without him. . . .
>
> The hypothetical student, still a mule, would drift around for a while. He would get another kind of education quite as valuable as the one he'd abandoned, in what used to be called the "school of hard knocks." Instead of wasting money and time as a high-status mule, he would now have to get a job as a low-status mule, maybe as a mechanic. Actually his real status would go up. He would be making a contribution for a change. Maybe that's what he would do for the rest of his life. Maybe he'd found his level. But don't count on it.
>
> In time . . . six months; five years, perhaps . . . a change could easily begin to take place. He would become less and less satisfied with a kind of dumb, day-to-

> day shopwork. His creative intelligence, stifled by too much theory and too many grades in college, would now become reawakened by the boredom of the shop. Thousands of hours of frustrating mechanical problems would have made him more interested in machine design. He would like to design machinery himself. He'd think he could do a better job. He would try modifying a few engines, meet with success, look for more success, but feel blocked because he didn't have the theoretical information. He would discover that when before he felt stupid because of his lack of interest in theoretical information, he'd now find a brand of theoretical information which he'd have a lot of respect for, namely, mechanical engineering.
>
> So he would come back to our degreeless and gradeless school, but with a difference. He'd no longer be a grade-motivated person. He'd be a knowledge-motivated person. He would need no external pushing to learn. His push would come from inside. He'd be a free man. He wouldn't need a lot of discipline to shape him up. In fact, if the instructors assigned him were slacking on the job he would be likely to shape them up by asking rude questions. He'd be there to learn something, would be paying to learn something and they'd better come up with it.

IV

In this last installment, I will discuss [Ford's] proposed agenda for dealing with abundance, and then present my own counter-agenda.

Ford uses the term "Luddite fallacy" for those who deny the possibility of technological unemployment in principle.

> This line of reasoning says that, while technological progress will cause some workers to lose their jobs as a result of outdated skills, any concern that advancing technology will lead to widespread, increasing unemployment is, in fact, a fallacy. In other words, machine automation will never lead to economy-wide, systemic unemployment. The reasoning offered by economists is that, as automation increases the productivity of workers, it leads to lower prices for products and services, and in turn, those lower prices result in increased consumer demand. As businesses strive to meet that increased demand, they ramp up production—and that means new jobs. (pp. 95-96)

The problem with their line of reasoning, as I argued here[1] and I think Ford would agree, is that it assumes demand is infinitely, upwardly elastic, and that some of the productivity increase won't be taken in the form of leisure.

My critique of Ford's scenario is from a perspective almost directly *opposite* what he calls the Luddite fallacy. I believe the whole concept of employment will become less meaningful as the falling cost of producer goods causes them to take on an increasingly tool-like character, and as the falling price of consumer goods reduces the need for wage income.

Ford refers to something like my perspective, among the hypothetical objections he lists at the end of the book: "In the future, wages/income may be very low because of job automation, but technology will also make everything plentiful and cheap—so low income won't matter" (pp. 220-221). Or as \I would put it, the reduced need for labor will be offset by labor's reduced need for employment.

Ford's response is that, first, manufactured goods are only a small percentage of the average person's total expenditures, and the costs of housing and healthcare would still require a significant income. Second, he points to "intellectual property" the source of prices that are above marginal cost, even at present, when technology has already lowered production costs, and argues that in the future "intel-

[1]<http://blog.p2pfoundation.net/abundance-creates-utility-but-destroys-exchange-value/2010/02/02>.

lectual property" will cause the prices of goods to exceed their marginal costs of production.

Ford's objections, ironically, point directly to my own agenda: to make housing and healthcare cheap as well by allowing asset prices to collapse, eliminate the artificial scarcities and cost floors that make healthcare expensive, and eliminate "intellectual property" as a source of artificially high prices.

Where Ford supports new government policies to maintain purchasing power, I propose eliminating existing government policies that put a floor under product prices, asset prices, and the cost of means of production.

Ford, like Fernhout and Arvidsson and many other post-scarcity thinkers, proposes various government measures to provide individuals with purchasing power independent of wage labor (p. 161). As a solution to the problem of externalities, he proposes a differential in government-provided income based on how socially responsible one's actions are—essentially Pigovian taxation in reverse (p. 177). He also proposes shifting the tax base for the social safety net from current payroll taxes to taxes on gross margins that remain stable regardless of employment levels (p. 142).

Such proposals have been common for solving the problems of overproduction and underconsumption, going back at least to Major Douglas and Social Credit. (I'm surprised Ford didn't hit on the same idea as Douglas, and dispense with the idea of taxation altogether—just create enough purchasing power out of thin air to fill the demand gap, and deposit it into people's bank accounts.) Something like it is also popular with many Georgists and Geolibertarians: tax the site value of land and other economic rents, resource extraction, and negative externalities like pollution and carbon emissions, and then use the revenue to fund a citizen's dividend or guaranteed minimum income.

Interestingly, some who propose such an agenda also favor leaving patent and copyright law in place and then taxing it as a rent to fund the basic income.

Ford raises the question, from a hypothetical critic, of whether this is not just "Robin Hood socialism": stealing from the productive in order to pay people to do nothing (p. 180). I'd attack it from the other side and argue that it's in fact the opposite of Robin Hood socialism: it leaves scarcity rents in place and then redistributes them, rather than allowing the competitive market to socialize the benefits of innovation through free goods.

I prefer just the opposite approach: where rents and inflated prices result, not from the market mechanism itself, but from government-enforced artificial scarcity, we should eliminate the artificial scarcity. And when negative externalities result from government subsidies to waste or insulation from the real market costs of pollution, we should simply eliminate the legal framework that promotes the negative externality in the first place. Rather than maintaining the purchasing power needed to consume present levels of output, we should reduce the amount of purchasing power required to consume those levels of output. We should eliminate all artificial scarcity barriers to meeting as many of our consumption needs as possible outside the wage economy.

And Ford seems to accept the conventional mass-consumption economy as a given. The problem, he says, "is really not that Americans have spent too much. The problem is that their spending has been sustained by borrowing rather than by growth in real income (p. 161)."

I disagree. The problem is that a majority of our spending goes to pay the embedded costs of subsidized waste and artificial scarcity rents. Overbuilt industry could run at full capacity, before the present downturn, only at the cost of landfills piled with mountains of discarded goods. Most of the money we spend is not on the necessary costs of producing the use-value we consume, but on the moral equivalent of superfluous steps in a Rube Goldberg machine: essentially digging holes and filling them back in. They include—among many other things—rents on copyright and patents, long-distance shipping costs, planned obsolescence, the costs of large inventories and high-pressure marketing associated with supply-push distribution, artificial scarcity rents on capital resulting from government restraints on competition in the supply of credit, and rents on artificial property in land (i.e. holding land out of use or charging tribute to the first user through government enforced titles to vacant and unimproved land).

The waste of resources involved in producing disposable goods for the landfill (after a brief detour through our living rooms), or shipping stuff across country that could be more efficiently produced in a small factory in the same town where it was consumed, was motivated by the same considerations of surplus disposal that, as Emmanuel Goldstein's "Book" described it in *1984*, caused the superpowers to sink millions of tons of industrial output to the bottom of the ocean or blast them into the stratosphere. It's motivated by the same considerations that caused Huxley's World-State to indoctrinate every consumer-citizen with tens of thousands of hypnopaedic injunctions that "ending is better than mending." Human beings have become living disposal units to prevent the wheels of industry from being clogged with unwanted output.

If all these artificial scarcity rents and subsidized inefficiencies were eliminated, and workers weren't deprived of part of the value of our labor by state-enforced unequal bargaining power, right now we could purchase all the consumption goods we currently consume with the wages of fifteen or twenty hours of labor a week.

What we need is not to guarantee sufficient purchasing power to absorb the output of overbuilt industry. It is to eliminate the excess capacity that goes to producing for planned obsolescence.

As with mass consumption, Ford seems to accept the job culture as a bulwark of social stability and purpose. What he has in mind, as I read it, is that the guaranteed income, as a source of purchasing power, be tied to some new "moral equivalent of jobs" that will maintain a sense of normalcy and fill the void left by the reduced need for wage labor (pp. 168-169). His agenda for decoupling purchasing power from wage income involves, rather than the basic income proposals of the Social Credit movement and some Geolibertarians, the use of government income subsidies as a targeted incentive or carrot to encourage favored kinds of behavior like continuing education, volunteering, and the like. "If we cannot pay people to work, then we must pay them to do something else that has value" (p. 194).

Again, I disagree. The loss of the job as an instrument of social control is a good thing.

I share Claire Wolfe's view of the job culture as unnatural from the standpoint of libertarian values, and as a historical anomaly. From an American historical perspective, the whole idea of the job was a radical departure from the previous mainstream in which most people were self-employed artisans and family farmers. It arose mainly because of the high cost of production machinery in the Industrial Revolution. From that perspective, the idea of the "job" as the main source of livelihood over the past 150-200 years—a situation in which the individual spends eight hours a day as a "poor relation" on someone else's property, and takes orders from an authority figure behind a desk in the same way that a schoolchild would from a teacher or a prisoner would from a guard, is just plain weird.

The generation after the American Revolution viewed standing armies as a threat to liberty, not primarily because of their potential for suppressing freedom by force, but because their internal culture inculcated authoritarian values that undermined the cultural atmosphere necessary for the preservation of political freedom in society at large. At the time, standing armies (along with perhaps the Post Office and ecclesiastical hierarchies like that of the Anglican Church) were just about the only large-scale hierarchical institutions around, in a society where most people were self-employed. As such, they were a breeding ground for a personality type fundamentally at odds with the needs of a republican society—people in the habit of taking orders from other people. And today, it seems self-evident that people who spend eight hours a day taking orders, and serving the values and goals of people who utterly unaccountable to them, are unlikely to resist the demands of any other form of authority in the portion of their lives where they're still theoretically "free."

The shift to the pre-job pattern of self-employment in the informal sector promises to eliminate this pathological culture in which one secures his livelihood by winning the approval of an authority figure. In my opinion, therefore, we should take advantage of the opportunity to eliminate this pattern of livelihood, instead of—as Ford proposes—replacing the boss with a bureaucrat as the authority figure on whose whims our livelihood depends. The sooner we destroy the idea of the "job" as a primary source of livelihood, and replace the idea of work as something we're given with the idea of work as something we do, the better. And then we should sow the ground with salt.

So here's my post-scarcity agenda:

1) Eliminating all artificial scarcity rents and mandated artificial levels of overhead for small-scale production, in order to reduce the overhead cost of everyday life, and to reduce the household revenue stream necessary to service it. That means, among other things:

1a) Eliminating "intellectual property" as a source of scarcity rents in informational and cultural goods, and embedded rents on patents as a component of the price of manufactured goods. See, for example, Tom Peters' enthusiastic description in The Tom Peters Seminar that ninety percent of the cost of his new Minolta camera was "intellect" or "ephemera" rather than parts and labor.

1b) An end to local business licensing, zoning laws, and spurious "safety" and "health" codes insofar as they prohibit operating microenterprises out of family residences, or impose arbitrary capital outlays and overhead on such microenterprises by mandating more expensive equipment than the nature of the case requires. It means, for example, eliminating legal barriers to running a microbakery out of one's own home using an ordinary kitchen oven and selling the bread out of one's home or at the Farmer's Market (such as, e.g., requirements to rent a stand-alone piece of commercial real estate, buy an industrial-size oven and dishwasther, etc.).

1c) Likewise, an end to local building codes whose main effect is to lock in conventional building techniques used by established contractors, and to criminalize innovative practices like the use of new low-cost building techniques and cheap vernacular materials.

1d) An end to occupational licensing, or at least an end to artificial restrictions on the number of licenses granted and licensing fees greater than necessary to fund the costs of administration. This would mean that, in place of a limited number of NYC cab medallions costing hundreds of thousands of dollars apiece, medallions would be issued to anyone who met the objective licensing requirements and the cost would be just enough to cover a driving record and criminal background check and a vehicle inspection.

2) An end to government policies aimed at propping up asset prices, allowing the real estate bubble to finish popping.
3) An increase in work-sharing and shorter work weeks to evenly distribute the amount of necessary work that remains. Ford also calls for job-sharing (pp. 185-186), and quotes Keynes 1930 essay on post-scarcity on the principle "spread the bread thinly on the butter—to make what work there is still to be done to be as widely shared as possible" (p. 190). Our disagreement seems to rely in this: I believe that, absent artificial scarcity rents to disrupt the link between effort and consumption, the average individual share of available work would provide sufficient income to purchase a comfortable standard of living. Ford explicitly denies that a part-time income would be sufficient to pay for the necessities of life (p. 191), but seems to operate on the assumption that most of the mechanisms of artificial scarcity would continue as before.
4) The decoupling of the social safety net from both wage employment and the welfare state, through 4a) an increase in extended family or multi-family income-pooling arrangements, cohousing projects, urban communes, etc., and 4b) a rapid expansion of mutuals (of the kind described by Kropotkin, E.P. Thompson, and Colin Ward) as mechanisms for pooling cost and risk. Ford also recognizes the imperative of decoupling the safety net from employment (p. 191), although he advocates government funding as a substitute. But libertarian considerations aside, government is increasingly subject to what James O'Connor called the "fiscal crisis of the state." And this crisis is exacerbated by the tendencies Douglas Rushkoff described in California, as the imploding capital costs required for production rendered most investment capital superfluous and destroyed the tax base. The whole gross margin from capital that Ford presupposes as a partial replacement for payroll taxes is for that reason becoming obsolete.
5) A shift of consumption wherever feasible, from the purchase of store goods with wage income, to subsistence production or production for barter in the household economy using home workshops, sewing machines, ordinary kitchen food prep equipment, etc. If every unemployed or underemployed person with a sewing machine and good skills put them to full use producing clothing for barter, and if every unemployed or underemployed person turned to such a producer as their first resort in obtaining clothing (and ditto for all other forms of common home production, like baking, daycare services, hairstyling, rides and running errands, etc.) the scale of the shift from the capitalist economy to the informal economy would be revolutionary;
6) A rapid expansion in local alternative currency and barter networks taking advantage of the latest network technology, as a source of liquidity of direct exchange between informal/household producers.

Putting it all together, the agenda calls for people to transfer as much of their subsistence needs out of the money economy as it's feasible to do right now, and to that extent to render themselves independent of the old laws of economic value; and where scarcity and exchange value and the need for purchases in the money economy persist, to restore the linkages of equity between effort and purchasing power.

Suppose that the amount of necessary labor, after technological unemployment, was only enough to give everyone a twenty-hour work week—but at the same time the average rent or mortgage payment fell to $150/month, anyone could join a neighborhood cooperative clinic (with several such cooperatives pooling their resources to fund a hospital out of membership fees) for a $50 monthly fee, the price of formerly patented drugs fell 95%, and a microfactory in the community was churning out quality manufactured goods for a fraction of their former price. For most people, myself included, I would call that a greatly improved standard of living.

4

Back to the Future

Even with the decentralizing potential of electrical power neglected and sidetracked into the paleotechnic framework, and even with the diversion of technical development into the needs of mass-production industry, small-scale production tools were still able to achieve superior productivity—even working with the crumbs and castoffs of Sloanist mass-production, and even at the height of Moloch's glory. Two models of production have arisen within the belly of the Sloanist beast, and between them offer the best hopes for replacing the mass-production model: 1) the informal and household economy; and 2) relocalized industry using general-purpose machinery to produce in small batches for the local market, frequently switching between production runs.

A. Home Manufacture

First, even at the height of mass-productionist triumphalism, the superior productivity of home manufacture was demonstrated in many fields. In the 1920s and 1930s, the zenith of mass production's supposed triumph, Ralph Borsodi showed that with electricity most goods could be produced in small shops and even in the home with an efficiency at least competitive with that of the great factories, once the greatly reduced distribution costs of small-scale production were taken into account. Borsodi's law—the tendency of increased distribution costs to offset reduced unit costs of production at a relatively small scale—applies not only to the relative efficiencies of large versus small factories, but also to the comparative efficiencies of factory versus home production. Borsodi argued that for most light goods like food, textiles, and furniture, the overall costs were actually lower to manufacture them in one's own home. The reason was that the electric motor put small-scale production machinery in the home on the same footing as large machinery in the factory. Although economies of large-scale machine production exist, most economies of machine production are captured with the bare adoption of the machinery itself, even with household electrical machinery. After that, the downward production cost curve is very shallow, while the upward distribution cost curve is steep.

Borsodi's study of the economics of home production began with the home-grown tomatoes his wife canned. Expressing some doubts as to Mrs. Borsodi's confidence that it "paid" to do it, he systematically examined all the costs going into the tomatoes, including the market value of the labor they put into growing them and canning them, the cost of the household electricity used, etc. Even with all these things factored in, Bordodi still found the home product cost 20-30% less than the canned tomatoes at the market. The reason? The home product, produced at the point of consumption, had zero distribution cost. The modest unit cost savings from large-scale machinery were insufficient to offset the enormous cost of distribution and marketing.[1]

Borsodi went on to experiment with home clothing production with loom and sewing machine, and building furniture in the home workshop.

> I discovered that more than two-thirds of the things which the average family now buys could be produced more economically at home than they could be bought factory made;
>
> —that the average man and woman could earn more by producing at home than by working for money in an office or factory and that, therefore, the less time they spent working away from home and the more time they spent working at home, the better off they would be;
>
> —finally, that the home itself was still capable of being made into a productive and creative institution and that an investment in a homestead equipped with efficient domestic machinery would yield larger returns per dollar of investment than investments in insurance, in mortgages, in stocks and bonds. . . .
>
> These discoveries led to our experimenting year after year with domestic appliances and machines. We began to experiment with the problem of bringing back into the house, and thus under our own direct control, the various machines which the textile-mill, the cannery and packing house, the flour-mill, the clothing and garment factory, had taken over from the home during the past two hundred years. . . .
>
> In the main the economies of factory production, which are so obvious and which have led economists so far astray, consist of three things: (1) quantity buying of materials and supplies; (2) the division of labor with each worker in industry confined to the performance of a single operation; and (3) the use of power to eliminate labor and permit the operation of automatic machinery. Of these, the use of power is unquestionably the most important. today, however, power is something which the home can use to reduce costs of production just as well as can the factory. The situation which prevailed in the days when water power and steam-engines furnished the only forms of power is at an end. As long as the only available form of power was centralized power, the transfer of machinery and production from the home and the individual, to the factory and the group, was inevitable. But with the development of the gas-engine and the electric motor, power became available in decentralized forms. The home, so far as power was concerned, had been put in position to compete with the factory.
>
> With this advantage of the factory nullified, its other advantages are in themselves insufficient to offset the burden of distribution costs on most products. . . .
>
> The average factory, no doubt, does produce food and clothing cheaper than we produce them even with our power-driven machinery on the Borsodi homestead. But factory costs, because of the problem of distribution, are only first costs. They cannot, therefore, be compared with home costs, which are final costs.[2]

Even the internal economies of the factory, it should be added, were offset by the overhead costs of administration, and the dividends and interest on capital. Profliferating departmentalization entails

> gang bosses, speed bosses, inspectors, repair bosses, planning department representatives and of course corresponding "office" supervisors: designers, planners, record keepers and cost clerks. . . . there are office managers, personnel managers, sales managers, advertising managers and traffic managers. . . . All tend to absorb the reductions in manufacturing costs which are made possible by the factory machinery and factory methods.

[1]Ralph Borsodi, *Flight From the City: An Experiment in Creative Living on the Land* (New York, Evanston, San Francisco, London: Harper & Row, 1933, 1972), pp. 10-15.

[2]*Ibid.*, pp. 17-19.

These are only the costs within the factory. Above the factory, in a firm of numerous factories and branch offices, comes an additional layer of administrative overhead for the corporate headquarters.

And on top of all that, there are the distribution costs of producing for a large market area: "wholesaling transportation and warehousing costs, wholesaling expenses, wholesaling profits, retailing transportation and warehousing costs, retailing expenses, retailing profits."[1]

Since Borsodi's time, the variety and sophistication of electrically powered small machinery has increased enormously. As we saw in Chapter One, after the invention of clockwork the design of machine processes for every conceivable function was nearly inevitable. Likewise once electrically powered machinery was introduced, the development of small-scale electrical machinery for every purpose followed as a matter of course.

Since first reading Borsodi's account I have encountered arguments that his experience was misleading or atypical, given that he was a natural polymath and therefore perhaps a quicker study than most, and therefore failed to include learning time in his estimate of costs. These objections cannot be entirely dismissed.

One of Borsodi's genuine shortcomings was his treatment of household production in largely autarkic terms. He generally argued that the homestead should produce for itself when it was economical to do so, and buy from the conventional money economy with wages when it was not, with little in between. The homesteader should not produce a surplus for the market, he said, because it could only be sold on disadvantageous terms in the larger capitalist economy and would waste labor that could be more efficiently employed either producing other goods for home consumption or earning wages on the market. He did mention the use of surpluses for gifting and hospitality, but largely ignored the possibility of a thriving informal and barter economy outside the capitalist system.

A relatively modest degree of division of labor in the informal and barter economy would be sufficient to overcome a great deal of the learning curve for craft production. Most neighborhoods probably have a skilled home seamstress, a baker famous for his homemade bread, a good home brewer, someone with a well-equipped woodworking or metal shop, and so forth. Present-day home hobbyists, producing for barter, could make use of their existing skills. What's more, in so doing they would optimize efficiency even over Borsodi's model: they would fully utilize the spare capacity of household equipment that would have been idle much of the time with entirely autarkic production, and spread the costs of such capital equipment over a number of households (rather than, as in Borsodi's model, duplicating it in each household).

One of the most important effects of licensing, zoning, and assorted "health" and "safety" codes, at the local level, is to prohibit production on a scale intermediate between individual production for home consumption, and production for the market in a conventional business enterprise. Such regulations criminalize the intermediate case of the household microenterprise, producing either for the market or for barter on a significant scale. This essentially mandates the level of autarky that Borsodi envisioned, and enables larger commercial enterprises to take advantage of the rents resulting from individual learning curves. Skilled home producers are prevented from taking advantage of the spare capacity of their capital equipment, and other households are forced either to acquire all the various specialty skills for themselves or to buy from a commercial enterprise.

[1]Borsodi, *This Ugly Civilization* (Philadelphia: Porcupine Press, 1929, 1975), pp. 34-38.

B. Relocalized Manufacturing

Borsodi's other shortcoming was his inadequate recognition of the possibility of scales of manufacturing below the mass production factory. In *Prosperity and Security*, he identified four scales of production: "(I) family production, (II) custom production, (III) factory production, and (IV) social production."[1] He confused factory production with mass-production. In fact, custom production fades into factory production, with some forms of small-scale factory production that bear as much (or more) resemblance to custom production than to stereotypically American mass-production. In arguing that large-scale factory production was more economical only for a handful of products—"automobiles, motors, electrical appliances, wire, pipe, and similar goods"—he ignored the possibility that even many of those goods could be produced more economically in a small factory using general-purpose machinery in short production runs.[2]

In making "serial production" the defining feature of the factory, as opposed to the custom shop, he made the gulf between factory production and custom production greater and more fixed than was necessary, and ignored the extent to which the line between them is blurred in reality.

> In the sense in which I use the term factory it applies only to places equipped with tools and machinery to produce "goods, wares or utensils" by a system involving serial production, division of labor, and uniformity of products.
>
> .... A garage doing large quantities of repair work on automobiles is much like a factory in appearance. So is a railroad repair shop. Yet neither of these lineal descendants of the roadside smithy is truly a factory.
>
> The distinctive attribute of the factory itself is the system of serial production. It is not, as might be thought, machine production nor even the application of power to machinery.... Only the establishment in which a product of uniform design is systematically fabricated with more or less subdivision of labor during the process is a factory.[3]
>
> .... But none of the economies of mass production, mass distribution, and mass consumption is possible if the finished product is permitted to vary in this manner. Serial production in the factory is dependent at all stages upon uniformities: uniformities, of design, material and workmanship. Each article exactly duplicates every other.... [4]

In arguing that some products ("of which copper wire is one example") could "best be made, or made most economically, by the factory," he neglected the question of whether such things as copper wire could be made more economically in much smaller factories with much less specialized machinery.[5] Elsewhere, citing the superior cost efficiency of milling grain locally or in the home using small electric mills rather than shipping bolted white flour from the mega-mills in Minneapolis, he appealed to the vision of a society of millions of household mills, along with "a few factories making these domestic mills and supplying parts and replacements for them.... "[6] This begs the question of whether a large, mass-production factory is best suited to the production of small appliances.

In fact the possibility of an intermediate model of industrial production has been well demonstrated in industrial districts like Emilia-Romagna. As we men-

[1]Borsodi, *Prosperity and Security: A Study in Realistic Economics* (New York and London: Harper & Brothers Publishers, 1938), p. 172.

[2]*Ibid.*, p. 181.

[3]Borsodi, *This Ugly Civilization*, pp. 56-57.

[4]*Ibid.*, p. 187.

[5]*Ibid.*, p. 78.

[6]*Ibid.*, p. 90.

tioned in Chapter One, Sabel's and Piore's "path not taken" (integrating flexible, electrically powered machinery into craft production) was in fact taken in a few isolated enclaves. In the late 1890s, for example, even after the tide had turned toward mass-production industry, "the German Franz Ziegler could still point to promising examples of the technological renovation of decentralized production in Remscheid, through the introduction of flexible machine tools, powered by small electric motors."[1]

But with the overall economy structured around mass-production industry, the successful industrial districts were relegated mainly to serving niche markets in the larger Sloanist economy. In some cases, like the Lyon textile district (see below), the state officially promoted the liquidation of the industrial district and its absorption by the mass-production economy. In the majority of cases, with the predominance of large-scale mass-production industry encouraged by the state and an economic environment artificially favorable to such forms of organization, flexible manufacturing firms in the industrial districts were "spontaneously" absorbed into a larger corporate framework. The government having created an economy dominated by large-scale, mass-production industry, the pattern of development of small-scale producers was distorted by the character of the overall system. Two examples of the latter phenomenon were the Sheffield and Birmingham districts, in which flexible manufacturers increasingly took on the role of supplying inputs to large manufacturers (they were drawn "ever more closely into the orbit of mass producers," in Piore's and Sabel's words), and as a result gradually lost their flexibility and their ability to produce anything but inputs for the dominant manufacturer. Their product became increasingly standardized, and their equipment more and more dedicated to the needs of a particular large manufacturer.[2] The small-scale machine tools of Remscheid, a decade after Ziegler wrote, were seen as doomed.[3]

But all this has changed with the decay of Mumford's "cultural pseudomorph," and the adoption of alternatives to mass production (as we saw in Chapter Three) as a response to economic crisis. Today, in both Toyota's "single minute exchange of dies" and in the flexible production in the shops of north-central Italy, factory production takes on many of the characteristics of custom production. With standardized, modular components and the ability to switch quickly between various combinations of features, production approaches a state of affairs in which every individual item coming out of the factory is unique. A small factory or workshop, frequently switching between products, can still obtain most of the advantages of Borsodi's "uniformity" through the simple expedient of modular design. Lean production is a synthesis of the good points of mass production and custom or craft production.

Lean production, broadly speaking, has taken two forms, typified respectively by the Toyota Production System and Emilia-Romagna. Robert Begg et al characterize them, respectively, as two ways of globally organizing flexible specialization: producer-driven commodity chains and consumer-driven commodity chains. The former, exemplified in the TPS and to some extent by most global manufacturing corporations, outsources production to small, networked supplier firms. Such firms usually bear the brunt of economic downturns, and have (because they must compete for corporate patronage) have little bargaining power against the corpo-

[1]Michael J. Piore and Charles F. Sabel, *The Second Industrial Divide: Possibilities for Prosperity* (New York: HarperCollins, 1984), p. 47.

[2]*Ibid.*, p. 37.

[3]*Ibid.*, p. 47.

rate purchasers of their output. The latter, exemplified by Emilia-Romagna, entail cooperative networks of small firms for which a large corporate patron most likely doesn't even exist, and production is driven by demand.[1] (Of course the large manufacturing corporations, in the former model, are far more vulnerable to bypassing by networked suppliers than the authors' description would suggest.)

The interesting thing about the Toyota Production System is that it's closer to custom production than to mass production. In many ways, it's Craft Production 2.0.

Craft production, as described by James Womack et al in *The Machine That Changed the World,* was characterized by

- A workforce that was highly skilled in design, machine operation, and fitting. . . .
- Organizations that were extremely decentralized, although concentrated within a single city. Most parts and much of the vehicle's design came from small machine shops. The system was coordinated by an owner/entrepreneur in direct contact with everyone involved—customers, employers, and suppliers.
- The use of general-purpose machine tools to perform drilling, grinding, and other operations on metal and wood.
- A very low production volume. . . .[2]

The last characteristic, low volume (Panhard et Levassor's custom automobile operation produced a thousand or fewer vehicles a year) resulted from the inability to standardize parts, which in turn resulted from the inability of machine tools to cut hardened steel. Before this capability was achieved, it would have been a waste of time to try producing to gauge; steel parts had to be cut and then hardened, which distorted them so that they had to be custom-fitted. The overwhelming majority of production time was taken up by filing and fitting each individual part to the other parts on (say) a car.

Most of the economies of speed achieved by Ford resulted, not from the assembly line (although as a secondary matter it may be useful for maintaining production flow), but from precision and interchangeability. Ford was the first to take advantage of recent advances in machine tools which enabled them to work on prehardened metal. As a result, he was able to produce parts to a standardized gauging system that remained constant throughout the manufacturing process.[3] In so doing, he eliminated the old job of fitter, which was the primary source of cost and delay in custom production.

But this most important innovation of Ford's—interchangeable parts produced to gauge—could have been introduced just as well into craft production, radically increasing the output and reducing the cost of craft industry. Ford managed to reduce task cycle time for assemblers from 514 minutes to 2.3 minutes by August 1913, before he ever introduced the moving assembly line. The assembly line itself reduced cycle time only from 2.3 to 1.19 minutes.[4]

With this innovation, a craft producer might still have used general-purpose machinery and switched frequently between products, while using precision machining techniques to produce identical parts for a set of standardized modular

[1]Robert Begg, Poli Roukova, John Pickles, and Adrian Smith, "Industrial Districts and Commodity Chains: The Garage Firms of Emilia-Romagna (Italy) and Haskovo (Bulgaria)," *Problems of Geography* (Sofia, Bulgarian Academy of Sciences), 1-2 (2005), p. 162.

[2]James P. Womack, Daniel T. Jones, and Daniel Roos, *The Machine That Changed the World* (New York, Toronto, London, Sydney: The Free Press, 1990 and 2007), p. 22.

[3]*Ibid.*, pp. 24-25.

[4]*Ibid.*, pp. 25-26.

designs. By radically reducing setup times and removing the main cost of fitting from craft production ("all filing and adjusting of parts had . . . been eliminated"), craft producers would have achieved many of the efficiencies of mass production with none of the centralization costs we saw in Chapter Two.

In a brilliant illustration of history's tendency to reappear as farce, by the way, GM's batch-and-queue production resurrected the old job of fitter, supposedly eliminated forever by production to gauge, to deal with the enormous output of defective parts. At GM's Framingham plant, besides the weeks' worth of inventory piled among the work stations, Waddell and his co-authors found workers "struggling to attach poorly fitting parts to the Oldsmobile Ciera models they were building."[1]

The other cost of craft production was setup time: the cost and time entailed in skilled machinists readjusting machine tools for different products. Ford reduced setup time through the use of product-specific machinery, foolproofed with simple jigs and gauges to ensure they worked to standard.[2] The problem was that this required batch production, the source of all the inefficiencies we saw in Chapter Two.

This second cost was overcome in the Toyota Production System by Taichi Ohno's "single-minute exchange of dies" (SMED), which reduced the changeover time between products by several orders of magnitude. By the time of World War II, in American-style mass production, manufacturers were dedicating a set of presses to specific parts for months or even years at a time in order to minimize the unit costs from a day or more of downtime to change dies.[3] Ohno, beginning in the late 1940s to experiment with used American machinery, by the late 1950s managed to reduce die-change time to three minutes. In so doing, he discovered that (thanks to the elimination of in-process inventories, and thanks to the fact that defects showed up immediately at the source) "it actually cost less per part to make small batches of stampings than to run off enormous lots."[4] In effect, he turned mass-production machinery into general-purpose machinery.

In industrial districts like Emilia-Romagna, the problem of setup and changeover time was overcome by the development of flexible general purpose machine tools, particularly the small numerically controlled machine tools which the microprocessor revolution permitted in the 1970s. Ford's innovations in precision cutting of pre-hardened metal to gauge, and the elimination of setup time with small CNC tools in the 1970s, between them made it possible for craft production to capture all the efficiencies of mass production.

Ohno's system was essentially a return to craft production methods, but with the speed of Ford's mass production assembly line. With the single-minute exchange of dies, factory machinery bore more of a functional resemblance to general-purpose machinery than to the dedicated and inflexible machinery of GM. But with precision cutting capabilities and a few standardized, modular designs, it achieved nearly the same economies of speed as mass production.

We already described, in Chapter Two, how Sloanism's "economies of speed" differ from those of the Toyota Production System. The irony, according to Waddell and Bodek, is that Toyota and other lean manufacturers reduce direct labor costs (supposedly the raison d'etre of Sloanism) "at rates that leave Sloan companies in the dust."

[1] *Ibid.*, p. 78.
[2] *Ibid.*, p. 33.
[3] *Ibid.*, p. 51.
[4] *Ibid.*, p. 52.

> The critical technology to cutting direct labor hours by fifty percent or more is better than sixty years old. Electric motors small enough and powerful enough to drive a machine tool had a negligible impact on productivity in America, but a huge impact in Japan.
>
> When belt drives came off of machines, and each machine was powered by its own electric motor the door opened up to a productivity improvement equal to that realized by Henry Ford with the advent of the assembly line. . . .
>
> . . . [T]he day came in the evolution of electrical technology that each machine could be equipped with its own motor. Motors were powerful enough, small enough and cheap enough for the belts and shafts to go by the wayside. . . .
>
> To American thinking, this was not much of an event. Sloan's system was firmly entrenched by the time the shafts and belts were eliminated. Economy was perceived to result exclusively from running machines as fast as possible, making big batches at a time. There was still one man to one machine, for the most part, and maximizing the output from that man's labor cost was the objective. Whether machines were lined up in rows, or scattered at random around the factory did not make much difference to the results of that equation.
>
> Shigeo Shingo presented a paper at a technical conference conducted by the Japan Management Association in 1946 entitled "Production Mechanism of Process and Operation." It was based on the principle that optimizing the overall production process . . . is the key to manufacturing. To quote Shingo, "Improvement of process must be accomplished prior to improvement of operation." While the Americans saw manufacturing as a set of isolated operations, all linked by sizeable inventories, the Japanese saw manufacturing as a flow. Where the machines are is a big deal to people concerned about flow while it matters little to people concerned only with isolated operations. To Shingo, the flexibility to put machines anywhere he wanted opened the door to fantastic productivity improvements.[1]

In other words, lean manufacturing—as Sabel and Piore put it—amounts to the discovery, after a century-long dead end, of how to integrate electrical power into manufacturing.

Emilia-Romagna is part of a larger phenomenon, the so-called "Third Italy" (as distinguished from the old industrial triangle of Milan-Turin-Genoa, and the cash crop plantation agriculture of the South):

> a vast network of very small enterprises spread through the villages and small cities of central and Northeast Italy, in and around Bologna, Florence, Ancona, and Venice. . . . These little shops range across the entire sprectrum of the modern industrial structure, from shoes, ceramics, textiles, and garments on one side to motorcycles, agricultural equipment, automotive parts, and machine tools on the other.[2]

Although these small shops (quite small on average, with ten workers or fewer not unusual) "perform an enormous variety of the operations associated with mass production," they do so using "artisans' methods rather than industrial techniques of production."[3]

A typical factory is housed on the ground floor of a building, with two or three floors of apartments above for the several extended families that own it.

[1]Waddell and Bodek, pp. 119-122.

[2]Piore and Sabel, "Italian Small Business Development: Lessons for U.S. Industrial Policy," in John Zysman and Laura Tyson, eds., *American Industry in International Competition: Governnment Policies and Corporate Strategies* (Ithaca and London: Cornell University Press, 1983).

[3]*Ibid*, pp. 392-393.

> The workrooms are clean and spacious. A number of hand operations are interspersed with the mechanized ones. The machinery, however, is fully modern technology and design; sometimes it is exactly the same as that found in a modern factory, sometimes a reduced version of a smaller machine. The work is laid out rationally: the workpieces flow along miniature conveyors, whose twists and turns create the impression of a factory in a doll house.[1]

At the smaller end of the scale, "production is still centered in the garage . . . "
Despite high productivity, the pace of work is typically relaxed, with production stopping daily for workers to retreat to their upstairs apartments for an extended lunch or siesta.[2]

> Some [factories] recall turn-of-the century sweatshops. . . . But many of the others are spotless; the workers extremely skilled and the distinction between them and their supervisors almost imperceptible; the tools the most advanced numerically controlled equipment of its type; the products, designed in the shop, sophisticated and distinctive enough to capture monopolies in world markets. If you had thought so long about Rousseau's artisan clockmakers at Neuchatel or Marx's idea of labor as joyful, self-creative association that you had begun to doubt their possibility, then you might, watching these craftsmen at work, forgive yourself the sudden conviction that something more utopian than the present factory system is practical after all.[3]

Production on the Emilia-Romagna model is regulated on a demand-pull basis: general-purpose machinery makes it possible to produce in small batches and switch frequently and quickly from one product line to another, as orders come in. Further, with the separate stages of production broken down in a networked relationship between producers, constant shifts in contractual relationships between suppliers and outlets are feasible at relatively low cost.[4]

While the small subcontractors in a sector are zealous of their autonomy and often vigorously competitive, they are also quite likely to collaborate as they become increasingly specialized, "subcontracting to each other or sharing the cost of an innovation in machine design that would be too expensive for one producer to order by himself." There is a tendency toward cooperation, especially, because the network relationships betgween specialized firms may shift rapidly with changes in demand, with the same firms alternately subcontracting to one another.[5] Piore and Sabel describe the fluidity of supply chains in an industrial district:

> The variability of demand meant that patterns of subcontracting were constantly rearranged. Firms that had underestimated a year's demand would subcontract the overflow to less well situated competitors scrambling to adapt to the market. But the next year the situation might be reversed, with winners in the previous round forced to sell off equipment to last year's losers. Under these circumstances, every employee could become a subcontractor, every subcontractor a manufacturer, every manufacturer an employee.[6]

The Chinese *shanzhai* phenomenon bears a striking resemblance to the Third Italy. The literal meaning of *shanzhai* is "mountain fortress," but it carries the connotation of a redoubt or stronghold outside the state's control, or a place of refuge for bandits or rebels (much like the Cossack communities on the fringes of the

[1] *Ibid.*, p. 394.
[2] *Ibid.*, p. 394.
[3] Piore and Sabel, "Italy's High-Technology Cottage Industry," *Transatlantic Perspectives* 7 (December 1982), p. 6.
[4] Piore and Sabel, *Second Industrial Divide*, pp. 29-30.
[5] Piore and Sabel, "Italian Small Business Development," pp. 400-401.
[6] Piore and Sabel, *Second Industrial Divide*, p. 32.

Russian Empire, or the Merry Men in Sherwood Forest). Andrew "Bunnie" Huang writes:

> The contemporary shanzhai are rebellious, individualistic, underground, and self-empowered innovators. They are rebellious in the sense that the shanzhai are celebrated for their copycat products; they are the producers of the notorious knock-offs of the iPhone and so forth. They individualistic in the sense that they have a visceral dislike for the large companies; many of the shanzhai themselves used to be employees of large companies (both US and Asian) who departed because they were frustrated at the inefficiency of their former employers. They are underground in the sense that once a shanzhai "goes legit" and starts doing business through traditional retail channels, they are no longer considered to be in the fraternity of the shanzai. They are self-empowered in the sense that they are universally tiny operations, bootstrapped on minimal capital, and they run with the attitude of "if you can do it, then I can as well".
>
> An estimate I heard places 300 shanzhai organizations operating in Shenzhen. These shanzai consist of shops ranging from just a couple folks to a few hundred employees; some just specialize in things like tooling, PCB design, PCB assembly, cell phone skinning, while others are a little bit broader in capability. The shanzai are efficient: one shop of under 250 employees churns out over 200,000 mobile phones per month with a high mix of products (runs as short as a few hundred units is possible); collectively an estimate I heard places shanzhai in the Shenzhen area producing around 20 million phones per month. That's an economy approaching a billion dollars a month. Most of these phones sell into third-world and emerging markets: India, Africa, Russia, and southeast Asia; I imagine if this model were extended to the PC space the shanzhai would easily accomplish what the OLPC failed to do. Significantly, the shanzai are almost universally bootstrapped on minimal capital with almost no additional financing—I heard that typical startup costs are under a few hundred thousand for an operation that may eventually scale to over 50 million revenue per year within a couple years.
>
> Significantly, they do not just produce copycat phones. They make original design phones as well. . . . These original phones integrate wacky features like 7.1 stereo sound, dual SIM cards, a functional cigarette holder, a high-zoom lens, or a built-in UV LED for counterfeit money detection. Their ability to not just copy, but to innovate and riff off of designs is very significant. They are doing to hardware what the web did for rip/mix/burn or mashup compilations. . . . Interestingly, the shanzhai employ a concept called the "open BOM"—they share their bill of materials and other design materials with each other, and they share any improvements made; these rules are policed by community word-of-mouth, to the extent that if someone is found cheating they are ostracized by the shanzhai ecosystem.
>
> To give a flavor of how this is viewed in China, I heard a local comment about how great it was that the shanzhai could not only make an iPhone clone, they could improve it by giving the clone a user-replaceable battery. US law would come down on the side of this activity being illegal and infringing, but given the fecundity of mashup on the web, I can't help but wonder out loud if mashup in hardware is all that bad. . . .
>
> In a sense, I feel like the shanzhai are brethren of the classic western notion of hacker-entrepreneurs, but with a distinctly Chinese twist to them. My personal favorite shanzhai story is of the chap who owns a house that I'm extraordinarily envious of. His house has three floors: on the top, is his bedroom; on the middle floor is a complete SMT manufacturing line; on the bottom floor is a retail outlet, selling the products produced a floor above and designed two floors above. How cool would it be to have your very own SMT line right in your home! It would certainly be a disruptive change to the way I innovate to own infrastructure like that—not only would I save on production costs, reduce my prototyping time, and turn inventory aggressively (thereby reducing inventory capital requirements), I would be able to cut out the 20-50% minimum retail margin typi-

> cally required by US retailers, assuming my retail store is in a high-traffic urban location.
>
> . . . . I always had a theory that at some point, the amount of knowledge and the scale of the markets in the area would reach a critical mass where the Chinese would stop being simply workers or copiers, and would take control of their own destiny and become creators and ultimately innovation leaders. I think it has begun—these stories I'm hearing of the shanzhai and the mashup they produce are just the beginning of a hockey stick that has the potential to change the way business is done, perhaps not in the US, but certainly in that massive, untapped market often referred to as the "rest of the world".[1]

And like the flexible manufacturing networks in the Third Italy, Huang says, the density and economic diversity of the environment in which shanzhai enterprises function promotes flow and adaptability.

> . . . [T]he retail shop on the bottom floor in these electronic market districts of China enables goods to actually flow; your neighbor is selling parts to you, the guy across the street sells your production tools, and the entire block is focused on electronics production, consumption or distribution in some way. The turnover of goods is high so that your SMT and design shop on the floors above can turn a profit.[2]

The success of shanzhai enterprises results not only from their technical innovativeness, according to Vassar professor Yu Zhou, but from "how they form supply chains and how rapidly they react to new trends."[3]

C. New Possibilities for Flexible Manufacturing

Considerable possibilities existed for increasing the efficiency of craft production through the use of flexible machinery, even in the age of steam and water power. The Jacquard loom, for example, used in the Lyon silk industry, was a much lower-tech precursor of Ohno's Single Minute Exchange of Dies (SMED). With the loom controlled by perforated cards, the setup time for switching to a new pattern was reduced substantially. In so doing, it made small-batch production profitable that would have been out of the question with costly, dedicated mass-production machinery.[4] Lyon persisted as a thriving industrial district, by the way, until the French government killed it off in the 1960s: official policy being to encourage conversion to a more "progressive," mass-production model through state-sponsored mergers and acquisitions, the local networked firms became subsidiaries of French-based transnational corporations.[5]

Such industrial districts, according to Piore and Sabel, demonstrated considerable "technological vitality" in the "speed and sophistication with which they adapted power sources to their needs."

> The large Alsatian textile firms not only made early use of steam power but also became—through their sponsorship of research institutes—the nucleus of a major theoretical school of thermodynamics. Small firms in Saint-Etienne experimented with compressed air in the middle of the nineteenth century, before turning, along with Remscheid and Solingen, to the careful study of small steam

[1]Bunnie Huang, "Tech Trend: Shanzhai," *Bunnie's Blog*, February 26, 2009 <http://www.bunniestudios.com/blog/?p=284>.

[2]Comment under *ibid.* <http://www.bunniestudios.com/blog/?p=284#comment-415355>.

[3]David Barboza, "In China, Knockoff Cellphones are a Hit," *New York Times*, April 28, 2009 <http://www.nytimes.com/2009/04/28/technology/28cell.html>.

[4]Piore and Sabel, p. 30.

[5]*Ibid.*, p. 36.

> and gasoline engines. After 1890, when the long-distance transmission of electric power was demontrated at Frankfurt, these three regions were among the first industrial users of small electric motors.[1]

With the introduction of electric motors, the downscaling of power machinery to virtually any kind of small-scale production was no longer a matter of technological possibilities. It was only a question of institutional will, in deciding whether to allocate research and development resources into large- or small-scale production. As we saw in Chapter One, the state tipped the balance toward large-scale mass-production industry, and production with small-scale power machinery was relegated to a few isolated industrial districts. Nevertheless, as we saw in earlier chapters, Borsodi demonstrated that small-scale production—even starved for developmental resources and with one hand tied behind its back—was able to surpass mass-production industry in efficiency.

For the decades of Sloanist dominance, local industrial districts were islands in a hostile sea.

But with the decay of the first stage of the paleotechnic pseudomorph, flexible manufacturing has become the wave of the future—albeit still imprisoned within a centralized corporate framework. And better yet, networked, flexible manufacturing shows great promise for breaking through the walls of the old corporate system and becoming the basis of a fundamentally different kind of society.

By the 1970s, anarchist Murray Bookchin was proposing small general-purpose machinery as the foundation of a decentralized successor to the mass-production economy.

In a 1970s interview with *Mother Earth News*, Borsodi repeated his general theme: that when distribution costs were taken into account, home and small shop manufacture were the most efficient way to produce some two-thirds of what we consume. But he conceded that some goods, like "electric wire or light bulbs," could not be produced "very satisfactorily on a limited scale."[2]

But as Bookchin and Kirkpatrick Sale pointed out, developments in production technology since Borsodi's experiments had narrowed considerably the range of goods for which genuine economies of scale existed. Bookchin proposed the adoption of multiple-purpose production machinery for frequent switching from one short production run to another.

> The new technology has produced not only miniaturized electronic components and smaller production facilities but also highly versatile, multi-purpose machines. For more than a century, the trend in machine design moved increasingly toward technological specialization and single purpose devices, underpinning the intensive division of labor required by the new factory system. Industrial operations were subordinated entirely to the product. In time, this narrow pragmatic approach has "led industry far from the rational line of development in production machinery," observe Eric W. Leaver and John J. Brown. "It has led to increasingly uneconomic specialization. . . . Specialization of machines in terms of end product requires that the machine be thrown away when the product is no longer needed. Yet the work the production machine does can be reduced to a set of basic functions—forming, holding, cutting, and so on—and these functions, if correctly analyzed, can be packaged and applied to operate on a part as needed."

[1] *Ibid.*, p. 31.

[2] "Plowboy Interview" (Ralph Borsodi), *Mother Earth News*, March-April 1974 <http://www.soilandhealth.org/03sov/0303critic/Brsdi.intrvw/The%20Plowboy-Borsodi%20 Interview.htm>.

> Ideally, a drilling machine of the kind envisioned by Leaver and Brown would be able to produce a hole small enough to hold a thin wire or large enough to admit a pipe. . . .
>
> The importance of machines with this kind of operational range can hardly be overestimated. They make it possible to produce a large variety of products in a single plant. A small or moderate-sized community using multi-purpose machines could satisfy many of its limited industrial needs without being burdened with underused industrial facilities. There would be less loss in scrapping tools and less need for single-purpose plants. The community's economy would be more compact and versatile, more rounded and self-contained, than anything we find in the communities of industrially advanced countries. The effort that goes into retooling machines for new products would be enormously reduced. Retooling would generally consist of changes in dimensioning rather than in design.[1]

And Sale, commenting on this passage, observed that many of Borsodi's stipulated exceptions could in fact now be produced most efficiently in a small community factory. The same plant could (say) finish a production run of 30,000 light bulbs, and then switch to wiring or other electrical products—thus "in effect becoming a succession of electrical factories." A machine shop making electric vehicles could switch from tractors to reapers to bicycles.[2]

Eric Husman, commenting on Bookchin's and Sale's treatment of multiple-purpose production technology, points out that they were 1) to a large extent reinventing the wheel, and 2) incorporating a large element of Sloanism into their model:

> *Human Scale* (1980) was written without reference to how badly the Japanese production methods . . . were beating American mass production methods at the time. . . . What Sale failed to appreciate is that the Japanese method (. . . almost diametrically opposed to the Sloan method that Sale is almost certainly thinking of as "mass production") allows the production of *higher* quality articles at *lower* prices. . . .
>
> . . . . Taichi Ohno would laugh himself silly at the thought of someone toying with the idea [of replacing large-batch production on specialized machinery with shorter runs on general-purpose machinery] 20 years after he had perfected it. Ohno's development of Toyota's Just-In-Time method was born exactly out of such circumstances, when Toyota was a small, intimate factory in a beaten country and could not afford the variety and number of machines used in such places as Ford and GM. Ohno pushed, and Shingo later perfected, the idea of Just-In-Time by using Single Minute Exchange of Dies (SMED), making a mockery of a month-long changeover. The idea is to use general machines (e.g. presses) in specialized ways (different dies for each stamping) and to vary the product mix on the assembly line so that you make some of every product every day.
>
> The Sale method (the slightly modified Sloan/GM method) would require extensive warehouses to store the mass-produced production runs (since you run a year's worth of production for those two months and have to store it for the remaining 10 months). If problems were discovered months later, the only recourse would be to wait for the next production run (months later). If too many light bulbs were made, or designs were changed, all those bulbs would be waste. And of course you can forget about producing perishables this way. The JIT method would be to run a few lightbulbs, a couple of irons, a stove, and a refrigerator every hour, switching between them as customer demand dictated. No warehouse needed, just take it straight to the customer. If problems are discovered, the next batch can be held until the problems are solved, and a new batch

[1]Murray Bookchin, *Post-Scarcity Anarchism* (Berkeley, Ca.: The Ramparts Press, 1971), pp. 110-111.

[2]Kirkpatrick Sale, *Human Scale* (New York: Coward, McCann, & Geoghegan, 1980), pp. 409-410.

> will be forthcoming later in the shift or during a later shift. If designs or tastes change, there is no waste because you only produce as customers demand.[1]

Since Bookchin wrote *Post-Scarcity Anarchism*, incidentally, Japanese technical innovations blurred even further the line between the production model he proposed above and the Japanese model of lean manufacturing. The numerically controlled machine tools of American mass-production industry, scaled down thanks to the microprocessor revolution, became suitable as a form of general-purpose machinery for the small shop. As developed by the Japanese, it was

> a new kind of machine tool: numerically controlled general-purpose equipment that is easily programmed and suited for the thousands of small and medium-sized job shops that do much of the batch production in metalworking. Until the mid-1970s, U.S. practice suggested that computer-controlled machine tools could be economically deployed only in large firms (typically in the aerospace industry); in these firms such tools were programmed, by mathematically sophisticated technicians, to manufacture complex components. But advances in the 1970s in semiconductor and computer technology made it possible to build a new generation of machine tools: numerically controlled (NC) or computer-numerical-control (CNC) equipment. NC equipment could easily be programmed to perform the wide range of simple tasks that make up the majority of machining jobs. The equipment's built-in microcomputers allowed a skilled metalworker to teach the machine a sequence of cuts simply by performing them once, or by translating his or her knowledge into a program through straightforward commands entered via a keyboard located on the shop floor.[2]

According to Piore and Sabel, CNC machinery offers the same advantages over traditional craft production—i.e., flexibility with reduced setup cost—that craft production offered over mass production.

> Efficiency in production results from adapting the equipment to the task at hand: the specialization of the equipment to the operation. With conventional technology, this adaptation is done by physical adjustments in the equipment; whenever the product is changed, the specialized machine must be rebuilt. In craft production, this means changing tools and the fixtures that position the workpiece during machining. In mass production, it means scrapping and replacing the machinery. With computer technology, the equipment (the hardware) is adapted to the operation by the computer program (the software); therefore, the equipment can be put to new uses without physical adjustments—simply by reprogramming.[3]

The more setup time and cost are reduced, and the lower the cost of redeploying resources, the less significant both economies of scale and economies of specialization become. Hence, the wider the range of products it is feasible to produce for the local or regional market.[4]

Interestingly, as recounted by David Noble, numeric control was first introduced for large-batch production with expensive machinery in heavy industry, and because of its many inefficiencies was profitable only with massive government subsidies. But the small-scale numerically controlled machine tools, made possible by the invention of the microprocessor, were ideally suited to small-batch production by small local shops.

[1]Eric Husman, "Human Scale Part II—Mass Production," *Grim Reader* blog, September 26, 2006 <http://www.zianet.com/ehusman/weblog/2006/09/human-scale-part-ii-mass-production.html>.

[2]Piore and Sabel, p. 218.

[3]*Ibid.*, p. 260.

[4]*Ibid.*, p. 277.

This is a perennial phenomenon, which we will examine at length in Chapter Seven: even when the state capitalist system heavily subsidizes the development of technologies specifically suited to large-scale, centralized production, decentralized industry takes the crumbs from under the table and uses them more efficiently than state capitalist industry. Consider, also, the role of the state in creating the technical prerequisites for the desktop and Internet revolutions, which are destroying the proprietary culture industries and proprietary industrial design. State capitalism subsidizes its gravediggers.

If Husman compared the Bookchin-Sale method to the Toyota Production System, and found it wanting, H. Thomas Johnson in turn has subjected the Toyota Production System to his own critique. As amazing as Ohno's achievements were at Toyota, introducing his lean production methods within the framework of a transnational corporation amounted to putting new wine in old bottles. Ohno's lean production methods, Johnson argued, are ideally suited to a relocalized manufacturing economy. (This is another example of the decay of the cultural pseudomorph discussed in the previous chapter—the temporary imprisonment of lean manufacturing techniques in the old centralized corporate cocoon.)

In his Foreword to Waddell's and Bodek's *The Rebirth of American Industry* (something of a bible for American devotees of the Toyota Production System), Johnson writes:

> Some people, I am afraid, see lean as a pathway to restoring the large manufacturing giants the United States economy has been famous for in the past half century.
>
> . . . The cheap fossil fuel energy sources that have always supported such production operations cannot be taken for granted any longer. One proposal that has great merit is that of rebuilding our economy around smaller scale, locally-focused organizations that provide just as high a standard living [sic] as people now enjoy, but with far less energy and resource consumption. Helping to create the sustainable local living economy may be the most exciting frontier yet for architects of lean operations. Time will tell.[1]

The "warehouses on wheels" (or "container ships") distribution model used by centralized manufacturing corporations, even "lean" ones like Toyota, is fundamentally at odds with the principles of lean production. Lean production calls for eliminating inventory by gearing production to orders on a demand-pull basis. But long distribution chains simply sweep the huge factory inventories of Sloanism under the rug, and shift them to trucks and ships. There's still an enormous inventory of finished goods at any given time—it's just in motion.

Husman, whom we have already seen is an enthusiastic advocate for lean production, has himself pointed to "warehouses on wheels" as just an outsourced version of Sloanist inventories:

> For another view of self-sufficiency—and I hate to beat this dead horse, but the parallel seems so striking—we have the lean literature on local production. In *Lean Thinking*, Womack et al discuss the travails of the simple aluminum soda can. From the mine to the smelter to the rolling mill to the can maker alone takes several months of storage and shipment time, yet there is only about 3 hours worth of processing time. A good deal of aluminum smelting is done in Norway and/or Sweden, where widely available hydroelectric power makes aluminum production from alumina very cheap and relatively clean. From there, the cans are shipped to bottlers where they sit for a few more days before being filled, shipped, stored, bought, stored, and drank. All told, it takes 319 days to go

[1]H. Thomas Johnson, "Foreword," William H. Waddell and Norman Bodek, *Rebirth of American Industry: A Study of Lean Management* (Vancouver, WA: PCS Press, 2005), p. xxi.

> from the mine to your lips, where you spend a few minutes actually using the can. The process also produces about 24% scrap (most of which is recycled at the source) because the cans are made at one location and shipped empty to the bottler and they get damaged in transit. It's an astounding tale of how wasteful the whole process is, yet still results in a product that—externalities aside—costs very little to the end user. Could this type of thing be done locally? After all, every town is awash in a sea of used aluminum cans, and the reprocessing cost is much lower than the original processing cost (which is why Reynolds and ALCOA buy scrap aluminum).
>
> Taking this problem to the obvious conclusion, Bill Waddell and other lean consultants have been trying to convince manufacturers that if they would only fire the MBAs and actually learn to manufacture, they could do so much more cheaply locally than they can by offshoring their production. Labor costs simply aren't the deciding factor, no matter what the local Sloan school is teaching: American labor may be more expensive then [sic] foreign labor, but it is also more productive. Further, all of the (chimerical) gains to be made from going to cheaper labor are likely to be lost in shipping costs. Think of that flotilla of shipping containers on cargo ships between here and Asia as a huge warehouse on the ocean, warehouses that not only charge rent, but also for fuel.[1]

Regarding the specific example of aluminum cans, Womack et al speculate that the slow acceptance of recycling results from evaluating its efficiencies as a discrete step, rather than in terms of its effects on the entire production stream. If the rate of recycling approached 100%,

> interesting possibilities would emerge for the entire value stream. Mini-smelters with integrated mini-rolling mills might be located near the can makers in England, eliminating in a flash most of the time, storage, and distances involved today in the steps above the can maker.[2]

A similar dynamic might result from the proliferation of mini-mills scaled to local needs, with most of the steel inputs for small-scale industry supplied from recycled local scrap.

As Womack et al point out, lean production—properly understood—requires not only the scaling of machinery to production flow within the factory. It also requires scaling the factory to local demand, and siting it as close as possible to the point of consumption, in order to eliminate as much as possible of the "inventory" in trucks and ships. It is necessary "to locate both design and physical production in the appropriate place to serve the customer."

> Just as many manufacturers have concentrated on installing larger and faster machines to eliminate the direct labor, they've also gone toward massive centralized facilities for product families . . . while outsourcing more and more of the actual component part making to other centralized factories serving many final assemblers. To make matters worse, these are often located on the wrong side of the world from both their engineering operations and their customers . . . to reduce the cost per hour of labor.

[1]Husman, "Human Scale Part III—Self-Sufficiency," *GrimReader* blog, October 2, 2006 <http://www.zianet.com/ehusman/weblog/2006/10/human-scale-part-iii-self-sufficiency.html>.

[2]James P. Womack and Daniel T. Jones, *Lean Thinking: Banish Waste and Create Wealth in Your Corporation* (Simon & Schuster, 1996), p. 43. In addition, recycling's slow takeoff may reflect a cost structure determined by the kind of standard, high-overhead bureaucratic organization which we saw dissected by Paul Goodman in Chapter Two. As recounted by Karl Hess and David Morris in *Neighborhood Power*, a neighborhood church group which set up a recycling center operated by local residents found they could sort out trash themselves and receive $20-50 a ton (this was in the mid-70s). Karl Hess and David Morris, *Neighborhood Power: The New Localism* (Boston: Beacon Press, 1975), p. 139.

> The production process in these remotely located, high-scale facilities may even be in some form of flow, but . . . the flow of the product stops at the end of the plant. In the case of bikes, it's a matter of letting the finished product sit while a whole sea container for a given final assembler's warehouse in North America is filled, then sending the filled containers to the port, where they sit some more while waiting for a giant container ship. After a few weeks on the ocean, the containers go by truck to one of the bike firm's regional warehouses, where the bikes wait until a specific customer order needs filling often followed by shipment to the customer's warehouse for more waiting. In other words, there's no flow except along a tiny stretch of the total value stream inside one isolated plant.
>
> The result is high logistics costs and massive finished unit inventories in transit and at retailer warehouses. . . . When carefully analyzed, these costs and revenue losses are often found to more than offset the savings in production costs from low wages, savings which can be obtained in any case by locating smaller flow facilities incorporating more of the total production steps much closer to the customer.[1]
>
> To achieve the scale needed to justify this degree of automation it will often be necessary to serve the entire world from a single facility, yet customers want to get exactly the product they want exactly when they want it. . . . It follows that oceans and lean production are not compatible. We believe that, in almost every case, locating smaller and less-automated production systems within the market of sale will yield lower total costs (counting logistics and the cost of scrapped goods no one wants by the time they arrive) and higher customer satisfaction.[2]

Husman, incidentally, describes a localized "open-source production" model, with numerous small local machine shops networked to manufacture a product according to open-source design specifications and then to manufacture replacement parts and do repairs on an as-needed basis, as "almost an ideally Lean manufacturing process. Dozens of small shops located near their customers, each building one at a time."[3]

The authors of *Natural Capitalism* devote a separate chapter to lean production. And perhaps not surprisingly, their description of the lean approach seems almost tailor-made for relocalized manufacturing on the Emilia-Romagna model:

> The essence of the lean approach is that in almost all modern manufacturing, the combined and often synergistic benefits of the lower capital investment, greater flexibility, often higher reliability, lower inventory cost, and lower shipping cost of much smaller and more localized production equipment will far outweigh any modest decreases in its narrowly defined "efficiency" per process step. It's more efficient overall, in resources and time and money, to scale production properly, using flexible machines that can quickly shift between products. By doing so, all the different processing steps can be carried out immediately adjacent to one another with the product kept in continuous flow. The goal is to have no stops, no delays, no backflows, no inventories, no expediting, no bottlenecks, no buffer stocks, and no *muda* [i.e., waste or superfluity].[4]

Decentralizing technologies undermined the rationale for large scale not only in mass-production industries, but in continuous-processing industries. In steel, for example, the introduction of the minimill with electric-arc furnace eliminated

[1] Womack, *Lean Thinking*, p. 64.

[2] *Ibid.*, p. 244.

[3] Husman, "Open Source Automobile," *GrimReader*, March 3, 2005 <http://www.zianet.com/ehusman/weblog/2005/03/open-source-automobile.html>.

[4] Paul Hawken, Amory Lovins, and L. Hunter Lovins, *Natural Capitalism: Creating the Next Industrial Revolution* (Boston, New York, London: Little, Brown and Company, 1999), pp. 129-130.

the need for operating on a large enough scale to keep a blast furnace in continuous operation. Not only did the minimill make it possible to scale steel production to the local industrial economy, but it processed scrap metal considerably more cheaply than conventional blast furnaces processed iron ore.[1]

[1]Piore and Sabel, p. 209.

SIDEBAR
MARXIST OBJECTIONS TO NON-CAPITALIST MARKETS: THE RELEVANCE OF THE DECENTRALIZED INDUSTRIAL MODEL

In opposing a form of socialism centered on cooperatives and non-capitalist markets, a standard argument of Marxists and other non-market socialists is that it would be unsustainable and degenerate into full-blown capitalism: "What happens to the losers?" Non-capitalist markets would eventually become capitalistic, through the normal operation of the laws of the market. Here's the argument as stated by Christian Siefkes, a German Marxist active in the P2P movement, on the Peer to Peer Research List:

> Yes, they would trade, and initially their trading wouldn't be capitalistic, since labor is not available for hire. But assuming that trade/exchange is their primary way of organizing production, capitalism would ultimately result, since some of the producers would go bankrupt, they would lose their direct access to the means of production and be forced to sell their labor power. If none of the other producers is rich enough to hire them, they would be unlucky and starve (or be forced to turn to other ways of survival such as robbery/thievery, prostitutioing—which is what we also saw as a large-scale phenomenon with the emergence of capitalism, and which we still see in so-called developing countries where there is not enough capital to hire all or most of the available labor power). But if there are other producers/people would *can* hire them, the seed of capitalism with it's capitalist/worker divide is laid.
>
> Of course, the emerging class of capitalists won't be just passive bystanders watching this process happen. Since they need a sufficiently large labor force, and since independent producers are unwanted competition for them, they'll actively try to turn the latter into the former. Means for doing so are enclosure/privatization laws that deprive the independent producers of their means of productions, technical progress that makes it harder for them to compete (esp. if expensive machines are required which they simple lack the money to buy), other laws that increase the overhead for independent producers (e.g. high bookkeeping requirements), creation of big sales points that non-capitalist producers don't have access to (department stores etc.), simple overproduction that drives small-scale producers (who can't stand huge losses) out of the market, etc. But even if they were passive bystanders (which is an unrealistic assumption), the conversion of independent producers into workers forced to sell their labor power would still take place through the simple laws of the market, which cause some producers to fail and go bankrupt.
>
> So whenever you start with trade as the primary way of production, you'll sooner or later end up with capitalism. It's not a contradiction, it's a process.[1]

One answer, in the flexible production model, is that there's no reason to have any permanent losers. First of all, the overhead costs are so low that it's possible to ride out a slow period indefinitely. Second, in low-overhead flexible production, in which the basic machinery for production is widely affordable and can be easily reallocated to new products, there's really no such thing as a "business" to go out of. The lower the capitalization required for entering the market, and the lower the overhead to be borne in periods of slow business, the more the labor market takes on a networked, project-oriented character—like, e.g., peer production of soft-

[1]Christian Siefkes, "[p2p-research] Fwd: Launch of Abundance: The Journal of Post-Scarcity Studies, preliminary plans," Peer to Peer Research List, February 25, 2009 <http://listcultures.org/pipermail/p2presearch_listcultures.org/2009-February/001555.html>.

ware. In free software, and in any other industry where the average producer owns a full set of tools and production centers mainly on self-managed projects, the situation is likely to be characterized not so much by the entrance and exit of discrete "firms" as by a constantly shifting balance of projects, merging and forking, and with free agents constantly shifting from one to another. The same fluidity prevails, according to Piore and Sabel, in the building trades and the garment industry.[1]

Another point: in a society where most people own the roofs over their heads and can meet a major part of their subsistence needs through home production, workers who own the tools of their trade can afford to ride out periods of slow business, and to be somewhat choosy in waiting to contract out to the projects most suited to their preference. It's quite likely that, to the extent some form of wage employment still existed in a free economy, it would take up a much smaller share of the total economy, wage labor would be harder to find, and attracting it would require considerably higher wages; as a result, self-employment and cooperative ownership would be much more prevalent, and wage employment would be much more marginal. To the extent that wage employment continued, it would be the province of a class of itinerant laborers taking jobs of work when they needed a bit of supplementary income or to build up some savings, and then periodically retiring for long periods to a comfortable life living off their own homesteads. This pattern—living off the common and accepting wage labor only when it was convenient—was precisely what the Enclosures were intended to stamp out.

For the same reason, the standard model of "unemployment" in American-style mass-production industry is in fact quite place-bound, and largely irrelevant to flexible manufacture in European-style industrial districts. In such districts, and to a considerable extent in the American garment industry, work-sharing with reduced hours is chosen in preference to layoffs, so the dislocations from an economic downturn are far less severe. Unlike the American presumption of a fixed and permanent "shop" as the central focus of the labor movement, the industrial district assumes the solidaristic craft community as the primary long-term attachment for the individual worker, and the job site at any given time as a passing state of affairs.[2]

And finally, in a relocalized economy of small-scale production for local markets, where most money is circulated locally, there is apt to be far less of a tendency toward boom-bust cycles or wild fluctuations in commodity prices. Rather, there is likely to be a fairly stable long-term matching of supply to demand.

In short, the Marxist objection assumes the high-overhead industrial production model as "normal," and judges cooperative and peer production by their ability to adapt to circumstances that almost certainly wouldn't exist.

[1]Piore and Sabel, *The Second Industrial Divide*, pp. 117-118.
[2]*Ibid.*, pp. 120-121.

5

The Small Workshop, Desktop Manufacturing, and Household Production

A. Neighborhood and Backyard Industry

A recurring theme among early writers on decentralized production and the informal and household economies is the community workshop, and its use in particular for repair and recycling. Even in the 1970s, when the price of the smallest machine tools was much higher in real terms, it was feasible by means of cooperative organization to spread the capital outlay cost over a large pool of users.

Kirkpatrick Sale speculated that neighborhood recycling and repair centers would put back into service the almost endless supply of defunct appliances currently sitting in closets or basements—as well as serving as "remanufacturing centers" for (say) diesel engines and refrigerators.[1]

Writing along similar lines, Colin Ward suggested "the pooling of equipment in a neighborhood group."

> Suppose that each member of the group had a powerful and robust basic tool, while the group as a whole had, for example, a bench drill, lathes and a saw bench to relieve the members from the attempt to cope with work which required these machines with inadequate tools of their own, or wasting their resources on under-used individually-owned plant. This in turn demands some kind of building to house the machinery: the Community Workshop.
>
> But is the Community Workshop idea nothing more than an aspect of the leisure industry, a compensation for the tedium of work?[2]

In other words, is it just a "hobby"? Ward argued, to the contrary, that it would bridge the growing gap between the worlds of work and leisure by making productive activity in one's free time a source of real use-value.

> Could [the unemployed] make a livelihood for themselves today in the community workshop? If the workshop is conceived merely as a social service for 'creative leisure' the answer is that it would probably be against the rules. . . . But if the workshop were conceived on more imaginative lines than any existing venture of this kind, its potentialities could become a source of livelihood in the truest sense. In several of the New Towns in Britain, for example, it has been found necessary and desirable to build groups of small workshops for individuals and small businesses engaged in such work as repairing electrical equipment or car bodies, woodworking and the manufacture of small components. The Commu-

[1]Kirkpatrick Sale, *Human Scale* (New York: Coward, McCann, & Geoghegan, 1980), p. 406.

[2]Colin Ward, *Anarchy in Action* (London: Freedom Press, 1982), p. 94.

nity Workshop would be enhanced by its cluster of separate workplaces for 'gainful' work. Couldn't the workshop become the community factory, providing work or a place for work for anyone in the locality who wanted to work that way, not as an optional extra to the economy of the affluent society which rejects an increasing proportion of its members, but as one of the prerequisites of the worker-controlled economy of the future?

> Keith Paton . . . , in a far-sighted pamphlet addressed to members of the Claimants' Union, urged them not to compete for meaningless jobs in the economy which has thrown them out as redundant, but to use their skills to serve their own community. (One of the characteristics of the affluent world is that it denies its poor the opportunity to feed, clothe, or house *themselves*, or to meet their own and their families' needs, except from grudgingly doled-out welfare payments). He explains that:
>
> > . . . [E]lectrical power and 'affluence' have brought a spread of intermediate machines, some of them very sophisticated, to ordinary working class communities. Even if they do not own them (as many claimants do not) the possibility exists of borrowing them from neighbours, relatives, ex-workmates. Knitting and sewing machines, power tools and other do-it-yourself equipment comes in this category. Garages can be converted into little workshops, home-brew kits are popular, parts and machinery can be taken from old cars and other gadgets. If they saw their opportunity, trained metallurgists and mechanics could get into advanced scrap technology, recycling the metal wastes of the consumer society for things which could be used again regardless of whether they would fetch anything in a shop. Many hobby enthusiasts could begin to see their interests in a new light.[1]

Karl Hess also discussed community workshops—or as he called them, "shared machine shops"—in *Community Technology*.

> The machine shop should have enough basic tools, both hand and power, to make the building of demonstration models or test facilities a practical and everyday activity. The shared shop might just be part of some other public facility, used in its off hours. Or the shop might be separate and stocked with cast-off industrial tools, with tools bought from government surplus through the local school system . . . Work can, of course, be done as well in home shops or in commercial shops of people who like the community technology approach. . . .
>
> Thinking of such a shared workshop in an inner city, you can think of its use . . . for the maintenance of appliances and other household goods whose replacement might represent a real economic burden in the neighborhood. . . .
>
> . . . The machine shop could regularly redesign cast-off items into useful ones. Discarded refrigerators, for instance, suggest an infinity of new uses, from fish tanks, after removing doors, to numerous small parts as each discarded one is stripped for its components, which include small compressors, copper tubing, heat transfer arrays, and so on. The same goes for washing machines. . . .[2]

Hess's choice of words, by the way, evidenced a failure to anticipate the extent to which flexible networked manufacturing would blur the line between "demonstration models" or test facilities and serial production.

Sharing is a way of maximizing the utilization of idle productive goods owned by individuals. Just about any tool or appliance you need for a current project, but lack, is probably gathering dust on the shelf of someone within a few blocks of where you live. If the pooling of such idle resources doesn't seem like much of a deal for the person with the unused appliances, keep in mind first that he isn't get-

[1]Keith Paton, *The Right to Work or the Fight to Live?* (Stoke-on-Trent, 1972), in Ward, *Anarchy in Action*, pp. 108-109.

[2]Karl Hess, *Community Technology*, pp. 96-97.

ting anything at all out of them now, second that he may trade access to them for access to other people's tools that *he* needs, and third that the arrangement may increase the variety of goods and services he has to choose from outside the wage system.

The same idea has appeared in the San Francisco Bay area, albeit in a commercial rather than communitarian form, as TechShop:[1]

> TechShop is a 15,000 square-foot membership-based workshop that provides members with access to tools and equipment, instruction, and a creative and supportive community of like-minded people so you can build the things you have always wanted to make. . . .
>
> TechShop provides you with access to a wide variety of machinery and tools, including milling machines and lathes, welding stations and a CNC plasma cutter, sheet metal working equipment, drill presses and band saws, industrial sewing machines, hand tools, plastic and wood working equipment including a 4' x 8' ShopBot CNC router, electronics design and fabrication facilities, Epilog laser cutters, tubing and metal bending machines, a Dimension SST 3-D printer, electrical supplies and tools, and pretty much everything you'd ever need to make just about anything.

Hess linked his idea for a shared machine shop to another idea, "[s]imilar in spirit," the shared warehouse:

> A community decision to share a space in which discarded materials can be stored, categorized, and made easily available is a decision to use an otherwise wasted resource. . . .
>
> The shared warehouse . . . should collect a trove of bits and pieces of building materials. . . . There always seems to be a bundle of wood at the end of any project that is too good to burn, too junky to sell, and too insignificant to store. Put a lot of those bundles together and the picture changes to more and more practical possibilities of building materials for the public space.
>
> Spare parts are fair game for the community warehouse. Thus it can serve as a parts cabinet for the community technology experimenter. . . .
>
> A problem common to many communities is the plight of more resources leaving than coming back in. . . . The shared work space and the shared warehouse space involve a community in taking a first look at this problem at a homely and nonideological level.[2]

This ties in closely with Jane Jacobs' recurring themes of the development of local, diversified economies through the discovery of creative uses for locally generated waste and byproducts, and the use of such innovative technologies to replace imports.[3]

E. F. Schumacher recounted his experiences with the Scott Bader Commonwealth, encouraging (often successfully) the worker-owners to undertake such ventures as a community auto repair shop, communally owned tools and other support for household gardening, a community woodworking shop for building and repairing furniture, and so forth. The effect of such measures was to take off some of the pressure to earn wages, so that workers might scale back their work hours.[4]

[1]<http://techshop.ws/>.

[2]Karl Hess, *Community Technology* (New York, Cambridge, Hagerstown, Philadelphia, San Francisco, London, Mexico City, Sao Paulo, Sydney: Harper & Row, Publishers, 1979), pp. 96-98.

[3]Jane Jacobs, *The Economy of Cities* (New York: Vintage Books, 1969, 1970)

[4]E. F. Schumacher, *Good Work* (New York, Hagerstown, San Fransisco, London: Harper & Row, 1979), pp. 80-83.

The potential for such common workspaces increases by an order of magnitude, of course, with the kinds of small, cheap, computerized machine tools we will consider later in this chapter.

The building, bottom-up, of local economies based on small-scale production with multiple-purpose machinery might well take place piecemeal, beginning with such small shops, at first engaged primarily in repair and remanufacture of existing machinery and appliances. As Peak Oil and the degradation of the national transportation system cause corporate logistic chains for spare parts to dry up, small garage and backyard machine shops may begin out of sheer necessity to take up the slack, custom-machining the spare parts needed to keep aging appliances in operation. From this, the natural progression would be to farming out the production of components among a number of such small shops, and perhaps designing and producing simple appliances from scratch. (An intermediate step might be "mass customization," the custom design of modular accessories for mass-produced platforms.) In this manner, networked production of spare parts by small shops might be the foundation for a new industrial revolution.

As Jacobs described it, the Japanese bicycle industry had its origins in just such networking between custom producers of spare parts.

> To replace these imports with locally made bicycles, the Japanese could have invited a big American or European bicycle manufacturer to establish a factory in Japan . . . Or the Japanese could have built a factory that was a slavish imitation of a European or American bicycle factory. They would have had to import most or all of the factory's machinery, as well as hiring foreign production managers or having Japanese production managers trained abroad. . . .
>
> . . . [Instead], shops to repair [imported bicycles] had sprung up in the big cities. . . . Imported spare parts were expensive and broken bicycles were too valuable to cannibalize the parts. Many repair shops thus found it worthwhile to make replacement parts themselves—not difficult if a man specialized in one kind of part, as many repairmen did. In this way, groups of bicycle repair shops were almost doing the work of manufacturing entire bicycles. That step was taken by bicycle assemblers, who bought parts, on contract, from repairmen: the repairmen had become "light manufacturers."[1]

Karl Hess and David Morris, in *Neighborhood Power*, suggested a progression from retail to repair to manufacturing as the natural model for a transition to relocalized manufacturing. They wrote of a process by which "repair shops begin to transform themselves into basic manufacturing facilities . . . "[2] Almost directly echoing Jacobs, they envisioned a bicycle collective's retail shop adding maintenance facilities, and then:

> After a number of people have learned the skills in repairs in a neighborhood, a factory could be initiated to produce a few vital parts, like chains or wheels or tires. Finally, if the need arises, full-scale production of bicycles could be attempted.

Interestingly enough, Don Tapscott and Anthony Williams describe just such a process taking place in micromanufacturing facilities (about which more below) which have been introduced in the Third World. Indian villagers are using fab labs

[1]Jacobs, *Economy of Cities*, pp. 63-64.

[2]Karl Hess and David Morris, *Neighborhood Power: The New Localism* (Boston: Beacon Press, 1975), p. 69.

(again, see below) "to make replacement gears for out-of date copying machines. . . ."[1]

The same process could be replicated in many areas of production. Retail collectives might support community-supported agriculture as a primary source of supply, followed by a small canning factory and then by a glass recycling center to trade broken bottles and jars for usable ones on an arrangement with the bottling companies.[2] Again, the parallels with Jane Jacobs are striking:

> Cities that replace imports significantly replace not only finished goods but, concurrently, many, many items of producers' goods and services. They do it in swiftly emerging, logical chains. For example, first comes the local processing of fruit preserves that were formerly imported, then the production of jars or wrappings formerly imported for which there was no local market of producers until the first step had been taken. Or first comes the assembly of formerly imported pumps for which, once the assembly step has been taken, parts are imported; then the making of parts for which metal is imported; then possibly even the smelting of metal for these and other import-replacements. The process pays for itself as it goes along. When Tokyo went into the bicycle business, first came repair work cannibalizing imported bicycles, then manufacture of some of the parts most in demand for repair work, then manufacture of still more parts, finally assembly of whole, Tokyo-made bicycles. And almost as soon as Tokyo began exporting bicycles to other Japanese cities, there arose in some of those customer cities much the same process of replacing bicycles imported from Tokyo, . . . as had happened with many items sent from city to city in the United States.[3]

A directly analogous process of import substitution can take place in the informal economy, with production for barter at the household and neighborhood level using household capital goods (about which more below) replacing the purchase of consumption goods in the wage economy.

Paul and Percival Goodman wrote, in *Communitas*, of the possibility of decentralized machining of parts by domestic industry, given the universal availability of power and the ingenuity of small machinery, coupled with assembly at a centralized location. It is, they wrote, "almost always cheaper to transport material than men."[4]

A good example of this phenomenon in practice is the Japanese "shadow factories" during World War II. Small shops attached to family homes played an important role in the Japanese industrial economy, according to Nicholas Wood. Many components and subprocesses were farmed out for household manufacture, in home shops consisting of perhaps a few lathes, drill presses or milling machines. In the war, the government had actively promoted such "shadow factories," distributing machine tools in workers' homes in order to disperse concentrated industry and reduce its vulnerability to American strategic bombing.[5] After the war, the government encouraged workers to purchase the machinery.[6] As late as the late fifties, such home manufacturers were still typically tied to particular companies, in what amounted to industrial serfdom. But according to Wood, by the time

[1]Don Tapscott and Anthony D. Williams, *Wikinomics: How Mass Collaboration Changes Everything* (New York: Portfolio, 2006), p. 213.

[2]Hess and Morris, p. 142.

[3]Jacobs, *Cities and the Wealth of Nations: Principles of Economic* Life (New York: Vintage Books, 1984), p. 38.

[4]p. 83.

[5]Nicholas Wood, "The 'Family Firm'—Base of Japan's Growing Economy," *The American Journal of Economics and Sociology*, vol. 23 no. 3 (1964), p. 316.

[6]*Ibid.*, p. 319.

of his writing (1964), many home manufacturers had become free agents, contracting out to whatever firm made the best offer.[1] The overhead costs of home production, after the war, were reduced by standardization and modular design. For example, household optical companies found it impossible at first to produce and stock the many sizes of lenses and prisms for the many different models. But subsequently all Japanese companies standardized their designs to a few models.[2]

A similar shadow factory movement emerged in England during the war, as described by Goodman: "Home manufacture of machined parts was obligatory in England during the last war because of the bombings, and it succeeded."[3]

The Chinese pursued a system of localized production along roughly similar lines in the 1970s. According to Lyman van Slyke, they went a long way toward meeting their small machinery needs in this way. This was part of a policy known as the "Five Smalls," which involved agricultural communes supplying their own needs locally (hydroelectric energy, agro-chemicals, cement, iron and steel smelting, and machinery) in order to relieve large-scale industry of the burden. In the case of machinery, specifically, van Slyke gives the example of the hand tractor:

> . . . [O]ne of the most commonly seen pieces of farm equipment is the hand tractor, which looks like a large rototiller. It is driven in the field by a person walking behind it. . . . This particular design is common in many parts of Asia, not simply in China. Now, at the small-scale level, it is impossible for these relatively small machine shops and machinery plants to manufacture all parts of the tractor. In general, they do not manufacture the engine, the headlights, or the tires, and these are imported from other parts of China. But the transmission and the sheet-metal work and many of the other components may well be manufactured at the small plants. Water pumps of a variety of types, both gasoline and electric, are often made in such plants, as are a variety of other farm implements, right down to simple hand tools. In addition, in many of these shops, a portion of plant capacity is used to build machine tools. That is, some lathes and drill presses were being used not to make the farm machinery but to make additional lathes and drill presses. These plants were thus increasing their own future capabilities at the local level. Equally important is a machinery-repair capability. It is crucial, in a country where there isn't a Ford agency just down the road, that the local unit be able to maintain and repair its own equipment. Indeed, in the busy agricultural season many small farm machinery plants close down temporarily, and the work force forms mobile repair units that go to the fields with spare parts and tools in order to repair equipment on the spot.
>
> Finally, a very important element is the training function played in all parts of the small-scale industry spectrum, but particularly in the machinery plants. Countless times we saw two people on a machine. One was a journeyman, the regular worker, and the second was an apprentice, a younger person, often a young woman, who was learning to operate the machine.[4]

It should be stressed that this wasn't simply a repeat of the disastrous Great Leap Forward, which was imposed from above in the late 1950s. It was, rather, an example of local ingenuity in filling a vacuum left by the centrally planned economy. If anything, in the 1970s—as opposed to the 1950s—the policy was considered a painful concession to necessity, to be abandoned as soon as possible, rather than

[1]*Ibid.*, p. 317.

[2]*Ibid.*, p. 318.

[3]*Paul Goodman, People or Personnel*, in *People or Personnel* and *Like a Conquered Province* (New York: Vintage Books, 1965, 1967, 1968), p. 95.

[4]Lyman P. van Slyke, "Rural Small-Scale Industry in China," in Richard C. Dorf and Yvonne L. Hunter, eds., *Appropriate Visions: Technology the Environment and the Individual* (San Francisco: Boyd & Fraser Publishing Company, 1978) pp. 193-194.

a vision pursued for its own sake. Van Slyke was told by those responsible for small-scale industry, "over and over again," that their goals were to move "from small to large, from primitive to modern, and from here-and-there to everywhere."[1] Aimin Chen reported in 2002 that the government was actually cracking down on local production under the "Five Smalls" in order to reduce idle capacity in the beleaguered state sector.[2] The centrally planned economy under state socialism, like the corporate economy, can only survive by suppressing small-scale competition.

The raw materials for such relocalized production are already in place in most neighborhoods, to a large extent, in the form of unused or underused appliances, power tools gathering dust in basements and garages, and the like. It's all just waiting to be integrated onto a local economy, as soon as producers can be hooked up to needs, and people realize that every need met by such means reduces their dependence on wage labor by an equal amount—and probably involves less labor and more satisfaction than working for the money. The problem is figuring out what's lying around, who has what skills, and how to connect supply to demand. As Hess and Morris put it,

> In one block in Washington, D.C., such a survey uncovered plumbers, electricians, engineers, amateur gardeners, lawyers, and teachers. In addition, a vast number of tools were discovered; complete workshops, incomplete machine-tool shops, and extended family relationships which added to the neighborhood's inventory—an uncle in the hardware business, an aunt in the cosmetics industry, a brother teaching biology downtown. The organizing of a directory of human resources can be an organizing tool itself.[3]

Arguably the neighborhood workshop and the household microenterprise (which we will examine later in this chapter) achieve an optimal economy of scale, determined by the threshold at which a household producer good is fully utilized, but the overhead for a permanent hired staff and a stand-alone dedicated building is not required.

The various thinkers quoted above wrote on community workshops at a time when the true potential of small-scale production machinery was just starting to emerge.

B. The Desktop Revolution and Peer Production in the Immaterial Sphere

Since the desktop revolution of the 1970s, computers have promised to be a decentralizing force on the same scale as electrical power a century earlier. The computer, according to Michel Piore and Charles Sabel, is "a machine that meets Marx's definition of an artisan's tool: it is an instrument that responds to and extends the productive capacities of the user."

> It is therefore tempting to sum the observations of engineers and ethnographers to the conclusion that technology has ended the domination of specialized machines over un- and semiskilled workers, and redirected progress down the path

[1] *Ibid.*, p. 196.

[2] Aimin Chen, "The structure of Chinese industry and the impact from China's WTO entry," *Comparative Economic Studies* (Spring 2002) <http://www.entrepreneur.com/tradejournals/article/print/86234198.html>.

[3] Hess and Morris, *Neighborhood Power*, p. 127.

of craft production. The advent of the computer restores human control over the production process; machinery again is subordinated to the operator.[1]

As Johan Soderberg argues, "[t]he universally applicable computer run on free software and connected to an open network . . . have [sic] in some respects leveled the playing field. Through the global communication network, hackers are matching the coordinating and logistic capabilities of state and capital."[2]

Indeed, the computer itself is the primary item of capital equipment in a growing number of industries, like music, desktop publishing and software design. The desktop computer, supplemented by assorted packages of increasingly cheap printing or sound editing equipment, is capable of doing what previously required a minimum investment of hundreds of thousands of dollars.

The growing importance of human capital, and the implosion of capital outlay costs required to enter the market, have had revolutionary implications for production in the immaterial sphere. In the old days, the immense outlay for physical assets was the primary basis for the corporate hierarchy's power, and in particular for its control over human capital and other intangible assets.

As Luigi Zingales observes, the declining importance of physical assets relative to human capital has changed this. Physical assets, "which used to be the major source of rents, have become less unique and are not commanding large rents anymore." And "the demand for process innovation and quality improvement . . . can only be generated by talented employees," which increases the importance of human capital.[3] This is even more true since Zingales wrote, with the rise of what has been variously called the wikified workplace,[4] the hyperlinked organization,[5] etc. What Niall Cook calls Enterprise 2.0[6] is the application of the networked platform technologies (blogs, wikis, etc.) associated with Web 2.0 to the internal organization of the business enterprise. It refers to the spread of self-managed peer network organization *inside* the corporation, with the internal governance of the corporation increasingly resembling the organization of the Linux developer community.

Tom Peters remarked in quite similar language, some six years earlier in *The Tom Peters Seminar*, on the changing balance of physical and human capital. Of *Inc.* magazine's 500 top-growth companies, which included a good number of information, computer technology and biotech firms, 34% were launched on initial capital of less than $10,000, 59% on less than $50,000, and 75% on less than $100,000.[7] The only reason those companies remain viable is that they control the value created by their human capital. And the only way to do that is through the ownership of artificial property rights like patents, copyrights and trademarks.

[1]Piore and Sabel, p. 261.

[2]Johan Soderberg, *Hacking Capitalism: The Free and Open Source Software Movement* (New York and London: Routledge, 2008), p. 2.

[3]Luigi Zingales, "In Search of New Foundations," *The Journal of Finance*, vol. lv, no. 4 (August 2000), pp. 1641-1642.

[4]Don Tapscott and Anthony D. Williams, *Wikinomics: How Mass Collaboration Changes Everything* (New York: Portfolio, 2006), pp. 239-267.

[5]Chapter Five, "The Hyperlinked Organization," in Rick Levine, Christopher Locke, Doc Searls and David Weinberger. *The Cluetrain Manifesto: The End of Business as Usual* (Perseus Books Group, 2001) <http://www.cluetrain.com/book/ index.html>.

[6]Niall Cook, *Enterprise 2.0: How Social Software Will Change the Future of Work* (Burlington, Vt.: Gower, 2008).

[7]Tom Peters. *The Tom Peters Seminar: Crazy Times Call for Crazy Organizations* (New York: Vintage Books, 1994), p. 35.

In many information and culture industries, the initial outlay for entering the market in the broadcast days was in the hundreds of thousands of dollars or more. The old broadcast mass media, for instance, were "typified by high-cost hubs and cheap, ubiquitous, reception-only systems at the end. This led to a limited range of organizational models for production: those that could collect sufficient funds to set up a hub."[1] The same was true of print periodicals, with the increasing cost of printing equipment from the mid-nineteenth century on serving as the main entry barrier for organizing the hubs. Between 1835 and 1850, the typical startup cost of a newspaper increased from $500 to $100,000—or from roughly $10,000 to $2.38 million in 2005 dollars.[2]

The networked economy, in contrast, is distinguished by "network architecture and the [low] cost of becoming a speaker."

> The first element is the shift from a hub-and-spoke architecture with unidirectional links to the end points in the mass media, to distributed architecture with multidirectional connections among all nodes in the networked information environment. The second is the practical elimination of communications costs as a barrier to speaking across associational boundaries. Together, these characteristics have fundamentally altered the capacity of individuals, acting alone or with others, to be active participants in the public sphere as opposed to its passive readers, listeners, or viewers.[3]

In the old days, the owners of the hubs—CBS News, the Associated Press, etc.—decided what you could hear. Today you can set up a blog, or record a podcast, and anybody in the world who cares enough to go to your URL can look at it free of charge (and anyone who agrees with it—or wants to tear it apart—can provide a hyperlink to his readers).

The central change that makes these things possible is that "the basic physical capital necessary to express and communicate human meaning is the connected personal computer."

> The core functionalities of processing, storage, and communications are widely owned throughout the population of users. . . . The high capital costs that were a prerequisite to gathering, working, and communicating information, knowledge, and culture, have now been widely distributed in the society. The entry barrier they posed no longer offers a condensation point for the large organizations that once dominated the information environment.[4]

The desktop revolution and the Internet mean that the minimum capital outlay for entering most of the entertainment and information industry has fallen to a few thousand dollars at most, and the marginal cost of reproduction is zero. If anything that overstates the cost of entry in many cases, considering how rapidly computer value depreciates and the relatively miniscule cost of buying a five-year-old computer and adding RAM.

The networked environment, combined with endless varieties of cheap software for creating and editing content, makes it possible for the amateur to produce output of a quality once associated with giant publishing houses and recording companies.[5] That is true of the software industry, desktop publishing, and to a

[1]Yochai Benkler, *The Wealth of Networks: How Social Production Transforms Markets and Freedom* (New Haven and London: Yale University Press, 2006), p. 179.

[2]*Ibid.*, p. 188.

[3]*Ibid.*, pp. 212-13.

[4]*Ibid.*, pp. 32-33.

[5]*Ibid.*, p. 54.

certain extent even to film (as witnessed by affordable editing technology and the success of *Sky Captain*).

In the case of the music industry, thanks to cheap equipment and software for high quality recording and sound editing, the costs of independently producing and distributing a high-quality album have fallen through the floor. Bassist Steve Lawson writes:

> . . . [T]he recording process—studio time and expertise used to be hugely expensive. But the cost of recording equipment has plummeted, just as the quality of the same has soared. Sure, expertise is still chargeable, but it's no longer a non-negotiable part of the deal. A smart band with a fast computer can now realistically make a release quality album-length body of songs for less than a grand. . . .
>
> What does this actually mean? Well, it means that for me—and the hundreds of thousands of others like me—the process of making and releasing music has never been easier. The task of finding an audience, of seeding the discovery process, has never cost less or been more fun. It's now possible for me to update my audience and friends (the cross-over between the two is happening on a daily basis thanks to social media tools) about what I'm doing—musically or otherwise—-and to hear from them, to get involved in their lives, and for my music to be inspired by them. . . .
>
> So, if things are so great for the indies, does that mean loads of people are making loads of money? Not at all. But the false notion there is that any musicians were before! We haven't moved from an age of riches in music to an age of poverty in music. We've moved from an age of massive debt and no creative control in music to an age of solvency and creative autonomy. It really is win/win.[1]

As Tom Coates put it, "the gap between what can be accomplished at home and what can be accomplished in a work environment has narrowed dramatically over the last ten to fifteen years."[2]

Podcasting makes it possible to distribute "radio" and "television" programming, at virtually no cost, to anyone with a broadband connection. As radio historian Jesse Walker notes, satellite radio's lackadaisical economic performance doesn't mean people prefer to stick with AM and FM radio; it means, rather, that the ipod has replaced the transistor radio as the primary portable listening medium, and that downloaded files have replaced the live broadcast as the primary form of content.[3]

A network of amateur contributors has peer-produced an encyclopedia, Wikipedia, which Britannica sees as a rival.

It's also true of news, with ever-expanding networks of amateurs in venues like Indymedia, with alternative new operations like those of Robert Parry, Bob Giordano and Greg Palast, and with natives and American troops blogging news firsthand from Iraq—all at the very same time the traditional broadcasting networks are relegating themselves to the stenographic regurgitation of press releases and press conference statements by corporate and government spokespersons, and "reporting" on celebrity gossip. Even conceding that the vast majority of shoe-leather reporting of original news is still done by hired professionals from a traditional journalistic background, blogs and other news aggregators are increasingly

[1]Steve Lawson, "The Future of Music is . . . Indie!" *Agit8*, September 10, 2009 <http://agit8.org.uk/?p=336>.

[2]Tom Coates, "(Weblogs and) The Mass Amateurisation of (Nearly) Everything . . . " Plasticbag.org, September 3, 2003 <http://www.plasticbag.org/archives/2003/09/weblogs_and_the_mass_ amateurisation_of_nearly_everything>.

[3]Jesse Walker, "The Satellite Radio Blues: Why is XM Sirius on the verge of bankruptcy?," *Reason*, February 27, 2009 <http://reason.com/news/show/131905.html>.

becoming the "new newspapers," making better use of reporter-generated content than the old, high-overhead news organizations. But in fact most of the traditional media's "original content" consists of verbatim conveyance of official press releases, which could just as easily be achieved by bloggers and news aggregators linking directly to the press releases at the original institutional sites. Genuine investigative reporting consumes an ever shrinking portion of news organizations' budgets.

The network revolution has drastically lowered the transaction costs of organizing education outside the conventional institutional framework. In most cases, the industrial model of education, based on transporting human raw material to a centrally located "learning factory" for processing, is obsolete. Over thirty years ago Ivan Illich, in *Deschooling Society*, proposed decentralized community learning nets that would put people in contact with the teachers they wished to learn from, and provide an indexed repository of learning materials. The Internet has made this a reality beyond Illich's wildest dreams. MIT's Open Courseware project was one early step in this direction. But most universities, even if they don't have a full database of lectures, at least have some sort of online course catalog with bare-bones syllabi and assigned readings for many individual courses.

A more recent proprietary attempt at the same thing is the online university StraighterLine.[1] Critics like to point to various human elements of the learning process that students are missing, like individualized attention to students with problems grasping the material. This criticism might be valid, if StraighterLine were competing primarily with the intellectual atmosphere of small liberal arts colleges, with their low student-to-instructor ratios. But StraighterLine's primary competition is the community college and state university, and its catalog[2] is weighted mainly toward the kinds of mandatory first- and second-year introductory courses that are taught by overworked grad assistants to auditoriums full of freshmen and sophomores.[3] The cost, around $400 per course,[4] is free of the conventional university's activity fees and all the assorted overhead that comes from trying to manage thousands of people and physical plant at a single location. What's more, StraighterLine offers the option of purchasing live tutorials.[5] *Washington Monthly* describes the thinking behind the business model:

> Even as the cost of educating students fell, tuition rose at nearly three times the rate of inflation. Web-based courses weren't providing the promised price competition—in fact, many traditional universities were charging extra for online classes, tacking a "technology fee" onto their standard (and rising) rates. Rather than trying to overturn the status quo, big, publicly traded companies like Phoenix were profiting from it by cutting costs, charging rates similar to those at traditional universities, and pocketing the difference.
>
> This, Smith explained, was where StraighterLine came in. The cost of storing and communicating information over the Internet had fallen to almost nothing. Electronic course content in standard introductory classes had become a low-cost commodity. The only expensive thing left in higher education was the labor, the price of hiring a smart, knowledgeable person to help students when only a person would do. And the unique Smarthinking call-center model made that much cheaper, too. By putting these things together, Smith could offer in-

[1]<http://www.straighterline.com/>.

[2]<http://www.straighterline.com/courses/>.

[3]Kevin Carey, "College for $99 a Month," *Washington Monthly*, September/October 2009 <http://www.washingtonmonthly.com/college_guide/feature/college_for_99_a_month.php>.

[4]<http://www.straighterline.com/costs/>.

[5]<http://smarthinking.com/static/sampleTutorials/>.

troductory college courses à la carte, at a price that seemed to be missing a digit or two, or three: $99 per month, by subscription. Economics tells us that prices fall to marginal cost in the long run. Burck Smith simply decided to get there first.

StraighterLine, he argues, threatens to do to universities what Craigslist did to newspapers. Freshman intro courses, with auditoriums stuffed like cattle cars and low-paid grad students presiding over the operation, are the cash cow that supports the expensive stuff—like upper-level and grad courses, not to mention a lot of administrative perks. If the cash cow is killed off by cheap competition, it will have the same effect on universities that Craigslist is having on newspapers.[1]

Of course StraighterLine is far costlier and less user-friendly than it might be, if it were peer-organized and open-source. Imagine a similar project with open-source textbooks (or which assigned, with a wink and a nudge, digitized proprietary texts available via a file-sharing network), free lecture materials like those of MIT's Open Courseware, and the creative use of email lists, blogs and wikis for the student community to help each other (much like the use of social networking tools for problem-solving among user communities for various kinds of computers or software).

For that matter, unauthorized course blogs and email lists created by students may have the same effect on StraighterLine that it is having on the traditional university—just as Wikipedia did to Encarta what Encarta did to the traditional encyclopedia industry.

The same model of organization can be extended to fields of employment outside the information and entertainment industries—particularly labor-intensive service industries, where human capital likewise outweighs physical capital in importance. The basic model is applicable in any industry with low requirements for initial capitalization and low or non-existent overhead. Perhaps the most revolutionary possibilities are in the temp industry. In my own work experience, I've seen that hospitals using agency nursing staff typically pay the staffing agency about three times what the agency nurse receives in pay. Cutting out the middleman, perhaps by means of some sort of cross between a workers' co-op and a longshoremen's union hiring hall, seems like a no-brainer. An AFL-CIO organizer in the San Francisco Bay area has attempted just such a project, as recounted by Daniel Levine.[2]

The chief obstacle to such attempts is non-competition agreements signed by temp workers at their previous places of employment. Typically, a temp worker signs an agreement not to work independently for any of the firm's clients, or work for them through another agency, for some period (usually three to six months) after quitting. Of course, this can be evaded fairly easily, if the new cooperative firm has a large enough pool of workers to direct particular assignments to those who aren't covered by a non-competition clause in relation to that particular client.

And as we shall see in the next section, the implosion of capital outlay requirements even for physical production has had a similar effect on the relative importance of human and physical capital, in a considerable portion of manufacturing, and on the weakening of firm boundaries.

These developments have profoundly weakened corporate hierarchies in the information and entertainment industries, and created enormous agency problems

[1]Carey, "College for $99 a Month."

[2]Daniel S. Levine, *Disgruntled: The Darker Side of the World of Work* (New York: Berkley Boulevard Books, 1998), p. 160.

as well. As the value of human capital increases, and the cost of physical capital investments needed for independent production by human capital decreases, the power of corporate hierarchies becomes less and less relevant. As the value of human relative to physical capital increases, the entry barriers become progressively lower for workers to take their human capital outside the firm and start new firms under their own control. Zingales gives the example of the Saatchi and Saatchi advertising agency. The largest block of shareholders, U.S. fund managers who controlled 30% of stock, thought that gave them effective control of the firm. They attempted to exercise this perceived control by voting down Maurice Saatchi's proposed increased option package for himself. In response, the Saatchi brothers took their human capital (in actuality the lion's share of the firm's value) elsewhere to start a new firm, and left a hollow shell owned by the shareholders.[1]

> Interestingly, in 1994 a firm like Saatchi and Saatchi, with few physical assets and a lot of human capital, could have been considered an exception. Not any more. The wave of initial public offerings of purely human capital firms, such as consultant firms, and even technology firms whose main assets are the key employees, is changing the very nature of the firm. Employees are not merely automata in charge of operating valuable assets but valuable assets themselves, operating with commodity-like physical assets.[2]

In another, similar example, the former head of Salomon Brothers' bond trading group formed a new group with former Salomon traders responsible for 87% of the firm's profits.

> . . . if we take the standpoint that the boundary of the firm is the point up to which top management has the ability to exercise power . . . , the group was not an integral part of Salomon. It merely rented space, Salomon's name, and capital, and turned over some share of its profits as rent.[3]

Marjorie Kelly gave the breakup of the Chiat/Day ad agency as an example of the same phenomenon.

> . . . What is a corporation worth without its employees?
>
> This question was acted out . . . in London, with the revolutionary birth of St. Luke's ad agency, which was formerly the London office of Chiat/Day. In 1995, the owners of Chiat/Day decided to sell the company to Omnicon—which meant layoffs were looming and Andy Law in the London office wanted none of it. He and his fellow employees decided to rebel. They phoned clients and found them happy to join the rebellion. And so at one blow, London employees and clients were leaving.
>
> Thus arose a fascinating question: What exactly did the "owners" of the London office now own? A few desks and files? Without employees and clients, what was the London branch worth? One dollar, it turned out. That was the purchase price—plus a percentage of profits for seven years—when Omnicon sold the London branch to Law and his cohorts after the merger. They renamed it St. Luke's. . . . All employees became equal owners . . . Every year now the company is re-valued, with new shares awarded equally to all.[4]

[1]Zingales, "In Search of New Foundations," p. 1641.

[2]*Ibid.*, p. 1641.

[3]Raghuram Rajan and Luigi Zingales, "The Governance of the New Enterprise," in Xavier Vives, ed., *Corporate Governance: Theoretical and Empirical Perspectives* (Cambridge: Cambridge University Press, 2000), pp. 211-212.

[4]Marjorie Kelly, "The Corporation as Feudal Estate" (an excerpt from *The Divine Right of Capital*) *Business Ethics*, Summer 2001. Quoted in *GreenMoney Journal*, Fall 2008 <http://greenmoneyjournal.com/article.mpl?articleid=60&newsletterid=15>.

David Prychitko remarked on the same phenomenon in the tech industry, the so-called "break-away" firms, as far back as 1991:

> Old firms act as embryos for new firms. If a worker or group of workers is not satisfied with the existing firm, each has a skill which he or she controls, and can leave the firm with those skills and establish a new one. In the information age it is becoming more evident that a boss cannot control the workers as one did in the days when the assembly line was dominant. People cannot be treated as workhorses any longer, for the value of the production process is becoming increasingly embodied in the intellectual skills of the worker. This poses a new threat to the traditional firm if it denies participatory organization.
>
> The appearance of break-away computer firms leads one to question the extent to which our existing system of property rights in ideas and information actually protects bosses in other industries against the countervailing power of workers. Perhaps our current system of patents, copyrights, and other intellectual property rights not only impedes competition and fosters monopoly, as some Austrians argue. Intellectual property rights may also reduce the likelihood of break-away firms in general, and discourage the shift to more participatory, cooperative formats.[1]

C. The Expansion of the Desktop Revolution and Peer Production into the Physical Realm

Although peer production first emerged in the immaterial realm—i.e., information industries like software and entertainment—its transferability to the realm of physical production is also a matter of great interest.

1. Open-Source Design: Removal of Proprietary Rents from the Design Stage, and Modular Design

One effect of the shift in importance from tangible to intangible assets is the growing portion of product prices that reflects embedded rents on "intellectual property" and other artificial property rights rather than the material costs of production.

The radical nature of the peer economy, especially as "intellectual property" becomes increasingly unenforceable, lies in its potential to cause the portion of existing commodity price that results from such embedded rents to implode.

Open source hardware refers, at the most basic level, to the development and improvement of designs for physical goods on an open-source basis, with no particular mode of physical production being specified. The design stage ceases to be a source of proprietary value, but the physical production stage is not necessarily affected. To take it in Richard Stallman's terms, 'free speech" only affects the portion of beer's price that results from the cost of a proprietary design phase: open source hardware means the design is free as in free speech, not free beer. Although the manufacturer is not hindered by patents on the design, he must still bear the costs of physical production. Edy Ferreira defined open-source hardware as

> any piece of hardware whose manufacturing information is distributed using a license that provides specific rights to users without the need to pay royalties to the original developers. These rights include freedom to use the hardware for any purpose, freedom to study and modify the design, and freedom to redistribute copies of either the original or modified manufacturing information. . . .

[1]David L Prychitko, *Marxism and Workers' Self-Management: The Essential Tension* (New York; London; Westport, Conn.: Greenwood Press, 1991), p. 121n.

> In the case of open source software (OSS), the information that is shared is software code. In OSH, what is shared is hardware manufacturing information, such as . . . the diagrams and schematics that describe a piece of hardware.[1]

At the simplest level, a peer network may develop a product design and make it publicly available; it may be subsequently built by any and all individuals or firms with the necessary production machinery, without coordinating their efforts with the original designer(s). A conventional manufacturer may produce open source designs, with feedback from the user community providing the main source of innovation.

Karim Lakhani describes this general phenomenon, the separation of open-source design from an independent production stage, as "communities driving manufacturers out of the design space," with

> users innovating and developing products that can out compete traditional manufacturers. But this effect is not just limited to software. In physical products . . . , users have been shown to be the dominant source of functionally novel innovations. Communities can supercharge this innovation mechanism. And may ultimately force companies out of the product design space. Just think about it—for any given company—there are more people outside the company that have smarts about a particular technology or a particular use situation then [sic] all the R&D engineers combined. So a community around a product category may have more smart people working on the product then [sic] the firm it self. So in the end manufacturers may end up doing what they are supposed to—manufacture—and the design activity might move . . . into the community.[2]

As one example, Vinay Gupta has proposed a large-scale library of open-source hardware designs as an aid to international development:

> An open library of designs for refrigerators, lighting, heating, cooling, motors, and other systems will encourage manufacturers, particularly in the developing world, to leapfrog directly to the most sustainable technologies, which are much cheaper in the long run. Manufacturers will be encouraged to use the efficient designs because they are free, while inefficient designs still have to be paid for. The library could also include green chemistry and biological solutions to industry challenges. . . . This library should be free of all intellectual property restrictions and open for use by any manufacturer, in any nation, without charge.[3]

One item of his own design, the Hexayurt, is "a refugee shelter system that uses an approach based on "autonomous building" to provide not just a shelter, but a comprehensive family support unit which includes drinking water purification, composting toilets, fuel-efficient stoves and solar electric lighting."[4] The basic construction materials for the floor, walls and roof cost about $200.[5]

[1] "Open Source Hardware," *P2P Foundation Wiki* <http://www.p2pfoundation.net/Open_Source_Hardware>.

[2] Karim Lakhana, "Communities Driving Manufacturers Out of the Design Space," *The Future of Communities Blog*, March 25, 2007 <http://www.futureofcommunities.com/2007/03/25/communities-driving-manufacturers-out-of-the-design-space/>.

[3] Vinay Gupta, "Facilitating International Development Through Free/Open Source," <http://guptaoption.com/5.open_source_development.php> Quoted from Beatrice Anarow, Catherine Greener, Vinay Gupta, Michael Kinsley, Joanie Henderson, Chris Page and Kate Parrot, Rocky Mountain Institute, "Whole-Systems Framework for Sustainable Consumption and Production." Environmental Project No. 807 (Danish Environmental Protection Agency, Ministry of the Environment, 2003), p. 24. <http://files.howtolivewiki.com/A%20Whole%20Systems%20Framework%20for%20Sustainable%20Production%20and%20Consumption.pdf>

[4] <http://www.p2pfoundation.net/Hexayurt>.

[5] <http://hexayurt.com/>.

Michel Bauwens, of the P2P foundation, provides a small list of some of the more prominent open-design projects:

- The Grid Beam Building System, at http://www.p2pfoundation.net/Grid_Beam_Building_Syste
- The Hexayurt, at http://www.p2pfoundation.net/Hexayurt
- Movisi Open Design Furniture, at http://www.p2pfoundation.net/Movisi_Open_Design_Furniture
- Open Cores, at http://www.p2pfoundation.net/Open_Cores and other Open Computing Hardware, at http://www.p2pfoundation.net/Open_Hardware
- Open Source Green Vehicle, at http://www.p2pfoundation.net/Open_Source_Green_Vehicle
- Open Source Scooter http://www.p2pfoundation.net/Open_Source_Scooter
- The Ronja Wireless Device at http://www.p2pfoundation.net/Twibright_Ronja_Open_Wireless_Networking_Device
- Open Source Sewing patterns, at http://www.p2pfoundation.net/Open_Source_Sewing_Patterns
- Velomobiles http://www.p2pfoundation.net/Open_Source_Velomobile_Development_Project
- Open Energy http://www.p2pfoundation.net/SHPEGS_Open_Energy_Project[1]

One of the most ambitious attempts at such an open design project is Open Source Ecology, which is developing an open-source, virally reproducible, vernacular technology-based "Open Village Construction Set" in its experimental site at Factor E Farm.[2] (Of course OSE is also directly involved in the physical implementation of its own designs; it is a manufacturing as well as a design network.)

A more complex scenario involves the coordination of an open source design stage with the production process, with the separate stages of production distributed and coordinated by the same peer network that created the design. Dave Pollard provides one example:

> Suppose I want a chair that has the attributes of an Aeron without the $1800 price tag, or one with some additional attribute (e.g. a laptop holder) the brand name doesn't offer? I could go online to a Peer Production site and create an instant market, contributing the specifications . . . , and, perhaps a maximum price I would be willing to pay. People with some of the expertise needed to produce it could indicate their capabilities and self-organize into a consortium that would keep talking and refining until they could meet this price. . . . Other potential buyers could chime in, offering more or less than my suggested price. Based on the number of 'orders' at each price, the Peer Production group could then accept orders and start manufacturing. . . .
>
> As [Erick] Schonfeld suggests, the intellectual capital associated with this instant market becomes part of the market archive, available for everyone to see, stripping this intellectual capital cost, and the executive salaries, dividends and corporate overhead out of the cost of this and other similar product requests and fulfillments, so that all that is left is the lowest possible cost of material, labour

[1]Michel Bauwens, "What kind of economy are we moving to? 3. A hierarchy of engagement between companies and communities," *P2P Foundation Blog*, October 5, 2007 <http://blog.p2pfoundation.net/what-kind-of-economy-are-we-moving-to-3-a-hierarchy-of-engagement-between-companies-and-communities/2007/10/05>.

[2]Marcin Jakubowski, "Clarifying OSE Vision," *Factor E Farm Weblog*, September 8, 2008 <http://openfarmtech.org/weblog/?p=325>.

and delivery to fill the order. And the order is exactly what the customer wants, not the closest thing in the mass-producer's warehouse.[1]

In any case, the removal of proprietary control over the implementation of designs means that the production phase will be subject to competitive pressure to adopt the most efficient production methods—a marked departure from the present, where "intellectual property" enables privileged producers to set prices as a cost-plus markup based on whatever inefficient production methods they choose.

The most ambitious example of an open-source physical production project is the open source car, or "OScar."

> Can open-source practices and approaches be applied to make hardware, to create tangible and physical objects, including complex ones? Say, to build a car? . . .
>
> Markus Merz believes they can. The young German is the founder and "maintainer" (that's the title on his business card) of the OScar project, whose goal is to develop and build a car according to open-source (OS) principles. Merz and his team aren't going for a super-accessorized SUV—they're aiming at designing a simple and functionally smart car. And, possibly, along the way, reinvent transportation.[2]

As of June 2009, the unveiling of a prototype—a two-seater vehicle powered by hydrogen fuel cells—was scheduled in London.[3]

Well, actually there's a fictional example of an open-source project even more ambitious than the OScar: the open-source moon project, a volunteer effort of a peer network of thousands, in Craig DeLancy's "Openshot." The project's ship (the *Stallman*), built largely with Russian space agency surplus, beats a corporate-funded proprietary project to the moon.[4]

A slightly less ambitious open-source manufacturing project, and probably more relevant to the needs of most people in the world, is Open Source Ecology's open-source tractor (LifeTrac). It's designed for inexpensive manufacture, with modularity and easy disassembly, for lifetime service and low cost repair. It includes, among other things, a well-drilling module, and is designed to serve as a prime mover for machinery like OSE's Compressed Earth Block Press and saw mill.[5]

When physical manufacturing is stripped of the cost of proprietary design and technology, and the consumer-driven, pull model of distribution strips away most of the immense marketing cost, we will find that the portion of price formerly made up of such intangibles will implode, and the remaining price based on actual production cost will be as much as an order of magnitude lower.

Just as importantly, open-source design reduces cost not only by removing proprietary rents from "intellectual property," but by the substantive changes in design that it promotes. Eliminating patents removes legal barriers to the competitive pressure for interoperability and reparability. And interoperability and repara-

[1]Dave Pollard, "Peer Production," *How to Save the World*, October 28, 2005 <http://blogs.salon.com/0002007/2005/10/28.html#a1322>.

[2]Bruno Giussani, "Open Source at 90 MPH," *Business Week*, December 8, 2006 <http://www.businessweek.com/innovate/content/dec2006/id20061208_509041.htm?>. See also the OS Car website, <http://www.theoscarproject.org/>.

[3]Lisa Hoover, "Riversimple to Unveil Open Source Car in London This Month," *Ostatic*, June 11, 2009 <http://ostatic.com/blog/riversimple-to-unveil-open-source-car-in-london-this-month>.

[4]Craig DeLancey, "Openshot," *Analog*, December 2006, pp. 64-74.

[5]"LifeTrac," Open Source Ecology wiki <http://openfarmtech.org/index.php?title=LifeTrac>.

bility promote the kind of modular design that is most conducive to networked production, with manufacture of components distributed among small shops producing a common design.

The advantages of modular design of physical goods are analogous to those in the immaterial realm.

> Current thinking says peer production is only suited to creating information-based goods—those made of bits, inexpensive to produce, and easily subdivided into small tasks and components. Software and online encyclopedias have this property. Each has small discrete tasks that participants can fulfill with very little hierarchical direction, and both can be created with little more than a networked computer.
>
> While it's true that peer production is naturally suited to bit products, it's also true that many of the attributes and advantages of peer production can be replicated for products made of atoms. If physical products are designed to be modular—i.e., they consist of many interchangeable parts that can be readily swapped in or out without hampering the performance of the overall product—then, theoretically at least, large numbers of lightly coordinated suppliers can engage in designing and building components for the product, much like thousands of Wikipedians add to and modify Wikipedia's entries.[1]

This is hardly mere theory, but is reflected in the real-world reality of China's motorcycle industry: "The Chinese approach emphasizes a modular motorcycle architecture that enables suppliers to attach component subsystems (like a braking system) to standard interfaces."[2] And in an open-source world, independent producers could make unauthorized modular components or accessories, as well.

Costs from outlays on physical capital are not a constant, and modular design is one factor that can cause those costs to fall significantly. It enables a peer network to break a physical manufacturing project down into discrete sub-projects, with many of the individual modules perhaps serving as components in more than one larger appliance. According to Christian Siefkes,

> Products that are modular, that can be broken down into smaller modules or components which can be produced independently before being assembled into a whole, fit better into the peer mode of production than complex, convoluted products, since they make the tasks to be handled by a peer project more manageable. Projects can build upon modules produced by others and they can set as their own (initial) goal the production of a specific module, especially if components can be used stand-alone as well as in combination. The Unix philosophy of providing lots of small specialized tools that can be combined in versatile ways is probably the oldest expression in software of this modular style. The stronger emphasis on modularity is another phenomenon that follows from the differences between market production and peer production. Market producers have to prevent their competitors from copying or integrating their products and methods of production so as not to lose their competitive advantage. In the peer mode, re-use by others is good and should be encouraged, since it increases your reputation and the likelihood of others giving something back to you. . . .
>
> Modularity not only facilitates decentralized innovation, but should also help to increase the longevity of products and components. Capitalism has developed a throw-away culture where things are often discarded when they break (instead of being repaired), or when one aspect of them is no longer up-to-date or in fashion. In a peer economy, the tendency in such cases will be to replace just a single component instead of the whole product, since this will generally be

[1]Tapscott and Williams, pp. 219-220.
[2]*Ibid.*, p. 222.

the most labor-efficient option (compared to getting a new product, but also to manually repairing the old one).[1]

Siefkes is wrong only in referring to producers under the existing corporate system as "market producers," since absent "intellectual property" as a legal bulwark to proprietary design, the market incentive would be toward designing products that were interoperable with other platforms, and toward competition in the design of accessories and replacement parts tailored to other companies' platforms. And given the absence of legal barriers to the production of such interoperable accessories, the market incentive would be to designing platforms as broadly interoperable as possible.

This process of modularization is already being promoted within corporate capitalism, although the present system is struggling mightily—and unsuccessfully—to keep itself from being torn apart by the resulting increase in productive forces. As Eric Hunting argues, the high costs of technical innovation, the difficulty of capturing value from it, and the mass customization or long tail market, taken together, create pressures for common platforms that can be easily customized between products, and for modularization of components that can be used for a wide variety of products. And Hunting points out, as we already saw in regard to flexible manufacturing networks in Chapters Four and Five, that the predominant "outsource everything" and "contract manufacturing" model increasingly renders corporate hubs obsolete, and makes it possible for contractees to circumvent the previous corporate principals and undertake independent production on their own account.

> Industrial ecologies are precipitated by situations where traditional industrial age product development models fail in the face of very high technology development overheads or very high demassification in design driven by desire for personalization/customization producing Long Tail market phenomenon [sic]. A solution to these dilemmas is modularization around common architectural platforms in order to compartmentalize and distribute development cost risks, the result being 'ecologies' of many small companies independently and competitively developing intercompatible parts for common product platforms—such as the IBM PC.
>
> The more vertical the market profile for a product the more this trend penetrates toward production on an individual level due [to] high product sophistication coupled to smaller volumes. . . . Competitive contracting regulations in the defense industry (when they're actually respected . . .) tend to, ironically, turn many kinds of military hardware into open platforms by default, offering small businesses a potential to compete with larger companies where production volumes aren't all that large to begin with. Consequently, today we have a situation where key components of some military vehicles and aircraft are produced on a garage-shop production level by companies with fewer than a dozen employees.
>
> All this represents an intermediate level of industrial demassification that is underway today and not necessarily dependent upon open source technology or peer-to-peer activity but which creates a fertile ground for that in the immediate future and drives the complementary trend in the miniaturization of machine tools.[2]

[1]Christian Siefkes, *From Exchange to Contributions: Generalizing Peer Production into the Physical World* Version 1.01 (Berlin, October 2007), pp. 104-105.

[2]Hunting comment under Michel Bauwens, "Phases for implementing peer production: Towards a Manifesto for Mutually Assured Production," P2P Foundation *Forum*, August 30, 2008 <http://p2pfoundation.ning.com/forum/topics/2003008:Topic:6275>.

In other words, the further production cost falls relative to the costs of design, the greater the economic incentive to modular design as a way of defraying design costs over as many products as possible.

In an email to the Open Manufacturing list, Hunting summed up the process more succinctly. Industrial relocalization

> compels the modularization of product design, which results in the replacement of designs by platforms and the competitive commoditization of their components. Today, automobiles are produced as whole products made with large high-capital-cost machinery using materials—and a small portion of pre-made components—transported long distances to a central production site from which the end product is shipped with a very poor transportation efficiency to local sales/distribution points. In the future automobiles may be assembled on demand in the car dealership from modular components which ship with far greater energy efficiency than whole cars and can come from many locations. By modularizing the design of the car to allow for this, that design is changed from a product to a platform for which many competitors, using much smaller less expensive means of production, can potentially produce parts to accommodate customers desire for personalization and to extend the capabilities of the automobile beyond what was originally anticipated. End-users are more easily able to experiment in customization and improvement and pursue entrepreneurship based on this innovation at much lower start-up costs. This makes it possible to implement technologies for the automobile—like alternative energy technology—earlier auto companies may not have been willing to implement because of a lack of competition and because their capital costs for their large expensive production tools and facilities take so long (20 years, typically) to amortize. THIS is the reason why computers, based on platforms for modular commodity components, have evolved so rapidly compared to every other kind of industrial product and why the single-most advanced device the human race has ever produced is now something most anyone can afford and which a child can assemble in minutes from parts sourced around the world.[1]

The beauty of modular design, Hunting writes elsewhere (in the specific case of modular prefab housing), is that the bulk of research and development man-hours are incorporated into the components themselves, which can be duplicated across many different products. The components are smart, but the combinations are dumbed-down and user friendly. A platform is a way to spread the development costs of a single component over as many products as possible.

> But underneath there are these open structural systems that are doing for house construction what the standardized architecture of the IBM PC did for personal computing, encoding a lot of engineering and pre-assembly labor into small light modular components created in an industrial ecology so that, at the high level of the end-user, it's like Lego and things go together intuitively with a couple of hand tools. In the case of the Jeriko and iT houses based on T-slot profiles, this is just about a de-facto public domain technology, which means a zillion companies around the globe could come in at any time and start making compatible hardware. We're tantalizingly close to factoring out the 'experts' in basic housing construction just like we did with the PC where the engineers are all down in the sub-components, companies don't actually manufacture computers they just do design and assemble-on-demand, and now kids can build computers in minutes with parts made all over the world. Within 20 years you'll be going to places like IKEA and Home Depot and designing your own home by picking parts out of catalogs or showrooms, having them delivered by truck, and then assembling

[1]Eric Hunting, "[Open Manufacturing] Re: Why automate? and opinions on Energy Descent?" *Open* Manufacturing, September 22, 2008 <http://groups.google.com/group/openmanufacturing/browse_thread/thread/1f40d031453b94eb>.

most of them yourself with about the same ease you put in furniture and home appliances.[1]

More recently, Hunting wrote of the role of modularized development for common platforms in this history of the computer industry:

> We commonly attribute the rapid shrinking in scale of the computer to the advance of integrated circuit technology. But that's just a small part of the story that doesn't explain the economy and ubiquity of computers. The real force behind that was a radically different industrial paradigm that emerged more-or-less spontaneously in response to the struggle companies faced in managing the complexity of the new technology. Put simply, the computer was too complicated for any one corporation to actually develop independently—not even for multi-national behemoths like IBM that once prided itself on being able to do everything. A radically new way of doing things was needed to make the computer practical.
>
> The large size of early computers was a result not so much of the primitive nature of the technology of the time but on the fact that most of that early technology was not actually specific to the application of computers. It was repurposed from electronic components that were originally designed for other kinds of machines. Advancing the technology to where the vast diversity of components needed could be made and optimized specifically for the computer demanded an extremely high development investment -more than any one company in the world could actually afford. There simply wasn't a big enough computer market to justify the cost of development of very sophisticated parts exclusively for computers. While performing select R&D on key components, early computer companies began to position themselves as systems integrators for components made by sub-contractracted suppliers rather than manufacturing everything themselves. While collectively the development of the full spectrum of components computers needed was astronomically expensive, individually they were quite within the means of small businesses and once the market for computers reached a certain minimum scale it became practical for such companies to develop parts for these other larger companies to use in their products. This was aided by progress in other areas of consumer, communications, and military digital electronics—a general shift to digital electronics—that helped create larger markets for parts also suited to computer applications. The more optimized for computer use subcomponents became, the smaller and cheaper the computer as a whole became and the smaller and cheaper the computer the larger the market for it, creating more impetus for more companies to get involved in computer-specific parts development. ICs were, of course, a very key breakthrough but the nature of their extremely advanced fabrication demanded extremely large product markets to justify. The idea of a microprocessor chip exclusive to any particular computer is actually a rather recent phenomenon even for the personal computer industry. Companies like Intel now host a larger family of concurrently manufactured and increasingly use-specialized microprocessors than was ever imaginable just a decade ago.
>
> For this evolution to occur the nature of the computer as a designed product had to be very different from other products common to industrial production. Most industrial products are monolithic in the sense that they are designed to be manufactured whole from raw materials and very elemental parts in one central mass production facility. But the design of a computer isn't keyed to any one resulting product. It has an 'architecture' that is independent of any physical form. A set of component function and interface standards that define the electronics of a computer system but not necessarily any particular physical configuration. Unlike other technologies, electronics is very mutable. There are an infi-

[1] Hunting, "[Open Manufacturing] Re:Vivarium," *Open Manufacturing*, March 28, 2009 <http://groups.google.com/group/openmanufacturing/browse_thread/thread/a891d6f72243436d/e58d837ac4022484?hl=en&q=vivarium+hunting#>.

nite variety of potential physical configurations of the same electronic circuit. This is why electronics engineering can be based on iconographic systems akin to mathematics—something seen in few other industries to a comparable level of sophistication. (chemical engineering) So the computer is not a product but rather a *platform* that can assume an infinite variety of shapes and accommodate an infinite diversity of component topologies as long as their electronic functions conform to the architecture. But, of course, one has to draw the line somewhere and with computer parts this is usually derived from the topology of standardized component connections and the most common form factors for components. Working from this a computer designer develops configurations of components integrated through a common motherboard that largely defines the overall shape possible for the resulting computer product. Though companies like Apple still defy the trend, even motherboards and enclosures are now commonly standardized, which has ironically actually encouraged diversity in the variety of computer forms and enclosure designs even if their core topological features are more-or-less standardized and uniform.

Thus the computer industry evolved into a new kind of industrial entity; an Industrial Ecology formed of a food-chain of interdependencies between largely independent, competitive, and globally dispersed companies defined by component interfaces making up the basis of computer platform architectures. This food chain extends from discrete electronics components makers, through various tiers of sub-system makers, to the computer manufacturers at the top—though in fact they aren't manufacturing anything in the traditional sense. They just cultivate the platforms, perform systems integration, customer support, marketing, and—decreasingly as even this is outsourced to contract job shops—assemble the final products.

For an Industrial Ecology to exist, an unprecedented degree of information must flow across this food chain as no discrete product along this chain can hope to have a market unless it conforms to interface and function standards communicated downward from higher up the chain. This has made the computer industry more open than any other industry prior to it. Despite the obsessions with secrecy, propriety, and intellectual property among executives, this whole system depends on an open flow of information about architectures, platforms, interfaces standards, software, firmware, and so on—communicated through technical reference guides and marketing material. This information flow exists to an extent seen nowhere else in the Industrial Age culture. . . .

Progressive modularization and interoperability standardization tends to consolidate and simplify component topologies near the top of the food chain. This is why a personal computer is, today, so simple to assemble that a child can do it—or for that matter an end-user or any competitor to the manufacturers at the top. All that ultimately integrates a personal computer into a specific physical form is the motherboard and the only really exclusive aspect of that is its shape and dimensions and an arrangement of parts which, due to the nature of electronics, is topologically mutable independent of function. There are innumerable possible motherboard forms that will still work the same as far as software is concerned. This made the PC an incredibly easy architecture to clone for anyone who could come up with some minor variant of that motherboard to circumvent copyrights, a competitive operating system, a work-around the proprietary aspects of the BIOS, and could dip into that same food chain and buy parts in volume. Once an industrial ecology reaches a certain scale, even the folks at the top become expendable. The community across the ecology has the basic knowledge necessary to invent platforms of its own, establish its own standards bottom-up, and seek out new ways to reach the end-user customer. And this is what happened to IBM when it stupidly allowed itself to become a bottleneck to the progress of the personal computer in the eyes of everyone else in its ecology. That ecology, for sake of its own growth, simply took the architecture of the PC from IBM and established its own derivative standards independent of IBM—and there was nothing even that corporate giant could ultimately do about it. . . .

> . . . Again, this is all an astounding revolution in the way things are supposed to work in the Industrial Age. A great demassification of industrial power and control. Just imagine what the car industry would be like if things worked like this—as well one should as this is, in fact, coming. Increasingly, the model of the computer industry is finding application in a steadily growing number of other industries. Bit by bit, platforms are superceding products and Industrial Ecologies are emerging around them.[1]

The size limitations of fabrication in the small shop, and the lack of facilities for plastic injection molding or sheet metal stamping of very large objects, constitute a further impetus to modular design.

> By virtue of the dimensional limits resulting from the miniaturization of fabrication systems, Post-Industrial design favors modularity following a strategy of maximum diversity of function from a minimum diversity of parts and materials—Min-A-Max. . . .
>
> Post-industrial artifacts tend to exhibit the characteristic of perpetual demountability, leading to ready adaptive reuse, repairability, upgradeability, and recyclability. By extension, they compartmentalize failure and obsolescence to discrete demountable components. A large Post-Industrial artifact can potentially live for as long as its platform can evolve -potentially forever.
>
> A scary prospect for the conventional manufacturer banking on the practice of planned obsolescence. . . . [2]

One specific example Hunting cites is the automobile. It was, more than anything, "the invention of pressed steel welded unibody construction in the 1930s," with its requirement for shaping sheet metal in enormous multi-story stamping presses, that ruled out modular production by a cooperative ecology of small manufacturers. Against that background, Hunting sets the abortive Africar project of the 1980s, with a modular design suitable for networked production in small shops.[3] The Africar had a jeeplike body design; but instead of pressed sheet metal, its surface was put together entirely from components capable of being cut from flat materials (sheet metal or plywood) using subtractive machinery like cutting tables, attached to a structural frame of cut or bent steel.

A more recent modular automobile design project is Local Motors. It's an open design community with all of its thousands of designs shared under Creative Commons licenses. All of them are designed around a common light-weight chassis, which is meant to be produced economically in runs of as little as two thousand. Engines, brakes, batteries and other components are modular, so as to be interchangeable between designs. Components are produced in networks of "microfactories." The total capital outlay required to produce a Local Motors design is a little over a million dollars (compared to hundreds of millions for a conventional auto plant), with minimal inventories and turnaround times a fifth those of conventional Detroit plants.[4]

[1]Hunting, "On Defining a Post-Industrial Style (1): from Industrial blobjects to post-industrial spimes," *P2P Foundation Blog*, November 2, 2009 <http://blog.p2pfoundation .net/on-defining-a-post-industrial-style-1-from-industrial-blobjects-to-post-industrial-spimes/ 2009/11/02>.

[2]"On Defining a Post-Industrial Style (2): some precepts for industrial design," *P2P Foundation Blog*, November 3, 2009 <http://blog.p2pfoundation.net/on-defining-a-post-industrial-style-2-some-precepts-for-industrial-design/2009/11/03>.

[3]"On Defining a Post-Industrial Style (3): Emerging examples," *P2P Foundation Blog*, November 4, 2009 <http://blog.p2pfoundation.net/on-defining-a-post-industrial-style-3-emerging-examples/2009/11/04>.

[4]"Jay Rogers: I Challenge You to Make Cool Cars," Alphachimp Studio Inc., November 10, 2009 <http://www.alphachimp.com/poptech-art/2009/11/10/jay-rogers-i-challenge-you-to-make-cool-cars.html>; Local Motors website at <http://www.local-motors.com>.

Michel Bauwens, in commenting on Hunting's remarks, notes among the "underlying trends . . . supporting the emergence of peer production in the physical world,"

> the 'distribution' of production capacity, i.e. lower capital requirements and modularisation making possible more decentralized and localized production, which may eventually be realized through the free self-aggregation of producers.[1]

Modular design is an example of stigmergic coordination. Stigmergy was originally a concept developed in biology, to describe the coordination of actions between a number of individual organisms through the individual response to markers, without any common decision-making process. Far from the stereotype of the "hive mind," ants—the classic example of biological stigmergy—coordinate their behavior entirely through the individual's reading of and reaction to chemical markers left by other individuals.[2] As defined in the Wikipedia entry, stigmergy is

> a mechanism of spontaneous, indirect coordination between agents or actions, where the trace left in the environment by an action stimulates the performance of a subsequent action, by the same or a different agent. Stigmergy is a form of self-organization. It produces complex, apparently intelligent structures, without need for any planning, control, or even communication between the agents. As such it supports efficient collaboration between extremely simple agents, who lack any memory, intelligence or even awareness of each other.[3]

The development of the platform is a self-contained and entirely self-directed action by an individual or a peer design group. Subsequent modules are developed with reference to the platform, but the design of each module is likewise entirely independent and self-directed; no coordination with the platform developer or the developers of other modules takes place. The effect is to break design down into numerous manageable units.

2. *Reduced Transaction Costs of Aggregating Capital*

We will consider the cheapening of actual physical tools in the next section. But even when the machinery required for physical production is still expensive, the reduction of transaction costs involved in aggregating funds is bringing on a rapid reduction in the cost of physical production. In addition, networked organization increases the efficiency of physical production by making it possible to pool more expensive capital equipment and make use of "spare cycles." This possibility was hinted at by proposals for pooling capital outlays through cooperative organization even back in the 1970s, as we saw in the first section. But the rise of network culture takes it to a new loevel (which, again, we will consider in the next section). As a result, Stallman's distinction between "free speech" and "free beer" is eroding even when tools themselves are costly. Michel Bauwens writes:

> P2P can arise not only in the immaterial sphere of intellectual and software production, but wherever there is access to distributed technology: spare computing cycles, distributed telecommunications and any kind of viral communicator meshwork.

[1]Michel Bauwens, "Contract manufacturing as distributed manufacturing," *P2P Foundation Blog*, September 11, 2008 <http://blog.p2pfoundation.net/contract-manufacturing-as-distributed-manufacturing/2008/09/11>.

[2]John Robb, "Stigmergic Leaning and Global Guerrillas," *Global Guerrillas*, July 14, 2004 <http://globalguerrillas.typepad.com/globalguerrillas/2004/07/stigmergic_syst.html>.

[3]"Stigmergy," *Wikipedia* <http://en.wikipedia.org/wiki/Stigmergy> (accessed September 29, 2009).

> P2P can arise wherever other forms of distributed fixed capital is [sic] available: such is the case for carpooling, which is the second mode of transportation in the U.S. . . .
>
> P2P can arise wherever financial capital can be distributed. Initiatives such as the ZOPA bank point in that direction. Cooperative purchase and use of large capital goods are a possibility. . . . [1]

As the reference to "distributed financial capital" indicates, the availability of crowdsourced and distributed means of aggregating dispersed capital is as important as the implosion of outlay costs for actual physical capital. A good example of such a system is the Open Source Hardware Bank, a microcredit network organized by California hardware hackers to pool capital for funding new open source hardware projects.[2]

The availability (or unavailability) of capital to working class people will have a significant effect on the rate of self-employment and small business formation. The capitalist credit system, in particular, is biased toward large-scale, conventional, absentee-owned firms. David Blanchflower and Andrew Oswald[3] found that childhood personality traits and test scores had almost no value in predicting adult entrepreneurship. On the other hand, access to startup capital was the single biggest factor in predicting self-employment. There is a strong correlation between self-employment and having received an inheritance or a gift.[4] NSS data indicate that most small businesses were begun not with bank loans but with own or family money. . . . "[5] The clear implication is that there are "undesirable impediments to the market supply of entrepreneurship."[6] In short, the bias of the capitalist credit system toward conventional capitalist enterprise means that the rate of wage employment is higher, and self-employment is lower, than their likely free market values. The lower the capital outlays required for self-employment, and the easier it is to aggregate such capital outside the capitalist credit system, the more self-employment will grow as a share of the total labor market.

Jed Harris, at *Anomalous Presumptions* blog, reiterates Bauwens' point that peer production makes it possible to produce without access to large amounts of capital. "The change that enables widespread peer production is that today, an entity can become self-sustaining, and even grow explosively, with very small amounts of capital. As a result it doesn't need to trade ownership for capital, and so it doesn't need to provide any return on investment."[7]

Charles Johnson adds that, because of the new possibilities the Internet provides for lowering the transaction costs entailed in networked mobilization of capital, peer production can take place even when significant capital investments are required—without relying on finance by large-scale sources of venture capital:

> it's not just a matter of projects being able to expand or sustain themselves with little capital. . . . It's also a matter of the way in which both emerging distributed

[1] Bauwens, "The Political Economy of Peer Production," *CTheory*, December 2005 <http://www.ctheory.net/articles.aspx?id=499>.

[2] Priya Ganapati, "Open Source Hardware Hackers Start P2P Bank," *Wired*, March 18, 2009 <http://www.wired.com/gadgetlab/2009/03/open-source-har/>.

[3] David G. Blanchflower and Andrew J. Oswald, "What Makes an Entrepreneur?" <http://www2.warwick.ac.uk/fac/soc/economics/staff/faculty/oswald/entrepre.pdf>. Later appeared in *Journal of Labor Economics*, 16:1 (1998), pp. 26-60.

[4] *Ibid.*, p. 2.

[5] *Ibid.*, p. 28.

[6] *Ibid.*, p. 3.

[7] Jed Harris, "Capitalists vs. Entrepreneurs," *Anomalous Presumptions*, February 26, 2007 <http://jed.jive.com/?p=23>.

technologies in general, and peer production projects in particular, facilitate the aggregation of dispersed capital—without it having to pass through a single capitalist chokepoint, like a commercial bank or a venture capital fund. . . . Meanwhile, because of the way that peer production projects distribute their labor, peer-production entrepreneurs can also take advantage of spare cycles on existing, widely-distributed capital goods—tools like computers, facilities like offices and houses, software, etc. which contributors own, which they still would have owned personally or professionally whether or not they were contributing to the peer production project. . . . So it's not just a matter of cutting total aggregate costs for capital goods . . . ; it's also, importantly, a matter of new models of aggregating the capital goods to meet whatever costs you may have, so that small bits of available capital can be rounded up without the intervention of moneymen and other intermediaries.[1]

So network organization not only lowers the transaction costs of aggregating capital for the purchase of physical means of production, but also increases the utilization of the means of production when they are expensive.

3. *Reduced Capital Outlays for Physical Production*

As described so far, the open-source model only removes proprietary rents from the portion of the production process—the design stage—that has no material cost, and from the process of aggregating capital. As Richard Stallman put it, to repeat, it's about "free speech" rather than "free beer." Simply removing proprietary rents from design, and removing all transaction costs from the free transfer of digital designs for automated production, will have a revolutionary effect by itself. Marcin Jakubowski, of Factor E Farm, writes:

> The unique contribution of the information age arises in the proposition that data at one point in space allows for fabrication at another, using computer numerical control (CNC) of fabrication. This sounds like an expensive proposition, but that is not so if open source fabrication equipment is made available. With low cost equipment and software, one is able to produce or acquire such equipment at approximately $5k for a fully-equipped lab with metal working, cutting, casting, and electronics fabrication, assisted by open source CNC.[2]

Or as Janne Kyttänen describes it:

> I'm trying to do for products what has already happened to music and digital photography, money, literature—to store them as information and be able to send the data files around the world to be produced. By doing this, you can reduce the waste of the planet, the labor cost, transportation . . . it's going to have a huge impact in the next couple of decades for the manufacturing of goods; we believe it's a new industrial revolution. We will be able to produce products without using the old mass production infrastructure that's been around for two hundred years and is fully out of date.[3]

Jakubowski's reference to the declining cost of fabrication equipment suggests that the revolution in open-source manufacturing goes beyond the design stage, and promises to change the way physical production itself is organized. Chris An-

[1]Charles Johnson, "Dump the rentiers off your back," *Rad Geek People's Daily*, May 29, 2008 <http://radgeek.com/gt/2008/05/29/dump_the/>.

[2]Marcin Jakubowski, "OSE Proposal—Towards a World Class Open Source Research and Development Facility," v0.12, January 16, 2008 <http://openfarmtech.org/OSE_Proposal.doc> (accessed August 25, 2009).

[3]Quoted in Diane Pfeiffer, "Digital Tools, Distributed Making and Design." Thesis submitted to the faculty of the Virginia Polytechnic Institute and State University in partial fulfillment of the requirements for Master of Science in Architecture, 2009, p. 36.

derson is not the first, and probably won't be the last, to point to the parallels between what the desktop computer revolution did to the information and culture industries, and what the desktop manufacturing revolution will do in the physical realm:

> The tools of factory production, from electronics assembly to 3-D printing, are now available to individuals, in batches as small as a single unit. Anybody with an idea and a little expertise can set assembly lines in China into motion with nothing more than some keystrokes on their laptop. A few days later, a prototype will be at their door, and once it all checks out, they can push a few more buttons and be in full production, making hundreds, thousands, or more. They can become a virtual micro-factory, able to design and sell goods without any infrastructure or even inventory; products can be assembled and drop-shipped by contractors who serve hundreds of such customers simultaneously.
>
> Today, micro-factories make everything from cars to bike components to bespoke furniture in any design you can imagine. The collective potential of a million garage tinkerers is about to be unleashed on the global markets, as ideas go straight into production, no financing or tooling required. "Three guys with laptops" used to describe a Web startup. Now it describes a hardware company, too.
>
> "Hardware is becoming much more like software," as MIT professor Eric von Hippel puts it. That's not just because there's so much software in hardware these days, with products becoming little more than intellectual property wrapped in commodity materials, whether it's the code that drives the off-the-shelf chips in gadgets or the 3-D design files that drive manufacturing. It's also because of the availability of common platforms, easy-to-use tools, Web-based collaboration, and Internet distribution.
>
> We've seen this picture before: It's what happens just before monolithic industries fragment in the face of countless small entrants, from the music industry to newspapers. Lower the barriers to entry and the crowd pours in. . . .
>
> A garage renaissance is spilling over into such phenomena as the booming Maker Faires and local "hackerspaces." Peer production, open source, crowdsourcing, user-generated content—all these digital trends have begun to play out in the world of atoms, too. The Web was just the proof of concept. Now the revolution hits the real world.
>
> In short, atoms are the new bits.[1]

The distinction, not only between being "in business" and "out of business," but between worker and owner, is being eroded. The whole concept of technological employment assumes the factory paradigm—in which means of production are extremely expensive, and the only access to work for most people is employment by those rich enough to own the machinery—will continue unaltered. But the imploding price of is making that paradigm obsolete. Neil Gerschenfeld, like Anderson, draws a parallel between hardware today and software thirty years ago:

> The historical parallel between personal computation and personal fabrication provides a guide to what those business models might look like. Commercial software was first written by and for big companies, because only they could afford the mainframe computers needed to run it. When PCs came along anyone could become a software developer, but a big company was still required to develop and distribute big programs, notably the operating systems used to run other programs. Finally, the technical engineering of computer networks combined with the social engineering of human networks allowed distributed teams of individual developers to collaborate on the creation of the most complex software. . . .

[1]Chris Anderson, "In the Next Industrial Revolution, Atoms Are the New Bits," *Wired*, January 25, 2010 <http://www.wired.com/magazine/2010/01/ff_newrevolution/all/1>.

> Similarly, possession of the means for industrial production has long been the dividing line between workers and owners. But if those means are easily acquired, and designs freely shared, then hardware is likely to follow the evolution of software. Like its software counterpart, open-source hardware is starting with simple fabrication functions, while nipping at the heels of complacent companies that don't believe personal fabrication "toys" can do the work of their "real" machines. That boundary will recede until today's marketplace evolves into a continuum from creators to consumers, servicing markets ranging from one to one billion.[1]

Diane Pfeiffer draws a comparison to the rise of desktop publishing in the 1980s.[2]

We already saw, in Chapter Three, what all this meant from the standpoint of investors: they're suffering from the superfluity of most investment capital, resulting from the emerging possibility of small producers and entrepreneurs owning their own factories. From the perspective of the small producer and entrepreneur, the same trend is a good thing because it enables them to own their own factories without any dependency on finance capital. Innovations not only in small-scale manufacturing technology, but in networked communications technology for distribution and marketing, are increasingly freeing producers from the need for large amounts of capital. Charles Hugh Smith writes:

> What I find radically appealing is not so much the technical aspects of desktop/workbench production of parts which were once out of financial reach of small entrepreneurs—though that revolution is the enabling technology—it is the possibility that entrepreneurs *can own the means of production without resorting to vulture/bank investors/loans.*
>
> Anyone who has been involved in a tech startup knows the drill—in years past, a tech startup required millions of dollars to develop a new product or the IP (intellectual property). To raise the capital required, the entrepreneurs had to sell their souls (and company) to venture capital (vulture capital) "investors" who simply took ORPM (other rich people's money) and put it to work, taking much of the value of new promising companies in trade for their scarce and costly capital.
>
> The only alternative were banks, who generally shunned "speculative investments" (unless they were in the billions and related to derivatives, heh).
>
> So entrepreneurs came up with the ideas and did all the hard work, and then vulture capital swooped in to rake off the profits, all the while crying bitter tears about the great risks they were taking with other rich people's spare cash.
>
> Now that these production tools are within reach of small entrepreneurs, the vulture capital machine will find less entrepreneural fodder to exploit. The entrepreneurs themselves can own/rent the *means of production.*
>
> That is a fine old Marxist phrase for the tools and plant which create value and wealth. Own that and you create your own wealth.
>
> In the post-industrial economies of the West and Asia, distribution channels acted as means of wealth creation as well: you want to make money selling books or music, for instance, well, you had to sell your product to the owners of the distribution channels: the record labels, film distributors, book publishers and retail cartels, all of whom sold product through reviews and adverts in the mainstream media (another cartel).
>
> **The barriers to entry were incredibly high.** It took individuals of immense wealth (Spielberg, et al.) to create a new film studio from scratch (DreamWorks) a few years ago. Now any artist can sell their music/books via the Web, completely bypassing the gatekeepers and distribution channels.

[1]Neil Gerschenfeld, *Fab: The Coming Revolution on Your Desktop—From Personal Computers to Personal Fabrication* (New York: Basic Books, 2005), pp. 14-15.

[2]Pfeiffer, "Digital Tools," pp. 33-35.

> **In a great irony, publishers and labels are now turning to the Web to sell their product.** If all they have is the Web, then what value can they add? I fully expect filmakers to go directly to the audience via the Web in coming years and bypass the entire film distribution cartel entirely.
>
> Why go to Wal-Mart to buy a DVD when you can download hundreds of new films off the Web?
>
> **Both the supply chain and distribution cartels are being blown apart by the Web.** Not only can entrepreneurs own/rent the means of production and arrange their own supply/assembly chains, they can also own their own distribution channels.
>
> **The large-scale factory/distribution model is simply no longer needed for many products.** As the barriers to owning the means of production and distribution fall, a Renaissance in small-scale production and wealth creation becomes not just possible but inevitable.[1]

Even without the latest generation of low-cost digital fabrication machinery, the kind of flexible manufacturing network that exists in Emilia-Romagna or Shenzen is ideally suited to the open manufacturing philosophy. Tom Igoe writes:

> There are some obvious parallels here [in the shanzhai manufacturers of China—see Chapter Four] to the open hardware community. Businesses like Spark Fun, Adafruit, Evil Mad Scientist, Arduino, Seeed Studio, and others thrive by taking existing tools and products, re-combining them and repackaging them in more usable ways. We borrow from each other and from others, we publish our files for public use, we improve upon each others' work, and we police through licenses such as the General Public License, and continual discussion between competitors and partners. We also revise products constantly and make our businesses based on relatively small runs of products tailored to specific audiences.[2]

The intersection of the open hardware and open manufacturing philosophies with the current model of flexible manufacturing networks will be enabled, Igoe argues, by the availability of

> **Cheap tools.** Laser cutters, lathes, and milling machines that are affordable by an individual or a group. This is increasingly coming true. The number of colleagues I know who have laser cutters and mills in their living rooms is increasing (and their asthma is worsening, no doubt). There are some notable holes in the open hardware world that exist partially because the tools aren't there. Cheap injection molding doesn't exist yet, but injection molding services do, and they're accessible via the net. But when they're next door (as in Shenzen), you've got a competitive advantage: your neighbor.[3]

(Actually hand-powered, small-scale injection molding machines are now available for around $1500, and Kenner marketed a fully functional "toy" injection molding machine for making toy soldiers, tanks, and the like back in the 1960s.)[4]

And the flexible manufacturing network, unlike the transnational corporate environment, is actively conducive to the sharing of knowledge and designs.

> **Open manufacturing information.** Manufacturers in this scenario thrive on adapting existing products and services. Call them knockoffs or call them new

[1] Charles Hugh Smith, "The Future of Manufacturing in the U.S." oftwominds, February 5, 2010 <http://charleshughsmith.blogspot.com/2010/02/future-of-manufacturing-in-us.html>.

[2] Tom Igoe, "Idle speculation on the shan zhai and open fabrication," *hello* blog, September 4, 2009 <http://www.tigoe.net/blog/category/environment/295/>.

[3] *Ibid.*

[4] Joseph Flaherty, "Desktop Injection Molding," *Replicator*, February 1, 2020 <http://replicatorinc.com/blog/2010/02/desktop-injection-molding>.

hybrids, they both involve reverse engineering something and making it fit your market. Reverse engineering takes time and money. When you're a mom & pop shop, that matters a lot more to you. If you've got a friend or a vendor who's willing to do it for you as a service, that helps. But if the plans for the product you're adapting are freely available, that's even better. In a multinational world, open source manufacturing is anathema. Why would Nokia publish the plans for a phone when they could dominate the market by doing the localization themselves? But in a world of networked small businesses, it spurs business. You may not have the time or interest in adapting your product for another market, but someone else will, and if they've got access to your plans, they'll be grateful, and will return the favor, formally or informally.[1]

The availability of modestly priced desktop manufacturing technology (about which we will see more immediately below), coupled with the promise of crowd-sourced means of aggregating capital, has led to a considerable shift in opinion in the peer-to-peer community, as evidenced by Michel Bauwens:

> I used to think that the model of peer production would essentially emerge in the immaterial sphere, and in those cases where the design phase could be split from the capital-intensive physical production sphere. . . .
>
> However, as I become more familiar with the advances in Rapid Manucturing [sic] . . . and Desktop Manufacturing . . . , I'm becoming increasingly convinced of the strong trend towards the distribution of physical capital.
>
> If we couple this with the trend towards the direct social production of money (i.e. the distribution of financial capital . . .) and the distribution of energy . . . ; and how the two latter trends are interrelated . . . , then I believe we have very strong grounds to see a strong expansion of p2p-based modalities in the physical sphere.[2]

The conditions of physical production have, in fact, experienced a transformation almost as great as that which digital technology has brought about on immaterial production. The "physical production sphere" itself has become far less capital-intensive. If the digital revolution has caused an implosion in the physical capital outlays required for the information industries, the revolution in garage and desktop production tools promises an analogous effect almost as great on many kinds of manufacturing. The radical reduction in the cost of machinery required for many kinds of manufacturing has eroded Stallman's distinction between "free speech" and "free beer." Or as Chris Anderson put it, "Atoms would like to be free, too, but they're not so pushy about it."[3]

The same production model sweeping the information industries, networked organization of people who own their own production tools, is expanding into physical manufacturing. A revolution in cheap, general purpose machinery, and a revolution in the possibilities for networked design made possible by personal computers and network culture, according to Johann Soderberg, is leading to

> an extension of the dream that was pioneered by the members of the Homebrew Computer Club [i.e., a cheap computer able to run on the kitchen table]. It is the vision of a universal factory able to run on the kitchen table. . . . [T]he desire for a 'desktop factory' amounts to the same thing as the reappropriation of the means of production.[4]

[1]Igoe, *op. cit.*

[2]Michel Bauwens post to Institute for Distributed Creativity email list, May 7, 2007. <https://lists.thing.net/pipermail/idc/2007-May/002479.html>

[3]Chris Anderson, *Free: The Future of a Radical Price* (New York: Hyperion, 2009), p. 241.

[4]Soderberg, *Hacking Capitalism*, pp. 185-186.

Clearly, the emergence of cheap desktop technology for custom machining parts in small batches will greatly lower the overall capital outlays needed for networked physical production of light and medium consumer goods.

We've already seen the importance of the falling costs of small-scale production machinery made possible by the Japanese development of small CNC machines in the 1970s. That is the technological basis of the flexible manufacturing networks we examined in the last chapter.

When it comes to the "Homebrew" dream of an actual desktop factory, the most promising current development is the Fab Lab. The concept started with MIT's Center for Bits and Atoms. The original version of the Fab Lab included CNC laser cutters and milling machines, and a 3-D printer, for a total cost of around $50,000.[1]

Open-source versions of the machines in the Fab Lab have brought the cost down to around $2-5,000.

One important innovation is the multimachine, an open-source, multiple-purpose machine tool that includes drill press, lathe and milling machine; it can be modified for computerized numeric control. The multimachine was originally developed by Pat Delaney, whose YahooGroup has grown into a design community and support network of currently over five thousand people.[2]

As suggested by the size of Delaney's YahooGroup membership, the multimachine has been taken up independently by open-source developers all around the world. The Open Source Ecology design community, in particular, envisions a Fab Lab which includes a CNC multimachine as "the central tool piece of a flexible workshop . . . eliminating thousands of dollars of expenditure requirement for similar abilities" and serving as "the centerpieces enabling the fabrication of electric motor, CEB, sawmill, OSCar, microcombine and all other items that require processes from milling to drilling to lathing."[3]

> It is a high precision mill-drill-lathe, with other possible functions, where the precision is obtained by virtue of building the machine with discarded engine blocks. . . .
>
> The central feature of the Multimachine is the concept that either the tool or the workpiece rotates when any machining operation is performed. As such, a heavy-duty, precision spindle (rotor) is the heart of the Multimachine—for milling, drilling and lathing applications. The precision arises from the fact that the spindle is secured within the absolutely precise bore holes of an engine block, so precision is guaranteed simply by beginning with an engine block.
>
> If one combines the Multimachine with a CNC XY or XYZ movable working platform—similar to ones being developed by the Iceland Fab Lab team[4], RepRap[5], CandyFab 4000[6] team, and others—then a CNC mill-drill-lathe is the result. At least Factor 10 reduction in price is then available compared to the competition. The mill-drill-lathe capacity allows for the subtractive fabrication of any allowable shape, rotor, or cylindrically-symmetric object. Thus, the CNC Multimachine can be an effective cornerstone of high precision digital fabrication—down to 2 thousandths of an inch.

[1]MIT Center for Bits and Atoms, "Fab Lab FAQ" <http://fab.cba.mit.edu/about/faq/> (accessed August 31, 2009).

[2]"Multimachine," *Wikipedia* <http://en.wikipedia.org/wiki/Multimachine> (accessed August 31, 2009>; <http://groups.yahoo.com/group/multimachine/>.

[3]"Multimachine & Flex Fab—Open Source Ecology" <http://openfarmtech.org/index.php?title=Multimachine_%26_Flex_Fab>.

[4]<http://smari.yaxic.org/blag/2007/11/14/the-routing-table/> (note in quoted text).

[5]<http://reprap.org/bin/view/Main/RepRap>. (note in quoted text).

[6]<http://www.makingthings.com/projects/CandyFab-4000> (note in quoted text).

Interesting features of the Multimachine are that the machines can be scaled from small ones weighing a total of ~1500 lb to large ones weighing several tons, to entire factories based on the Multimachine system. The CNC XY(Z) tables can also be scaled according to the need, if attention to this point is considered in development. The whole machine is designed for disassembly. Moreover, other rotating tool attachments can be added, such as circular saw blades and grinding wheels. The overarm included in the basic design is used for metal forming operations.

Thus, the Multimachine is an example of appropriate technology, where the user is in full control of machine building, operation, and maintenance. Such appropriate technology is conducive to successful small enterprise for local community development, via its low capitalization requirement, ease of maintenance, scaleability and adaptability, and wide range of products that can be produced. This is relevant both in the developing world and in industrialized countries.[1]

The multimachine, according to Delaney, "can be built by a semi-skilled mechanic using just common hand tools," from discarded engine blocks, and can be scaled from "a closet size version" to "one that would weigh 4 or 5 tons."[2]

In developing countries, in particular, the kinds of products that can be built with a multimachine include:

AGRICULTURE:

Building and repairing irrigation pumps and farm implements.

WATER SUPPLIES:

Making and repairing water pumps and water-well drilling rigs.

FOOD SUPPLIES:

Building steel-rolling-and-bending machines for making fuel efficient cook stoves and other cooking equipment.

TRANSPORTATION:

Anything from making cart axles to rebuilding vehicle clutch, brake, and other parts. . . .

JOB CREATION:

A group of specialized but easily built MultiMachines can be combined to form a small, very low cost, metal working factory which could also serve as a trade school. Students could be taught a single skill on a specialized machine and be paid as a worker while learning other skills that they could take elsewhere.[3]

More generally, a Fab Lab (i.e. a digital flexible fabrication facility centered on the CNC multimachine along with a CNC cutting table and open-source 3-D printer like RepRap) can produce virtually anything—especially when coupled with the ability of such machinery to run open-source design files.

Flexible fabrication refers to a production facility where a small set of non-specialized, general-function machines (the 5 items mentioned [see below]) is capable of producing a wide range of products if those machines are operated by skilled labor. It is the opposite of mass production, where unskilled labor and specialized machinery produce large quantities of the same item (see section *II, Economic Base*). When one adds *digital fabrication* to the flexible fabrication mix—then the skill level on part of the operator is reduced, and the rate of production is increased.

Digital fabrication is the use of computer-controlled fabrication, as instructed by data files that generate tool motions for fabrication operations. Digi-

[1]Jakubowski, "OSE Proposal."
[2]<http://groups.yahoo.com/group/multimachine/?yguid=234361452>.
[3]<http://opensourcemachine.org/node/2>.

tal fabrication is an emerging byproduct of the computer age. It is becoming more accessible for small scale production, especially as the influence of open source philosophy is releasing much of the know-how into non-proprietary hands. For example, the Multimachine is an open source mill-drill-lathe by itself, but combined with computer numerical control (CNC) of the workpiece table, it becomes a digital fabrication device.

It should be noted that open access to digital design—perhaps in the form a global repository of shared open source designs—introduces a unique contribution to human prosperity. This contribution is the possibility that data at one location in the world can be translated immediately to a product in any other location. This means anyone equipped with flexible fabrication capacity can be a producer of just about any manufactured object. The ramifications for localization of economies are profound, and leave the access to raw material feedstocks as the only natural constraint to human prosperity.[1]

Open Source Ecology, based on existing technology, estimates the cost of producing a CNC multimachine with their own labor at $1500.[2] The CNC multimachine is only one part of a projected "Fab Lab," whose total cost of construction will be a few thousand dollars.

1. CNC Multimachine—Mill, drill, lathe, metal forming, other grinding/cutting. This constitutes a robust machining environment that may be upgraded for open source computer numerical control by OS software, which is in development.[3]
2. XYZ-controlled torch and router table—can accommodate an acetylene torch, plasma cutter, router, and possibly CO_2 laser cutter diodes
3. Metal casting equipment—all kinds of cast parts from various metals
4. Plastic extruder—extruded sheet for advanced glazing, and extruded plastic parts or tubing
5. Electronics fabrication—oscilloscope, circuit etching, others—for all types of electronics from power control to wireless communications.

This equipment base is capable of producing just about anything—electronics, electromechanical devices, structures, and so forth. The OS Fab Lab is crucial in that it enables the self-replication of all the 16 technologies.[4]

(The "16 technologies" refers to Open Source Ecology's entire line of sixteen products, including not only construction and energy generating equipment, a tractor, and a greenhouse, but using the Fab Lab to replicate the five products in the Fab Lab itself. See the material on OSE in the Appendix.)

Another major component of the Fab Lab, the 3-D printer, sells at a price starting at over $20,000 for commercial versions. The RepRap, an open-source 3-D printer project, has reduced the cost to around $500.[5] MakerBot[6] is a closely related commercial 3-D printer project, an offshoot of RepRap that shares much of

[1]Jakubowski, "OSE Proposal."

[2]Marcin Jakubowski, "Rapid Prototyping for Industrial Swadeshi," Factor E Farm Weblog, August 10, 2008 <http://openfarmtech.org/weblog/?p=293>. "Open Source Fab Lab," Open Source Ecology wiki (accessed August 22, 2009) <http://openfarmtech.org/index.php?title=Open_Source_Fab_Lab>.

[3]Open source CNC code is being developed by Smari McCarthy of the Iceland Fab Lab, <http://smari.yaxic.org/blag/2007/11/14/the-routing-table/>.

[4]Jakubowski, "OSE Proposal."

[5]RepRap site <http://reprap.org/bin/view/Main/WebHome>; "RepRap Project," Wikipedia <http://en.wikipedia.org/wiki/RepRap_Project> (accessed August 31, 2009).

[6]<http://makerbot.com/>

its staff in common.[1] Makerbot has a more streamlined, finished (i.e., commercial-looking) appearance. Unlike RepRap, it doesn't aim at total self-replicability; rather, most of its parts are designed to be built with a laser cutter.[2]

3-D printers are especially useful for making casting molds. Antique car enthusiast Jay Leno, in a recent issue of *Popular Mechanics*, described the use of a combination 3-D scanner/3-D printer to create molds for out-of-production parts for old cars like his 1907 White Steamer.

> The 3D printer makes an exact copy of a part in plastic, which we then send out to create a mold. . . .
>
> The NextEngine scanner costs $2995. The Dimension uPrint Personal 3D printer is now under $15,000. That's not cheap. But this technology used to cost 10 times that amount. And I think the price will come down even more.[3]

Well, yeah—especially considering RepRap can *already* be built for around $500 in parts. Even the Desktop Factory, a commercial 3-D printer, sells for about $5,000.[4]

Automated production with CNC machinery, Jakubowski argues, holds out some very exciting possibilities for producing at rates competitive with conventional industry.

> It should be pointed out that a particularly exciting enterprise opportunity arises from automation of fabrication, such as arises from computer numerical control. For example, the sawmill and CEB discussed above are made largely of DfD, bolt-together steel. This lends itself to a fabrication procedure where a CNC XYZ table could cut out all the metal, including bolt holes, for the entire device, in a fraction of the time that it would take by hand. As such, complete sawmill or CEB kits may be fabricated and collected, ready for assembly, on the turn-around time scale of days. . . .
>
> The digital fabrication production model may be equivalent in production rates to that of any large-scale, high-tech firms.[5]
>
> The concept of a CNC XYZ table is powerful. It allows one to prepare all the metal, such as that for a CEB press or the boundary layer turbine, with the touch of a button if a design file for the toolpath is available. This indicates on-demand fabrication capacity, at production rates similar to that of the most highly-capitalized industries. With modern technology, this is doable at low cost. With access to low-cost computer power, electronics, and open source blueprints, the capital needed for producing a personal XYZ table is reduced merely to structural steel and a few other components: it's a project that requires perhaps $1000 to complete.[6]

(Someone's actually developed a CNC XYZ cutting table for $100 in materials, although the bugs are not yet completely worked out.)[7]

Small-scale fabrication facilities of the kind envisioned at Factor E Farm, based on CNC multimachines, cutting tables and 3D printers, can even produce

[1]Keith Kleiner, "3D Printing and Self-Replicating Machines in Your Living Room—Seriously," *Singularity Hub*, April 9, 2009 <http://singularityhub.com/2009/04/09/3d-printing-and-self-replicating-machines-in-your-living-room-seriously/>.

[2]"What is the relationship between RepRap and Makerbot?" *Hacker News* <http://news.ycombinator.com/item?id=696785>.

[3]Jay Leno, "Jay Leno's 3-D Printer Replaces Rusty Old Parts," *Popular Mechanics*, July 2009 <http://www.popularmechanics.com/automotive/jay_leno_garage/4320759.html?page=1>.

[4]<http://www.desktopfactory.com/>.

[5]Jakubowski, "OSE Proposal."

[6]*Ibid.*

[7]"CNC machine v2.0—aka 'Valkyrie'," *Let's Make Robots*, July 14, 2009 <http://letsmakerobots.com/node/9006>.

motorized vehicles like passenger cars and tractors, when the heavy engine block is replaced with light electric motor. Such electric vehicles, in fact, are part of the total product package at Factor E Farm.

> The central part of a car is its propulsion system. *Fig. 6* shows a *fuel* source feeding a *heat generator*, which heats a flash steam generator *heat exchanger*, which drives a *boundary layer turbine*, which drives a *wheel motor* operating as an electrical generator. The electricity that is generated may either be fed into *battery storage*, or controlled by *power electronics* to drive 4 separate wheel motors. This constitutes a hybrid electric vehicle, with 4 wheel drive in this particular implementation.
>
> This hybrid electric vehicle is one of intermediate technology design that may be fabricated in a small-scale, flexible workshop. The point is that a complicated power delivery system (clutch-transmission-drive shaft-differential) has been replaced by four electrical wires going to the wheel electrical motors. This simplification results in high localization potential of car manufacturing.
>
> The first step in the development of open source, Hypercar-like vehicles is the propulsion system, for which the boundary layer turbine hybrid system is a candidate. Our second step will be structural optimization for lightweight car design.[1]

The CubeSpawn project is also involved in developing a series of modular desktop machine tools. The first stage is a cubical 3-axis milling machine (or "milling cell"). The next step will be to build a toolchanger and head changer so the same cubical framework and movement controls can be used for a 3-D printer.[2]

> It starts by offering a simple design for a 3 axis, computer controlled milling machine.
>
> With this resource, you have the ability to make a significant subset of all the parts in existence! So, parts for additional machines can be made on the mill, allowing the system to add to itself, all based on standards to promote interoperability. . . .
>
> The practical consequence is a self expanding factory that will fit in a workshop or garage. . . .
>
> Cross pollenization with other open source projects is inevitable and beneficial although at first, commercial products will be used if no open source product exists. This has already begun, and CubeSpawn uses 5 other open source+ projects as building blocks in its designs These are electronics from the Sanguino / RepRap specific branch of the Arduino project, Makerbeam for cubes of small dimensions, and the EMC control software for an interface to individual cells. There is an anticipated use of SKDB for part version and cutting geometry file retrieval, with Debian Linux as a central host for the system DB. . . .
>
> By offering a standardized solution to the problems of structure, power connections, data connections, inter-cell transport, and control language, we can bring about an easier to use framework to collaborate on. The rapid adoption of open source hardware should let us build the "better world" industry has told us about for over 100 years.[3]

With still other heads, the same framework can be used as a cutting table.

If these examples are not enough, the P2P Foundation's "Product Hacking" page provides, under the heading of "Production/Machinery," a long list of open-source CNC router, cutting table, 3-D printer, modular electronics, and other pro-

[1]Jakubowski, "OSE Proposal."

[2]<http://www.cubespawn.com/>.

[3]"CubeSpawn, An open source, Flexible Manufacturing System (FMS)" <http://www.kickstarter.com/projects/1689465850/cubespawn-an-open-source-flexible-manufacturing>.

jects.[1] DIYLILCNC is a cheap homebrew 3-axis milling machine that can be built with "basic shop skills and tool access."[2]

One promising early attempt at distributed garage manufacturing is 100kGarages, which we will examine in some detail in the Appendix. 100kGarages is a joint effort of the ShopBot 3-axis router company and the Ponoko open design network (which itself linked a library of designs to local Makers with CNC laser cutters).

Besides Ponoko, a number of other commercial firms have appeared recently which offer production of custom parts to the customer's digital design specifications, at a modest price, using small-scale, multipurpose desktop machinery. Two of the most prominent are Big Blue Saw[3] and eMachineShop.[4] The way the latter works, in particular, is described in a *Wired* article:

> The concept is simple: Boot up your computer and design whatever object you can imagine, press a button to send the CAD file to Lewis' headquarters in New Jersey, and two or three weeks later he'll FedEx you the physical object. Lewis launched eMachineShop a year and a half ago, and customers are using his service to create engine-block parts for hot rods, gears for home-brew robots, telescope mounts—even special soles for tap dance shoes.[5]

Another project of the same general kind was just recently announced: CloudFab, which offers access to a network of job-shops with 3-D printers.[6] Also promising is mobile manufacturing (Factory in a Box).[7]

Building on our earlier speculation about networked small machine shops and hobbyist workshops, new desktop manufacturing technology offers an order of magnitude increase in the quality of work that can be done for the most modest expense.

Kevin Kelly argues that the actual costs of physical production are only a minor part of the cost of manufactured goods.

> . . . . material industries are finding that the costs of duplication near zero, so they too will behave like digital copies. Maps just crossed that threshold. Genetics is about to. Gadgets and small appliances (like cell phones) are sliding that way. Pharmaceuticals are already there, but they don't want anyone to know. It costs nothing to make a pill.[8]

If, as Kelley suggests, the cheapness of digital goods reflects the imploding cost of copying them, it follows that the falling cost of "copying" physical goods will follow the same pattern.

There is a common thread running through all the different theories of the interface between peer production and the material world: as technology for physical

[1]<http://p2pfoundation.net/Product_Hacking#Production.2FMachinery>.

[2]<http://diylilcnc.org/>.

[3]<http://www.bigbluesaw.com/saw/>.

[4]<http://www.emachineshop.com/> (see also <www.barebonespcb.com/!BB1.asp>).

[5]Clive Thompson, "The Dream Factory," *Wired*, September 2005 <http://www.wired.com/wired/archive/13.09/fablab_pr.html>.

[6]"The CloudFab Manifesto," *Ponoko Blog*, September 28, 2009 <http://blog.ponoko.com/2009/09/28/the-cloudfab-manifesto/>.

[7]Carin Stillstrom and Mats Jackson, "The Concept of Mobile Manufacturing," *Journal of Manufacturing Systems* 26:3-4 (July 2007) <http://www.sciencedirect.com/science?_ob=ArticleURL&_udi=B6VJD-4TK3FG8-6&_user=108429&_rdoc=1&_fmt=&_orig=search&_sort=d&view=c&_version=1&_urlVersion=0&_userid=108429&md5=bf6e603b5de29cdfd026d5d00379877c>.

[8]Kevin Kelly, "Better Than Free," *The Technium*, January 31, 2008 <http://www.kk.org/thetechnium/archives/2008/01/ better_than_fre.php>.

production becomes feasible on increasingly smaller scales and at less cost, and the transaction costs of aggregating small units of capital into large ones fall, there will be less and less disconnect between peer production and physical production.

It's worth repeating one last time: the distinction between Stallman's "free speech" and "free beer" is eroding. To the extent that embedded rents on "intellectual property" are a significant portion of commodity prices, "free speech" (in the sense of the free use of ideas) will make our "beer" (i.e., the price of manufactured commodities) at least a lot cheaper. And the smaller the capital outlays required for physical production, the lower the transaction costs for aggregating capital, and the lower the overhead, the cheaper the beer becomes as well.

If, as we saw Sabel and Piore say above, the computer is a textbook example of an artisan's tool—i.e., an extension of the user's creativity and intellect—then small-scale, computer-controlled production machinery is a textbook illustration of E. F. Schumacher's principles of appropriate technology:

- cheap enough that they are accessible to virtually everyone;
- suitable for small-scale application; and
- compatible with man's need for creativity.

D. The Microenterprise

We have already seen, in Chapter Four, the advantages of low overhead and small batch production that lean, flexible manufacturing offers over traditional mass-production industry. The household microenterprise offers these advantages, but increased by another order of magnitude. As we saw Charles Johnson suggest above, the use of "spare cycles" of capital goods people own anyway results in enormous cost efficiencies.

Consider, for example, the process of running a small, informal brew pub or restaurant out of your home, under a genuine free market regime. Buying a brewing vat and a few small fermenters for your basement, using a few tables in a remodeled spare room as a public restaurant area, etc., would require a small bank loan for at most a few thousand dollars. And with that capital outlay, you could probably make payments on the debt with the margin from one customer a day. A few customers evenings and weekends, probably found mainly among your existing circle of acquaintances, would enable you to initially shift some of your working hours from wage labor to work in the restaurant, with the possibility of gradually phasing out wage labor altogether or scaling back to part time, as you built up a customer base. In this and many other lines of business (for example a part-time gypsy cab service using a car and cell phone you own anyway), the minimal entry costs and capital outlay mean that the minimum turnover required to pay the overhead and stay in business would be quite modest. In that case, a lot more people would be able to start small businesses for supplementary income and gradually shift some of their wage work to self employment, with minimal risk or sunk costs.

But that's *illegal*. You have to buy an extremely expensive liquor license, as well as having an industrial sized stove, dishwasher, etc. You have to pay rent on a separate, dedicated commercial building. And that level of capital outlay can only be paid off with a large dining room and a large kitchen/waiting staff, which means you have to keep the place filled or the overhead costs will eat you alive—in other words, Chapter Eleven. These high entry costs and the enormous overhead are the reason you can't afford to start out really small and cheap, and the reason restaurants have such a high failure rate. It's illegal to use the surplus capacity of the ordinary household items we have to own anyway but remain idle most of the time (including small-scale truck farming): e.g. RFID chip requirements and bans on unpasteurized milk, high fees for organic certification, etc., which make it prohibitively expensive to sell a few hundred dollars surplus a month from the household economy. As Roderick Long put it,

> In the absence of licensure, zoning, and other regulations, how many people would start a restaurant today if all they needed was their living room and their kitchen? How many people would start a beauty salon today if all they needed was a chair and some scissors, combs, gels, and so on? How many people would start a taxi service today if all they needed was a car and a cell phone? How many people would start a day care service today if a bunch of working parents could simply get together and pool their resources to pay a few of their number to take care of the children of the rest? These are not the sorts of small businesses that receive SBIR awards; they are the sorts of small businesses that get hammered down by the full strength of the state whenever they dare to make an appearance without threading the lengthy and costly maze of the state's permission process.[1]

Shawn Wilbur, an anarchist writer with half a lifetime in the bookselling business, describes the resilience of a low-overhead business model: "My little store was enormously efficient, in the sense that it could weather long periods of low sales, and still generally provide new special order books in the same amount of time as a Big Book Bookstore." The problem was that, with the state-imposed paperwork burden associated with hiring help, it was preferable—i.e. less complicated—to work sixty-hour weeks.[2] The state-imposed administrative costs involved in the cooperative organization of labor amount to an entry barrier that can only be hurdled by the big guy. After some time out of the business of independent bookselling and working a number of wage-labor gigs in chain bookstores, Wilbur has recently announced the formation of Corvus—a micropublishing operation that operates on a print-on-demand basis.[3] In response to my request for information on his business model, Wilbur wrote:

> In general . . . , Corvus Editions is a hand-me-down laptop and a computer that should probably have been retired five years ago, and which has more than paid for itself in my previous business, some software, all of which I previously owned and none of which is particularly new or spiffy, a $20 stapler, a $150 laser printer, a handful of external storage devices, an old flatbed scanner, the usual computer-related odds and ends, and the fruits of thousands of hours of archival research and sifting through digital sources (all of which fits on a single portable harddrive.) The online presence did not involve any additional expense, beyond the costs of the free archive, except for a new domain name. My hosting costs, including holding some domain registrations for friendly projects, total around $250/year, but the Corvus site and shop could be hosted for $130.
>
> Because Portland has excellent resources for computer recycling and the like, I suspect a similar operation, minus the archive, using free Linux software tools, could almost certainly be put together for less than $500, including a small starting stock of paper and toner—and perhaps more like $300.

[1]Roderick Long, "Free Market Firms: Smaller, Flatter, and More Crowded," *Cato Unbound*, November 25, 2008 <http://www.cato-unbound.org/2008/11/25/roderick-long/free-market-firms-smaller-flatter-and-more-crowded/>.

[2]Comment under Shawn Wilbur, "Who benefits most economically from state centralization" *In the Libertarian Labyrinth*, December 9, 2008 <http://libertarian-labyrinth.blogspot.com/2008/12/who-benefits-most-economically-from.html>.

[3]Shawn Wilbur, "Taking Wing: Corvus Editions," *In the Libertarian Labyrinth*, July 1, 2009 <http://libertarian-labyrinth.blogspot.com/2009/07/taking-wing-corvus-editions.html>; Corvus Distribution website <http://www.corvusdistribution.org/shop/>.

The cost of materials is some 20% of Wilbur's retail price on average, with the rest of the price being compensation free and clear for his labor: "the service of printing, folding, stapling and shipping. . . . " There are no proprietary rents because the pdf files are themselves free for download; Wilbur makes money entirely from the convenience-value of his doing those printing, etc., services for the reader.[1]

As an example of a more purely service-oriented microenterprise, Steve Herrick describes the translators' cooperative he's a part of:

> . . . We effectively operate as a job shop. Work comes in from clients, and our coordinator posts the offer on email. People offer to take it as they're available. So far, the supply and demand have been roughly equal. When multiple people are available, members take priority over associates, and members who have taken less work recently take priority over those who have taken more.
>
> We have seven members, plus eight or ten associates, who have not paid a buy-in and who are not expected to attend meetings. They do, however, make the same pay for the same work.
>
> Interpreting and translating are commonly done alone. So, why have a co-op? First, we all hate doing the paperwork and accounting. We'd rather be doing our work. A co-op lets us do that. The other reason is branding/marketing/reputation. Clients can't keep track of the contact info for a dozen people, but they can remember the email and phone number for our coordinator, who can quickly contact us all. Also, with us, they get a known entity, even if it's a new person. (Unlike most other services an organization might contract for, clients don't usually know how well their interpreters are doing for their pay. With us, they worry about that a lot less.)
>
> We keep our options open by taking many kinds of work. We don't compete with the local medical and court interpreter systems (and some of us also work in them), but that leaves a lot of work to do: we work for schools and universities, non-profits, small businesses, individuals, unions, and so on. We've pondered whether there are clients we would refuse to work for, but so far, that hasn't been an issue.
>
> We have almost no overhead. We are working on getting an accountant, but we don't anticipate having to pay more than a few hours a month for that. Our books aren't that complicated. We also pay rent to the non-profit we spun off from, but that's set up as a percentage of our income, not a fixed amount, so it can't put us under water. It also serves as an incentive for them to send us work! Other than that, we really have no costs. As a co-op, taxes are "pass-through," meaning the co-op itself pays no taxes; we pay taxes on our income from the co-op. We will be doing some marketing soon, but we're investigating very low-cost ways to reach our target market, like in-kind work. And we have no capital costs, apart from our interpreting mic and earpieces, which we inherited from the non-profit. Occasionally, we have to buy batteries, but I'm going to propose we buy rechargables, so even that won't be a recurring cost. And finally, we're looking in to joining our local Time Bank.
>
> What this means is that we can operate at a very low volume. As a ballpark figure, I'd say we average an hour of work per member per week. That's not much more than a glorified hobby. Even so, 2009 brought in considerably more work than 2008, which saw twice as much work as 2007 (again, with essentially no marketing). We're not looking for it to increase too rapidly, because each of us has at least one other job, and six of the seven of us have kids (ranging from

[1]Shawn Wilbur, "Re: [Anarchy-List] Turnin' rebellion into money (or not . . . your choice)," email to *Anarchy List*, July 17, 2009 <http://lists.anarchylist.org/private.cgi/anarchy-list-anarchylist.org/2009-July/003406.html>.

mine at three weeks to one member with school-age grandkids). A slow, steady increase would be great.[1]

More generally, this business model applies to a wide range of service industries where overhead requirements are minimal. An out of work plumber or electrician can work out of his van with parts from the hardware store, and cut his prices by the amount that formerly went to commercial rent, management salaries and office staff, and so forth—not to mention working for a "cash discount." Like Herrick's translator cooperative, one of the main functions of a nursing or other temporary staffing agency is branding—providing a common reference point for accountability to clients. But the actual physical capital requirements don't go much beyond a phone line and mail drop, and maybe a scanner/fax. The business consists, in essence, of a personnel list and a way of contacting them. The main entry barrier to cooperative self-employment in this field is non-competition agreements (when you work for a client of a commercial staffing agency, you agree not to work for that client either directly or through another agency for some period—usually three months—after your last assignment there). But with a large enough pool of workers in the cooperative agency, it should be possible to direct assignments to those who haven't worked for a particular client, until the non-competition period expires.

The lower capital outlays and fixed costs fall, the more meaningless the distinction between being "in business" and "out of business" becomes.

Another potential way to increase the utilization of capacity of capital goods in the informal and household economy is through sharing networks of various kinds. The sharing of tools through neighborhood workshops, discussed earlier, is one application of the general principle. Other examples include ride-sharing, time-sharing one another's homes during vacations, gift economies like FreeCycle, etc. Regarding ride-sharing in particular, *Dilbert* cartoonist Scott Adams speculates quite plausibly on the potential for network technologies like the iPhone to facilitate sharing in ways that previous technology could not, by reducing the transaction costs of connecting participants. The switch to network connections by mobile phone increases flexibility and capability for short-term changes and adjustments to plans by an order of magnitude over desktop computers. Adams describes how such a system might work:

> . . . [T]he application should use GPS to draw a map of your location, with blips for the cars available for ridesharing. You select the nearest blip and a bio comes up telling you something about the driver, including his primary profession, age, a photo, and a picture of the car. If you don't like something about that potential ride, move on to the next nearest blip. Again, you have a sense of control. Likewise, the driver could reject you as a passenger after seeing your bio.
>
> After you select your driver, and he accepts, you can monitor his progress toward your location by the moving blip on your iPhone. . . .
>
> I also imagine that all drivers would have to pass some sort of "friend of a friend" test, in the Facebook sense. In other words, you can only be a registered rideshare driver if other registered drivers have recommended you. Drivers would be rated by passengers after each ride, again by iPhone, so every network of friends would carry a combined rating. That would keep the good drivers from recommending bad drivers because the bad rating would be included in their own network of friends average. . . . And the same system could be applied to potential passengers. As the system grew, you could often find a ride with a

[1]Steve Herrick, private email, December 10, 2009.

tential passengers. As the system grew, you could often find a ride with a friend of a friend.[1]

Historically the prevalence of such enterprises has been associated with economic downturn and unemployment.

The shift to value production outside the cash nexus in the tech economy has become a common subject of discussion in recent years. We already discussed at length, in Chapter Three, how technological innovation has caused the floor to drop out from beneath capital outlay costs, and thereby rendered a great deal of venture capital superfluous. Although this was presented as a negative from the standpoint of capitalism's crisis of overaccumulation, we can also see it as a positive from the standpoint of opportunities for the growth of a new economy outside the cash nexus.

Michel Bauwens describes the way most innovation, since the collapse of the dotcom bubble, has shifted to the social realm and become independent of capital.

> To understand the logic of this promise, we can look to a less severe, but nevertheless serious crisis: that of the internet bubble collapse in 2000-1. As an internet entrepreneur, I personally experienced both the manic phase, and the downturn, and the experience was life changing because of the important discovery I and others made at that time. All the pundits where [sic] predicting, then as now, that without capital, innovation would stop, and that the era of high internet growth was over for a foreseeable time. In actual fact, the reality was the very opposite, and something apparently very strange happened. In fact, almost everything we know, the Web 2.0, the emergence of social and participatory media, was born in the crucible of that downturn. In other words, innovation did not slow down, but actually increased during the downturn in investment. This showed the following new tendency at work: capitalism is increasingly being divorced from entrepreneurship, and entrepreneurship becomes a networked activity taking place through open platforms of collaboration.
>
> The reason is that internet technology fundamentally changes the relationship between innovation and capital. Before the internet, in the Schumpeterian world, innovators need capital for their research, that research is then protected through copyright and patents, and further funds create the necessary factories. In the post-schumpeterian world, creative souls congregate through the internet, create new software, or any kind of knowledge, create collaboration platforms on the cheap, and paradoxically, only need capital when they are successful, and the servers risk crashing from overload. As an example, think about Bittorrent, the most important software for exchanging multimedia content over the internet, which was created by a single programmer, surviving through a creative use of some credit cards, with zero funding. But the internet is not just for creative individual souls, but enables large communities to cooperate over platforms. Very importantly, it is not limited to knowledge and software, but to everything that knowledge and software enables, which includes manufacturing. Anything that needs to be physically produced, needs to be 'virtually designed' in the first place.
>
> This phenomena [sic] is called social innovation or social production, and is increasingly responsible for most innovation. . . .
>
> But what does this all mean for the Asian economic crisis and the plight of the young people that we touched upon at the beginning? The good news is this: first, the strong distinction between working productively for a wage, and idly waiting for one, is melting. All the technical and intellectual tools are available to allow young people, and older people for that matter, to continue being engage [sic] in value production, and hence also to continue to build their experience (knowledge capital), their social life (relationship capital) and reputation.

[1]Scott Adams, "Ridesharing in the Future," *Scott Adams Blog*, January 21, 2009 <http://dilbert.com/blog/entry/ ridesharing_in_the_future/>.

All three of which will be crucial in keeping them not just employable, but will actually substantially increase their potential and capabilities. The role of business must be clear: it can, on top of the knowledge, software or design commons created by social production, create added value services that are needed and demanded by the market of users of such products (which includes other businesses), and can in turn sustain the commons from which it benefits, making the ecology sustainable. While the full community of developers create value for businesses to build upon, the businesses in term help sustain the infrastructure of cooperation which makes continued development possible.[1]

The shift of value-creation outside the cash nexus provoked an interesting blogospheric discussion between Tyler Cowen and John Quiggin. Cowen raised the possibility that much of the productivity growth in recent years has taken place "outside of the usual cash and revenue-generating nexus."[2] Quiggin, in an article appropriately titled "The end of the cash nexus," took the idea and ran with it:

> There has been a huge shift in the location of innovation, with much of it either deriving from, or dependent on, public goods produced outside the market and government sectors, which may be referred to as social production. . . .
>
> If improvements in welfare are increasingly independent of the market, it would make sense to shift resources out of market production, for example by reducing working hours. The financial crisis seems certain to produce at least a temporary drop in average hours, but the experience of the Depression and the Japanese slowdown of the 1990s suggest that the effect may be permanent. . . .[3]

Michel Bauwens, as we saw in Chapter Three, draws a parallel between the current crisis of realization in capitalism and previous crises like that of the Roman slave economy. When the system hits limits to extensive development, it instead turns to intensive development in ways that lead to a phase transition. But there is another parallel, Bauwens argues: each systemic decline and phase transition is associated with an "exodus" of labor:

> **The first transition: Rome to feudalism**
>
> At some point in its evolution (3rd century onwards?), the Roman empire ceases to expand (the cost of of maintaining empire and expansion exceeds its benefits). No conquests means a drying up of the most important raw material of a slave economy, i.e. the slaves, which therefore become more 'expensive'. At the same time, the tax base dries up, making it more and more difficult to maintain both internal coercion and external defenses. It is in this context that Perry Anderson mentions for example that when Germanic tribes were about to lay siege to a Roman city, they would offer to free the slaves, leading to an exodus of the city population. This exodus and the set of difficulties just described, set of a reorientation of some slave owners, who shift to the system of coloni, i.e. serfs. I.e. slaves are partially freed, can have families, can produce from themselves and have villages, giving the surplus to the new domain holders.
>
> Hence, the phase transition goes something like this: 1) systemic crisis ; 2) exodus 3) mutual reconfiguration of the classes. . . .
>
> Hypothesis of a third transition: capitalism to peer to peer

[1]Michel Bauwens, "Asia needs a Social Innovation Stimulus plan," *P2P Foundation Blog*, March 23, 2009 <http://blog.p2pfoundation.net/asia-needs-a-social-innovation-stimulus-plan/2009/03/23>.

[2]Tyler Cowen, "Was recent productivity growth an illusion?" *Marginal Revolution*, March 3, 2009 <http://www.marginalrevolution.com/marginalrevolution/2009/03/was-recent-productivity-growth-an-illusion.html>.

[3]John Quiggin, "The End of the Cash Nexus," *Crooked Timber*, March 5, 2009 <http://crookedtimber.org/2009/03/05/the-end-of-the-cash-nexus>.

Again, we have a system faced with a crisis of extensive globalization, where nature itself has become the ultimate limit. It's way out, cognitive capitalism, shows itself to be a mirage.

What we have then is an exodus, which takes multiple forms: precarity and flight from the salaried conditions; disenchantement with the salaried condition and turn towards passionate production. The formation of communities and commons are shared knowledge, code and design which show themselves to be a superior mode of social and economic organization.

The exodus into peer production creates a mutual reconfiguration of the classes. A section of capital becomes netarchical and 'empowers and enables peer production', while attempting to extract value from it, but thereby also building the new infrastructures of cooperation.[1]

If, as we saw in earlier chapters, economic downturns tend to accelerate the expansion of the custom industrial periphery at the expense of the mass-production core, such downturns also accelerate the shift from wage labor to self-employment or informal production outside the cash nexus. James O'Connor described the process in the economic stagnation of the 1970s and 1980s: "the accumulation of stocks of means and objects of reproduction within the household and community took the edge off the need for alienated labor."

> Labor-power was hoarded through absenteeism, sick leaves, early retirement, the struggle to reduce days worked per year, among other ways. Conserved labor-power was then expended in subsistence production. . . . The living economy based on non- and anti-capitalist concepts of time and space went underground: in the reconstituted household; the commune; cooperatives; the single-issue organization; the self-help clinic; the solidarity group. Hurrying along the development of the alternative and underground economies was the growth of underemployment . . . and mass unemployment associated with the crisis of the 1980s. "Regular" employment and union-scale work contracted, which became an incentive to develop alternative, localized modes of production. . . .
>
> . . . New social relationships of production and alternative employment, including the informal and underground economies, threatened not only labor discipline, but also capitalist markets. . . . Alternative technologies threatened capital's monopoly on technological development . . . Hoarding of labor-power threatened capital's domination of production. Withdrawal of labor-power undermined basic social disciplinary mechanisms. . . .[2]

And back in the recession of the early eighties, Samuel Bowles and Herbert Gintis speculated that the "reserve army of the unemployed" was losing some of its power to depress wages. They attributed this to the "*partial deproletarianization of wage labor*" (i.e. the reduced profile of wage labor alone as the basis of household subsistence). Bowles and Gintis identified this reduced dependency largely on the welfare state, which seems rather quaint for anyone who since lived through the Reagan and Clinton years.[3] But the partial shift in value creation from paid employment to the household and social economies, which we have seen in the past decade, fully accords with the same principle.

Dante-Gabryell Monson speculated on the possibility that the open manufacturing movement was benefiting from the skills of corporate tech people under-

[1] Michel Bauwens, "Three Times Exodus, Three Phase Transitions," *P2P Foundation Blog*, May 2, 2010 <http://blog.p2pfoundation.net/three-times-exodus-three-phase-transitions/2010/05/02>.

[2] James O'Connor, *Accumulation Crisis* (New York: Basil Blackwell, 1984), pp. 184-186.

[3] Samuel Bowles and Herbert Gintis, "The Crisis of Liberal Democratic Capitalism: The Case of the United States," *Politics and Society* 11:1 (1982), pp. 79-84.

employed in the current downturn, or even from their deliberate choice to hoard labor:

> *Is there a potential scenario for a brain drain from corporations to intentional peer producing networks ?*. . . .
>
> Can part-time, non-paid (in mainstream money) "hobby" work in open, diy, collaborative convergence spaces become an *argument for long term material security of the participating peer* towards he's/her family ?
>
> Hacker spaces seem to be convergence spaces for open source programmers, and possibly more and more other artists, open manufacturing, diy permaculture, . . . ?
>
> Can we expect a "Massive Corporate Dropout" . . . to drain into such diy convergence and interaction spaces ?
>
> *Can "Corporate Dropouts" help financing new open p2p infrastructures ?*
>
> Is there an increase of part-time "Corporates", working part time in open p2p ?
>
> Would such a transition, potentially part time "co-working / co-living " space be a convergence "model" and scenario some of us would consider working on ? . . .
>
> I personally observe some of my friends working for money as little as possible, sometimes on or two months a year, and spend the rest of their time working on their own projects.[1]

The main cause for the apparently stabilizing level of unemployment in the present recession, despite a decrease in the number of employed, is that so many "discouraged workers" have disappeared from the unemployment rolls altogether. At the same time, numbers for self-employment are continuing to rise.

> We [Canadians] lost another 45,000 jobs in July, but the picture is much worse on closer examination. There were 79,000 fewer workers in paid jobs compared to June, while self-employment rose by 35,000. This was on top of another big jump in self-employment of 37,000 last month.
>
> Put it all together and the picture is of large losses in paid jobs, with the impact on the headline unemployment rate cushioned by workers giving up the search for jobs or turning to self-employment.[2]

A recent article in the Christian Science Monitor discussed the rapid growth of the informal economy, even as the formal economy and employment within it shrink (Friedrich Schneider, a scholar who specializes in the shadow economy, expects it to grow at least five percent this year). Informal enterprise is mushrooming among the unemployed and underemployed of the American underclass: street vendors of all kinds (including clothing retail), unlicensed moving services consisting of a pickup truck and cell phone, people selling food out of their homes, etc.

And traditional small businesses in permanent buildings resent the hell out of it (if you ever saw that episode of *The Andy Griffith Show* where established retailer Ben Weaver tries to shut down Emmett's pushcart, you get the idea).

> "Competition is competition," says Gene Fairbrother, the lead small-business adviser in Dallas for the National Association for the Self-Employed. But competition from producers who don't pay taxes and licensing fees isn't fair to the many struggling small businesses who play by the rules.

[1]Dante-Gabryell Monson, "[p2p-research] trends ? : 'Corporate Dropouts' towards Open diy ? . . . ," *P2P Research*, October 13, 2009 <http://listcultures.org/pipermail/p2presearch_listcultures.org/2009-October/005128.html>.

[2]Andrew Jackson, "Recession Far From Over," *The Progressive Economics Forum*, August 7, 2009 <http://www.progressive-economics.ca/2009/08/07/recession-far-from-over/>.

> Mr. Fairbrother says he's seen an increase in the number of callers to his Shop Talk show who ask about starting a home-based business, and many say they're working in a salon and would rather work out of their homes or that they want to start selling food from their kitchens. Businesses facing this price pressure should promote the benefits of regulation, he advises, instead of trying to get out from under it.

Uh huh. Great "benefits" if you're one of the established businesses that uses the enormous capital outlays for rent on dedicated commercial real estate, industrial-sized ovens and dishwashers, licensing fees, etc., to crowd out competitors. Not so great if you're one of the would-be microentrepreneurs forced to pay artificially inflated overhead on such unnecessary costs, or one of the consumers who must pay a price with such overhead factored in. Parasitism generally has much better benefits for the tapeworm than for the owner of the colon.

Fortunately, in keeping with our themes of agility and resilience throughout this book, microentrepreneurs tend to operate on a small scale beneath the radar of the government's taxing, regulatory and licensing authorities. In most cases, the cost of catching a small operator with a small informal client network is simply more than it's worth.

> The Internal Revenue Service or local tax authorities would have to track down thousands of elusive small vendors and follow up for payment to equal, by one estimate, the $100 million a year that the US could gain by taxing several hundred holders of Swiss and other foreign bank accounts.[1]

So we can expect the long-term structural reduction in employment and the shortage of liquidity, in the current Great Recession or Great Malaise, to lead to rapid growth of an informal economy based on the kinds of household microenterprises we described above. Charles Hugh Smith, after considering the enormous fixed costs of conventional businesses and the inevitability of bankruptcy for businesses with such high overhead in a period of low sales, draws the conclusion that businesses with low fixed costs are the wave of the future. Here is his vision of the growing informal sector of the future:

> **The recession/Depression will cut down every business paying high rent and other fixed costs like a razor-sharp scythe hitting dry corn stalks. . . .**
>
> **. . . [H]igh fixed costs will take down every business which can't remake itself into a low-fixed-cost firm. . . .**
>
> **For the former employees, the landscape is bleak: there are no jobs anywhere, at any wage. . . .**
>
> **So how can anyone earn a living in The End of Work? Look to Asia for the answer.** The MSM snapshot of Asia is always of glitzy office towers in Shanghai or a Japanese factory or the docks loaded with containers: the export machine.
>
> But if you actually wander around Shanghai (or any city in Japan, Korea, southeast Asia, etc.) then you find the number of people working in the glitzy office tower is dwarfed by the number of people making a living operating informal businesses.
>
> **Even in high-tech, wealthy Japan, tiny businesses abound.** Wander around a residential neighborhood and you'll find a small stall fronting a house staffed by a retired person selling cigarettes, candy and soft drinks. Maybe they only sell a few dollars' worth of goods a day, but it's something, and in the meantime the proprietor is reading a magazine or watching TV.

[1]Taylor Barnes, "America's 'shadow economy' is bigger than you think—and growing," *Christian Science Monitor*, November 12, 2009 <http://features.csmonitor.com/economyrebuild/2009/11/12/americas-shadow-economy-is-bigger-than-you-think-and-growing>.

In old Shanghai, entire streets are lined with informal vendors. Some are the essence of enterprise: a guy buys a melon for 40 cents, cuts it into 8 slices and then sells the slices for 10 cents each. Gross profit, 40 cents.

In Bangkok, such areas actually have two shifts of street vendors: one for the morning traffic, the other for the afternoon/evening trade. The morning vendors are up early, selling coffee, breakfasts, rice soup, etc. to workers and school kids. By 10 o'clock or so, they've folded up and gone home.

That clears the way for the lunch vendors, who have prepared their food at home and brought it to sell. In some avenues, a third shift comes in later to sell cold drinks, fruit and meat sticks as kids get out of school and workers head home.

Fixed costs of these thriving enterprises: a small fee to some authority, an old cart and umbrella—and maybe a battered wok or ice chest.

So this is what I envision happening as the Depression drives standard-issue high-fixed cost "formal" enterprises out of business in the U.S.:

1. The mechanic who used to tune your (used) vehicle for $300 at the dealership (now gone) tunes it up in his home garage for $120—parts included.

2. The gal who cut your hair for $40 at the salon now cuts it at your house for $10.

3. The chef who used to cook at the restaurant that charged $60 per meal now delivers a gourmet plate to your door for $10 each.

4. The neighbor kids' lemonade stand is now a permanent feature; you pay 50 cents for a lemonade or soft drink instead of $3 at Starbucks.

5. Used book sellers spread their wares on the sidewalk, or in fold-up booths; for reasons unknown, one street becomes the "place to go buy used books."

6. The neighborhood jazz guy/gal sets up and plays with his/her pals in the backyard; donations welcome.

7. The neighborhood chips in a few bucks each to make it worth a local Iraqi War vet's time to keep an eye on things.

8. When your piece-of-crap Ikea desk busts, you call a guy who can fix it for $10 (glue, clamps, a few ledger strips and screws) rather than go blow $50 on another particle board P.O.C. which will bust anyway. (oh, and you don't have the $50 anyway.)

9. The guy with a Dish runs cables to the other apartments in his building for a few bucks each.

10. One person has an "unlimited" Netflix account, and everyone pays him/her a buck a week to get as many movies as they want (he/she burns a copy of course).

11. The couple with the carefully tended peach or apple tree bakes 30 pies and trades them for vegtables, babysitting, etc.[1]

The crushing costs of formal business (State and local government taxes and junk fees rising to pay for unaffordable pensions, etc.) and the implosion of the debt-bubble economy will drive millions into the informal economy of barter, trade and "underground" (cash) work.

As small businesses close their doors and corporations lay off thousands, the unemployed will of necessity shift their focus from finding a new formal job (essentially impossible for most) to fashioning a livelihood in the informal economy.

One example of the informal economy is online businesses—people who make a living selling used items on eBay and other venues. Such businesses can be operated at home and do not require storefronts, rent to commercial landlords, employees, etc., and because they don't require a formal presence then they also fly beneath all the government junk fees imposed on formal businesses.

[1]Charles Hugh Smith, "End of Work, End of Affluence III: The Rise of Informal Businesses," *Of Two Minds*, December 10, 2009 <http://www.oftwominds.com/blogdec08/informal12-08.html>.

I have mentioned such informal businesses recently, and the easiest way to grasp the range of possibilities is this: whatever someone did formally, they can do informally.

Chef had a high fixed-cost restaurant which bankrupted him/her? Now he/she prepares meals at home and delivers them to neighbors/old customers for cash. No restaurant, no skyhigh rent, no employees, no payroll taxes, no business licenses, inspection fees, no sales tax, etc. Every dime beyond the cost of food and utilities to prepare the meals stays in Chef's pocket rather than going to the commercial landlords and local government via taxes and fees.

All the customers who couldn't afford $30 meals at the restaurant can afford $10. Everybody wins except commercial landlords (soon to be bankrupt) and local government (soon to be insolvent). How can you bankrupt all the businesses and not go bankrupt yourself?

As long as Chef reports net income on Schedule C, he/she is good to go with Federal and State tax authorities. [And if Chef doesn't, fuck 'em.]

Now run the same scenario for mechanics, accountants, therapists, even auto sales—just rent a house with a big yard or an apartment with a big parking lot and away you go; the savvy entrepreneur who moves his/her inventory can stock a few vehicles at a time. No need for a huge lot, high overhead, employees or junk fees. It's cash and carry.

Lumber yard? Come to my backyard lot. Whatever I don't have I can order from a jobber and have delivered to your site.

This is the result of raising the fixed costs of starting and running a small business to such a backbreaking level that few formal businesses can survive.[1]

[1]Smith, "Trends for 2009: The Rise of Informal Work," *Of Two Minds*, December 30, 2009 <http://www.oftwominds.com/blogdec08/rise-of-informal12-08.html>.

Appendix
Case Studies in the Coordination of Networked Fabrication and Open Design

1. *Open Source Ecology/Factor e Farm*

Open Source Ecology, with its experimental demo site at Factor e Farm, is focused on developing the technological building blocks for a resilient local economy.

> We are actively involved in demonstrating the world's first replicable, post-industrial village. We take the word *replicable* very seriously—we do not mean a top-down funded showcase—but one that is based on ICT, open design, and digital fabrication—in harmony with its natural life support systems. As such, this community is designed to be self-reliant, highly productive, and sufficiently transparent so that it can truly be replicated in many contexts—whether it's parts of the package or the whole. Our next frontier will be education to train Village Builders—just as we're learning how to do it from the ground up.[1]
>
> Open Source Ecology's latest core message is "Building the world's first replicable, open source, modern off-grid global village—to transcend survival and evolve to freedom." . . .
>
> Replicable means that the entire operation can be copied and 'replicated' at another location at low cost.
>
> Open source means that the knowledge of how it works and how to make it is documented to the point that others can "make it from scratch." It can also be changed and added to as needed. . . .
>
> Permafacture: A car is a temporarily useful consumer product—eventually it breaks down and is no longer useful as a car. The same is true for almost any consumer product—they are temporary, and when they break down they are no longer useful for their intended purpose. They come from factories that use resources from trashing ecosystems and using lots of oil. Even the "green" ones. Most consumer food is grown on factory farms using similar processes, and resulting in similar effects. When the resources or financing for those factories and factory farms dries up they stop producing, and all the products and food they made stop flowing into the consumer world. Consumers are dependent on these products and food for their very survival, and every product and food they buy from these factories contributes to the systems that are destroying the ecosystems that they will need to survive when finances or resources are interrupted. The more the consumers buy, the more dependent they are on the factories consuming and destroying the last of the resources left in order to maintain their current easy and dependent survival. These factories are distributed all over the world, and need large amounts of cheap fuel to move the products to market through the global supply and production chain, trashing ecosystems all along the way. The consumption of the products and food is completely disconnected from their production and so consumers do not actually see any of these connections or their interruptions as the factories and supply chains try hard to keep things flowing smoothly, until things reach their breaking point and the supply of products to consumers is suddenly interrupted. Open Source Ecology aims to create the means of production and reuse on a small local scale, so that we can produce the machines and resources that make survival trivial without being de-

[1]Marcin Jakubowski, "Clarifying OSE Vision," *Factor e Farm Weblog*, September 8, 2008 <http://openfarmtech.org/weblog/?p=325>.

pendent on global supply and production chains, trashing ecosystems, and cheap oil.[1]

The focus of OSE is to secure "right livelihood," according to founder Marcin Jakubowski, who cites Vinay Gupta's "The Unplugged" as a model for achieving it:

> The focus of our *Global Village Construction program* is to deploy *communities* that live according to the intention of *right livelihood*. We are considering the *ab initio* creation of nominally 12 person communities, by networking and marketing this *Buy Out at the Bottom* (BOAB) package, at a fee of approximately $5k to participants. *Buying Out at the Bottom* is a term that I borrowed from Vinay Gupta in his article about The Unplugged—where *unplugging* means the creation of *an independent life-support infrastructure and financial architecture—a society within society—which allowed anybody who wanted to "buy out" to "buy out at the bottom" rather than "buying out at the top."*
>
> Our Global Village Construction program is an implementation of The Unplugged lifestyle. With 12 people buying out at $5k each, that is $60k seed infrastructure capital.
>
> We have an option to stop feeding invading colonials, from our own empire-building governments to slave goods from China. Structurally, the more self-sufficient we are, the less we have to pay for our own enslavement—through education that dumbs us down to producers in a global workforce—through taxation that funds rich peoples' wars of commercial expansion—through societal engineering and PR that makes the quest for an honest life dishonorable if we can't keep up with the Joneses. [2]

Several of the most important projects interlock to form an "OSE Product Ecology."[3] For example the LifeTrac Open Source Tractor acts as prime mover for Fabrication (i.e., the machine shop, in which the Multi-Machine features prominently), and the Compressed Earth Block Press and the Sawmill, which in turn are the basic tools for housing construction. The LifeTrac also functions, of course, as a tractor for hauling and powering farm machinery.

Like LifeTrac, the PowerCube—a modular power-transmission unit—is a multi-purpose mechanism designed to work with several of the other projects.

> Power Cube is our open source, self-contained, modular, interchangeable, hydraulic power unit for all kinds of power eguipment. It has an 18 hp gasoline engine coupled to a hydraulic pump, and it will later be be powered by a flexible-fuel steam engine. Power Cube will be used to power MicroTrac (under construction) and it is the power source for the forthcoming CEB Press Prototype 2 adventures. It is designed as a general power unit for all devices at Factor e Farm, from the CEB press, power take-off (PTO) generator, heavy-duty workshop tools, even to the LifeTrac tractor itself. Power Cube will have a quick attachment, so it can be mounted readily on the quick attach plate of LifeTrac. As such, it can serve as a backup power source if the LifeTrac engine goes out. . . .
>
> The noteworthy features are modularity, hydraulic quick-couplers, lifetime design, and design-for-disassembly. Any device can be plugged in readily through the quick couplers.
>
> It can be maintained easily because of its transparency of design, ready access to parts, and design for disassembly. It is a major step towards realizing the

[1]Jeremy Mason, "What is Open Source Ecology?" *Factor e Farm Weblog*, March 20, 2009 <http://openfarmtech.org/weblog/?p=595>.

[2]"Organizational Strategy," Open Source Ecology wiki, February 11, 2009 <http://openfarmtech.org/index.php?title=Organizational_Strategy> (accessed August 28, 2009).

[3]Marcin Jakubowski, "CEB Proposal—Community Supported Manufacturing," *Factor e Farm* weblog, October 23, 2008 <http://openfarmtech.org/weblog/?p=379>.

> true, life-size Erector Set or Lego Set of heavy-duty, industrial machinery in the style of Industrial Swadeshi.[1]
>
> A universal mechanical power source is one of the key components of the Global Village Construcgtion Set—the set of building blocks for creating resilient communities. The basic concept is that instead of using a dedicated engine on a particular powered device—which means hundreds of engines required for a complete resilient community, you need one (or a few) power unit. If this single power unit can be coupled readily to the powered device of interest, then we have the possibility of this single power unit being interchangeable between an unlimited number of devices. Our implementation of this is the hydrauilic PowerCube—whose power can be tapped simply by attaching 2 hydraulic hoses to a device of interest. A 3/4" hydraulic hose . . . can transfer up to 100 horsepower in the form of usable hydraulic fluid flow.[2]

Among projects that have reached the prototype stage, the foremost is the Compressed Earth Block Press, which can be built for $5000—some 20% of the price of the cheapest commercial competitor.[3] In field testing, the CEB Press demonstrated the capability of producing a thousand blocks in eight-hours, on a day with bad weather (the expected norm in good weather is 1500 a day).[4] On August 20, 2009, Factor e Farm announced completion of a second model prototype, its most important new feature being an extendable hopper that can be fed directly by a tractor loader. Field testing is expected to begin shortly.[5]

> The speed of the CEB Press was recently augmented by the prototyping of a complementary product, the Soil Pulverizer.
>
> Initial testing achieved 5 ton per hour soil throughput, while The Liberator CEB press requires about 1.5 tons of soil per hour. . . .
>
> Stationary soil pulverizers comparable in throughput to ours cost over $20k. Ours cost $200 in materials—which is not bad in terms of 100-fold price reduction. The trick to this feat is modular design. We are using components that are already part of our LifeTrac infrastructure. The hydraulic motor is our power take-off (PTO) motor, the rotor is the same tiller that we made last year—with the tiller tines replaced by pulverizer tines. The bucket is the same standard loader bucket that we use for many other applications. . . .
>
> It is interesting to compare this development to our CEB work from last year—given our lesson that soil moving is the main bottleneck in earth building. It takes 16 people, 2 walk-behind rototillers, many shovels and buckets, plus backbreaking labor—to load our machine as fast as it can produce bricks. We can now replace this number of people with 1 person—by mechanizing the earth moving work with the tractor-mounted pulverizer. In a sample run, it took us about 2 minutes to load the pulverizer bucket—with soil sufficient for about 30 bricks. Our machine produces 5 bricks per minute—so we have succeeded in removing the soil-loading bottleneck from the equation.
>
> This is a major milestone for our ability to do CEB construction. Our results indicate that we can press 2500 bricks in an 8 hour day—with 3 people.[6]

[1]Jakubowski, "Power Cube Completed," *Factor e Farm Weblog*, June 29, 2009 <http://openfarmtech.org/weblog/?p=814>.

[2]Jakubowski, "PowerCube on LifeTrak," *Factor e Farm Weblog* , April 26, 2010 <http://openfarmtech.org/weblog/?p=1761>.

[3]Jakubowski, "CEB Phase 1 Done," *Factor e Farm Weblog*, December 26, 2007 <http://openfarmtech.org/weblog/?p=91>.

[4]Jakubowski, "The Thousandth Brick: CEB Field Testing Report," *Factor e Farm Weblog*, Nov. 16, 2008 <http://openfarmtech.org/weblog/?p=422>.

[5]Jakubowski, "CEB Prototype II Finished," *Factor e Farm Weblog*, August20, 2009 <http://openfarmtech.org/weblog/?p=1025>.

[6]Jakubowski, "Soil Pulverizer Annihilates Soil Handling Limits," *Factor e Farm Weblog*, September 7, 2009 <http://openfarmtech.org/weblog/?p=1063>.

In October Jakubowski announced plans to release the CEB Beta Version 1.0 on November 1, 2009. The product as released will have a five block per minute capacity and include automatic controls (the software for which is being released on an open-source basis).[1] The product was released, on schedule, on November 1.[2] Shortly thereafter, OSE was considering options for commercial production of the CEB Press as a source of revenue to fund new development projects.[3]

The MicroTrac, a walk-behind tractor, has also been prototyped. Its parts, including the Power Cube, wheel, quick-attach motor and cylinder are interchangeable with LifeTrac and other machines. "We can take off the wheel motor from MicroTrac, and use it to power shop tools."[4]

OSE's planned facilities for replication and machining are especially exciting, including a 3-D printer and a Multi-Machine with added CNC controls.

> There is a significant set of open source technologies available for rapid prototyping in small workshops. By combining 3D printing with low-cost metal casting, and following with machining using a computer controlled Multimachine, the capacity arises to make rapid prototypes and products from plastic and metal. This still does not address the feedstocks used, but it is a practical step towards the post-centralist, participatory, distributive economy with industrial swadeshi on a regional scale. . . .
>
> The interesting part is that the budget is $500 for RepRap, $200 for the casting equipment, and $1500 for a Multimachine with CNC control added. Using available knowhow, this can be put together in a small workshop for a total of about $2200—for full, LinuxCNC computer controlled rapid fabrication in plastic and metal. Designs may be downloaded from the internet, and local production can take place based on global design.
>
> This rapid fabrication package is one of our near-term (one year) goals. The research project in this area involves the fabrication and integration of the individual components as described. . . .
>
> Such a project is interesting from the standpoint of localized production in the context of the global economy—for creating significant wealth in local economies. This is what we call industrial swadeshi. For example, I see this as the key to casting and fabricating low-cost steam engines ($300 for 5 hp) for the Solar Turbine—as one example of Gandhi's mass production philosophy.[5]

The entire Fab Lab project aims to produce "the following equipment infrastructure, in order of priorities . . . ":

- 300 lb/hour steel melting Foundry—$1000
- Multimachine-based Lathe, mill, and drill, with addition of CNC control—$1500
- CNC Torch Table (plasma and oxyacetylene), adaptable to a router table
- RepRap or similar 3D printer for printing casting molds—$400
- Circuit fabrication—precise xyz router table

[1]Jakubowski, "Exciting Times: Nearing Product Release," *Factor e Farm Weblog*, October 10, 2009 <http://openfarmtech.org/weblog/?p=1168>.

[2]Jakubowski, "Product," *Factor e Farm Weblog*, November 4, 2009 <http:// openfarmtech.org/weblog/?p=1224>.

[3]Jakubowski, "CEB Sales: Rocket Fuel for Post-Scarcity Economic Development?" *Factor e Farm Weblog*, November 28 2009 <http://openfarmtech.org/weblog/?p=1331>.

[4]Jakubowski, "MicroTrac Completed," *Factor e Farm Weblog*, July 7, 2009 <http:// openfarmtech.org/weblog/?p=852>.

[5]Jakubowski, "Rapid Prototyping for Industrial Swadeshi," *Factor e Farm Weblog*, August 10, 2008 <http://openfarmtech.org/weblog/?p=293>.

- Open Source Wire Feed Welder[1]

In August 2009, Lawrence Kincheloe moved to Factor e Farm in August 2009 under contract to build the torch table in August and September.[2] He ended his visit in October with work on the table incomplete, owing to "a host of fine tuning and technical difficulties which all have solutions but were not addressable in the time left."[3] Nevertheless, the table was featured in the January issue of *MAKE Magazine* as RepTab (the name reflects the fact that—aside from motors and microcontrollers—it can replicate itself):

> One of the interesting features of RepTab is that the cutting head is interchangeable (router, plasma, oxyacetylene, laser, water jet, etc.), making it versatile and extremely useful.
>
> "Other machines make that difficult without major modifications," says Marcin Jakubowski, the group's founder and director. "We can make up to 10-foot-long windmill blades if we modify the table as a router table. That's pretty useful."[4]

Since then, Factor e Farm has undertaken to develop an open-source lathe, as well as a 100-ton ironworker punching/shearing/bending machine; Jakubowski estimates an open-source version can be built for a few hundred dollars in materials, compared to $10,000 for a commercial version.[5]

In December 2009 Jakubowski announced that a donor had committed $5,000 to a project for developing an open-source induction furnace for smelting, and solicited bids for the design contract.

> You may have heard us talk about recasting civilization from scrap metal. Metal is the basis of advanced civilization. Scrap metal in refined form can be mined in abundance from heaps of industrial detritus in junkyards and fence rows. This can help us produce new metal in case of any unanticipated global supply chain disruptions. This will have to do until we can take mineral resources directly and smelt them to pure metal.
>
> I look forward to the day when our induction furnace chews up our broken tractors and cars—and spits them out in fluid form. This leads to casting useful parts, using molds printed by open source ceramic printers—these exist. This also leads to hot metal processing, the simplest of which is bashing upon an anvil—and the more refined of which is rolling. Can we do this to generate metal bar and sheet in a 4000 square foot workshop planned for Factor e Farm? We better. Technology makes that practical, though this is undeard-of outside of centralized steel mills. We see the induction furnace, hot rolling, forging, casting, and other processes critical to the fabrication component of the Global Village Construction Set.

[1]"Open Source Fab lab," Open Source Ecology wiki (accessed August 22, 2009) <http://openfarmtech.org/index.php?title=Open_Source_Fab_Lab>.

[2]Marcin Jakubowski, "Moving Forward," Factor e Farm Weblog, August 20, 2009 <http://openfarmtech.org/weblog/?p=1020>; "Lawrence Kincheloe Contract," OSE Wiki <http://openfarmtech.org/index.php?title=Lawrence_Kincheloe_Contract>; "Torch Table Build," *Open Source Ecology* wiki (accessed August 22, 2009 <http:// openfarmtech.org/ index.php?title=Torch_Table_Build>.

[3]Lawrence Kincheloe, "First Dedicated Project Visit Comes to a Close," *Factor e Farm Weblog*, October 25, 2009 <http://openfarmtech.org/weblog/?p=1187> (see especially comment no. 5 under the post).

[4]Abe Connally, "Open Source Self-Replicator," *MAKE Magazine*, No. 21 <http://www.make-digital.com/make/vol21/?pg=69>.

[5]Jakubowski, "CEB Sales"; "Ironworkers," Open Source Ecology Wiki <http:// openfarmtech.org/index.php?title=Ironworkers>. Accessed December 10, 2009.

We just got a $5k commitment to open-source this technology.[1]

In January, Jacubowski reported initial efforts to build a lathe-drill-mill multimachine (not CNC, apparently) powered by the LifeTrac motor.[2]

In addition to the steel casting functions of the Foundry, Jakubowski ultimately envisions the production of aluminum from clay as a key source of feedstock for relocalized production. As an alternative to "high-temperature, energy-intensive smelting processes" involving aluminum oxide (bauxite), he proposes "extracting aluminum from clays using baking followed by an acid process."[3]

OSE's flexible and digital fabrication facility is intended to produce a basic set of sixteen products, five of which are the basic set of means of fabrication themselves:

- *Boundary layer turbine*—simpler and more efficient alternative to most external and internal combustion engines and turbines, such as gasoline and diesel engines, Stirling engines, and air engines. The only more efficient energy conversion devices are bladed turbines and fuel cells.
- *Solar concentrators*—alternative heat collector to various types of heat generators, such as petrochemical fuel combustion, nuclear power, and geothermal sources
- *Babington[4] and other fluid burners*--alternative heat source to solar energy, internal combustion engines, or nuclear power
- *Flash steam generators*—basis of steam power
- *Wheel motors*—low-speed, high-torque electric motors
- *Electric generators*—for generating the highest grade of usable energy: electricity
- *Fuel alcohol* production systems—proven biofuel of choice for temperate climates
- *Compressed wood gas*—proven technology; cooking fuel; usable in cars if compressed
- *Compressed Earth Block* (CEB) press—high performance building material
- *Sawmill*—production of dimensional lumber
- *Aluminum from clay*—production of aluminum from subsoil clays

Means of fabrication:

- *CNC Multimachine[5]*—mill, drill, lathe, metal forming, other grinding/cutting
- *XYZ*-controlled torch and router table—can accommodate an acetylene torch, plasma cutter, router, and possibly CO_2 laser cutter diodes
- *Metal casting* equipment—various metal parts
- *Plastic extruder[6]*—plastic glazing and other applications
- *Electronics fabrication*—oscilloscope, multimeter, circuit fabrication; specific power electronics products include battery chargers, inverters, converters,

[1]Jakubowski, "Open Source Induction Furnace," *Factor e Farm Weblog*, December 15, 2009 <http://openfarmtech.org/weblog/?p=1373>.

[2]Jakubowski, "Initial Steps to the Open Source Multimachine," *Factor e Farm Weblog*, January 26, 2010 <http://openfarmtech.org/weblog/?p=1408>.

[3]Jakubowski, "OSE Proposal—Towards a World Class Open Source Research and Development Facility" v0.12, January 16, 2008 <http://openfarmtech.org/OSE_Proposal.doc> (accessed August 25, 2009).

[4]http://www.aipengineering.com/babington/Babington_Oil_Burner_HOWTO.html

[5]<http://opensourcemachine.org/>.

[6]See Extruder_doc.pdf at <http://www.fastonline.org/CD3WD_40/CD3WD/INDEX.HTM>.

> transformers, solar charge controllers, PWM DC motor controllers, multipole motor controllers.[1]

The Solar Turbine, as it was initially called, uses the sun's heat to power a steam-driven generator, as an alternative to photovoltaic electricity.[2] It has since been renamed the Solar Power Generator, because of the choice to use a simple steam engine as the heat engine instead of a Tesla turbine.[3]

The Steam Engine, still in the design stage, is based on a simple and efficient design for a 3kw engine, with an estimated bill of parts of $250.[4]

The Sawmill, which can be built with under $2000 in parts (a "Factor 10 cost reduction"), has "the highest production rate of any small, portable sawmills."[5]

OSE's strategy is to use the commercial potential of the first products developed to finance further development. As we saw earlier, Jakubowski speculates that a fully equipped digital fabrication facility could turn out CEB presses or sawmills with production rates comparable to those of commercial manufacturing firms, cutting out all the metal parts for the entire product with a turn-around time of days. The CEBs and sawmills could be sold commercially, in that case, to finance development of other products.[6]

And in fact, Jakubowski has made a strategic decision to give priority to developing the CEB Press as rapidly as possible, in order to leverage the publicity and commercial potential as a source of future funding for the entire project.[7]

OSE's goal of replicability, once the first site is completed with a full range of production machinery and full product line, involves hosting interns who wish to replicate the original experiment at other sites, and using fabrication facilities to produce duplicate machinery for the new sites.[8] Jakubowski recently outlined a more detailed timeline:

> Based on our track record, the schedule may be off by up to twenty years. Thus, the proposed timeline can be taken as either entertainment or a statement of intent—depending on how much one believes in the project.
>
> 2008—modularity and low cost features of open source products have been demonstrated with LifeTrac and CEB Press projects
>
> 2009—First product release
>
> 2010—TED Fellows or equivalent public-relations fellowship to propel OSE to high visibility
>
> 2011—$10k/month funding levels achieved for scaling product development effort
>
> 2012—Global Village Construction Set finished
>
> 2013—First true post-scarcity community built

[1]Jakubowski, "OSE Proposal" [Note—OSE later decided to replace the boundary layer turbine with a simple steam engine as their primary heat engine. Also "Babington oil burner, compressed fuel gas production, and fuel alcohol production have now been superseded by pelletized biomass-fueled steam engines." (Marcin Jakubowski, private email, January 22, 2010)]

[2]"Solar Turbine—Open Source Ecology" <http://openfarmtech.org/index.php?title=Solar_Turbine>.

[3]Marcin Jakubowski, "Factor e Live Distillations—Part 8—Solar Power Generator," *Factor e Farm Weblog*, February 3, 2009 <http://openfarmtech.org/weblog/?p=507>.

[4]Nick Raaum, "Steam Dreams," *Factor e Farm* weblog, January 22, 2009 <http://openfarmtech.org/weblog/?p=499>.

[5]Jeremy Mason, "Sawmill Development," *Factor e Farm* weblog, January 22, 2009 <http://openfarmtech.org/weblog/?p=498>.

[6]Jakubowski, "OSE Proposal."

[7]*Ibid.*

[8]"Organizational Strategy."

2014—OSE University (immersion training) established, to be competitive with higher education but with an applied focus

2015—OSE Fellows program started (the equivalent of TED Fellows, but with explicit focus of solving pressing world issues)

2016—First productive recursion completed (components can be produced locally anywhere)

2017—Full meterial [sic] recursion demostrated (all materials become producible locally anywhere)

2018—Ready self-replicability of resilient, post-scarcity communities demonstrated

2019—First autonomous republic created, along the governance principles of Leashless

2020—Ready replicability of autonomous republics demonstrated[1]

In August 2009, some serious longtime tensions came to a head at OSE, as the result of personality conflicts beyond the scope of this work, and the subsequent departure of members Ben De Vries and Jeremy Mason.

Since then, the project has given continuing signs of being functional and on track. As of early October 2009, Lawrence Kincheloe had completed torch table Prototype 1 (pursuant to his contract described above), and was preparing to produce a debugged Prototype 2 (with the major portion of its components produced with Prototype 1).[2] As recounted above, OSE also went into serial production of the CEB Press and has undertaken new projects to build the open-source lathe and ironworker.

2. *100kGarages*

Another very promising open manufacturing project, besides OSE's, is 100kGarages—a joint effort of ShopBot and Ponoko. ShopBot is a maker of CNC routers.[3] Ponoko is both a network of designers and a custom machining service, that produces items as specified in customer designs uploaded via Internet, and ships them by mail, and also has a large preexisting library of member product designs available for production.[4] 100kGarages is a nationwide American network of fabbers aimed at "distributed production in garages and small workshops"[5]: linking separate shops with partial tool sets together for the division of labor needed for networked manufacturing, enabling shops to contract for the production of specific components, or putting customers in contact with fabbers who can produce their designs. Ponoko and ShopBot, in a joint announcement, described it as helping 20,000 creators meet 6,000 fabricators, and specifically putting them in touch with fabricators in their own communities.[6] As described at the 100kGarages site:

> 100kGarages.com is a place for anyone who wants to get something made ("Makers") to link up with those having tools for digital fabrication ("Fabbers")

[1] Jakubowski, ""TED Fellows," *Factor e Farm Weblog*, September 22, 2009 <http://openfarmtech.org/weblog/?p=1121>.

[2] Lawrence Kincheloe, "One Month Project Visit: Take Two," *Factor e Farm Weblog*, October 4, 2009 <http://openfarmtech.org/weblog/?p=1146>.

[3] <http://www.shopbottools.com/>.

[4] <http://www.ponoko.com/>.

[5] "What's Digital Fabrication?" 100kGarages website <http://100kgarages.com/digital_fabrication.html>.

[6] Ted Hall (ShopBot) and Derek Kelley (Ponoko), "Ponoko and ShopBot announce partnership: More than 20,000 online creators meet over 6,000 digital fabricators," joint press release, September 16, 2009. Posted on Open Manufacturing email list, September 16, 2009 <http://groups.google.com/group/openmanufacturing/browse_thread/thread/fdb7b4d562f5e59d?hl=en>.

used to make parts or projects. . . . At the moment, the structure is in place to for [sic] Makers to find Fabbers and to post jobs to the Fabber community. . . . We're working hard to provide software and training resources to help those who want to design for Fabbers, whether doing their own one-off projects or to use the network of Fabbers for distributed manufacturing of products (as done by the current gallery of designers on the Ponoko site).

In the first few weeks there have been about 40 Fabbers who've joined up. In the beginning, we are sticking to Fabbers who are ShopBotters. This makes it possible to have some confidence in the credibilty and capability of the Fabber, without wasting enormous efforts on certification. . . . But before long, we expect to open up 100kGarages.com to all digital fabrication tools, whether additive or subtractive. We're hoping to grow to a couple of hundred Fabbers over the next few months, and this should provide a geographical distribution that brings fabrication capabilities pretty close to everyone and helps get the system energized.[1]

As we all are becoming environmentally aware, we realize that our environment just can't handle transporting all our raw materials across the country or around the world, just to ship them back as finished products. These new technologies make practical and possible doing more of our production and manufacturing in small distributed facilities, as small as our garages, and close to where the product is needed. Most importantly our new methods for collaboration and sharing means that we don't have to do it all by ourselves . . . that designers with creative ideas but without the capability to see their designs become real can work with fabricators that might not have the design skills that they need but do have the equipment and the skills and orientation that's needed to turn ideas into reality . . . that those who just want to get stuff made or get their ideas realized can work with the Makers/designers who can help them create the plans and the local fabricators who fulfill them.

To get this started ShopBot Tools, Makers of popular tools for digital fabrication and Ponoko, who are reinventing how goods are designed, made and distributed, are teaming-up to create a network of workshops and designers, with resources and infrastructure to help facilitate "rolling up our sleeves and getting to work." Using grass roots enterprise and ingenuity this community can help get us back in action, whether it's to modernize school buildings and infrastructure, develop energy-saving alternatives, or simply produce great new products for our homes and businesses.

There are thousands of ShopBot digital-fabrication (CNC) tools in garages and small shops across the country, ready to locally fabricate the components needed to address our energy and environmental challenges and to locally produce items needed to enhance daily living, work, and business. Ponoko's web methodologies offer people who want to get things made an environment that integrates designers and inventors with ShopBot fabricators. Multiple paths for getting from idea to object, part, component, or product are possible in a dynamic network like this, where ideas can be realized in immediate distributed production and where production activities can provide feedback to improve designs.[2]

Although all ShopBot CNC router models are quite expensive compared to the reverse-engineered stuff produced by hardware hackers (most models are in the $10-20,000 range, and the two cheapest are around $8,000), ShopBot's recent open-sourcing of its CNC control code received much fanfare in the open manufacturing community.[3]

[1]100KGarages founder Ted Hall, "100kGarages is Open: A Place to Get Stuff Made," *Open Manufacturing* email list, September 15, 2009 <http://groups.google.com/group/openmanufacturing/browse_thread/thread/ae45b45de1d055a7?hl=en#>.

[2]"Our Big Idea!" 100kGarages site <http://100kgarages.com/our_big_idea.html>.

[3]Gareth Branwyn, "ShopBot Open-Sources Their Code," *Makezine*, April 13, 2009 <http://blog.makezine.com/archive/2009/04/shopbot_open-sources_their_code.html>.

And as the 100kGarages site says, they plan to open up the network to machines other than routers, and to "home-brew routers" other than ShopBot, as the project develops. Ponoko already had a similar networking project among owners of CNC laser cutters.[1] As a first step toward its intention to "expand to all kinds of digital fabrication tools," in October ShopBot ordered a MakerBot kit with a view to investigating the potential for incorporating additive fabrication into the mix.[2] 100kGarages announced in January 2010 it had signed up 150 Fabbers, and was still developing plans to add other digital tools like cutting tables and 3-D printers to its network.[3] In February they elaborated on their plans, specifying that 100kGarages would add the owners of other digitally controlled tools, with the same certification mechanism for reliability they already used for the ShopBot:

> The plan we've come up with is to work with other Digital Fabrication Equipment manufacturers and let them do the same sort of ownership verification steps that ShopBot has done with the original Fabbers. If a person with a Thermwood (or an EZRouter, Universal Laser, etc) wants to join 100kGarages they can have the manufacturer of their tool verify that they are an owner. We'll work out a simple process for this verification and will work to develop relationships with other manufacturers over time to make the process as painless as possible and to let them get involved if they would like.

Plans to incorporate homebrew tools are also in the works, although much less far along than plans for commercially manufactured tools.

> It also leaves a question of the home-made and home-brew Fabbers. We appreciate that some of these tools can be pretty good. There may be other kinds of user organizations for some types of tools that could help with certification, but we've got to admit that we don't know exactly how we'll deal with it yet. It may be as simple as "send us a picture of yourself, your machine, and a portfolio of work", or we may have to develop some sort of certification method involving cutting a sample. We'll let you know when we come up with something, but we'll try to make it as painless for you (and for us) as possible.[4]

Interestingly, this was almost identical to the relocalized manufacturing model described by John Robb:

> It is likely that by 2025, the majority of the "consumer" goods you purchase/acquire, will be manufactured locally. However, this doesn't likely mean what you think it means. The process will look like this:
>
> 1. You will purchase/trade for/build a design for the product you desire through online trading/sharing systems. That design will be in a standard file format and the volume of available designs for sale, trade, or shared openly will be counted in the billions.
>
> 2. You or someone you trust/hire will modify the design of the product to ensure it meets your specific needs (or customize it so it is uniquely yours). Many products will be smart (in that they include hardware/software that makes them responsive), and programmed to your profile.
>
> 3. The refined product design will be downloaded to a small local manufacturing company, co-operative, or equipped home for production. Basic feedstock materials will be used in its construction (from metal to plastic powders derived from generic sources, recycling, etc.). Delivery is local and nearly costless.

[1] "What's Digital Fabrication?"

[2] "100kGarages is Building a MakerBot," *100kGarages*, October 17, 2009 <http://blog.100kgarages.com/2009/10/17/100kgarages-is-building-a-makerbot/>.

[3] "What are we working on?" *100kGarages*, January 8, 2010 <http://blog.100kgarages.com/2010/01/08/what-are-we-working-on/>.

[4] "What's Next for 100kGarages?" *100kGarages News*, February10, 2010 <http://blog.100kgarages.com/2010/02/10/whats-next-for-100kgarages/>.

The relocalization of manufacturing will be promoted among other things, Robb says, by the fact that

> [l]ocal fabrication will get cheap and easy. The cost of machines that can print, lathe, etch, cut materials to produce three dimensional products will drop to affordable levels (including consumer level versions). This sector is about to pass out of its "home brew computer club phase" and rocket to global acceptance.[1]

It's impossible to underestimate the revolutionary significance of this development. As Lloyd Alter put it, "This really does change everything."[2]

Back in January, Eric Hunting considered the slow takeoff in the open manufacturing/Making movement on the Open Manufacturing email list.

> There seem to be a number of re-occurring questions that come up—openly or in the back of peoples minds- seeming to represent key obstacles or stumbling blocks in the progress of open manufacturing or Maker culture. . . .
>
> Why are Makers still fooling around with toys and mash-ups and not making serious things? (short answer; like early computer hackers lacking off-the-shelf media to study, they're still stuck reverse- engineering the off-the-shelf products of existing industry to learn how the technology works and hacking is easier than making something from scratch)
>
> Why are Makers rarely employing many of the modular building systems that have been around since the start of the 20th century? Why do so few tech-savvy people seem to know what T-slot is when it's ubiquitous in industrial automation? Why little use of Box Beam/Grid Beam when its cheap, easy, and has been around since the 1960s? Why does no one in the world seem to know the origin and name of the rod and clamp framing system used in the RepRap? (short answer: no definitive sources of information)
>
> Why are 'recipes' in places like Make and Instructibles most [sic] about artifacts and rarely about tools and techniques? (short answer; knowledge of these are being disseminated ad hoc)
>
> Why is it so hard to collectivize support and interest for open source artifact projects and why are forums like Open Manufacture spending more time in discussion of theory rather than nuts & bolts making? (short answer; no equivalent of Source Forge for a formal definition of hardware projects—though this is tentatively being developed—and no generally acknowledged definitive channel of communication about open manufacturing activity)
>
> Why are Fab Labs not self-replicating their own tools? (short answer; no comprehensive body of open source designs for those tools and no organized effort to reverse-engineer off-the-shelf tools to create those open source versions)
>
> Why is there no definitive 'users manual' for the Fab Lab, its tools, and common techniques? (short answer; no one has bothered to write it yet)
>
> Why is there no Fab Lab in my neighborhood? Why so few university Fab Labs so far? Why is it so hard to find support for Fab Lab in certain places even in the western world? (short answer; 99% of even the educated population still doesn't know what the hell a Fab Lab is or what the tools it's based on are)
>
> Why do key Post-Industrial cultural concepts remain nascent in the contemporary culture, failing to coalesce into a cultural critical mass? Why are entrepreneurship, cooperative entrepreneurship, and community support networks still left largely out of the popular discussion on recovery from the current economic crash? Why do advocates of Post-Industrial culture and economics still often hang their hopes on nanotechnology when so much could be done with

[1]John Robb, "The Switch to Local Manufacturing," *Global Guerrillas*, July 8, 2009 <http://globalguerrillas.typepad.com/globalguerrillas/2009/07/journal-the-switch-to-local-manufacturing.html>.

[2]Lloyd Alter, "Ponoko + ShopBot = 100kGarages: This Changes Everything in Downloadable Design," *Treehugger*, September 16, 2009 <http://www.treehugger.com/files/2009/09/ponoko-shopbot.php>.

the technology at-hand? (short answer; no complete or documented working models to demonstrate potential with)

Are you, as I am, starting to see a pattern here? It seems like there's a Missing Link in the form of a kind of communications or media gap. There is Maker media—thanks largely to the cultural phenomenon triggered by Make magazine. But it's dominated by ad hoc individual media produced and published on-line to communicate the designs for individual artifacts while largely ignoring the tools. People are learning by making, but they never seem to get the whole picture of what they potentially could make because they aren't getting the complete picture of what the tools are and what they're capable of.

We seem to basically be in the MITS Altair, Computer Shack, Computer Faire, Creative Computing, 2600 era of independent industry. A Hacker era. Remember the early days of the personal computer? You had these fairs, users groups, and computer stores like Computer Shack basically acting like ad hoc ashrams of the new technology because there were no other definitive sources of knowledge. This is exactly what Maker fairs, Fab Labs, and forums like this one are doing. . . .

There are a lot of parallels here to the early personal computer era, except for a couple of things; there's no equivalent of Apple (yet.), no equivalent of the O'Reily Nutshell book series, no "##### For Dummies" books.[1]

100kGarages is a major step toward the critical mass Hunting wrote about. Although there's as yet no Apple of CNC tools (in the sense of the CAD file equivalent of a user-friendly graphic user interface), there is now an organized network of entrepreneurs with a large repository of open designs. As Michel Bauwens puts it, "Suddenly, anyone can pick one of 20,000 Ponoko Designs (or build one themselves) and get it cut out and built just about anywhere."[2] This is essentially what Marcin Jakubowski referred to above, when he speculated on distributed open source manufacturing shops linked to a "global repository of shared open source designs." To get back to Lloyd Alter's theme ("This changes everything"):

Ponoko is the grand idea of digital design and manufacture; they make it possible for designers to meet customers, "where creators, digital fabricators, materials suppliers and buyers meet to make (almost) anything." It is a green idea, producing only when something is wanted, transporting ideas instead of physical objects.

Except there wasn't a computerized router or CNC machine on every block, no 3D Kinko's where you could go and print out your object like a couple of photocopies. Until now, with the introduction of 100K Garages, a joint venture between Ponoko and ShopBot, a community of over six thousand fabricators.

Suddenly, anyone can pick one of 20,000 Ponoko Designs (or build one themselves) and get it cut out and built just about anywhere.[3]

The answer to Hunting's question about cooperative entrepreneurship seems to have come to a large extent from outside the open manufacturing movement, as such. And ShopBot and Ponoko, if not strictly speaking part of the committed open manufacturing movement, have grafted it onto their business model. This is an extension to the physical realm of a phenomenon Bauwens remarked on in the realm of open-source software:

[1]Eric Hunting, "Toolbook and the Missing Link," *Open Manufacturing*, January 30, 2009 <http://groups.google.com/group/openmanufacturing/msg/2fccdde02f402a5b>.

[2]Michel Bauwens, "A milestone for distributed manufacturing: 100kGarages," *P2P Foundation Blog*, September 19, 2009 <http://blog.p2pfoundation.net/a-milestone-for-distributed-manufacturing-100k-garages/2009/09/19>.

[3]Alter, *op. Cit.*

> . . . [M]ost peer production allies itself with an ecology of businesses. It is not difficult to understand why this is the case. Even at very low cost, communities need a basic infrastructure that needs to be funded. Second, though such communities are sustainable as long as they gain new members to compensate the loss of existing contributors; freely contributing to a common project is not sustainable in the long term. In practice, most peer projects follow a 1-10-99 rule, with a one percent consisting of very committed core individuals. If such a core cannot get funded for its work, the project may not survive. At the very least, such individuals must be able to move back and forth from the commons to the market and back again, if their engagement is to be sustainable.
>
> Peer participating individuals can be paid for their work on developing the first iteration of knowledge or software, to respond to a private corporate need, even though their resulting work will be added to the common pool. Finally, even on the basis of a freely available commons, many added value services can be added, that can be sold in the market. On this basis, cooperative ecologies are created. Typical in the open source field for example, is that such companies use a dual licensing strategy. Apart from providing derivative services such as training, consulting, integration etc., they usually offer an improved professional version with certain extra features, that are not available to non-paying customers. The rule here is that one percent of the customers pay for the availability of 99% of the common pool. Such model also consists of what is called benefit sharing practices, in which open source companies contribute to the general infrastructure of cooperation of the respective peer communities.
>
> Now we know that the world of free software has created a viable economy of open source software companies, and the next important question becomes: Can this model be exported, wholesale or with adaptations, to the production of physical goods?[1]

I think it's in process of being done right now.

Jeff Vail expressed some misgivings about Ponoko, wondering whether it could go beyond the production of trinkets and produce primary goods essential to daily living. 100kGarages' partnership with PhysicalDesignCo[2] (a group of MIT architects who design digitally prefabricated houses), announced in early October, may go a considerable way toward addressing that concern. PhysicalDesignCo will henceforth contract the manufacture of all its designs to 100kGarages.[3]

3. Assessment

Franz Nahrada, of the Global Village movement, has criticized Factor e Farm in terms of its relationship to a larger, surrounding networked economy. However, he downplayed the importance of autarky compared to that of cross-linking between OSE and the rest of the resilient community movement.

> I really think we enter a period of densification and intensive cross-linking between various projects. I would like to consider Factor_E_Farm the flagship project for the Global Village community even though I am not blind to some shortcomings. I talked to many people and they find and constantly bring up some points that are easy to critisize [sic]. But I want to make clear: I also see these points and they all can be dealt with and are IMHO of minor importance.
>
> • the site itself seems not really being locally embedded in regional development initiatives, but rather a "spaceship from Mars" for the surrounding popu-

[1]Bauwens, "The Emergence of Open Design and Open Manufacturing," *We Magazine*, vol. 2 <http://www.we-magazine.net/we-volume-02/the-emergence-of-open-design-and-open-manufacturing/>.

[2]<http://www.physicaldesignco.com/>.

[3]"PhysicalDesignCo teams up with 100kGarages," *100kGarages News*, October 4, 2009 <http://blog.100kgarages.com/2009/10/04/physicaldesignco-teams-up-with-100kgarages/>.

> lation. The same occured to me in Tamera 10 years ago when I stayed at a neighboring farmhouse with a very benevolent Portuguese lady who spoke perfect German (because she was the widow of a German diplomat). She was helpful im [sic] mediating, but still I saw the community through the "lenses of outsiders" and I saw how much damage too much cultural isolation can do to a village building effort and how many opportunities are missed that way. We must consider the local and the regional as equally important as the global, in fact the global activates the local and regional potential. It makes us refocus on our neighbors because we bring in a lot of interesting stuff for them—and they might do the same for us. . . .
>
> • the overall OSE project is radically geared towards local autonomy—something which sometimes seemingly cuts deeply into efficiency and especially life quality. I think that in many respects the Factor e Farm zeal, the backbreaking heroism of labor, the choice of the hard bottom-up approach, is more a symbolic statement—and the end result will differ a lot. In the end, we might have regional cooperatives, sophisticated regional division of labor and a size of operations that might still be comparable to small factories; especially when it comes to metal parts, standard parts of all kinds, modules of the toolkit etc. But the statement "we can do it ourselves" is an important antidote to todays absolutely distorted system of technology and competences.
>
> We cannot really figure out what is the threshold where this demonstration effort becomes unmanageable; I think that it is important to start with certain aspects of autarky, with the idea of partial autarky and self-reliance, but not with the idea of total self-sufficiency. This demonstration of aspectual autarky is important in itself and gives a strong message: we can build our own tractor. we can produce our own buidling materials. we can even build most of our own houses.[1]

So OSE is performing a valuable service in showing the outer boundaries of what can be done within a resilient, self-sufficient community. In a total systemic collapse, without (for example) any microchip foundries, the CNC tools in the Fab Lab will—obviously—be unsustainable on a long-term basis. But assuming that such resilient communities are part of a larger network with some of Nahrada's "regional division of labor" and "small factories" (including, perhaps, a decentralized, recycling-based rubber industry), OSE's toolkit will result in drastic increases in the *degree* of local independence and the length of periods a resilient local economy can weather on its own resources.

100kGarages and OSE may be converging toward a common goal from radically different starting points. That is, 100kGarages may be complementary to OSE in terms of Nahrada's criticism. If 100kGarages' networked distributed manufacturing infrastructure is combined with OSE's open-source design ecology, with designs aimed specifically at bootstrapping technologies for maximum local resilience and economy autonomy, the synergies are potentially enormous. Imagine if OSE products like the LifeTrac tractor/prime mover, sawmill, CEB, etc., were part of the library of readily available designs that could be produced through 100kGarages.

[1]Quoted in Michel Bauwens, "Strategic Support for Factor e Farm and Open Source Ecology," *P2P Foundation Blog*, June 19, 2009 <http://blog.p2pfoundation.net/strategic-support-for-factor-e-farm-and-open-source-ecology/2009/06/19>.

6

Resilient Communities and Local Economies

We already saw, in Chapter Five, the economy of networked micromanufacturing that's likely to emerge from the decline of the state capitalist system. We further saw in Chapter Three that there is a cyclical tendency of industrial production to shift from the mass-production core to the craft periphery in economic downturns. And we've witnessed just such a long-term structural shift during the stagnation of the past thirty years.

There is a similar historic connection between severe economic downturns, with significant periods of unemployment, and the formation of barter networks and resilient communities. If the comparison to manufacturing holds, given the cumulative effect of all of state capitalism's crises of sustainability which we examined in Paper No. 4, we can expect to see a long-term structural shift toward resilient communities and relocalized exchange. John Robb suggests that, given the severity of the present "Great Recession," it may usher in a phase transition in which the new society crystallizes around resilient communities as a basic building block; resilient communities will play the same role in resolving the current "Time of Troubles" that the Keynesian state did in resolving the last one.

> Historically, economic recessions that last longer than a year have durations/severities that can be plotted as power law distributions. . . . Given that we are already over a year into this recession, it implies that we are really into black swan territory (unknown and extreme outcomes) in regards to our global economy's current downturn and that no estimates of recovery times or ultimate severity based on historical data of past recessions apply anymore. This also means that the system has exceeded its ability to adapt using standard methods (that shouldn't be news to anyone).
>
> It may be even more interesting than that. The apparent non-linearity and turbulence of the current situation suggests we may be at a phase transition (akin to the shift in the natural world from ice to water). . . .
>
> As a result, a new control regime may emerge. To get a glimpse of what is in store for us, we need to look at the sources of emerging order (newly configured dissipative and self-organizing systems/networks/orgs that are better adapted to the new non-linear dynamics of the global system).
>
> In [the Great Depression] the sources of emerging organizational order were reconfigured nation-states that took a more active role in economics (total war economies during peacetime). In this situation, we are seeing emerging order at the local level: small resilient networks/communities reconfigured to handle this level of systemic environmental non-linearity and survive/thrive. . . . Further, it appears that these emerging communities and networks are well suited to drawing on a great behavioral shift occurring at the individual level, already evident in all economic statistics, that emphasis thrift/investment rather than consumption/gambling (the middle class consumer is becoming extinct).

> So what does this mean? These new communities will eventually start to link up, either physically or virtually . . . , into network clusters. IF the number of links in the largest cluster reaches some critical proportion of the entire system's nodes . . . , there will be a phase transition as entire system shifts to the new mode of operation. In other words, resilient communities *might* become the new configuration of the global economic system.[1]

Robb's phase transition resembles Jeff Vail's description of the gradually shifting correlation of forces between the old legacy system and his "Diagonal Economy":

> The diagonal economy might rise amidst the decline of our current system—the "Legacy System." Using America as an example (but certainly translatable to other regions and cultures), more and more people will gradually realize that there the "plausible promise" once offered by the American nation-state is no longer plausible. A decent education and the willingness to work 40 hours a week will no longer provide the "Leave it to Beaver" quid pro quo of a comfortable suburban existence and a secure future for one's children. As a result, our collective willingness to agree to the conditions set by this Legacy System (willing participation in the system in exchange for this once "plausible promise") will wane. Pioneers—and this is certainly already happening—will reject these conditions in favor of a form of networked civilizational entrepreneurship. While this is initially composed of professionals, independent sales people, internet-businesses, and a few market gardeners, it will gradually transition to take on a decidedly "third world" flavor of local self-sufficiency and import-replacement (leveraging developments in distributed, open-source, and peer-to-peer manufacturing) in the face of growing ecological and resource pressures. People will, to varying degrees, recognize that they cannot rely on the cradle-to-cradle promise of lifetime employment by their nation state. Instead, they will realize that they are all entrepreneurs in at least three—and possibly many more—separate enterprises: one's personal brand in interaction with the Legacy System (e.g. your conventional job), one's localized self-sufficiency business (ranging from a back yard tomato plant to suburban homesteads and garage workshops), and one's community entrepreneurship and network development. As the constitutional basis of our already illusory Nation-State system . . . erode further, the focus on #2 (localized self-sufficiency) and #3 (community/networking) will gradually spread and increase in importance, though it may take much more than my lifetime to see them rise to general prominence in replacement of the Nation-State system.[2]

In this chapter we will examine the general benefits of resilient local economies, consider some notable past examples of the phenomenon, and then survey some current experiments in resilient community which are especially promising as building blocks for a post-corporate society.

A. Local Economies as Bases of Independence and Buffers Against Economic Turbulence

One virtue of the local economy is its insulation from the boom-bust cycle of the larger money economy.

Paul Goodman wrote that a "tight local economy" was essential for maintaining "a close relation between production and consumption,"

[1]John Robb, "Viral Resilience," *Global Guerrillas*, January 12, 2009 <http://globalguerrillas.typepad.com/globalguerrillas/2009/01/journal-phase-t.html>.

[2]Jeff Vail, "Diagonal Economy 1: Overview," *JeffVail.Net*, August 24, 2009 <http://www.jeffvail.net/2009/08/diagonal-economy-1-overview.html>.

> for it means that prices and the value of labor will not be so subject to the fluctuations of the vast general market. A man's work, meaningful during production, will somewhat carry through the distribution and what he gets in return. That is, within limits, the nearer a system gets to simple household economy, the more it is an economy of specific things and services that are bartered, rather than an economy of generalized money.[1]

The greater the share of consumption needs met through informal (barter, household and gift) economies, the less vulnerable individuals are to the vagaries of the business cycle, and the less dependent on wage labor as well.

The ability to meet one's own consumption needs with one's own labor, using one's own land and tools, is something that can't be taken away by a recession or a corporate decision to offshore production to China (or just to downsize the work force and speed up work for the survivors). The ability to trade one's surplus for other goods, with a neighbor also using his own land and tools, is also much more secure than a job in the capitalist economy.

Ralph Borsodi described the cumulative effect of the concatenation of uncertainties in an economy of large-scale factory production for anonymous markets:

> Surely it is plain that no man can afford to be dependent upon some other man for the bare necessities of life without running the risk of losing all that is most precious to him. Yet that is precisely and exactly what most of us are doing today. Everybody seems to be dependent upon some one else for the opportunity to acquire the essentials of life. The factory-worker is dependent upon the man who employs him; both of them are dependent upon the salesmen and retailers who sell the goods they make, and all of them are dependent upon the consuming public, which may not want, or may not be able, to buy what they may have made.[2]

Imagine, on the other hand, an organic truck farmer who barters produce for clothing from a home seamstress living nearby. Neither the farmer nor the seamstress can dispose of her full output in this manner, or meet all of her subsistence needs. But both together have a secure and reliable source for all their sewing *and* vegetable needs, and a reliable outlet for the portion of the output of each that is consumed by the other. The more trades and occupations brought into the exchange system, the greater the portion of total consumption needs of each that can be reliably met within a stable sub-economy. At the same time, the less dependent each person is on outside wage income, and the more prepared to weather a prolonged period of unemployment in the outside wage economy.

Subsistence, barter, and other informal economies, by reducing the intermediate steps between production and consumption, also reduce the contingency involved in consumption. If the realization of capital follows a circuit, as described by Marx in *Capital*, the same is also true of labor. And the more steps in the circuit, the more likely the circuit is to be broken, and the realization of labor (the transformation of labor into use-value, through the indirect means of exchanging one's own labor for wages, and exchanging those wages for use-value produced by someone else's labor) is to fail. Marx, in *The Poverty of Philosophy*, pointed out long ago that the disjunction of supply from demand, which resulted in the boom-bust cycle, was inevitable given the large-scale production under industrial capitalism:

[1]Paul and Percival Goodman, *Communitas: Means of Livelihood and Ways of Life* (New York: Vintage Books, 1947, 1960), p. 170.

[2]Ralph Borsodi. *Flight from the City: An Experiment in Creative Living on the Land* (New York, Evanston, San Francisco, London: Harper & Row, 1933, 1972), p. 147.

> . . . [This true proportion between supply and demand] was possible only at a time when the means of production were limited, when the movement of exchange took place within very restricted bounds. With the birth of large-scale industry this true proportion had to come to an end, and production is inevitably compelled to pass in continuous succession through vicissitudes of prosperity, depression, crisis, stagnation, renewed prosperity, and so on.
>
> Those who . . . wish to return to the true proportion of production, while preserving the present basis of society, are reactionary, since, to be consistent, they must also wish to bring back all the other conditions of industry of former times.
>
> What kept production in true, or more or less true, proportions? It was demand that dominated supply, that preceded it. Production followed close on the heels of consumption. Large-scale industry, forced by the very instruments at its disposal to produce on an ever-increasing scale, can no longer wait for demand. Production precedes consumption, supply compels demands.[1]

In drawing the connection between supply-push distribution and economic crisis, Marx was quite perceptive. Where he went wrong was his assumption that large-scale industry, and production that preceded demand on the push model, were necessary for a high standard of living ("the present basis of society").

Leopold Kohr, in the same vein, compared local economies to harbors in a storm in their insulation from the business cycle and its extreme fluctuations of demand.[2]

Ebenezer Howard, in his vision of Garden Cities, argued that the overhead costs of risk and distribution (as well as rent, given the cheap rural land on which the new towns would be built) would be far lower for both industry and retailers serving the less volatile local markets.

> They might even sell considerably below the ordinary rate prevailing elsewhere, but yet, having an assured trade and being able very accurately to gauge demand, they might turn their money over with remarkable frequency. Their working expenses, too, would be absurdly small. They would not have to advertise for customers, though they would doubtless make announcements to them of any novelties; but all that waste of effort and of money which is so frequently expended by tradesmen in order to secure customers or to prevent their going elsewhere, would be quite unnecessary.[3]

His picture of the short cycle time and minimal overhead resulting from the gearing of supply to demand, by the way, is almost a word-for-word anticipation of lean principles.

We saw, in previous chapters, the way that lean production overcomes bottlenecks in supply by scaling production to demand and siting production as close as possible to the market. The small neighborhood shop and the household producer apply the same principle, on an even higher level. So the more decentralized and relocalized the scale of production, the easier it is to overcome the divorce of production from demand—the central contradiction of mass production. These remarks by Gandhi are relevant:

> Question : "Do you feel, Gandhiji, that mass production will raise the standard of living of the people?"

[1]Karl Marx, *The Poverty of Philosophy*, Marx and Engels *Collected Works*, vol. 6 (New York: International Publishers, 1976).

[2]Leopold Kohr, *The Overdeveloped Nations: The Diseconomies of Scale* (New York: Schocken Books, 1977), p. 110.

[3]Ebenezer Howard, *To-Morrow: A Peaceful Path to Real Reform*. Facsimile of original 1898 edition, with introduction and commentary by Peter Hall, Dennis Hardy and Colin Ward (London and New York: Routledge, 2003), pp. 100, 102 [facsimile pp. 77-78].

"I do not believe in it at all, there is a tremendous fallacy behind Mr. Ford's reasoning. Without simultaneous distribution on an equally mass scale, the production can result only in a great world tragedy."

"Mass production takes no note of the real requirement of the consumer. If mass production were in itself a virtue, it should be capable of indefinite multiplication. But it can be definitely shown that mass production carries within it its own limitations. If all countries adopted the system of mass production, there would not be a big enough market for their products. Mass production must then come to a stop."

"I would categorically state my conviction that the mania for mass production is responsible for the world crises. If there is production and distribution both in the respective areas where things are required, it is automatically regulated, and there is less chance for fraud, none for speculation." . . .

Question : Have you any idea as to what Europe and America should do to solve the problem presented by too much machinery?

"You see," answered Gandhiji, "that these nations are able to exploit the so-called weaker or unorganized races of the world. Once those races gain this elementary knowledge and decide that they are no more going to be exploited, they will simply be satisfied with what they can provide themselves. Mass production, then, at least where the vital necessities are concerned, will disappear." . . .

Question : "But even these races will require more and more goods as their needs multiply."

"They will them [sic] produce for themselves. And when that happens, mass production, in the technical sense in which it is understood in the West, ceases."

Question : "You mean to say it becomes local?"

"When production and consumption both become localized, the temptation to speed up production, indefinitely and at any price, disappears.

Question : If distribution could be equalized, would not mass production be sterilized of its evils?

"No," The evil is inherent in the system. Distribution can be equalized when production is localized; in other words, when the distribution is simultaneous with production. Distribution will never be equal so long as you want to tap other markets of the world to dispose of your goods.

Question : Then, you do not envisage mass production as an ideal future of India ?

"Oh yes, mass production, certainly," "But not based on force. After all, the message of the spinning wheel is that. It is mass production, but mass production in people's own homes. If you multiply individual production to millions of times, would it not give you mass production on a tremendous scale? But I quite understand that your 'mass production' is a technical term for production by the fewest possible number through the aid of highly complicated machinery. I have said to myself that that is wrong. My machinery must be of the most elementary type which I can put in the homes of the millions." Under my system, again, it is labour which is the current coin, not metal. Any person who can use his labour has that coin, has wealth. He converts his labour into cloth, he converts his labour into grain. If he wants paraffin oil, which he cannot himself produce, he uses his surplus grain for getting the oil. It is exchange of labour on free, fair and equal terms—hence it is no robbery. You may object that this is a reversion to the primitive system of barter. But is not all international trade based on the barter system?

Concentration of production ad infinitum can only lead to unemployment.[1]

Gandhi's error was assuming that localized and household production equated to low-tech methods, and that technological advancement was inevitably

[1]"Mahatma Gandhi on Mass Production" (1936), *TinyTech Plants* <http://www.tinytechindia.com/gandhiji2.html> (punctuation in original).

associated with large scale and capital intensiveness. As we saw in Chapter Five, nothing could be further from the truth.

Communities of locally owned small enterprises are much healthier economically than communities that are colonized by large, absentee-owned corporations. For example, a 1947 study compared two communities in California: one a community of small farms, and the other dominated by a few large agribusiness operations. The small farming community had higher living standards, more parks, more stores, and more civic, social and recreational organizations.[1]

Bill McKibben made the same point in *Deep Economy*. Most money that's spent buying stuff from a national corporation is quickly sucked out of the local economy, while money that's spent at local businesses circulates repeatedly in the local economy and leaks much more slowly to the outside. According to a study in Vermont, substituting local production for only ten percent of imported food would create $376 million in new economic output, including $69 million in wages at over 3600 new jobs. A similar study in Britain found the multiplier effect of ten pounds spent at a local business benefited the local economy to the tune of 25 pounds, compared to only 14 for the same amount spent at a chain store.

> The farmer buys a drink at the local pub; the pub owner gets a car tune-up at the local mechanic; the mechanic brings a shirt to the local tailor; the tailor buys some bread at the local bakery; the baker buys wheat for bread and fruit for muffins from the local farmer. When these businesses are not owned locally, money leaves the community at every transaction.[2]

B. Historical Models of Resilient Community

The prototypical resilient community, in the mother of all "Times of Troubles," was the Roman villa as it emerged in the late Empire and early Dark Ages. In Republican times, villas had been estates on which the country homes of the Senatorial class were located, often self-sufficient in many particulars and resembling villages in their own right. During the stresses of the "long collapse" in the fifth century, and in the Dark Ages following the fall of the Western Empire, the villas became stockaded fortresses, often with villages of peasants attached.

Since the rise of industrial capitalism, economic depression and unemployment have been the central motive forces behind the creation of local exchange systems and the direct production for barter by producers.

A good example is the Owenites' use of the social economy as a base of independence from wage labor. According to E. P. Thompson, "[n]ot only did the benefit societies on occasion extend their activities to the building of social clubs or alms-houses; there are also a number of instances of pre-Owenite trade unions when on strike, employing their own members and marketing the product."[3] G. D. H. Cole describes the same phenomenon:

> As the Trade Unions grew after 1825, Owenism began to appeal to them, and especially to the skilled handicraftsmen. . . . Groups of workers belonging to a particular craft began to set up Co-operative Societies of a different type—societies of producers which offered their products for sale through the Co-

[1] L. S. Stavrianos, *The Promise of the Coming Dark Age* (San Francisco: W. H. Freeman and Company, 1976), p. 41.

[2] Bill McKibben, *Deep Economy: The Wealth of Communities and the Durable Future* (New York: Times Books, 2007), p. 165.

[3] E. P. Thompson, *The Making of the English Working Class* (New York: Vintage Books, 1963, 1966), p. 790.

operative Stores. Individual Craftsmen, who were Socialists, or who saw a way of escape from the exactions of the middlemen, also brought their products to the stores to sell."[1]

... [This pattern of organization was characterized by] societies of producers, aiming at co-operative production of goods and looking to the Stores to provide them with a market. These naturally arose first in trades requiring comparatively little capital or plant. They appealed especially to craftsmen whose independence was being threatened by the rise of factory production or subcontracting through capitalist middlemen.

The most significant feature of the years we are discussing was the rapid rise of this ... type of Co-operative Society and the direct entry of the Trades Unions into Co-operative production. Most of these Societies were based directly upon or at least very closely connected with the Unions of their trades, ... which took up production as a part of their Union activity—especially for giving employment to their members who were out of work or involved in trade disputes. ... [2]

The aims and overall vision of such organization were well expressed in the rules of the Ripponden Co-operative Society, formed in 1832 in a weaving village in the Pennines:

> The plan of co-operation which we are recommending to the public is not a visionary one but is acted upon in various parts of the Kingdom; we all live by the produce of the land, and exchange labour for labour, which is the object aimed at by all Co-operative societies. We labourers do all the work and produce all the comforts of life;—why then should we not labour for ourselves and strive to improve our conditions.[3]

Cooperative producers' need for an outlet led to Labour Exchanges, where workmen and cooperatives could directly exchange their product so as "to dispense altogether with either capitalist employers or capitalist merchants." Exchange was based on labor time. "Owen's Labour Notes for a time not only passed current among members of the movement, but were widely accepted by private shopkeepers in payment for goods."[4]

The principle of labor-based exchange was employed on a large-scale. In 1830 the London Society opened an Exchange Bazaar for exchange of products between cooperative societies and individuals.[5] The Co-operative Congress, held at Liverpool in 1832, included a long list of trades among its participants (the B's alone had eleven). The National Equitable Labour Exchange, organized in 1832-33 in Birmingham and London, was a venue for the direct exchange of products between craftsmen, using Labour Notes as a medium of exchange.[6]

The Knights of Labor, in the 1880s, undertook a large-scale effort at organizing worker cooperatives. Their fate is an illustration of the central role of capital outlay requirements in determining the feasibility of self-employment and cooperative employment.

The first major wave of worker cooperatives, according to John Curl, was under the auspices of the National Trades' Union in the 1830s.[7] Like the Owenite trade union cooperatives in Britain, they were mostly undertaken in craft employ-

[1]G.D.H. Cole, *A Short History of the British Working Class Movement (1789-1947)* (London: George Allen & Unwin, 1948), p. 76.

[2]*Ibid.*, p. 78.

[3]*Ibid.*, pp. 793-794.

[4]*Ibid.*, pp. 78-79.

[5]*Ibid.*, p. 76.

[6]Thompson, *Making of the English Working Class*, p. 791.

[7]John Curl, *For All the People: Uncovering the Hidden History of Cooperation, Cooperative Movements, and Communalism in America* (Oakland, CA: PM Press, 2009), p.4

ments for which the basic tools of the trade were relatively inexpensive. From the beginning, worker cooperatives were a frequent resort of striking workers. In 1768 twenty striking journeyman tailors in New York, the first striking wage-workers in American history, set up their own cooperative shop. Journeyman carpenters striking for a ten-hour day in Philadelphia, in 1761, formed a cooperative (with the ten-hour day they sought) and undercut their master's price by 25%; they disbanded the cooperative when they went back to work. The same was done by shoemakers in Baltimore, 1794, and Philadelphia, 1806.[1] This was a common pattern in early American labor history, and the organization of cooperatives moved from being purely a strike tactic to providing an alternative to wage labor.[2] It was feasible because most forms of production were done by groups of artisan laborers using hand tools.

By the 1840s, the rise of factory production with expensive machinery had largely put an end to this possibility. As the prerequisites of production became increasingly unafforable, the majority of the population was relegated to wage labor with machinery owned by someone else.[3]

Most attempts at worker-organized manufacturing, after the rise of the factory system, failed on account of the capital outlays required. For example, when manufacturers refused to sell farm machinery to the Grangers at wholesale prices, the Nebraska Grange undertook its own design and manufacturing of machinery. (How's that for a parallel to modern P2P ideas?) Its first attempt, a wheat head reaper, sold at half the price of comparable models and drove down prices on farm machinery in Nebraska. The National Grange planned a complete line of farm machinery, but most Grange manufacturing enterprises failed to raise the large sums of capital needed.[4]

The Knights of Labor cooperatives were on shaky ground in the best of times. Many of them were founded during strikes, started with "little capital and obsolescent machinery," and lacked the capital to invest in modern machinery. Subjected to economic warfare by organized capital, the network of cooperatives disintegrated during the post-Haymarket repression.[5]

[1]*Ibid.*, p. 33.
[2]*Ibid.*, p. 34.
[3]*Ibid.*, pp. 35, 47.
[4]*Ibid.*, p. 77.
[5]*Ibid.*, p. 107.

The fate of the KofL cooperatives, resulting from the high capitalization requirements for production, is a useful contrast to the potential for small-scale production today. The economy today is experiencing a revolution as profound as the corporate transformation of the late 19th century. The main difference today is that, for material reasons, the monopolies on which corporate rule depends are becoming unenforceable. Another revolution, based on P2P and micromanufacturing, is sweeping society on the same scale as did the corporate revolution of 150 years ago. But the large corporations today are in the same position that the Grange and Knights of Labor were in the Great Upheaval back then, fighting a desperate, futile rearguard action, and doomed to be swept under by the tidal wave of history.

The worker cooperatives organized in the era of artisan labor paralleled, in many ways, the forms of work organization that are arising today. Networked organization, crowdsourced credit and the implosion of capital outlays required for physical production, taken together, are recreating the same conditions that made artisan cooperatives feasible in the days before the factory system. In the artisan manufactories that prevailed into the early 19th century, most of the physical capital required for production was owned by the work force; artisan laborers could walk out and essentially take the firm with them in all but name. Likewise, today, the collapse of capital outlay requirements for production in

Ebenezer Howard's Garden Cities were a way of "buying out at the bottom" (a phrase coined by Vinay Gupta—about whom more later): building the cities on cheap rural land and using it with maximum efficiency. The idea was that workers would take advantage of the rent differential between city and country, make more efficient use of underused land than the great landlords and capitalists could, and use the surplus income from production in the new cities (collected as a single tax on the site value of land) for quickly paying off the original capital outlays.[1] Howard also anticipated something like counter-economics: working people living within his garden cities, working through building societies, friendly societies, mutuals, consumer and worker cooperatives, etc., would find ways to employ themselves and each other outside the wage system.

> It is idle for working-men to complain of this self-imposed exploitation, and to talk of nationalizing the entire land and capital of this country under an executive of their own class, until they have first been through an apprenticeship at the humbler task of organising men and women with their own capital in constructive work of a less ambitious character. . . . The true remedy for capitalist oppression where it exists, is not the strike of no work, but the strike of true work, and against this last blow the oppressor has no weapon. If labour leaders spent half the energy in co-operative organisation that they now waste in co-operative disorganisation, the end of our present unjust system would be at hand.[2]

Howard, heavily influenced by Kropotkin's vision of the decentralized production made possible by small-scale electrically powered machinery,[3] wrote that "[t]own and country must be married, and out of this joyous union will spring a new hope, a new life, a new civilization."[4] Large markets, warehouses, and industry would be located along a ring road on the outer edge of each town, with markets and industry serving the particular ward in which its customers and workers lived.[5] A cluster of several individual towns (the "social city" of around a quarter million population in an area of roughly ten miles square) would ultimately be linked together by "[r]apid railway transit," much like the old mixed-use railroad suburbs which today's New Urbanists propose to resurrect and link together with light rail. Larger industries in each town would specialize in the production of commodities for the entire cluster, in which greater economies of scale were necessary.

In the Great Depression, the same principles used by the Owenites and Knights of Labor were applied in the Homestead Unit project in the Dayton area, an experiment with household and community production in which Borsodi played a prominent organizing role. Despite some early success, it was eventually

the cultural and information fields (software, desktop publishing, music, etc.) has created a situation in which human capital is the source of most book value for many firms; consequently, workers are able to walk out with their human capital and form "breakaway firms," leaving their former employers as little more than hollow shells. And the rise of cheap garage manufacturing machinery (a Fab Lab with homebrew CNC tools costing maybe two months' wages for a semi-skilled worker) is, in its essence, a return to the days when low physical capital costs made worker cooperatives a viable alternative to wage labor.

The first uprising against corporate power, in the late 19th century, was defeated by the need for capital. The present one will destroy the old system by making capital superfluous.

[1] Howard, *To-Morrow*, pp. 32, 42 [facsimile pp. 13, 20-21].

[2] *Ibid.*, pp. 108, 110 [facsimile pp. 85-86].

[3] Colin Ward, Commentator's introduction to *Ibid.*, p. 3.

[4] *Ibid.*, p. 28 [facsimile p. 10].

[5] *Ibid.*, p. 14 [facsimile p. 34].

killed off by Harold Ickes, a technocratic liberal who wanted to run the homestead project along the same centralist lines as the Tennessee Valley Authority. The Homestead Units were built on cheap land in the countryside surrounding Dayton, with a combination of three-acre family homesteads and some division of labor on other community projects. The family homestead included garden, poultry and other livestock, and a small orchard and berry patch. The community provided woodlot and pasture, in addition.[1] A Unit Committee vice president in the project described the economic security resulting from subsistence production:

> There are few cities where the independence of a certain sort of citizen has not been brought into relief by the general difficulties of the depression. In the environs of all cities there is the soil-loving suburbanite. In some cases these are small farmers, market gardeners and poultry raisers who try to make their entire living from their little acres. More often and more successful there is a combination of rural and city industry. Some member of the family, while the others grow their crops, will have a job in town. A little money, where wages are joined to the produce of the soil, will go a long way. . . .
>
> When the depression came most of these members of these suburban families who held jobs in town were cut in wages and hours. In many cases they entirely lost their jobs. What, then, did they do?. . . . The soil and the industries of their home provided them . . . work and a living, however scant. Except for the comparatively few dollars required for taxes and a few other items they were able, under their own sail, to ride out the storm. The sailing was rough, perhaps; but not to be compared with that in the wreck-strewn town. . . .
>
> Farming as an exclusive business, a full means of livelihood, has collapsed. . . . Laboring as an exclusive means of livelihood has also collapsed. The city laborer, wholly dependent on a job, is of all men most precariously placed. Who, then, is for the moment safe and secure? The nearest to it is this home and acres-owning family in between, which combines the two.[2]

An interesting experiment in restoring the "circuit of labor" through barter exchange was Depression-era organizations like the Unemployed Cooperative Relief Organization and Unemployed Exchange Association:

> . . . The real economy was still there—paralyzed but still there. Farmers were still producing, more than they could sell. Fruit rotted on trees, vegetables in the fields. In January 1933, dairymen poured more than 12,000 gallons of milk into the Los Angeles City sewers every day.
>
> The factories were there too. Machinery was idle. Old trucks were in side lots, needing only a little repair. All that capacity on the one hand, legions of idle men and women on the other. It was the financial casino that had failed, not the workers and machines. On street corners and around bare kitchen tables, people started to put two and two together. More precisely, they thought about new ways of putting two and two together. . . .
>
> In the spring of 1932, in Compton, California, an unemployed World War I veteran walked out to the farms that still ringed Los Angeles. He offered his labor in return for a sack of vegetables, and that evening he returned with more than his family needed. The next day a neighbor went out with him to the fields. Within two months 500 families were members of the Unemployed Cooperative Relief Organization (UCRO).

[1]Ralph Borsodi, *The Nation*, April 19, 1933; reproduced in *Flight From the City*, pp. 154-59. Incidentally, the New Town project in Great Britain was similarly sabotaged, first under the centralizing social-democratic tendencies of Labour after WWII, and then by Thatcherite looting (er, "privatization") in the 1980s. Ward commentary, Howard, *To-Morrow*, p. 45.

[2]Editorial by Walter Locke in *The Dayton News*, quoted by Borsodi in *Flight From the City*, pp. 170-71.

> That group became one of 45 units in an organization that served the needs of some 150,000 people.
>
> It operated a large warehouse, a distribution center, a gas and service station, a refrigeration facility, a sewing shop, a shoe shop, even medical services, all on cooperative principles. Members were expected to work two days a week, and benefits were allocated according to need. . . .
>
> The UCRO was just one organization in one city. Groups like it ultimately involved more than 1.3 million people, in more than 30 states. It happened spontaneously, without experts or blueprints. Most of the participants were blue collar workers whose formal schooling had stopped at high schools. Some groups evolved a kind of money to create more flexibility in exchange. An example was the Unemployed Exchange Association, or UXA, based in Oakland, California. . . . UXA began in a Hooverville . . . called "Pipe City," near the East Bay waterfront. Hundreds of homeless people were living there in sections of large sewer pipe that were never laid because the city ran out of money. Among them was Carl Rhodehamel, a musician and engineer.
>
> Rhodehamel and others started going door to door in Oakland, offering to do home repairs in exchange for unwanted items. They repaired these and circulated them among themselves. Soon they established a commissary and sent scouts around the city and into the surrounding farms to see what they could scavenge or exchange labor for. Within six months they had 1,500 members, and a thriving sub-economy that included a foundry and machine shop, woodshop, garage, soap, factory, print shop, wood lot, ranches, and lumber mills. They rebuilt 18 trucks from scrap. At UXA's peak it distributed 40 tons of food a week.
>
> It all worked on a time-credit system. . . . Members could use credits to buy food and other items at the commissary, medical and dental services, haircuts, and more. A council of some 45 coordinators met regularly to solve problems and discuss opportunities.
>
> One coordinator might report that a saw needed a new motor. Another knew of a motor but the owner wanted a piano in return. A third member knew of a piano that was available. And on and on. It was an amalgam of enterprise and cooperation—the flexibility and hustle of the market, but without the encoded greed of the corporation or the stifling bureaucracy of the state. . . . The members called it a "reciprocal economy.". . . .[1]

Stewart Burgess, in a 1933 article, described a day's produce intake by the warehouse of Unit No. 1 in Compton. It included some fifteen different kinds of fruits and vegetables, including two tons of cabbage and seventy boxes of pears, all the way down to a single crate of beets—not to mention a sack of salt. The production facilities and the waste materials it used as inputs foreshadow the ideas of Colin Ward, Kirkpatrick Sale and Karl Hess on community warehouses and workshops, discussed in the last chapter:

> In this warehouse is an auto repair shop, a shoe-repair shop, a small printing shop for the necessary slips and forms, and the inevitable woodpile where cast-off railroad ties are sawed into firewood. Down the street, in another building, women are making over clothing that has been bartered in. In another they are canning vegetables and fruit—Boy Scouts of the Burbank Unit brought in empty jars by the wagon-load.[2]

Such ventures, like the Knights of Labor cooperatives, were limited by the capital intensiveness of so many forms of production. The bulk of the labor performed within the barter networks was either in return for salvage goods in need of repair, for repairing such goods, or in return for unsold inventories of conven-

[1]Jonathan Rowe, "Entrepreneurs of Cooperation," *Yes!*, Spring 2006 <http://www.yesmagazine.org/article.asp?ID=1464>.

[2]J. Stewart Burgess, "Living on a Surplus," *The Survey* 68 (January 1933), p. 6.

tional businesses. When the supply of damaged machinery was exhausted by house-to-house canvassing, and local businesses disposed of their accumulated inventory, barter associations reached their limit. They could continue to function at a fairly low volume, directly undertaking for barter such low-capital forms of production as sewing, gardening on available land, etc., and trading labor for whatever percentage of output from otherwise idle capacity that conventional businesses were willing to barter for labor. But that level was quite low compared to the initial gains from absorbing excess inventory and salvageable machinery in the early days of the system. At most, once barter reached its sustainable limits, it was good as a partial mitigation of the need for wage labor.

But as production machinery becomes affordable to individuals independently of large employers, such direct production for barter will become increasingly feasible for larger and larger segments of the workforce.

The Great Depression was a renaissance of local barter currencies or "emergency currencies," adopted around the world, which enabled thousands of communities to weather the economic calamity with "the medium of exchange necessary for their activities, to give each other work."[1]

The revival of barter on the Internet coincides with a new economic downturn, as well. A Craigslist spokesman reported in March 2009 that bartering had doubled on the site over the previous year.

> Proposed swaps listed on the Washington area Craigslist site this week included accounting services in return for food, and a woman offering a week in her Hilton Head, S.C., vacation home for dental work for her husband.

Barter websites for exchanging goods and services without cash are proliferating around the world.

> With unemployment in the United States and Britain climbing, some people said bartering is the only way to make ends meet.
>
> "I'm using barter Web sites just to see what we can do to survive," said Zedd Epstein, 25, who owned a business restoring historic houses in Iowa until May, when he was forced to close it as the economy soured.
>
> Epstein, in a telephone interview, said he has not been able to find work since, and he and his wife moved to California in search of jobs.
>
> Epstein said he has had several bartering jobs he found on Craigslist. He drywalled a room in exchange for some tools, he poured a concrete shed floor in return for having a new starter motor installed in his car, and he helped someone set up their TV and stereo system in return for a hot meal.
>
> "Right now, this is what people are doing to get along," said Epstein, who is studying for an electrical engineering degree.
>
> "If you need your faucet fixed and you know auto mechanics, there's definitely a plumber out there who's out of work and has something on his car that needs to be fixed," he said.[2]

[1]Bernard Lietaer, *The Future of Money: A New Way to Create Wealth, Work and a Wiser World* (London: Century, 2001), p. 148. In pp. 151-157, he describes examples from all over the world, including "several thousand examples of local scrip from every state in the Union."

[2]Kevin Sullivan, "As Economy Plummets, Cashless Bartering Soars on the Internet," *Washington Post*, March 14, 2009 <http://www.washingtonpost.com/wp-dyn/content/article/2009/03/13/AR2009031303035_pf.html>.

C. Resilience, Primary Social Units, and Libertarian Values

As the crisis progresses, and with it the gradually increasing underemployment and unemployment and the partial shift of value production from wage labor to the informal sector, we can probably expect to see several converging trends: a long-term decoupling of health care and the social safety net from both state-based and employer-based provision of benefits; shifts toward shorter working hours and job-sharing; and the growth of all sorts of income-pooling and cost-spreading mechanisms in the informal economy.

These latter possibilities include a restored emphasis on mutual aid organizations of the kind described by left-libertarian writers like Pyotr Kropotkin and E. P. Thompson. As Charles Johnson wrote:

> It's likely also that networks of voluntary aid organizations would be strategically important to individual flourishing in a free society, in which there would be no expropriative welfare bureaucracy for people living with poverty or precarity to fall back on. Projects reviving the bottom-up, solidaritarian spirit of the independent unions and mutual aid societies that flourished in the late 19th and early 20th centuries, before the rise of the welfare bureaucracy, may be essential for a flourishing free society, and one of the primary means by which workers could take control of their own lives, without depending on either bosses or bureaucrats.[1]

More fundamentally, they are likely to entail people coalescing into primary social units at the residential level (extended family compounds or multi-family household income-pooling units, multi-household units at the neighborhood level, urban communes and other cohousing projects, squats, and stand-alone intentional communities), as a way of pooling income and reducing costs. As the state's social safety nets come apart, such primary social units and extended federations between them are likely to become important mechanisms for pooling cost and risk and organizing care for the aged and sick. One early sign of a trend in that direction: multi-generational or extended family households are at a fifty-year high, growing five percent in the first year of the Great Recession alone.[2] Here's how John Robb describes it:

> My solution is to form a tribal layer. Resilient communities that are connected by a network platform (a darknet). A decentralized and democratic system that can provide you a better interface with the dominant global economic system than anything else I can think of. Not only would this tribe protect you from shocks and predation by this impersonal global system, it would provide you with the tools and community support necessary to radically improve how you and your family does [sic] across all measures of consequence.[3]

[1]Charles Johnson, "Liberty, Equality, Solidarity: Toward a Dialectical Anarchism," in Roderick T. Long and Tibor R. Machan, eds., *Anarchism/Minarchism: Is a Government Part of a Free Country?* (Hampshire, UK, and Burlington, Vt.: Ashgate Publishing Limited, 2008). Quoted from textfile provided by author.

[2]Donna St. George, "Pew report shows 50-year high point for multi-generational family households," *Washington Post*, March 18, 2010 <http://www.washingtonpost.com/wp-dyn/content/article/2010/03/18/AR2010031804510.html>.

[3]John Robb, "You Are In Control," *Global Guerrillas*, January 3, 2010 <http://globalguerrillas.typepad.com/ globalguerrillas/2010/01/you-are-in-control.html>. For a wonderful fictional account of the growth of a society of resilient communities linked in a darknet, and its struggle with the host society, I strongly recommend two novels by Daniel Suarez: *Daemon* (Signet, 2009), and its sequel *Freedom(TM)* (Dutton, 2010). I reviewed them here: Kevin Carson, "Daniel Suarez: Daemon and Freedom, *P2P Foundation*

Poul Anderson, in the fictional universe of his Maurai series, envisioned a post-apocalypse society in the Pacific Northwest coalescing around the old fraternal lodges, with the Northwestern Federation (a polity extending from Alaska through British Columbia down to northern California) centered on lodges rather than geographical subdivisions as the component units represented in its legislature. The lodge emerged as the central social institution during the social disintegration following the nuclear war, much as the villa became the basic social unit of the new feudal society in the vacuum left by the fall of Rome. It was the principal and normal means for organizing benefits to the sick and unemployed, as well as the primary base for providing public services like police and fire protection.[1]

It's to be hoped that, absent a thermonuclear war, the transition will be a bit less abrupt. Upward-creeping unemployment, the exhaustion of the state's social safety net, and the explosion of affordable technologies for small-scale production and network organization, taken together, will likely encounter an environment in which the incentives for widespread experimentation are intense. John Robb speculates on one way these trends may come together:

> In order to build out resilient communities there needs to be a business mechanism that can financially power the initial roll-out. Here are some markets that may be serviced by resilient community formation:
>
> - An already large and growing group of people that are looking for a resilient community within which to live if the global or US system breaks down (ala the collapse of the USSR/Argentina or worse). Frankly, a viable place to live is a lot better than investing in gold that may not be valuable (gold assumes people are willing to part with what they have).
> - A larger and growing number of prospective students that want to learn how to build and operate resilient communities (rather than campus experiments and standard classroom blather).
> - A large and growing group of young people that want to work and live within a resilient community. A real job after school ends.
>
> Triangulating these markets yields the following business opportunity:
>
> - The ability of prospective residents of resilient communities to invest a portion of their IRA/401K and/or ongoing contributions in the construction and operation of a resilient community in exchange for home and connections to resilient systems (food, energy, local manufacturing, etc.) within that community.
> - An educational program, like Gaia University's collaboration with Factor e Farm, that allows students to get a degree while building out a resilient community (active permaculture/acquaculture plots, micro manufactories, local energy production, etc.). This allows access to government sponsored student debt.
> - A work study program that allows students of the University to pay off their student debt and make a living doing over a 5 year (flexible) period. IF they want to do that.
>
> I suspect there is a good way to construct a legal business framework that allows this to happen. What would make this even more interesting would be to combine this with a "Freedom" network/darknet that allows ideas to flow freely via an open source approach between active resilient communities on the network. The network would also allow goods and services to flow between sites (via an internal trading mechanism) and also allow these goods and intellectual property (protected by phalanxes of lawyers) to be sold to the outside world (via an

Blog, April 26, 2010 <http://blog.p2pfoundation.net/daniel-suarez-daemon-and-freedom/2010/4/26>.

[1]Poul Anderson, *Orion Shall Rise* (New York: Pocket Books, 1983).

> Ali Baba approach). At some point, if it is designed correctly, this network could become self-sustaining and able to generate the income necessary to continue a global roll-out by itself.[1]

(All except the "intellectual property" part.)

An article by Reihan Salam in *Time Magazine*, of all places, put a comparatively upbeat spin on the possibilities:

> Imagine a future in which millions of families live off the grid, powering their homes and vehicles with dirt-cheap portable fuel cells. As industrial agriculture sputters under the strain of the spiraling costs of water, gasoline and fertilizer, networks of farmers using sophisticated techniques that combine cutting-edge green technologies with ancient Mayan know-how build an alternative food-distribution system. Faced with the burden of financing the decades-long retirement of aging boomers, many of the young embrace a new underground economy, a largely untaxed archipelago of communes, co-ops, and kibbutzim that passively resist the power of the granny state while building their own little utopias.
>
> Rather than warehouse their children in factory schools invented to instill obedience in the future mill workers of America, bourgeois rebels will educate their kids in virtual schools tailored to different learning styles. Whereas only 1.5 million children were homeschooled in 2007, we can expect the number to explode in future years as distance education blows past the traditional variety in cost and quality. The cultural battle lines of our time, with red America pitted against blue, will be scrambled as Buddhist vegan militia members and evangelical anarchist squatters trade tips on how to build self-sufficient vertical farms from scrap-heap materials. To avoid the tax man, dozens if not hundreds of strongly encrypted digital currencies and barter schemes will crop up, leaving an underresourced IRS to play whack-a-mole with savvy libertarian "hacktivists."
>
> Work and life will be remixed, as old-style jobs, with long commutes and long hours spent staring at blinking computer screens, vanish thanks to ever increasing productivity levels. New jobs that we can scarcely imagine will take their place, only they'll tend to be home-based, thus restoring life to bedroom suburbs that today are ghost towns from 9 to 5. Private homes will increasingly give way to cohousing communities, in which singles and nuclear families will build makeshift kinship networks in shared kitchens and common areas and on neighborhood-watch duty. Gated communities will grow larger and more elaborate, effectively seceding from their municipalities and pursuing their own visions of the good life. Whether this future sounds like a nightmare or a dream come true, it's coming.
>
> This transformation will be not so much political as antipolitical. The decision to turn away from broken and brittle institutions, like conventional schools and conventional jobs, will represent a turn toward what military theorist John Robb calls "resilient communities," which aspire to self-sufficiency and independence. The left will return to its roots as the champion of mutual aid, cooperative living and what you might call "broadband socialism," in which local governments take on the task of building high-tech infrastructure owned by the entire community. Assuming today's libertarian revival endures, it's easy to imagine the right defending the prerogatives of state and local governments and also of private citizens—including the weird ones. This new individualism on the left and the right will begin in the spirit of cynicism and distrust that we see now, the sense that we as a society are incapable of solving pressing problems. It will

[1]John Robb, "An Entrepreneur's Approach to Resilient Communities," *Global Guerrillas*, February 22, 2010 <http://globalguerrillas.typepad.com/globalguerrillas/2010/02/turning-resilient-communities-into-a-business-opportunity.html>.

evolve into a new confidence that citizens working in common can change their lives and in doing so can change the world around them.[1]

I strongly suspect that, in whatever form of civil society stabilizes at the end of our long collapse, the typical person will be born into a world where he inherits a possessory right to some defined share in the communal land of an extended family or cohousing unit, and to some minimal level of support from the primary social unit in times of old age and sickness or unemployment in return for a customarily defined contribution to the common fund in his productive years. It will be a world in which the Amish barn-raiser and the sick benefit societies of Kropotkin and E.P. Thompson play a much more prominent role than Prudential or the anarcho-capitalist "protection agency."

Getting from here to there will involve a fundamental paradigm shift in how most people think, and the overcoming of centuries worth of ingrained habits of thought. This involves a paradigm shift from what James Scott, in *Seeing Like a State*, calls social organizations that are primarily "legible" to the state, to social organizations that are primary legible or transparent to the people of local communities organized horizontally and opaque to the state.[2]

The latter kind of architecture, as described by Kropotkin, was what prevailed in the networked free towns and villages of late medieval Europe. The primary pattern of social organization was horizontal (guilds, etc.), with quality certification and reputational functions aimed mainly at making individuals' reliability transparent to one another. To the state, such local formations were opaque.

With the rise of the absolute state, the primary focus became making society transparent (in Scott's terminology "legible") from above, and horizontal transparency was at best tolerated. Things like the systematic adoption of family surnames that were stable across generations (and the 20th century followup of citizen ID numbers), the systematic mapping of urban addresses for postal service, etc., were all for the purpose of making society transparent to the state. To put it crudely, the state wants to keep track of where its stuff is, same as we do—and we're its stuff.

Before this transformation, for example, surnames existed mainly for the convenience of people in local communities, so they could tell each other apart. Surnames were adopted on an ad hoc basis for clarification, when there was some danger of confusion, and rarely continued from one generation to the next. If there were multiple Johns in a village, they might be distinguished by trade ("John the Miller"), location ("John of the Hill"), patronymic ("John Richard's Son"), etc. By contrast, everywhere there have been family surnames with cross-generational continuity, they have been imposed by centralized states as a way of cataloguing and tracking the population—making it legible to the state, in Scott's terminology.[3]

To accomplish a shift back to horizontal transparency, it will be necessary to overcome a powerful residual cultural habit, among the general public, of thinking of such things through the mind's eye of the state. E.g., if "we" didn't have some way of verifying compliance with this regulation or that, some business somewhere might be able to get away with something or other. We must overcome six hundred years or so of almost inbred habits of thought, by which the state is the all-

[1]Reihan Salam, "The Dropout Economy," *Time*, March 10, 2010 <http://www.time.com/time/specials/packages/printout/0,29239,1971133_1971110_1971126,00.html>.

[2]James Scott, *Seeing Like a State* (New Haven and London: Yale University Press, 1998).

[3]*Ibid.*, pp. 64-73.

seeing guardian of society protecting us from the possibility that someone, somewhere might do something wrong if "the authorities" don't prevent it.

In place of this habit of thought, we must think instead of *ourselves* creating mechanisms on a networked basis, to make us as transparent as possible to *each other* as providers of goods and services, to prevent businesses from getting away with poor behavior by informing *each other*, to prevent *each other* from selling defective merchandise, to protect *ourselves* from fraud, etc. In fact, the creation of such mechanisms—far from making us *transparent* to the regulatory state—may well require active measures to render us *opaque* to the state (e.g. encryption, darknets, etc.) for protection *against* attempts to suppress such local economic self-organization against the interests of corporate actors.

In other words, we need to lose the centuries-long habit of thinking of "society" as a hub-and-spoke mechanism and viewing the world from the perspective of the hub, and instead think of it as a horizontal network in which we visualize things from the perspective of individual nodes. We need to lose the habit of thought by which transparency from above ever even became perceived as an issue in the first place.

This will require, more specifically, overcoming the hostility of conventional liberals who are in the habit of reacting viscerally and negatively, and on principle, to anything not being done by "qualified professionals" or "the proper authorities."

Arguably conventional liberals, with their thought system originating as it did as the ideology of the managers and engineers who ran the corporations, government agencies, and other giant organizations of the late 19th and early 20th century, have played the same role for the corporate-state nexus that the *politiques* did for the absolute states of the early modern period.

This is reflected in a common thread running through writers like Andrew Keene, Jaron Lanier, and Chris Hedges, as well as documentary producers like Michael Moore. They share a nostalgia for the "consensus capitalism" of the early postwar period, in which the gatekeepers of the Big Three networks controlled what we were allowed to see and it was just fine for GM to own the whole damned economy—just so long as everyone had a lifetime employment guarantee and a UAW contract.

Paul Fussell, in *Bad*, ridicules the whole Do-it-Yourself ethos as an endless Sahara of the Squalid, with blue collar schmoes busily uglifying their homes by taking upon themselves projects that should be left to—all together now—the Properly Qualified Professionals.

Keith Olbermann routinely mocks exhortations to charity and self-help, reaching for shitkicking imagery of the nineteenth century barnraiser for want of any other comparision to sufficiently get across just how backward and ridiculous that kind of thing really is. Helping your neighbor out directly, or participating in a local self-organized friendly society or mutual, is all right in its own way, if nothing else is available. But it carries the inescapable taint, not only of the quaint, but of the provincial and the picayune—very much like the perception of homemade bread and home-grown veggies promoted in corporate advertising in the early twentieth century, come to think of it. People who help each other out, or organize voluntarily to pool risks and costs, are to be praised—grudgingly and with a hint of condescension—for doing the best they can in an era of relentlessly downscaled social services. But that people are forced to resort to such expedients, rather than meeting all their social safety net needs through one-stop shopping at the Ministry of Central Services office in a giant monumental building with a statue of winged victory in the lobby, a la *Brazil*, is a damning indictment of any civilized society.

The progressive society is a society of comfortable and well-fed citizens, competently managed by properly credentialed authorities, happily milling about like ants in the shadows of miles-high buildings that look like they were designed by Albert Speer. And that kind of H.G. Wells utopia simply has no room for the barn-raiser or the sick benefit society.

Aesthetic sensibilities aside, such critics are no doubt motivated to some extent by genuine concern that networked reputational and certifying mechanisms just won't take up the slack left by the disappearance of the regulatory state. Things like *Consumer Reports*, Angie's List and the Better Business Bureau are all well and good, for educated people like themselves who have the sense and know-how to check around. But Joe Sixpack, God love him, will surely just go out and buy magic beans from the first disreputable salesman he encounters—and then likely put them right up his nose.

Seriously, snark aside, such reputational systems really are underused, and most people really do take inadequate precautions in the marketplace on the assumption that the regulatory state guarantees some minimum acceptable level of quality. But liberal criticism based on this state of affairs reflects a remarkably static view of society. It ignores the whole idea of crowding out, as well as the possibility that even the Great Unwashed may be capable of changing their habits quite rapidly in the face of necessity. Because people are not presently in the habit of automatically consulting such reputational networks to check up on people they're considering doing business with, and *are* in the habit of unconsciously assuming the government will protect them, conventional liberals assume that people will not shift from one to the other in the face of changing incentives, and scoff at the idea of a society that relies primarily on networked rating systems.

But in a society where people are aware that most licensing and safety/quality codes are no longer enforceable, and "caveat emptor" is no longer just a cliche, it would be remarkable if things like Angie's list, reputational certification by local guilds, customer word of mouth, etc., did *not* rapidly grow in importance for most people. They were, after all, at one time the main reputational mechanism that people *did* rely on before the rise of the absolute state, and as ingrained a part of ordinary economic behavior as reliance on the regulatory state is today.

People's habits change rapidly. Fifteen years ago, when even the most basic survey of a research topic began with an obligatory painful crawl through the card catalog, Reader's Guide and Social Science Index—and when the average person's investigations were limited to the contents of his $1000 set of Britannica—who could have foreseen how quickly Google and SSRN searches would become second nature?

In fact, if anything the assumption that "they couldn't sell it if it wasn't OK, because it's illegal" leaves people especially vulnerable, because it creates an unjustified confidence and complacency regarding what they buy. The standards of safety and quality, based on "current science," are set primarily by the regulated industries themselves, and those industries are frequently able to criminalize voluntary safety inspections with more stringent standards—or advertising that one adheres to such a higher standard—on the grounds that it constitutes disparagement of the competitor's product. For example, Monsanto frequently goes after grocers who label their milk rBGH free, and some federal district courts have argued that it's an "unfair competitive practice" to test one's beef cattle for Mad Cow Disease more frequently than the mandated industry standard. We have people slathering themselves with lotion saturated with estrogen-mimicing parabens, on the assumption that "they couldn't sell it if it was dangerous." So in many cases, this all-seeing central authority we count on to protect us is like a shepherd that puts the wolves in charge of the flock.

As an individualist anarchist, I'm often confronted with issues of how societies organized around such primary social units would affect the libertarian values of self-ownership and nonaggression.

First, it's extremely unlikely in my opinion that the collapse of centralized state and corporate power will be driven by, or that the post-corporate state society that replaces it will be organized according to, any single libertarian ideology (although I am hopeful, for reasons discussed later in this section, that there will be a significant number of communities organized primarily around such values, and that those values will have a significant leavening effect on society as a whole).

Second, although the kinds of communal institutions, mutual aid networks and primary social units into which people coalesce may strike the typical right-wing flavor of free market libertarian as "authoritarian" or "collectivist," a society in which such institutions are the dominant form of organization is by no means necessarily a violation of the substantive values of self-ownership and nonaggression.

I keep noticing, without ever really being able to put it in just the right words, that most conventional libertarian portrayals of an ideal free market society, and particularly the standard anarcho-capitalist presentation of a conceptual framework of individual self-ownership and non-aggression, seem implicitly to assume an atomized society of individuals living (at most) in nuclear families, with allodial ownership of a house and quarter-acre lot, and with most essentials of daily living purchased via the cash nexus from for-profit business firms.

But it seems to me that the libertarian concepts of self-ownership and nonaggression are entirely consistent with a wide variety of voluntary social frameworks, while at the same time the practical application of those concepts would vary widely. Imagine a society like most of the world before the rise of the centralized territorial state, where most ultimate (or residual, or reversionary) land ownership was vested in village communes, even though there might be a great deal of individual possession. Or imagine a society like the free towns that Kropotkin described in the late Middle Ages, where people organized social safety net functions through the guild or other convivial associations. Now, it might be entirely permissible for an individual family to sever its aliquot share of land from the peasant commune, and choose not to participate in the cooperative organization of seasonal labor like spring plowing, haying or the harvest. It might be permissible, in an anarchist society, for somebody to stay outside the guild and take his chances on unemployment or sickness. But in a society where membership in the primary social unit was universally regarded as the best form of insurance, such a person would likely be regarded as eccentric, like the individualist peasants in anarchist Spain who withdrew from the commune, or the propertarian hermits in Ursula LeGuin's *The Dispossessed*. And for the majority of people who voluntarily stayed in such primary social units, most of the social regulations that governed people's daily lives would be irrelevant to the Rothbardian conceptual framework of self-ownership vs. coercion.

By way of comparison, for the kinds of mainstream free market libertarians conventionally assigned to the Right, the currently predominating model of employment in a business firm is treated as the norm. Such libertarians regard the whole self-ownership vs. aggression paradigm as irrelevant to life within that organizational framework so long as participation in the framework is itself voluntary. Aha! but by the same token, when people are born into a framework in which they are guaranteed a share in possession of communal land and are offered social safety net protections in the event of illness or old age, in return for observance of communal social norms, the same principle applies.

And for most of human history, before the state started actively suppressing voluntary association, and discouraged a self-organized social safety net based on voluntary cooperation and mutual aid, membership in such primary social units was the norm. Going all the way back to the first *homo sapiens* hunter-gatherer groups, altruism was very much consistent with rational utility maximization as a form of insurance policy. When there's no such thing as unemployment compensation, food stamps, or Social Security, it makes a whole lot of sense for the most skillful or lucky hunter, or the farmer with the best harvest, to share with the old, sick and orphaned—and not to be a dick about it or rub it in their faces. Such behavior is almost literally an insurance premium to guarantee your neighbors will take care of you when you're in a similar position. Consider Sam Bowles' treatment of the altruistic ethos in the "weightless" forager economy:

> Network wealth is the contribution made by your social connections to your well-being. This could be measured by your number of connections, or by your centrality in different networks. A simple way to think about this is the number of people who will share food with you. . . .
>
> The culture of the foraging band emphasizes generosity and modesty. There are norms of sharing. You depricate what you catch, describing it as "not as big as a mouse", or "not even worth cooking", even when you've killed a large animal. In the Ache people of Eastern Paraguay, hunters are prohibited from eating their own catch. There's complex sanctioning of individually assertive behavior, particularly those that disturb or disrupt cooperation and group stability. This makes sense—if hunters can't expect that they'll be fed by other hunters—particularly by a hunter who suddenly develops a taste for eating his own catch—the society collapses rapidly.[1]

Before states began creating social safety nets, functions comparable to unemployment compensation, food stamps, and Social Security were almost universally organized through primary social units like the clan, the village commune, or the guild.

The irony is that the mainstream of market anarchism, particularly right-leaning followers of Murray Rothbard, are pushing for a society where there's no state to organize unemployment compensation, food stamps or Social Security. I suppose they just assume this function will be taken over by Prudential, but I suspect that what fills the void after the disintegration of the state will be a lot closer to Poul Anderson's above-mentioned society of lodges in the Northwest Federation.

It seems likely the Rothbardians are neglecting the extent to which the kinds of commercialized business relations they use as a preferred social model are, themselves, a product of the statism that they react against. The central state that they want to do away with played a large role in dismantling organic social institutions like clans, village communes, extended families, guilds, friendly societies, and so forth, and replacing them with an atomized society in which everybody sells his labor, buys consumables from the store, and is protected either by the department of human services or Prudential.

[1]Ethan Zuckerman, "Samuel Bowles Introduces Kudunomics," *My Heart's in Accra*, November 17, 2009 <http://www.ethanzuckerman.com/blog/2009/11/17/samuel-bowles-introduces-kudunomics/>.

Gary Chartier (a professor of ethics and philosophy at La Sierra University), in discussing some of these issues with me, raised some serious questions about my comparison between the right-libertarian view of civil rights in the employment relation, and the rights of the individual in the kinds of communal institutions I brought up. One of the central themes of "thick" libertarianism is that a social environment can have an unlibertarian character, and that nominally private and primary forms of exploitation and unfairness can exist, even when no formal injustice has taken place in terms of violation of the nonaggression principle.[1]

Cultural authoritarianism in the workplace, especially, is a central focus for many thick libertarians. Claire Wolfe, a writer with impeccable libertarian credentials and Gadsden Flag-waver nonpareil, has pointed out just how inconsistent the authoritarian atmosphere of the workplace is with libertarian cultural values.[2] At the other end of the spectrum are people like Hans Hermann Hoppe, who actively celebrate the potential for cultural authoritarianism when every square foot of the Earth has been appropriated and there is no such thing as a right of way or any other form of public space. Their ideal world is one in which the letter of self-ownership and nonaggression is adhered to, but in which one cannot move from Point A to Point B anywhere in the world without encountering a request for "Ihre Papiere, bitte!" from the private gendarmerie, or stopping for the biometric scanners, of whoever owns the bit of space they're standing on at any given momemt.

So could not an organic local community and its communal institutions, likewise, create an environment that would be considered authoritarian by thick libertarian norms, even when self-ownership and nonaggression were formally respected? Chartier continues:

> I think the interesting question, for a left libertarian who's interested in minimizing negative social pressure on minority groups of various sorts and who doesn't want to see people pushed around, is, What kinds of social arrangements would help to ensure that "the social regulation that governed people's daily lives" didn't replicate statism in a kindler, gentler fashion? ("Want access to the communal water supply? I'd better not see you working in your field on the Sabbath . . . .") Ostracism is certainly a hell of a lot better than jail, but petty tyrannies are still petty tyrannies. What's the best way, do you think, to keep things like zoning regulations from creeping in the back door via systems of persistent social pressure? I'd rather not live in a Hoppe/Tullock condominium community.
>
> One way of getting at this might be to note that, as [Michael] Taylor plausibly suggests, small scale communities are probably good at preventing things like workplace injustices and the kinds of abuses that are possible when there are vast disparities in wealth and so in social influence. But I'm less clear that they're good at avoiding abuses, not in the economic realm, but in the social or cultural realm. I'm more of a localist than a number of the participants in the recent discussions of these matters, but I think people like Aster [Aster Francesca, pen name of Jeanine Ring, a prolific and incisive writer on issues of social and cultural freedom] are surely right that the very solidarity that can prevent people

[1]See, for example, Roderick Long and Charles Johnson, "Libertarian Feminism: Can This Marriage Be Saved?" May 1, 2005 <http://charleswjohnson.name/essays/libertarian-feminism/>; Johnson, "Libertarianism Through Thick and Thin," *Rad Geek People's Daily*, October 3, 2008 <http://radgeek.com/gt/2008/10/03/libertarianism_through/>; Matt MacKenzie, "Exploitation: A Dialectical Anarchist Perspective," *Upaya: Skillful Means to Liberation*, March 20, 2007 <http://upaya.blogspot.com/2007/03/exploitation.html>. (link defunct—retrieved through Internet Archive).

[2]Claire Wolfe, "Insanity, the Job Culture, and Freedom," *Loompanics Catalog* 2005 <http://www.loompanics.com/Articles/insanityjobculture.html>.

in a close-knit community from going hungry or being arbitrarily fired can also keep them from being open about various kinds of social non-conformity. (My own social world includes a lot of people who need to avoid letting others with whom they work or worship know that they drink wine at dinner or learn about their sexual behavior; a generation ago, they'd have also avoided letting anyone know they went to movies.)

Self-ownership vs. aggression needn't be immediately relevant to community life any more than it might be to the firm. But the same sorts of objections to intra-firm hierarchy would presumably still apply to some kinds of social pressure at the community level, yes?[1]

One thing that's relevant is suggested by Michael Taylor's[2] treatment of hippie communalism as a way of reinventing community. To the extent that a reaction against the centralized state and corporate power is motivated by anti-authoritarian values, and rooted in communities like file-sharers, pot-smokers, hippie back-to-the-landers, etc. (and even to the extent that it takes place in a milieu "corrupted" by the American MYOB ethos), there will be at least a sizeable minority of communities in a post-state panarchy where community is seen as a safety net and a place for voluntary interaction rather than a straitjacket. And in America, at least, the majority of communities will also probably be leavened to some extent by the MYOB ethos, and by private access to the larger world via a network culture that it's difficult for the community to snoop on. (I've seen accounts of the monumental significance of net-connected cell phones to Third World teens who live in traditional patriarchal cultures without even their own private rooms—immensely liberating).

The best thing left-libertarians can do is probably try to strengthen ties between local resilience movements of various sorts and culturally left movements like open-source/filesharing, the greens, and all the other hippie-dippy stuff. The biggest danger from that direction is that, as in the rather unimaginatively PC environments of a lot of left-wing urban communes and shared housing projects today, people might have to hide the fact that they ate a non-vegan dinner.

As for communities that react against state and corporate power from the direction of cultural conservatism, the Jim Bob Duggar types (a revolt of "Jihad" against "McWorld"), probably the best we can hope for is 1) the leavening cultural effects of the American MYOB legacy and even surreptitious connection to the larger world, 2) the power of exit as an indirect source of voice, and 3) the willingness of sympathetic people in other communities to intervene on behalf of victims of the most egregious forms of bluestockingism and Mrs. Grundyism.

D. LETS Systems, Barter Networks, and Community Currencies

Local currencies, barter networks and mutual credit-clearing systems are a solution to a basic problem: "a world in which there is a lot of work to be done, but there is simply no money around to bring the people and the work together."[3]

Unconventional currencies are buffers against unemployment and economic downturn. Tsutomu Hotta, the founder of the Hureai Kippu ("Caring Relationship

[1]Gary Chartier, private email, January 15, 2010. The discussion took place in the context of my remarks on Michael Taylor's book *Community, Anarchy and Liberty* (Cambridge, UK: Cambridge University Press, 1982). To put the references to the Sabbath and other issues of personal morality in context, Chartier is from a Seventh-day Adventist backgrounds and teaches at a university affiliated with that denomination.

[2]Taylor, pp. 161-164 (see note immediately above).

[3]Lietaer, p. 112.

Tickets," a barter system in which participants accumulate credits in a "healthcare time savings account" by volunteering their own time), estimated that such unconventional currencies would replace a third to a half of conventional monetary functions. "As a result, the severity of any recession and unemployment will be significantly reduced."[1]

One barrier to local barter currencies and crowdsourced mutual credit is a misunderstanding of the nature of money. For the alternative economy, money is not primarily a store of value, but an accounting system to facilitate exchange. Its function is not to store accumulated value from past production, but to provide liquidity to facilitate the exchange of present and future services between producers.

The distinction is a very old one, aptly summarized by Schumpeter's contrast between the "money theory of credit" and the "credit theory of money." The former, which Schumpeter dismisses as entirely fallacious, assumes that banks "lend" money (in the sense of giving up use of it) which has been "withdrawn from previous uses by an entirely imaginary act of saving and then lent out by its owners. It is much more realistic to say that the banks 'create credit.,' than to say that they lend the deposits that have been entrusted to them."[2] The credit theory of money, on the other hand, treats finances "as a clearing system that cancels claims and debts and carries forward the difference. . . . "[3]

Thomas Hodgskin, criticizing the Ricardian "wage fund" theory from a perspective something like Schumpeter's credit theory of money, utterly demolished any moral basis for the creative role of the capitalist in creating a wage fund through "abstention," and instead made the advancement of subsistence funds from *existing* production a function that workers could just as easily perform for one another through mutual credit, were the avenues of doing so not preempted.

> The only advantage of circulating capital is that by it the labourer is enabled, he being assured of his present subsistence, to direct his power to the greatest advantage. He has time to learn an art, and his labour is rendered more productive when directed by skill. Being assured of immediate subsistence, he can ascertain which, with his peculiar knowledge and acquirements, and with reference to the wants of society, is the best method of labouring, and he can labour in this manner. Unless there were this assurance there could be no continuous thought, an invention, and no knowledge but that which would be necessary for the supply of our immediate animal wants. . . .
>
> The labourer, the real maker of any commodity, derives this assurance from a knowledge he has that the person who set him to work will pay him, and that with the money he will be able to buy what he requires. He is not in possession of any stock of commodities. Has the person who employs and pays him such a stock? Clearly not. . . .
>
> A great cotton manufacturer . . . employs a thousand persons, whom he pays weekly: does he possess the food and clothing ready prepared which these persons purchase and consume daily? Does he even know whether the food and clothing they receive are prepared and created? In fact, are the food and clothing which his labourers will consume prepared beforehand, or are other labourers busily employed in preparing food and clothing while his labourers are making cotton yarn? Do all the capitalists of Europe possess at this moment one week's food and clothing for all the labourers they employ? . . .

[1]*Ibid.*, pp. 23-24.

[2]*Joseph Schumpeter, History of Economic Analysis.* Edited from manuscript by Elizabeth Boody Schumpeter (New York: Oxford University Press, 1954), p. 1114.

[3]*Ibid.*, p. 717.

> . . . As far as food, drink and clothing are concerned, it is quite plain, then, that no species of labourer depends on any previously prepared stock, for in fact no such stock exists; but every species of labourer does constantly, and at all times, depend for his supplies on the co-existing labour of some other labourers.[1]
>
> . . . When a capitalist therefore, who owns a brew-house and all the instruments and materials requisite for making porter, pays the actual brewers with the coin he has received for his beer, and they buy bread, while the journeymen bakers buy porter with their money wages, which is afterwards paid to the owner of the brew-house, is it not plain that the real wages of both these parties consist of the produce of the other; or that the bread made by the journeyman baker pays for the porter made by the journeyman brewer? But the same is the case with all other commodities, and labour, not capital, pays all wages. . . .
>
> In fact it is a miserable delusion to call capital something saved. Much of it is not calculated for consumption, and never is made to be enjoyed. When a savage wants food, he picks up what nature spontaneously offers. After a time he discovers that a bow or a sling will enable him to kill wild animals at a distance, and he resolves to make it, subsisting himself, as he must do, while the work is in progress. He saves nothing, for the instrument never was made to be consumed, though in its own nature it is more durable than deer's flesh. This example represents what occurs at every stage of society, except that the different labours are performed by different persons—one making the bow, or the plough, and another killing the animal or tilling the ground, to provide subsistence for the makers of instruments and machines. To store up or save commodities, except for short periods, and in some particular cases, can only be done by more labour, and in general their utility is lessened by being kept. The savings, as they are called, of the capitalist, are consumed by the labourer, and there is no such thing as an actual hoarding up of commodities.[2]

What political economy conventionally referred to as the "labor fund," and attributed to past abstention and accumulation, resulted rather from the present division of labor and the cooperative distribution of its product. "Capital" is a term for a right of property in organizing and disposing of this present labor. The same basic cooperative functions could be carried out just as easily by the workers themselves, through mutual credit. Under the present system, the capitalist monopolizes these cooperative functions, and thus appropriates the productivity gains from the social division of labor.

> Betwixt him who produces food and him who produces clothing, betwixt him who makes instruments and him who uses them, in steps the capitalist, who neither makes nor uses them, and appropriates to himself the produce of both. With as niggard a hand as possible he transfers to each a part of the produce of the other, keeping to himself the large share. Gradually and successively has he insinuated himself betwixt them, expanding in bulk as he has been nourished by their increasingly productive labours, and separating them so widely from each other that neither can see whence that supply is drawn which each receives through the capitalist. While he despoils both, so completely does he exclude one from the view of the other that both believe they are indebted him for subsistence.[3]

Franz Oppenheimer made a similar argument in "A Post Mortem on Cambridge Economics":

[1]Thomas Hodgskin, *Labour Defended Against the Claims of Capital* (New York: Augustus M. Kelley, 1969 [1825]), pp. 36-40.

[2]Hodgskin, *Popular Political Economy: Four Lectures Delivered at the London Mechanics' Institution* (New York: Augustus M. Kelley, 1966 [1827]), p. 247.

[3]Hodgskin, *Labour Defended*, p. 71.

> THE JUSTIFICATION OF PROFIT, to repeat, rests on the claim that the entire stock of instruments of production must be "saved" during one period by private individuals in order to serve during a later period. This proof, it has been asserted, is achieved by a chain of equivocations. In short, the material instruments, for the most part, are not saved in a former period, but are manufactured in the same period in which they are employed. What is saved is capital in the other sense, which may be called for present purposes "money capital." But this capital is not necessary for developed production.
>
> Rodbertus, about a century ago, proved beyond doubt that almost all the "capital goods" required in production are created in the same period. Even Robinson Crusoe needed but one single set of simple tools to begin works which, like the fabrication of his canoe, would occupy him for several months. A modern producer provides himself with capital goods which other producers manufacture simultaneously, just as Crusoe was able to discard an outworn tool, occasionally, by making a new one while he was building the boat. On the other hand, money capital must be saved, but it is not absolutely necessary for developed technique. It can be supplanted by co-operation and credit, as Marshall correctly states. He even conceives of a development in which savers would be glad to tend their savings to reliable persons without demanding interest, even paying something themselves for the accommodation for security's sake. Usually, it is true, under capitalist conditions, that a certain personally-owned money capital is needed for undertakings in industry, but certainly it is never needed to the full amount the work will cost. The initial money capital of a private entrepreneur plays, as has been aptly pointed out, merely the rôle of the air chamber in the fire engine; it turns the irregular inflow of capital goods into a regular outflow.[1]

Oscar Ameringer illustrated the real-world situation in a humorous socialist pamphlet, "Socialism for the Farmer Who Farms the Farm," written in 1912. A river divided the nation of Slamerica into two parts, one inhabited by farmers and the other by makers of clothing. The bridge between them was occupied by a fat man named Ploot, who charged the farmers four pigs for a suit of clothes and the tailors four suits for a pig. The difference was compensation for the "service" he provided in letting them across the bridge and providing them with work. When a radical crank proposed the farmers and tailors build their own bridge, Ploot warned that by depriving him of his share of their production they would drive capital out of the land and put themselves out of work three-quarters of the time (while getting the same number of suits and pigs, of course).[2]

Schumpeter distinction between money theories of credit and credit theories of money is useful here. Critiquing the former, he wrote that it was misleading to treat bank credit as the lending of funds which had been "withdrawn from previous uses by an entirely imaginary act of saving and then lent out by their owners. It is much more realistic to say that the banks 'create credit . . . ,' than to say that they lend the deposits that have been entrusted to them."[3] The latter, in contrast, treat finances "as a clearing system that cancels claims and carries forward the difference."[4]

E. C. Riegel argues that issuing money is a function of the individual within the market, a side-effect of his normal economic activities. Currency is issued by

[1]Franz Oppenheimer, "A Post Mortem on Cambridge Economics (Part Three)," *The American Journal of*
Economics and Sociology, vol. 3, no. 1 (1944), pp, 122-123, [115-124]

[2]Oscar Ameriger. "Socialism for the Farmer Who Farms the Farm." Rip-Saw Series No. 15 (Saint Louis: The National Rip-Saw Publishing Co., 1912).

[3]Schumpeter, *History of Economic Analysis*, p. 1114.

[4]*Ibid.*, p. 717.

the buyer by the very act of buying, and it's backed by the goods and services of the seller.

> Money can be issued only in the act of buying, and can be backed only in the act of selling. Any buyer who is also a seller is qualified to be a money issuer. Government, because it is not and should not be a seller, is not qualified to be a money issuer.[1]

Money is simply an accounting system for tracking the balance between buyers and sellers over time.[2]

And because money is issued by the buyer, it comes into existence as a debit. The whole point of money is to create purchasing power where it did not exist before: " . . . [N]eed of money is a condition precedent to the issue thereof. To issue money, one must be without it, since money springs only from a debit balance on the books of the authorizing bank or central bookkeeper."[3]

> IF MONEY is but an accounting instrument between buyers and sellers, and has no intrinsic value, why has there ever been a scarcity of it? The answer is that the producer of wealth has not been also the producer of money. He has made the mistake of leaving that to government monopoly.[4]

Money is "simply number accountancy among private traders."[5] Or as Riegel's disciple Thomas Greco argues, currencies are not "value units" (in the sense of being stores of value). They are means of payment *denominated* in value units.[6]

In fact, as Greco says, "barter" systems are more accurately conceived as "credit clearing" systems. In a mutual credit clearing system, rather than cashing in official state currency for alternative currency notes (as is the case in too many local currency systems), participating businesses *spend the money into existence* by incurring debits for the purchase of goods within the system, and then earning credits to offset the debits by selling their own services within the system. The currency functions as a sort of IOU by which a participant monetizes the value of his future production.[7] It's simply an accounting system for keeping track of each member's balance:

> Your purchases have been indirectly paid for with your sales, the services or labor you provided to your employer.
>
> In actuality, everyone is both a buyer and a seller. When you sell, your account balance increases; when you buy, it decreases.

It's essentially what a checking account does, except a conventional bank does not automatically provide overdraft protection for those running negative balances, unless they pay a high price for it.[8]

[1]E. C. Riegel, *Private Enterprise Money: A Non-Political Money System* (1944), Introduction <http://www.newapproachtofreedom.info/pem/introduction.html>.

[2]*Ibid.*, Chapter Seven <http://www.newapproachtofreedom.info/pem/chapter07.html>.

[3]Riegel, *The New Approach to Freedom: together with Essays on the Separation of Money and State.* Edited by Spencer Heath MacCallum (San Pedro, California: The Heather Foundation, 1976), Chapter Four <http://www.newapproachtofreedom.info/naf/chapter4.html>.

[4]Riegel, "The Money Pact, in *Ibid.* <http://www.newapproachtofreedom.info/naf/essay1.html>.

[5]Spencer H. MacCallum, "E. C. Riegel on Money" (January 2008) <http://www.newapproachtofreedom.info/documents/AboutRiegel.pdf>.

[6]Thomas Greco, *Money and Debt: A Solution to the Global Crisis* (1990), Part III: Segregated Monetary Functions and an Objective, Global, Standard Unit of Account <http://circ2.home.mindspring.com/Money_and_Debt_Part3_lo.PDF>.

[7]Greco, *The End of Money and the Future of Civilization* (White River Junction, Vermont: Chelsea Green Publishing, 2009), p. 82.

[8]*Ibid.*, p. 102.

> There's no reason businesses cannot maintain a mutual credit-clearing system between themselves, without the intermediary of a bank or any other third party currency or accounting institution. The businesses agree to accept each other's IOUs in return for their own goods and services, and periodically use the clearing process to settle their accounts.[1]

And again, since some of the participants run negative balances for a time, the system offers what amounts to interest-free overdraft protection. As such a system starts out, members are likely to resort to fairly frequent settlements of account, and put fairly low limits on the negative balances that can be run, as a confidence building measure. Negative balances might be paid up, and positive balances cashed out, every month or so. But as confidence increases, Greco argues, the system should ideally move toward a state of affairs where accounts are never settled, so long as negative balances are limited to some reasonable amount.

> An account balance increases when a sale is made and decreases when a purchase is made. It is possible that some account balances may always be negative. That is not a problem so long as the account is actively trading and the negative balance does not exceed some appropriate limit. What is a reasonable basis for deciding that limit? . . . Just as banks use your income as a measure of your ability to repay a loan, it is reasonable to set maximum debit balances based on the amount of revenue flowing through an account. . . . [One possible rule of thumb is] that a negative account balance should not exceed an amount equivalent to three months' average sales.[2]

It's interesting how Greco's proposed limit on negative balances dovetails with the credit aspect of the local currency system. His proposed balance limit, a de facto interest-free loan, is sufficient to fund the minimum capital outlays for many kinds of low-overhead micro-enterprise. Even at the average wages of unskilled labor, three months' income is sufficient to acquire the basic equipment for a Fab Lab (at least the open-source versions described in Chapter Six). And it's far more than sufficient to meet the capital outlays needed for a microbakery or microcab.

Greco recounts an experiment with one such local credit clearing system, the Tucson Traders. It's fairly typical of his experience: initial enthusiasm, followed by gradual decline and dwindling volume, as the dwindling number of goods and services and the inconvenience of traveling between the scattered participating businesses take their toll.[3]

The reason for such failure, in normal economic times, is that local currency systems are crowded out by the official currency and the state-supported banking system.

For a credit clearing system to thrive, it must offer a valued alternative to those who lack sources of money in the conventional economy. That means it must have a large variety of participating goods and services, participating businesses must find it a valuable source of business that would not otherwise exist in the conventional economy, and unemployed and underemployed members must find it a valuable alternative for turning their skills into purchasing power they would not otherwise have. So we can expect LETS or credit clearing systems to increase in significance in periods of economic downturn, and even more so in the structural decline of the money and wage economy that is coming.

[1]*Ibid.* pp. 106-107

[2]*Ibid.*, p. 134.

[3]Greco, *The End of Money*, pp. 139-141.

Karl Hess and David Morris cite Alan Watts' illustration of the absurdity of saying it's impossible for willing producers, faced with willing consumers, to produce for exchange because "there's not enough money going around":

> Remember the Great Depression of the Thirties? One day there was a flourishing consumer economy, with everyone on the up-and-up; and the next: poverty, unemployment and breadlines. What happened? The physical resources of the country—the brain, brawn, and raw materials—were in no way depleted, but there was a sudden absence of money, a so-called financial slump. Complex reasons for this kind of disaster can be elaborated at lengths by experts in banking and high finance who cannot see the forest for the trees. But it was just as if someone had come to work on building a house and, on the morning of the Depression, that boss had to say, "Sorry, baby, but we can't build today. No inches." "Whaddya mean, no inches? We got wood. We got metal. We even got tape measures." "Yeah, but you don't understand business. We been using too many inches, and there's just no more to go around."[1]

The point of the mutual credit clearing system, as Greco describes it, is that two people who have goods and services to offer—but no money—are able to use their goods and services to buy other goods and services, even when there's "no money."[2] So we can expect alternative currency systems to come into play precisely at those times when people feel the lack of "inches." Based on case studies in the WIR system and the Argentine social money movement, Greco says, "complementary currencies will take hold most easily when they are introduced into markets that are starved for exchange media."[3] The widespread proliferation of local currencies in the Depression suggests that when this condition holds, the scale of adoption will follow as a matter of course. And as we enter a new, long-term period of stagnation in the conventional economy, it seems likely that local currency systems will play a growing role in the average person's strategy for economic survival.

There has been a new revival of local currency systems starting in the 1990s with the Ithaca Hours system and spreading to a growing network of LETS currencies.

But Ted Trainer, a specialist on relocalized economies who writes at "The Simpler Way" site, points out that LETS systems are, by themselves, largely worthless. The problem with LETS systems, by themselves, is that

> most people do not have much they can sell, i.e., they do not have many productive skills or the capital to set up a firm. It is therefore not surprising that LETSystems typically do not grow to account for more than a very small proportion of a town's economic activity. . . . What is needed and what LETSystems do not create is productive capacity, enterprises. It will not set up a cooperative bakery in which many people with little or no skill can be organised to produce their own bread.
>
> So the crucial element becomes clear. *Nothing significant can be achieved unless people acquire the capacity to produce and sell things that others want.* Obviously, unless one produces and sells to others one can't earn the money with which to purchase things one needs from others. So the question we have to focus on is how can the introduction of a new currency facilitate this *setting up of firms that will enable those who had no economic role to start producing, selling, earning and buying*. The crucial task is to create productive roles, not to

[1]Karl Hess and David Morris, *Neighborhood Power: The New Localism* (Boston: Beacon Press, 1975), pp. 154-155.

[2]Greco, *The End of Money*, p. 116.

[3]*Ibid.*, p. 158.

> create a currency. The new currency should be seen as little more than an accounting device, necessary but not the crucial factor.
>
> It is obvious here that what matters in local economic renewal is not redistribution of income or purchasing power. What matters is *redistribution of production power.* [1]
>
> It is ridiculous that millions of people are been unable to trade with each other simply because they do not have money, i.e., tokens which enable them to keep track of who owes what amount of goods and work to whom. LETS is a great solution to this elementary problem.
>
> However it is very important to understand that a LETSystem is far from sufficient. In fact a LETS on its own will not make a significant difference to a local economy. The evidence is that on average LETS transactions make up less than 5% of the economic activity of the average member of a scheme, let alone of the region. (See R. Douthwaite, Short Circuit, 1996, p. 76.) [[Look up]]
>
> LETS members soon find that they can only meet a small proportion of their needs through LETS, i.e., that there is not that much they can buy with their LETS credits, and not that much they can produce and sell. Every day they need many basic goods and services but very few of these are offered by members of the system. This is the central problem in local economic renewal; the need for ways of increasing the capacity of local people to produce things local people need. The core problem in other words is how to set up viable firms. . . .
>
> The core task in town economic renewal is to enable, indeed create a whole new sector of economic activity involving the people who were previously excluded from producing and earning and purchasing. This requires much more than just providing the necessary money; it requires the establishment of firms in which people a can produce and earn.[2]

As he writes elsewhere, the main purpose of local currency systems is "to contribute to getting the unused productive capacity of the town into action, i.e., stimulating/enabling increase in output to meet needs." Therefore the creation of a local currency system is secondary to creating firms by which the unemployed and underemployed can earn the means of exchange.[3]

For that reason, Trainer proposes Community Development Cooperatives as a way to promote the kinds of new enterprises that enable people to earn local currency outside the wage system.

> The economic renewal of the town will not get far unless its CDC actively works on this problem of establishing productive ventures *within the new money sector* which will enable that sector to sell things to the old firms in the town. In the case of restaurants the CDC's best option would probably be to set up or help others set up gardens to supply the restaurants with vegetables. Those who run the gardens would pay the workers in new money, sell the vegetables to the restaurants for new money, and use their new money incomes to buy meals from the restaurants.
>
> The Community Development Cooperative must work hard to find and set up whatever other ventures it can because the capacity of the previously poor and unemployed group of people in the town to *purchase from* normal/old firms is strictly limited by the volume that that group is able to sell to those firms. Getting these productive ventures going is by far the most important task of the Community Development Cooperative, much more important than just organising a new currency in which the exchanges can take place.

[1]Ted Trainer, "Local Currencies" (September 4, 2008), *The Simpler Way* <http://ssis.arts.unsw.edu.au/tsw/localcurrency.html>.

[2]Trainer, "We Need More Than LETS," *The Simpler Way* <http://ssis.arts.unsw.edu.au/tsw/D11WeNdMreThLETS2p.html>.

[3]Trainer, "The Transition Towns Movement; its huge significance and a friendly criticism," *(We) can do better*, July 30, 2009 <http://candobetter.org/node/1439>.

> The other very important thing the Community Development Cooperative must do is enable low skilled and low income people to cooperative [sic] produce many things for themselves. A considerable proportion of people in any region do not have the skills to get a job in the normal economy. This economy will condemn them to poverty and boredom. Yet they could be doing much useful work, especially work to produce many of the things they need. But again this will not happen unless it is organised. Thus the Community Development Cooperative must organise gardens and workshops and enterprises (such as furniture repair, house renovation and fuel wood cutting) whereby this group of people can work together to produce many of the things they need. They might be paid in new money according to time contributions, or they might just share goods and income from sales of surpluses.[1]

Trainer's critique of stand-alone LETS systems makes a lot of sense. When people earn official dollars in the wage economy, and then trade them in for local currency notes at the local bank that can only be spent in local businesses, they're trading dollars they already have for something that's *less* useful; local currency, in those circumstances, becomes just another greenwashed yuppie lifestyle choice financed by participation in the larger capitalist economy. As Greco puts it,

> a community currency that is issued on the basis of payment of a national currency (e.g., a local currency that is sold for dollars), amounts to a "gift certificate" or localized "traveler's check." It amounts to prepayment for the goods or services offered by the merchants that agree to accept the currency. That approach provides some limited utility in encouraging the holder of the currency to buy locally . . . [But] that sort of issuance requires that someone have dollars in order for the community currency to come into existence.[2]

Local currency should be a tool that's *more* useful than the alternative, giving people who are outside the wage system and who lack official dollars a way to transform their skills into purchasing power they would otherwise not have. A unit of local currency shouldn't be something one obtains by earning official money through wage employment and then trading it in for feel-good money at the bank to spend on establishment Main Street businesses. It should be an accounting unit for barter by the unemployed or underemployed person, establishing *new* microenterprises out of their own homes and exchanging goods and services directly with one another.

Trainer's main limitation is his focus on large-scale capital investment in conventional enterprises as the main source of employment. In examining the need for capital for setting up viable firms, he ignores the enormous amounts of capital that already exist.

The capital exists in the form of the ordinary household capital goods that most people already own, sitting idle in their own homes: the ordinary kitchen ovens that might form the basis of household microbakeries producing directly for credit in the barter network; the sewing machines that might be used to make clothes for credit in the network; the family car and cell phone that might be used to provide cab service for the network in exchange for credit toward other members' goods and services; etc. The unemployed or underemployed carpenter, plumber, electrician, auto mechanic, etc., might barter his services for credit to purchase tomatoes from a market gardener within the network, for the microbaker's bread or the seamstress's shirts, and so forth. The "hobbyist" with a well-equipped workshop in his basement or back yard might custom machine replacement parts to keep the home appliances of the baker, market gardener, and

[1]Trainer, "We Need More Than LETS."
[2]Greco, *The End of Money*, p. 81.

seamstress working, in return for their goods and services. Eventually "hobbyist" workshops and small local machine shops might begin networked manufacturing for the barter network, perhaps even designing their own open-source products with CAD software and producing them with CNC machine tools.

Hernando de Soto, in *The Mystery of Capital*, pointed to the homes and plots of land, to which so many ordinary people in the Third World hold informal title, as an enormous source of unrealized investment capital. Likewise, the spare capacity of people's ordinary household capital goods is a potentially enormous source of "plant and equipment" for local alternative economies centered on the informal and household sector.

There is probably enough idle oven capacity in the households of the average neighborhood or small town to create the equivalent of a hundred cooperative bakeries. Why waste the additional outlay cost, and consequent overhead, for relocating this capital to a stand-alone building?

Another thing to remember is that, even when a particular kind of production requires capital investment beyond the capabilities of the individual of average means, new infrastructures for crowdsourced, distributed credit—microcredit—make it feasible to aggregate sizable sums of investment capital from many dispersed small capitals, without paying tribute to a capitalist bank for performing the service. That's why it's important for a LETS system to facilitate not only the exchange of present goods and services, but the advance of credit against future goods and services.

Such crowdsourced credit might be used by members of a barter network to form their own community or neighborhood workshops in cheap rental space, perhaps (again) contributing the unused tools sitting in their garages and basements.

Of course the idle capacity of conventional local businesses shouldn't be entirely downplayed. Conventional enterprises with excess capacity can often use the spare capacity to produce at marginal costs a fraction of the normal cost, for barter against similar surpluses of other businesses. For instance, vacant hotel rooms in the off-season might be exchanged for discounted meals at restaurants during the slow part of the day, matinee tickets at the theater, etc. And local nonprofit organizations might pay volunteers in community currency units good for such surplus production at local businesses. In Minneapolis, for example, volunteers are paid in Community Service Dollars, which can be used for up to half the price of a restaurant meal before 7 p.m., or 90% of a matinee movie ticket. This enables local businesses to utilize idle capacity to produce goods sold at cost, and enables the unemployed to turn their time into purchasing power.[1]

As we already saw above, barter associations like UXA frequently exchanged their members' skills for the surplus inventory of conventional businesses.

E. Community Bootstrapping

The question of economic development in apparently dead-end areas has been of widespread interest for a long time. Of one such area, the so-called Arkansas Delta region (the largely rural, black, cash crop southeastern portion of the state) was recently the subject of a column by John Brummett:

[1]Lietaer, pp. 207-209.

Back when then-Gov. Mike Huckabee was trying to consolidate high schools for better educational opportunities, I was among dozens openly agreeing with him.

People in small towns cried out that losing their high schools would mean losing their towns. Only once did I work up the nerve to write that a town had no inalienable right to exist and that it wasn't much of a town if all it had was a school.

This comment was not well-received in some quarters. I was called an elitist enemy of the wholesome rural life.

But that wasn't so. I wasn't an enemy of the blissful advantages of a bucolic eden; I was only against inefficiently small schools getting propped up illogically in little incorporated spots on the road, anachronistic remnants of an olden time.

So imagine my reaction last week when I read Rex Nelson's idea. It is to abandon, more or less, whole towns in the Delta and consolidate people from those towns in other towns that Nelson termed "worth saving" on account of having "critical mass."

Presumably you'd go into Gould and Marianna and Marvell and Elaine and Clarendon and Holly Grove and say something like this: "Y'all need to get out; come on, get packed; get to Pine Bluff or Helena or Forrest City, because that's where the government money for schools and hospitals and infrastructure and such is going to go from now on. We can't afford to keep messing with this dead little town that doesn't have any remote hope of getting better. We don't have enough money to send a doctor around to your little health clinic once a week. We've got to get you over to the town where he lives and where they have a hospital that can provide him equipment and a living. This is for your own good."

Nelson, former press aide to Tommy Robinson and Huckabee but a decent sort anyway, has just left a Republican-rewarded patronage job with the Delta Regional Authority. That's an eight-state compact spending federal grants in the fast-dying Delta region along both sides of the Mississippi River.

Newly relocated to an advertising agency in Little Rock, Nelson gave an interview to a friendly newspaper columnist and, after some discussion of his liking Southern food and culture, shared his valedictory thoughts on what in the wide world we might do for the Delta.

So here's the idea: You pick out communities with hospitals and schools and decent masses of population and give them more federal grants than you give all these proliferating and tiny dead communities. You try to correct all this chronic dissipation of effort and resources.

It's school consolidation writ large. It's an attempt at redistribution of the population. It's eminent domain on steroids.

It's cold. It's difficult. And it's absolutely right.

What we call the Delta region of eastern Arkansas is a mechanized farm region, vast acreage of soybeans and rice, with pointless towns dotted at every crossroad. These one-time commerce centers thrived before farming was mechanized. Jobs for humans were to be had through the first half of the last century. Now they're home to boarded windows and people trapped in tragic cycles of poverty without hope of jobs because none is left and none is coming.[1]

Despite Brummet's assumptions, there is no shortage of examples of building an alternative economy almost scratch, a bit at a time, in an impoverished area. The Antigonish movement in Nova Scotia and the Mondragon cooperatives in Spain are two such examples. Both movements were sparked by radical Catholic priests serving impoverished areas, and heavily influenced by the Distributist ideas of G.K. Chesterton and Hilaire Belloc. The Antigonish movement, founded by Fr. Moses Coady, envisioned starting with credit unions and consumer retail coopera-

[1]John Brummett, "Delta Solution: Move," *The Morning News of Northwest Arkansas*, June 14, 2009 <http://arkansasnews.com/2009/06/14/delta-solution-move/>.

tives, which would obtain goods from cooperative wholesale societies, and which would in turn be supplied by factories owned by the whole movement. The result would be an integrated cooperative economy as a base of independence from capitalism.[1] In the specific example of Larry's River, the community began by building a cooperative sawmill; they went on to build a cooperative lobster cannery, a credit union, a cooperative store, a blueberry cannery, and a fish processing plant.[2] Mondragon—founded in the Basque country by Fr. Don Jose Maria Arizmendiarrietta—started similarly with a small factory, gradually adding a trade school, a credit union, and another factory at a time, until it became an enormous federated system with its own finance arm and tens of thousands of member-owners employed in its enterprises.[3]

More recently, the people of the Salinas region of the Ecuadorian Andes created a similar regional economy by essentially the same process, as recounted by Massimo de Angelis of *the editor's blog*.[4] The Salinas area, a region centering on the village of the same name, includes some thirty communities comprising a total of around six thousand people. The area economy is a network of cooperative enterprises, commonly called "the organization," that includes some 95% of the population.

> The "organization" is in reality a quick name for several associations, foundations, consortia and cooperatives, ranging from cheese producers to textile, ceramic and chocolate making, herbal medicine and trash collection, a radio station an hotel, a hostel, and a "office of community tourism".

The origin of "the organization" is reminiscent of a couple of Antigonish and Mondragon. The Salinas area was originally the typical domain of a patron, under the Latin American *hacienda* system. Most land belonged to the Cordovez family, who collected rents pursuant to a Spanish crown grant, and the Cordovez family's salt mine was the main non-agricultural employer. Like Antigonish and Mondragon, the organization started out with a single cooperative enterprise and from there grew by mitosis into an entire federated network of cooperatives. The first cooperative, formed in the 1970s, was a credit union created as a source of independence from the loan sharks who preyed on the poor. (This initial nucleus, like—again—Antigonish and Mondragon, was the project of an activist Catholic priest, the Italian immigrant Fr. Antonio Polo). The credit cooperative offered to buy the Cordovez family lands. With the encouragement of Fr. Polo, the village subsequently organized one cooperative enterprise after another to provide employment after the salt mine closed.

A significant social safety net operates in the village, funded by the surpluses of various cooperative enterprises, on a gift economy basis. And it's possible to earn exchange value outside of wage labor by contributing to something like a time bank.

> However, at the end of the year, the monetary surplus [of the cheese factory] is not distributed among coop members on the basis of their milk contribution, but is shared among them for common projects: either buying new equipment, or transferred to community funds. This way, as our guide told us, "the farmer who has 10 cows is helping the farmer that has only one cow", allowing for some re-

[1]Race Matthews, *Jobs of Our Own: Building a Stakeholder Society—Alternatives to the Market & the State* (Annandale, NSW, Australia: Pluto Press, 1999), pp. 125-172.

[2]*Ibid.* , pp. 151-152; p. 47.

[3]*Ibid.*, pp. 173-190.

[4]Massimo de Angelis, "Branding + Mingas + Coops = Salinas," *the editor's blog*, March 26, 2010 <http://www.commoner.org.uk/blog/?p=239>.

> distribution. Another example is the use of Mingas. Minga is a quechua word used by various ethnical groups throughout the Andes and refer to unwaged community work, in which men, women and children all participate in pretty much convivial ways and generally ends up in big banquets. Infrastructure work such as road maintenance, water irrigation, planting, digging, but also garbage collection and cleaning up the square are all type of work that calls for a Minga of different size and are used in Salinas. Yet another example is the important use of foundations, that channel funds earned in social enterprises for projects for the community.

Angelis, despite his admiration, has serious doubts as to whether the project is relevant or replicable. For one thing, this mixed commons/market system may be less sustainable when more capital-intensive forms of production are undertaken, and may accordingly be more vulnerable to destabilization and decay into exploitative capitalism. He raises the example of the new factory for turning wool into thread, to be vertically integrated with the household production of sweaters and other woolens. The large capital outlay, he says, means a break even point can only be achieved with fairly large batch production.

For another, de Angelis says, the success of the Salinas model arguably depends on its uniqueness, so that it can serve wide-open global niche markets without a lot of global competition from other local economies pursuing the same development model.

And finally, debt financing of capital investment leads to a certain degree of self-exploitation to service that debt.

De Angelis analyzes the cumulative implications of these problems:

> I have mixed feelings about this Salinas' experience. There is no doubt that the 69 agro-industrial and 38 service communities enterprises are quite a means for the local population to meet reproduction needs in ways that shield them from the most exploitative practices of other areas in the region and make them active participants in commoning processes centred on dignity. But the increasing reliance on, and strong preoccupation with, global export circuits and on the markets seems excessive, with the risk that experiments like these really become the vehicles for commons co-optation.

The newest venture along these lines is the Evergreen Cooperative Initiative in the decaying rust belt city of Cleveland—aka "the Mistake by the Lake," where the poverty rate is 30%.[1]

The Evergreen Cooperative Initiative is heavily influenced by the example of Mondragon.[2] The project had its origins in a study trip to Mondragon sponsored by the Cleveland Foundation, and is described as "the first example of a major city trying to reproduce Mondragon."[3] Besides the cooperative development fund, its umbrella of support organizations includes Evergreen Business services, which provides "back-office services, management expertise and turn-around skills should a co-op get into trouble down the road." Member enterprises are expected

[1]<http://www.evergreencoop.com/>

[2]Guy Alperowitz, Ted Howard, and Thad Williamson, "The Cleveland Model," *The Nation*, February 11, 2010 <http://www.thenation.com/doc/20100301/alperowitz_et_al/single>.

[3]Andrew MacLeod, "Mondragon—Cleveland—Sacramento," *Cooperate and No One Gets Hurt*, October 10, 2009 <http://coopgeek.wordpress.com/2009/10/10/mondragon-cleveland-sacramento/>; Ohio Employee Ownership Center, "Cleveland Goes to Mondragon," *Owners at Work* (Winter 2008-2009), pp.10-12 <http://dept.kent.edu/oeoc/OEOCLibrary/OaW_Winter08_Cleveland_Goes_to_Mondragon.pdf>.

to plow ten percent of pre-tax profits back into the development fund to finance investment in new cooperatives.[1]

The Evergreen Cooperative Laundry[2] was the first of some twenty cooperative enterprises on the drawing board, followed by Ohio Cooperative Solar[3] (which carries out large-scale installation of solar power generating equipment on the roofs of local government and non-profit buildings). A second and third enterprise, a cooperative greenhouse[4] and the *Neighborhood Voice* newspaper, are slated to open in the near future.

The Initiative is backed by stakeholders in the local economy, local government and universities. The primary focus of the new enterprises, besides marketing to individuals in the local community, is on serving local "anchor institutions"—the large hospitals and universities—that will provide a guaranteed market for a portion of their services. The Cleveland Foundation and other local foundations, banks, and the municipal government are all providing financing. The Evergreen Cooperative Development Fund is currently capitalized at $5 million, and expects to raise at least $10-12 million more.[5]

Besides the Cleveland Foundation, other important stakeholders are the Cleveland Roundtable and the Democracy Collaborative. The Roundtable is a project of Community-Wealth.org[6]; Community-Wealth[7], in turn, is a project of the Democracy Collaborative at the University of Maryland, College Park.[8] All three organizations are cooperating intensively to promote the Evergreen Cooperative Initiative.

> On December 7—8, 2006, The Democracy Collaborative, the Ohio Employee Ownership Center, and the Aspen Institute Nonprofit Sector Research Fund convened a Roundtable in Cleveland, Ohio. The event, titled "Building Community Wealth: New Asset-Based Approaches to Solving Social and Economic Problems in Cleveland and Northeast Ohio," brought together national experts, local government representatives, and more than three-dozen community leaders in Cleveland to discuss community wealth issues and identify action steps toward developing a comprehensive strategy.
>
> The fifty participants included representatives of the Federal Reserve Bank of Cleveland, the Ohio Public Employees Retirement System, universities, and employee-owned firms; directors of nonprofit community and economic development organizations such as community development corporations, housing land trusts, and community development financial institutions; the economic development director of the City of Cleveland and members of his staff; a director of the new veterans administration hospital to be established in the city; the treasurer of Cuyahoga County; and others of the public, private, philanthropic, faith-based and non-profit communities. Funding and other support for the meeting was provided by the Gund Foundation, the Cleveland Foundation, and the Sisters of Charity Foundation.[9]

[1]Alperovitz et al., "The Cleveland Model."

[2]<http://www.evergreencoop.com/Laundry/index.html>

[3]<http://www.evergreencoop.com/OhioSolar/index.html>

[4]<http://www.evergreencoop.com/GreenCity/greencity.html>

[5]Alperowitz et al., "The Cleveland Model."

[6]<http://www.community-wealth.org/strategies/cw-roundtables.html#cleveland>.

[7]<http://www.community-wealth.org/about/index.html>.

[8]<http://www.community-wealth.org/about/about-us.html>.

[9]"Community Wealth Building Conference in Cleveland, OH," *GVPT News*, February 2007, p. 14 <http://www.bsos.umd.edu/gvpt/newsletter/February07.news.pdf>.

This is one of the largest and most promising experiments in cooperative economics ever attempted in the United States, with an unprecedented number of local stakeholders at the table.

What do Antigonish, Mondragon, Salinas and Cleveland have in common? They all take the conventional commercial enterprise using existing production technology as a given, and simply tinker around with applying the cooperative principle and economic localism to such enterprises.

Most of Brummett's hits on the economic viability of small towns in the Delta are based on the technocratic liberal assumption that enormous capital outlays are required to accomplish particular economic functions. That's an assumption shared by technocratic liberals of the same stripe who promoted a Third World economic development model based on maximizing economies of scale by concentrating available capital in a few giant, capital-intensive enterprises rather than integrating intermediate production technologies into village economies.[1] That's true of most Progressive(TM) versions of community economic development—Obama's "green jobs" programs, alternative energy projects, and the like. Typically they entail "private-public partnerships," based on attracting colonization by "progressive" or "green" corporations with capital-intensive business models, and the capture of profits from new technology on the pattern of "cognitive capitalism": a sort of mashup of the Gates Foundation, Warren Buffett and Bono.

And the government's criteria for aiding such development efforts usually manage to exclude low-capital, bottom-up efforts by self-organized locals.[2]

> And de Angelis's critique of the Salinas experiment comes from a similar set of assumptions: namely, that capital-intensive forms of production, with the requirement for high capital outlays and debt finance, and an export-oriented economic model for servicing that debt and fully utilizing the expensive plant and equipment, are simply a given.

But as we saw in the previous chapter, decent standards of living no longer depend on building communities around enormous concentrations of capital assets housed in large buildings. Thanks to technical change, the capital outlays required to support a comfortable standard of living are scalable to smaller and smaller population units. So Muhammad no longer need go to the mountain.

This has enormous liberatory significance for experiments in cooperative local economies like Salinas. As production tools become cheaper and cheaper, for an

[1]See Chapter One, Appendix A, "Economy of Scale in Development Economics," in Kevin Carson, *Organization Theory: A Libertarian Perspective* (Booksurge, 2008), pp. 24 et seq.

[2]Keith Taylor, who is doing dissertation work on how wind farms relate to alternative models of economic development. The structure of refundable tax credits for "green energy" investment, in particular, massively empowers conventional corporate wind farms against electric power cooperatives. Making credits conditional on paying at least some taxes seems at first glance to be a fairness issue, ensuring that only people who pay taxes can get credits, and thus making refundable credits a bit less welfare-like. But the ostensible fairness is only superficial: Once the threshold of paying any taxes at all is triggered, the scale of the credit need bear no proportion at all to the amount of taxes paid. So a refundable credit which is available only to for-profit, tax-paying entities is equivalent to a $20 million welfare check that's available to anyone who paid a dollar in taxes, but not to the unemployed. And the refundable green energy investment tax credits are in effect a massive subsidy that is available only to for-profit corporations. Likewise, the Obama administration's "smart grid" policies are suited primarily to the interests of corporate wind farm mega-projects, situated far from the point of consumption, like those T. Boone Pickens is so busy promoting.

ever increasing range of products, the more feasible it is to produce more and more of the things the local population consumes in small shops scaled to the local market, without high capital outlays and overhead creating pressure to maximize batch size and amortize costs. This will also mean less indebtedness from capital investment, less pressure to self-exploitation, and less pressure to compete in a global marketplace instead of serving the local economy.

That means that manufacturing can move toward the kind of local subsistence model that de Angelis desires for the Salinas economy, and envisions as its idealized "better self": "a means for the local population to meet reproduction needs in ways that shield them from the most exploitative practices of other areas in the region . . . "

In general, the promise of low-cost production tools dovetails perfectly with the goals of the cooperative and relocalization movements. As we will see in more detail in the next chapter, the lower the cost of production tools, the less of a bottleneck investment capital becomes for local economic development, and the less dependent the local economy becomes on outside investors. The imploding cost of production machinery is a revolutionary reinforcement for the kind of process that Jane Jacobs regarded as the best approach to community economic development: import replacement by using local resources and putting formerly waste resources to use. Every technological change that reduces the capital outlays required for producing local consumption needs is a force multiplier, not only making import substitution more feasible but increasing its cost-effectiveness, and enabling local economies to do more with less. When the masters of the corporate state realize the full revolutionary significance of micromanufacturing technology in liberating local economies from corporate power, we'll be lucky if the people in the Fab Labs don't wind up being waterboarded at Gitmo.

Low capital outlays and other fixed costs, and the resulting low overhead burden to be serviced, are the key to the counter-economy's advantages as a path to community economic development.

The Indian villages Neil Gerschenfeld described in *Fab* (quoted extensively in the next chapter) one illustration of the possibilities for economically depressed, resource-poor areas using the latest generation of technology to bootstrap development and leapfrog previous generations of high-cost, capital-intensive technology.

Sam Kronick recently challenged members of the Open Manufacturing email list on the relevance of their pet micromanufacturing technology as a lifeline for dying rust belt communities like Braddock, Pennsylvania.

> The state has classified it a "distressed municipality"—bankrupt, more or less—since the Reagan administration. The tax base is gone. So are most of the residents. The population, about 18,000 after World War II, has declined to less than 3,000. Many of those who remain are unemployed. Real estate prices fell 50 percent in the last year.
>
> "Everyone in the country is asking, 'Where's the bottom?' " said the mayor, John Fetterman. "I think we've found it."
>
> Mr. Fetterman is trying to make an asset out of his town's lack of assets, calling it "a laboratory for solutions to all these maladies starting to knock on the door of every community." One of his first acts after being elected mayor in 2005 was to set up, at his own expense, a Web site to publicize Braddock—if you can call pictures of buildings destroyed by neglect and vandals a form of promotion.

He has encouraged the development of urban farms on empty lots, which employ area youths and feed the community. He started a nonprofit organization to save a handful of properties.[1]

This, Kronick says, "is as close as you'll get to an open invitation by a government to experiment with some of these ideas in the real world."

What could be done in the next week/month/year/decade? . . .

. . . [H]ow could a community fablab/hackerspace affect a place like this in the short term?[2]

Several other list members replied by pointing out the negative points of Braddock as a site for a Fab Lab or hackerspace: the high rates of crime and vandalism, the deteriorating buildings, etc. One member argued that micromanufacturing was about "building from abundance," not "trying to rebuild from scratch" in the worst-off areas. Kronick, nonplussed, rejoined that they had "made the case for Braddock as the prototypical challenge to many of your ideas."

If your post-scarcity dreams don't have a chance there, I don't know how much hope I have for them in the rest of the world. . . .

Vandalism is, I would argue, a key indicator of abundance or, put more simply, "free time." Vandalism can be an outlet for creativity and intelligence (and I don't just mean artistic graffiti. Some tend to venerate the bourgeois urban explorers with their ropes and headlamps and cameras but not the kids who risk arrest or injury climbing buildings or billboards to throw up a quick tag). I won't argue that you /should/ move there because of this, but try to understand how useless or upsetting your own pasttimes might seem to others. Buying cheap distressed property can lead to what many might call "gentrification," a prospect some find more terrifying to their way of life than broken windows and scribbles on the walls. It's a matter of perspective.

But I will not digress further; I will attempt to sustain my disbelief that this mailing list isn't really just a thin guise for endless theoretical musings on Utopia and return to the subject I originally asked about: what implications could "open manufacturing" have in a small town that is actively seeking out new ideas? . . .

What might the priorities be in a Braddock communal workshop? An army of Repraps? A few old Bridgeports? A safe, sound building that can be used year-round? Community show-and-tell nights to get the whole town interested in what's being built? Connections to the schools? Connections to local manufacturers? Initiatives that would bring in government "green jobs" money? Production of profitable items to bring cash into the community? Production of necessary items for people in the community? A focus on urban gardening, bicycle transportation, alternative energy, building rehabilitation, permaculture, electronics, EV's, biodiesel, art, music, etc etc etc?

I guess I see plenty of options and directions that the tools of "open manufacturing" could bring (though I appreciate those working on creating more/better tools, more options); now I want to know how their application would fare in a place that would provides both clear challenges and opportunities. I think this is what people like the openfarmtech people are doing already, but why not experiment in another situation?[3]

[1]David Streitfeld, "Rock Bottom for Decades, but Showing Signs of Life," *New York Times*, February 1, 2009 <http://www.nytimes.com/2009/02/01/us/01braddock.html>.

[2]Sam Kronick, "[Open Manufacturing] Re: How will laws be changed just by the existence of self-sufficient people?" *Open Manufacturing*, January 16, 2010 <http://groups.google.com/group/openmanufacturing/msg/8014b08692f05f8e>.

[3]Kronick, "[Open Manufacturing] Regenerating Braddock (was Re: How will laws be changed . . .)," *Open Manufacturing*, January 17, 2010 <http://groups.google.com/group/openmanufacturing/msg/12bd8bdf36290535>.

As I argued on-list, my position is midway between those of Kronick and the skeptics. It seems to me that depressed areas like Braddock, the Arkansas Delta, and a good many Rust Belt communities in the former Ohio Valley have a lot in common with the economic problems facing Indian villages, as described by Neil Gerschenfeld in *Fab*. Gerschenfeld's examples (which, again, we will examine in the next chapter) of rural hardware hackers reverse-engineering homebrew versions of proprietary tractors for a small fraction of the cost, or of village cable systems using cheap reverse-engineered satellite receivers, seems like something that would be relevant to American communities with high unemployment, collapsing asset values and eroding tax bases. Those villages in India that Gerschenfeld describes couldn't exactly be described as building from abundance, except in the sense that imploding fixed costs are creating potential abundance *ex nihilo* everywhere.

And as I also argued, it seems to me that stigmergic organization (see especially the discussion in the next chapter) is relevant to the problem. In my opinion micromanufacturing will benefit communities like Braddock and the Arkansas Delta a lot sooner than most people think. But the fastest way to get from here to there, from the perspective of those currently involved in the movement, is for them to develop and expand the technology as fast as they can from where they are right now. Those currently engaged in micromanufacturing should feel under no moral pressure to abandon the capital assets they've built up where they are to start over somewhere else, as some sort of missionary effort. The faster Fab Labs, hacker spaces and garage factories proliferate and drop in price, the more of a demonstration effect they'll create. And the cheaper and more demonstratedly feasible the technology becomes, the more it builds up an models of complete industrial ecologies in communities where it already exists, and the more it shows itself as benefiting those local economies by filling the void left by deindustrialization of old-style mass production employers, the more attractive it will be in places where it hasn't yet been tried. The more this happens, in turn, the more people there will be like Kronick's friend in Braddock (his suggestion to Kronick that it might be a useful site for a micromanufacturing effort after Kronick's graduation was what sparked the whole discussion), who are eager to experiment with it locally. And at the same time, the more people there will be in the existing fab/hackerspace movement who are willing to take a gamble in acting as micromanufacturing missionaries in the Rust Belt. Likewise, the more prominent a part of economic life it becomes in areas where it already exists, and the more public awareness it creates as a credible path to economic development in depressed levels, the more open people like the unconventional mayor of Braddock will be toward trying it out.

In keeping with Eric Raymond's stigmergic model, the people who are best suited to tackle particular problems do so, and put all their effort into doing what they're best at where they are. These contributions create a demonstration effect and go into the network culture's pool of common knowledge, for free adoption by anyone who finds them to be what they need. So the more everybody does their own thing, the more they're facilitating the eventual adoption of the benefits of their work in areas like Braddock.

Everything Kronick said of Braddock is true of Cleveland in spades; it's an unprecedented opportunity for micromanufacturing enthusiasts to put their ideas into operation. The micromanufacturing and open hardware movements are actively engaged in building the technological basis for the libertarian, decentralized manufacturing economy of the future. And right now Cleveland is engaged in the biggest experimental project around for building a relocalized cooperative economy. An alliance between the micromanufacturing movement and the Cleveland model would seem to be the opportunity of a century. As I asked in an article at P2P Foundation Blog on the Evergreen Cooperative Initiative:

> There is enormous potential for fruitful collaboration between the Cleveland experiment and the micromanufacturing, Fab Lab and hackerspace movements.
>
> What local resources exist in Cleveland right now for a networked micromanufacturing economy? Perhaps someone in our readership knows of someone in Cleveland with CNC tools who would be interested in joining the 100kGarages micromanufacturing network. Or someone in the Cleveland area with the appropriate skills might be interested in organizing a hackerspace.
>
> The university is one of the leading stakeholders in the effort. Universities like Stanford, MIT and UT Austin have played a central role in creating the leading tech economies in other parts of the country, and the flagship project of the Fab Lab movement is the Austin Fab Lab created under the auspices of UT. Perhaps the engineering department at one of the universities involved in building the Cleveland Model would be interested in supporting local micromanufacturing projects. Or maybe some high school shop classes, or community college machining classes, would be interested in collaborating to build a local Fab Lab.
>
> From the other direction, is anyone involved in networked manufacturing projects like 100kGarages, or in the Fab Lab and hackerspace movement, interested in feeling out some of the stakeholders in the Cleveland initiative?[1]

Counter-economic development initiatives in decaying American cities like Cleveland can achieve synergies not only with the micromanufacturing movement, but also with the microenterprise movement.

Micromanufacturing is a force multiplier because new, cheaper production technologies free local economies from dependence on external capital finance for organizing the local production of local needs. The microenterprise, on the other hand, is a force multiplier because it puts existing underutilized capital equipment to full use. The household microenterprise operates on extremely low overhead because it uses idle capacity ("spare cycles") of the ordinary capital goods that most households already own.

The Cleveland initiative could achieve very high bang for the buck, in building a resilient and self-sufficient local economy, by eliminating all the local regulatory barriers to microenterprises operating out of people's homes.

Such relocalization movements can also achieve synergies and get more bang from the buck in another way: by eliminating barriers to cheap subsistence by the homeless and unemployed. No matter how large a share of the goods and services we consume can be produced and exchanged in the counter-economy, most people still bear one significant fixed cost that can't be met outside the wage system: their rent or mortgage payment. And most of the possibilities for informal production go right out the window when a household lacks sufficient employment income to pay the rent or mortgage, and people consequently lose the roofs over their heads.

So the problem of "informal housing" needs to be addressed in some way as part of the larger agenda. This means efforts like those discussed later in this chapter: for law enforcement to de-prioritize foreclosure evictions and the eviction squatters, for local governments to open unused public buildings as barebones shelters (with group toilets, water taps and hot plates), and similarly to open va-

[1]Kevin Carson, "The Cleveland Model and Micromanufacturing," *P2P Foundation Blog*, April 6, 2010 <http://blog.p2pfoundation.net/the-cleveland-model-and-micromanufacturing-an-opportunity-for-collaboration/2010/04/06>.

cant public land as camping grounds with communal water taps and portable toilets.

F. Contemporary Ideas and Projects

To some extent Factor e Farm and 100kGarages, which we examined in the previous chapter, are local economy projects of sorts. Rather than duplicating the material in the last chapter, we refer you back to it.

1. *Jeff Vail's "Hamlet Economy."*

This is a system of networked villages based on an idealized version of the historical "lattice network of Tuscan hill towns" numbering in the hundreds (which became the basis of a modern regional economy based largely on networked production). The individual communities in Vail's network must be large enough to achieve self-sufficiency by leveraging division of labor, as well as providing sufficient redundancy to absorb systemic shock. When larger-scale division of labor is required to support some industry, Vail writes, this is not to be achieved through hierarchy, with larger regional towns becoming centers of large industry. Rather, it is to be achieved by towns of roughly similar size specializing in producing specialized surplus goods for exchange, via fairs and other horizontal exchange relationships.[1]

The Hamlet relies on a "design imperative," in an age of Peak Oil, for extracting the maximum quality of life from reduced energy inputs. The Tuscan hill towns Vail points to as a model are decentralized, open source and vernacular.

> How is the Tuscan village decentralized? Production is localized. Admittedly, everything isn't local. Not by a long shot. But compared to American suburbia, a great percentage of food and building materials are produced and consumed in a highly local network. A high percentage of people garden and shop at local farmer's markets.
>
> How is the Tuscan village open source? Tuscan culture historically taps into a shared community pool of technics in recognition that a sustainable society is a non-zero-sum game. Most farming communities are this way—advice, knowledge, and innovation is shared, not guarded. Beyond a certain threshold of size and centralization, the motivation to protect and exploit intellectual property seems to take over (another argument for decentralization). There is no reason why we cannot share innovation in technics globally, while acting locally—in fact, the internet now truly makes this possible, leveraging our opportunity to use technics to improve quality of life.
>
> How is the Tuscan village vernacular? You don't see many "Colonial-Style" houses in Tuscany. Yet strangely, in Denver I'm surrounded by them. Why? They make no more sense in Denver than in Tuscany. The difference is that the Tuscans recognize (mostly) that locally-appropriate, locally-sourced architecture improves quality of life. The architecture is suited to their climate and culture, and the materials are available locally. Same thing with their food—they celebrate what is available locally, and what is in season. Nearly every Tuscan with the space has a vegetable garden. And finally (though the pressures of globalization are challenging this), their culture is vernacular. They celebrate local festivals, local harvests, and don't rely on manufactured, mass-marketed, and global trends for their culture nearly as much as disassociated suburbanites—their

[1]Jeff Vail, "Re-Post: Hamlet Economy," *Rhizome*, July 28, 2008 <http://www.jeffvail.net/2008/07/re-post-hamlet-economy.html>.

strong sense of community gives prominence to whatever "their" celebration is over what the global economy tells them it should be.[1]

2. *Global Ecovillage Network*

GEN was based on, and in some cases went on to incorporate, a number of "apparently simultaneous ideas arising in different locations at about the same time."[2] It seems to have been a direct outgrowth of the "planetary village" movement, centered on the Findhorne community in Scotland, founded in 1962.[3]

> In 1975 the magazine *Mother Earth News* began constructing experimental energy systems, novel buildings, and organic gardens near its business office in Hendersonville, North Carolina, and in 1979, began calling this educational center an "eco-village."
>
> At about the same time in Germany, during the political resistance against disposal of nuclear waste in the town of Gorleben, anti-nuclear activists attempted to build a small, ecologically based village at the site, which they called an okodorf (literally ecovillage). In the largest police action seen in Germany since the Second World War, their camp was ultimately removed, but the concept lived on, and small okodorf experiments continued in both eastern and western Germany. The magazine *Okodorf Informationen* began publishing in 1985 and later evolved into *Eurotopia*. After reunification of Germany, the movement coalesced and became part of the International ecovillage movement.
>
> About the same time in Denmark, a number of intentional communities began looking beyond the social benefits of cohousing and other cooperative forms of housing towards the ecological potentials of a more thorough redesign of human habitats. In 1993 a small group of communities inaugurated the Danish ecovillage network, *Landsforeningen for Okosamfund*, the first network of its kind and a model for the larger ecovillage movement that was to follow. . . .
>
> Throughout the 1980s and early 1990, on Bainbridge Island near Seattle, Robert and Diane Gilman used their journal, In Context, to publish stories and interviews describing ecovillages as a strategy for creating a more sustainable culture. When Hildur Jackson, a Danish attorney and social activist, discovered In Context, the ecovillage movement suddenly got traction.
>
> Ross Jackson, Hildur's husband, was a Canadian computer whiz who had been working in the financial market, writing programs to predict shifts in international currencies. When he took his algorithms public as Gaia Corporation, his models made a fortune for his investors, but Ross, being a deeply spiritual man, wanted little of it for himself. Searching for the best way to use their prosperity, Ross and Hildur contacted the Gilmans and organized some gatherings of visionaries at Fjordvang, the Jackson's retreat in rural Denmark, to mull over the needs of the world. . . .
>
> Ross Jackson was also interested in utilizing the new information technology that was just then emerging: email and electronic file exchanges between universities and research centers (although it would still be a few years before the appearance of shareware browsers and the open-to-all World Wide Web).
>
> Ross and Hildur Jackson created a charitable foundation, the Gaia Trust, and endowed it with 90 percent of their share of company profits. In 1990, Gaia Trust asked *In Context* to produce a report, *Ecovillages and Sustainable Communities*, in order to catalog the various efforts at sustainable community living underway around the world, and to describe the emerging philosophy and princi-

[1]Vail, "The Design Imperative," *JeffVail.Net*, April 8, 2007 <http://www.jeffvail.net/2007/04/design-imperative.html>.

[2]Albert Bates, "Ecovillage Roots (and Branches): When, where, and how we re-invented this ancient village concept," *Communities Magazine* No. 117 (2003).

[3]Ross Jackson, "The Ecovillage Movement," *Permaculture Magazine* No. 40 (Summer 2004), p. 25.

ples in greater detail. The report was released in 1991 as a spiral bound book (now out of print).

In September 1991, Gaia Trust convened a meeting in Fjordvang to bring together people from eco-communities to discuss strategies for further developing the ecovillage concept. This led to a series of additional meetings to form national and international networks of ecovillages, and a decision, in 1994, to formalize networking and project development under the auspices of a new organization, the Global Ecovillage Network (GEN).

By 1994 the Internet had reached the point where access was becoming available outside the realm of university and government agencies and contractors. Mosaic was the universal browser of the day, and the first Internet cafes had begun to appear in major cities. Ross Jackson brought in a young Swedish web technician, Stephan Wik, who'd had a computer services business at Findhorn, and the Ecovillage Information Service was launched from Fjordvang at www.gaia.org. With Stephan and his co-workers gathering both the latest in hardware advances and outstanding ecovillage content from around the world, gaia.org began a steady growth of "hits," increasing 5 to 15 percent per month, that would go on for the next several years, making the GEN database a major portal for sustainability studies.

In October 1995, Gaia Trust and the Findhorn Foundation co-sponsored the first international conference "Ecovillages and Sustainable Communities—Models for the 21st Century," held at Findhorn in Scotland. After the conference, GEN held a formative meeting and organized three worldwide administrative regions: Europe and Africa; Asia and Oceania; and the Americas. Each region was to be overseen by a secretariat office responsible for organizing local ecovillage networks and developing outreach programs to encourage growth of the movement. A fourth secretariat was established in Copenhagen to coordinate all the offices, seek additional funding, and oversee the website. The first regional secretaries, chosen at the Findhorn meeting, were Declan Kennedy, Max Lindegger, and myself. Hamish Stewart was the first international secretary.[1]

According to Ross Jackson, the GEN was founded "to link the hundreds of small projects that had sprung up around the world. . . . "[2] The Gaia Trust website adds:

> The projects identified varied from well-established settlements like Solheimer in Iceland, Findhorn in Scotland, Crystal Waters in Australia, Lebensgarten in Germany to places like The Farm in Tennessee and the loosely knit inner-city Los Angeles Ecovillage project to places like the Folkecenter for Renewable Energy in Thy and many smaller groups that were barely started, not to mention the traditional villages of the South.[3]

Following the foundation of GEN, Albert Bates continues, "[w]ith generous funding from Gaia Trust for this new model, the ecovillage movement experienced rapid growth."

> Kibbutzim that re-vegetated the deserts of Palestine in the 20th century developed a new outlook with the formation of the Green Kibbutz Network. The Russian Ecovillage Network was inaugurated. Permaculture-based communities in Australia such as Crystal Waters and Jarlanbah pioneered easy paths to more environmentally sensitive lifestyles for the mainstream middle class. GEN-Europe hosted conferences attended by ecovillagers from dozens of countries, and national networks sprang up in many of them. In South and North America, nine representatives were designated to organize ecovillage regions by geography and language. By the turn of the 21st century GEN had catalogued thousands

[1]Bates, "Ecovillage Roots (and Branches)."

[2]Ross Jackson, "The Ecovillage Movement."

[3]"What is an Ecovillage?" Gaia Trust website <http://www.gaia.org/gaia/ecovillage/whatis/>.

of ecovillages, built "living and learning centers" in several of them, launched ecovillage experiments in universities, and sponsored university-based travel semesters to ecovillages on six continents. . . .

Ecovillages today are typically small communities with a tightly-knit social structure united by common ecological, social, or spiritual views. These communities may be urban or rural, high or low technologically, depending on circumstance and conviction. Okodorf Seiben Linden is a zero-energy cohousing settlement for 200 people in a rural area of eastern Germany. Los Angeles EcoVillage is a neighborhood around an intersection in inner Los Angeles. Sasardi Village is in the deep rainforest of Northern Colombia. What they share is a deep respect for nature, with humans as an integral part of natural cycles. Ecovillages address social, environmental, and economic dimensions of sustainability in an integrated way, with human communities as part of, not apart from, balanced ecologies. . . .[1]

The best concise description of an ecovillage that I've seen comes from what is apparently an older version of the Gaia Trust website, preserved on an article at *Permaculture Magazine*:

Ecovillages are urban or rural communities that strive to combine a supportive social environment with a low-impact way of life. To achieve this, they integrate various aspects of ecological design, permaculture, ecological building, green production, alternative energy, community building practices, and much more.

These are communities in which people feel supported by and responsible to those around them. They provide a deep sense of belonging to a group and are small enough for everyone to be seen and heard and to feel empowered. People are then able to participate in making decisions that affect their own lives and that of the community on a transparent basis.

Ecovillages allow people to experience their spiritual connection to the living earth. People enjoy daily interaction with the soil, water, wind, plants and animals. They provide for their daily needs—food, clothing, shelter—while respecting the cycles of nature.

They embody a sense of unity with the natural world, with cultural heritage around the world and foster recognition of human life and the Earth itself as part of a larger universe.

Most ecovillages do not place an emphasis on spiritual practices as such, but there is often a recognition that caring for one's environment does make people a part of something greater than their own selves. Observing natural cycles through gardening and cultivating the soil, and respecting the Earth and all living beings on it, ecovillages tend to maintain, recreate or find cultural expressions of human connectedness with nature and the universe.

Respecting this spirituality and culture manifests in many ways in different traditions and places.[2]

The typical ecovillage has 50-400 people. Many ecovillages, particularly in Denmark, are linked to a cohousing project of some sort.[3] Such projects lower the material cost of housing (construction materials, heating, etc.) per person, and reduce energy costs by integrating the home with workplace and recreation.[4] Neighborhood-based ecovillages in some places have influenced the liberalization of local zoning laws and housing codes, and promoted the adoption of new building techniques by the construction industry. Ecovillage practices include peripheral

[1]Bates, "Ecovillage Roots (and Branches)."

[2]"What is an Ecovillage?" (sidebar), Agnieszka Komoch, "Ecovillage Enterprise," *Permaculture Magazine* No. 32 (Summer 2002), p. 38.

[3]Jackson, p. 26.

[4]Jackson, p. 28.

parking, common open spaces and community facilities, passive solar design, vernacular materials, and composting toilets.[1]

The ecovillage movement is a loose and liberally defined network. According to Robert and Diane Giulman, in *Ecovillages and Sustainable Communities* (1991), an ecovillage is "A human-scale, full-featured settlement in which human activities are harmlessly integrated into the natural world in a way that is supportive of healthy human development and can be successfully continued into the indefinite future." The GEN refuses to police member communities or to enforce any centralized standard of compliance. At a 1998 GEN board meeting in Denmark, the Network affirmed "that a community is an ecovillage if it specifies an ecovillage mission, such as in its organizational documents, community agreements, or membership guidelines, and makes progress in that direction. The Network promotes the Community Sustainability Assessment Tool, a self-administered auditing survey, as a way to measure progress toward the same general set of goals.[2] The Ecological portion of the checklist, for example, includes detailed survey questions on

1. Sense of Place—community location & scale; restoration & preservation of nature
2. Food Availability, Production & Distribution
3. Physical Infrastructure, Buildings & Transportation—materials, methods, designs
4. Consumption Patterns & Solid Waste Management
5. Water—sources, quality & use patterns
6. Waste Water & Water Pollution Management
7. Energy Sources & Uses[3]

Question 2, "Food Availability," includes questions on the percentage of food produced within the community, what is done with food scraps, and whether greenhouses and rooftop gardens are used for production year-round.[4]

Such liberality of standards is arguably necessary, given the diversity of starting points of affiliate communities. An ecovillage based in an inner city neighborhood, it stands to reason, will probably have much further to go in achieving sustainability than a rural-based intentional community. Urban neighborhoods, of necessity, must be "vertically oriented," and integrate the production of food and other inputs on an incremental basis, often starting from zero.[5]

3. *The Transition Town Movement*

This movement, which began with the town of Totnes in the UK, is described by John Robb as an "open-source insurgency": a virally replicable, open-source model for resilient communities capable of surviving the Peak Oil transition. As of April 2008, some six hundred towns around the world had implemented Transition Town projects.[6]

The Transition Towns Wiki[7] includes, among many other things, a *Transition Initiatives Primer* (a 51 pp. pdf file), a guide to starting a Transition Town initiative

[1] Jackson, p. 29.

[2] Linda Joseph and Albert Bates, "What Is an 'Ecovillage'?" *Communities Magazine* No. 117 (2003).

[3] <http://gen.ecovillage.org/activities/csa/English/toc.html>.

[4] <http://gen.ecovillage.org/activities/csa/English/eco/eco2.php>.

[5] Joseph and Bates.

[6] John Robb, "Resilient Communities: Transition Towns," *Global Guerrillas*, April 7, 2008 <http://globalguerrillas.typepad.com/globalguerrillas/2008/04/transition-town.html>.

[7] <http://transitiontowns.org/>.

in a local community.[1] It has also published a print book, *The Transition Handbook.*[2]

Totnes is the site of Rob Hopkins' original Transition Town initiative, and a model for the subsequent global movement.

> The thinking behind [Transition Town Totnes] is simply that a town using much less energy and resources than currently consumed could, if properly planned for and designed, be more resilient, more abundant and more pleasurable than the present.
>
> Given the likely disruptions ahead resulting from Peak Oil and Climate Change, a resilient community—a community that is self-reliant for the greatest possible number of its needs—will be infinitely better prepared than existing communities with their total dependence on heavily globalised systems for food, energy, transportation, health and housing.
>
> Through 2007, the project will continue to develop an Energy Descent Action Plan for Totnes, designing a positive timetabled way down from the oil peak.[3]

The most complete Energy Descent Action Plan is that of Kinsale. It assumes a scenario in which Kinsale in 2021 has half the energy inputs as in 2005. It includes detailed targets and step-by-step programs, for a wide range of areas of local economic life, by which energy consumption per unit of output may be reduced and local inputs substituted for outside imports on a sustainable basis. In the area of food, for example, it envisions a shift to local market gardening as the primary source of vegetables and a large expansion in the amount of land dedicated to community-supported agriculture. By 2021, the plan says, most ornamental landscaping will likely be replaced with fruit trees and other edible plants, and the lawnmower will be as obsolete as the buggy whip. In housing, the plan calls for a shift to local materials, vernacular building techniques, and passive solar design. The plan also recommends the use of local currency systems, skill exchange networks, volunteer time banks, and barter and freecycling networks as a way to put local producers and consumers in contact with one another.[4]

4. Global Villages

These are designed to generate 80% of their income internally and 20% externally, with internally generated wealth circulating five times before it leaves the community. As described by Claude Lewenz, *in How to Build a Village:*

> The local economy is layered, built on a foundation that provides the basic needs independent of the global economy—if it melts down the Villagers will survive. The local economy is diversified. . . . The local economy must provide conditions that encourage a wide diversity of businesses and officers to operate. Then when some collapse or move away, the local economy only suffers a bit—it remains healthy.[5]

[1]Ben Brangwyn and Rob Hopkins, *Transition Initiatives Primer: becoming a Transition Town, City, District, Village, Community or even Island* (Version 26—August 12, 2008) <http://transitionnetwork.org/Primer/TransitionInitiativesPrimer.pdf>.

[2]Rob Hopkins, *The Transition Handbook: From Oil Dependency to Local Resilience* (Green Books) <http://transitiontowns.org/TransitionNetwork/TransitionHandbook>.

[3]*Ibid.*, p. 10.

[4]*Kinsale 2021: An Energy Descent Action Plan*. Version.1. 2005. By Students of Kinsale Further Education College. Edited by Rob Hopkins <http://transitionculture.org/wp-content/uploads/members/KinsaleEnergyDescentActionPlan.pdf>.

[5]Claude Lewenz, *How to Build a Village* (Auckland, New Zealand: Village Forum Press and Jackson House Publishing Company, 2007), p. 73.

Lewenz's Village is also essentially the kind of "resilient community" John Robb and Jeff Vail have in mind:

> . . . [E]conomies can collapse and first-world people can starve if systems fail. We have now built a food system almost entirely dependent on diesel fuelled tractors, diesel delivery trucks and a long-distance supermarket delivery system. More recently, we shifted to an economic and communication system entirely dependent on computers—a system that only runs if the electrical grid supplies power. In the Great Depression in the USA, poor people say they hardly noticed—in those days they kept gardens because the USA was predominantly rural and village. The potential for economic collapse always looms, especially as the global economic system becomes more complex and vulnerable. Prudence would dictate that in planning for a local economy, it include provisions to assure the Village sustained its people, and those of the surrounding region, in such adverse conditions.
>
> The challenge is to maintain a direct rural and farm connection for local, good food, and establish an underlying local economy that can operate independent of the larger economy and which can put unemployed people to work in hard times.[1]

The Global Villages network[2] has had fairly close ties with Marcin Jakubowski and Factor e Farm, which we considered in the previous chapter.

5. *Venture Communism*

Venture communism is a project developed by Dmytri Kleiner. The basic principle—purchasing undeveloped land and resources cheaply from the capitalist economy, and then financing itself internally from the rents on that land as development by venture communist enterprises causes it to appreciate in value—is reminiscent of Ebenezer Howard's original vision for the Garden City movement.

> Starting from the belief that political change can only follow a change in the mode of production, venture communism is an attempt to create a mode of production that will expand socialism by reducing the labour available to be exploited by property. . . .
>
> Socialism is defined as a mode of production where the workers own the means of production, and especially the final product.By withholding our labour from Capitalists and instead forming our own worker-owned enterprises we expand Socialism.
>
> The more labour withheld from Capitalists, the less they are able to exploit.[3]

In an extended passage from the P2P Foundation Wiki, Kleiner describes the actual functioning of a venture commune:

> "A Venture Commune is a joint stock corporation, much like the Venture Capital Funds of the Capitalist class, however it has four distinct properties which transform it into an effective vehicle for revolutionary worker's struggle.
>
> 1—A Share In The Venture Commune Can Only Be Acquired By Contributions Of Labour, and Not Property.
>
> In other words only by working is ownership earned, not by contributing Land, Capital or even Money. Only Labour.
>
> It is this contributed labour which represents the initial Investment capacity of the Commune.

[1]*Ibid*., p. 77.

[2]See also the Global Villages site maintained by Frahz Nahrada, another leading figure in the movement. <http://www.globalvillages.info>.

[3]Luca, "TeleKommunisten" (interview with Dmytri Kleiner), ecopolis, May 21, 2007 <http://www.ecopolis.org/ telekommunisten/>.

The Commune Issues its own currency, based on the value of the labour pledges it has.

It then invests this currency into the private enterprises which it intends to purchase or fund, these Enterprises thus become owned by the Commune, in the same way that Enterprises which receive Venture Capital become owned by a Venture Capital Fund.

2—The Venture Commune's Return On Investment From Its Enterprises Is Derived From Rent and Not Income.

As condition of investment, the Enterprise agrees to not own its own property, neither Land nor Capital, but rather to rent Land and Capital from the Commune.

The Commune, unlike a Venture Capital Fund, never takes a share of the income of the Enterprise nor of any of its workers.

The Commune finances the acquisition of Land and Capital by issuing Bonds, and then Rents the Land and Capital to its Enterprises, or an Enterprise can sell whatever Land and Capital it acquires through other means to the Commune, and in turn Rent it.

In this way Property is always owned Mutually by all the members of the Commune, however all workers and the Enterprises that employ them retain the entire product of their labour.

3—The Venture Commune Is Owned Equally By All Its Members.

Each member can have one share, and only one share. Thus although each worker is able to earn different prices for their labour from the Enterprises, based on the demand for their labour, each worker may never earn any more than one share in the ownership of the Commune itself, and therefore can never accumulate a disproportionate share of the proceeds of Property.

Ownership of Property can therefore never be concentrated in fewer and fewer hands and used to exploit the worker as in Capitalist corporations.

4—All Those Who Apply Their Labour To the Property of the Commune Must Be Eligible For Membership In The Commune.

The Commune may not refuse membership to any Labour employed by any of its enterprises that work with the Land and Capital controlled by the commune. In this way Commune members can not exploit outside wage earners, and the labour needs of the Enterprise will ensure that each Commune continues to grow and accept new members."

Discussion

Dmytri Kleiner:

"I see venture communism in two initial phases, in the first phase proto-venture-communist enterprises must break the Iron law and then join together to found a venture commune.

In a mature venture commune, cost-recovery is simply achieved by using rent-sharing to efficiently allocate property to its most productive use, thereby ensuring mutual accumulation. Rent sharing works by renting the property for it's full market value to member enterprises and then distributing the proceeds of this rent equally among all commune members.

Investment, when required by exogenous exchange, is funded by selling bonds at auction. Endogenous liquidity is achieved through the use of mutual credit.

However in the initial phase there is no property to rent-share and the demand for the bonds is likely to be insufficient, thus the only way the enterprise can succeed is to break the iron law and somehow capitalize and earn more than subsistence costs, making mutual accumulation possible.

IMO, there are two requirements for breaking the iron law:

a) The enterprise must have highly skilled creative labour, so that the labour itself can capture scarcity rents, i.e. artists, software developers.

b) Production must be based on what I call "commodity capital," that is Capital that is a common input to most, if not all, industries, and therefore is often subsidized by public and private foundations and available on the market for below it's actual cost. Examples of this are telecommunications and transportation infrastructure, both of which have been heavily subsidized.

Also, a third requirement for me, although not implied by the simple economic logic, is that the initial products are of general use to market segments I believe are most directly agents for social change, i.e. other peer producers, activists, diasporic/translocal communities and the informal economy broadly.

Also, I would like to note that while the initial enterprises depend on complex labour and should focus on products of strategic benefit, a mature venture commune can incorporate all types of labour and provide all types of goods and services once the implementation of rent-sharing, bond-auction and mutual-credit is achieved." (Oekonux mailing list, January 2008)[1]

The Telekommunisten collective is one such initial enterprise for raising money. "Venture Communism," Kleiner writes, "is a form of worker's self organization which provides a model of sharing property and forming mutual capital that is compatible with anti-capitalist ideals."

However, venture communism does not provide a means of acquiring such property in the first place. Telekommunisten is intended to realize possibilities in forming the privative mutual property required to initiate venture communism.

The lack of any initial financing, most forms of which would be incompatible with the venture communist principal of ownership as a reward for labour not wealth, present twin challenges for a proto-venture-communist enterprise to overcome: Forming capital and finding customers. The first challenge in essence requires breaking the Iron Law of Wages, the implications of which are that worker's can never form capital because they can never earn any more than their subsistence cost from wages alone.

The primitive accumulation theory of Telekommunisten proposes to break the Iron Law by exploiting it's boundary conditions, namely that some labour is scarce, and therefore captures a form of scarcity rent in addition to wages and that some forms of capital are themselves commodities, and therefore can not even capture interest, more to the point, often these forms of capital are common inputs to production and are subsidized by private and public funds and are available on the market for below their own reproduction costs.

Therefore, the Iron Law can be broken if you are able to invest scarce labour and employ commodity capital in production. An obvious example of such commodity capital is basic telephone and internet infrastructure, which connects the farthest reaches of the globe together, built almost entirely with public money and available to be exploited for far less than it's real cost. And likewise, an obvious example of the needed scarce labour investment is the IT and media skills required to derive new products from basic internet and telephone service.

Thus, Telekommunisten propose to form the primitive mutual property required to initiate venture communism by collective investment in the form of IT and media labour using only commonly available internet resources to derive marketable products. The first of these products is Dialstation, which allow any land line or mobile telephone to make very inexpensive international phone calls.

The second challenge, finding customers without any initial financing for marketing, is addressed by linking the artistic and political nature of the project very closely with our products, therefore we promote products such as Dialstation as a matter of course in our artistic production and our participation in the activist and hacker communities. Our basic premise is that people will use and

[1]"Venture Communism," *P2P Foundation Wiki* <http://p2pfoundation.net/Venture_Communism> (accessed August 8, 2009.

> promote our products if they identify with our artistic and political practices, and in turn the economy generated can support and expand these practices.[1]

It is most notable for its Dialstation project, an international long-distance service.[2]

6. Decentralized Economic and Social Organization (DESO)

This is a project in development by Reed Kinney. It's a continuation of the work of his late father, Mark Kinney, among other things a writer on alternative currency systems and an associate of Thomas Greco.[3] Kinney's book on DESO is forthcoming. Here's a brief summary of the project:

> This is a miniscule explanation of Decentralized Economic Social Organization, DESO.
>
> The text has required five years to research and write. As of July 2009, I'm now editing it. The text categorically unfolds every DESO structure, component, department, and its accompanying philosophies. It is a substantial work and will require a conventional publisher by October 2009. As a favor to Kevin Carson, I can offer this very brief overview.
>
> The content of this text is an object of dialogue. The assertion made here is that the base of human intercourse is structurally embedded. And that each type of socioeconomic structure generates a corresponding form of social intercourse. The stated objective here is the development of the socioeconomic pattern that best meets the real needs of its members and that generates the maximum and the fullest mental health among them. This content is derived from many contributors, like Paulo Freire, whom each create equally important components that are here molded into coherent functioning form. True dialogue is the soil, water and sunlight needed to germinate DESO.
>
> DESO is the creation of viable, independent communities within which the humanity of each person is supported through humanistic education and participatory decision making processes. The autonomous DESO economy is designed to both support and further cultivate those objectives.
>
> DESO's economic organization, its educational organization, and its civic organization are designed to interpenetrate and to be interdependent. From their incipience each DESO community develops those three fundamental DESO spheres concurrently. DESO culture is the consequence of inter-community networks. However, it is structured to maintain and perpetuate decentralization.
>
> DESO creates stable, regional economies that resemble the self-sustaining ecosystems of nature. DESO independence is proportional to its population. Structurally, DESO is designed to expand exponentially through mass centrist society, MCS, which it depopulates with astonishing rapidity. Ultimately, DESO curbs the destructive momentum of MCS.[4]
>
> . . .
>
> **Stateless Society**
>
> To refer to a society as stateless does not imply an absence of socioeconomic organization.
>
> To build an equitable society two basic interrelated tactics are used. First, the dissection, the "deconstruction," of the structures of mass centrist society, MCS, reveal what their opposite structures would be, then, second, all of the

[1] "Telekommunisten: The Revolution is Coming" <http://telekommunisten.net/about> Accessed October 19, 2009.

[2] <http://www.dialstation.com/>.

[3] See, for example, Mark Kinney's pamphlet "In Whose Interest?" (n.d) <http://www.appropriate-economics.org/materials/in_whose_interest.pdf>. It briefly sets forth a view of money much like Greco's. His work is quoted several times in Greco's body of work.

[4] Reed Kinney, private email.

known requirements, the conditions needed, for fermenting full, human **psychic** health are evaluated. These two known factors are then used to mold the functions of the structures of decentralized economic social organization, DESO.

The interpersonal relations born of genuine dialogical based organization (mutualism), both in civic and familial spheres, develops the self-realization of all members (The Knowledge of Man, A Philosophy of the Interhuman, by Martin Buber, Edited with an Introduction by Maurice Friedman, Harper & Row). When that is combined with education *through* art (Herbert Read), then, genuine individuation develops. These combined conditions must be met to ferment full, human **psychic** health.

DESO is member managed and is structured to be perpetually decentralized and networked. Each sovereign community is semi-self sufficient; organization is dialogical.

DESO uses technology to reduce the cost, time and space required for production. A production based economy is neither consumer nor profit-based.

Since DESO is a production based economy its production slows as the basic needs of its members are met; slows, levels off, and is then maintained. Its economy does not pose a threat to the life support systems of the planet.

Member objectives are not materialistic per se, although prosperity is generalized. Rather, the objectives of its members orbit their dialogical interpersonal relationships and their mutual self-development through all art, aesthetics, and all knowledge. Art and knowledge are not viewed as commodities, but rather as integral aspects of culture.

Unavoidably, incipient DESO grows alongside and through MCS. It purchases productive facilities from MCS and adopts from it what is useful for DESO. Nonetheless, the DESO objective is independence from MCS. Its independence is ever-augmented through the expansion of its own infrastructures. (Its internal monetized organization is an interest-free civic service.)

[The "political" implications are somewhat self-evident. MCS is not disrupted by internal modifications within its own context. However, people that live in a "humanistic," independent socioeconomic organization, one that is expansive and *competitive*, represent an external force that can curb the self-destructive momentum of MCS; not through direct confrontation per se, but, rather, by infiltrating and "depopulating" MCS.]

In this other DESO context . . . within its own circumstances . . . the indispensable and dynamic drama of equilibrium between individuation and mutualism can be maintained indefinitely.

The DESO scenario does not resemble anything that MCS produces; neither an economy of scarcity nor the alienated mind. No, rather, what you have in DESO is an economy of abundance and a post-alienated population of whole human beings; whole in all their dichotomies.[1]

7. The Triple Alliance

This is an interesting proposal for building a resilient community through social production by the urban underemployed and unemployed. The idea was originally sparked by a blog post by Dougald Hine: "Social Media vs the Recession."

> Looked at very simply: hundreds of thousands of people are finding or are about to find themselves with a lot more time and a lot less money than they are used to. The result is at least three sets of needs:
>
> - practical/financial (e.g. how do I pay the rent/avoid my house being repossessed?)
> - emotional/psychological (e.g. how do I face my friends? where do I get my identity from now I don't have a job?)
> - directional (e.g. what do I do with my time? how do I find work?) . . .

[1]Reed Kinney, personal email, April 8, 2010.

Arguably the biggest thing that has changed in countries like the UK since there was last a major recession is that most people are networked by the internet and have some experience of its potential for self-organisation . . . There has never been a major surge in unemployment in a context where these ways of "organising without organisations" were available.

As my School of Everything co-founder Paul Miller has written, London's tech scene is distinctive for the increasing focus on applying these technologies to huge social issues . . . Agility and the ability to mobilise and gather momentum quickly are characteristics of social media and online self-organisation, in ways that government, NGOs and large corporations regard with a healthy envy.

So, with that, the conversations I've been having keep coming back to this central question: is there a way we can constructively mobilise to respond to this situation in the days and weeks ahead? . . .

- Information sharing for dealing with practical consequences of redundancy or job insecurity. You can see this happening already on a site like the Sheffield Forum.
- Indexes of local resources of use to the newly-unemployed—including educational and training opportunities—built up in a user-generated style.
- Tools for reducing the cost of living. These already exist—LiftShare, Freecycle, etc.—so it's a question of more effective access and whether there are quick ways to signpost people towards these, or link together existing services better.
- An identification of skills, not just for potential employers but so people can find each other and organise, both around each other and emergent initiatives that grow in a fertile, socially-networked context.

If the aim is to avoid this recession creating a new tranche of long-term unemployed (as happened in the 1980s), then softening the distinction between the employed and unemployed is vital. In social media, we've already seen considerable softening of the line between producer and consumer in all kinds of areas, and there must be lessons to draw from this in how we view any large-scale initiative.

As I see it, such a softening would involve not only the kind of online tools and spaces suggested above, but the spread of real world spaces which reflect the collaborative values of social media. Examples of such spaces already exist:

- Media labs on the model of Access Space or the Brasilian Pontos de Cultura programme, which has applied this approach on a national scale
- Fab Labs for manufacturing, as already exist from Iceland to Afghanistan
- studio spaces like TenantSpin, the micro-TV station in Liverpool based in a flat in a towerblock—and like many other examples in the world of Community Media

Again, if these spaces are to work, access to them should be open, not restricted to the unemployed. (If, as some are predicting, we see the return of the three day week, the value of spaces like this open to all becomes even more obvious!)[1]

This was the direct inspiration for Nathan Cravens, of *Appropedia* and sometime Open Source Ecology collaborator, in outlining his Triple Alliance:

The Triple Alliance describes a network of three community supported organizations necessary to meet basic needs and comforts.

- The Open Cafe, a place to have a meal in good company without a price tag
- The CSA or community supported farm

[1]Dougald Hine, "Social Media vs the Recession," *Changing the World*, January 28, 2009 <http://otherexcuses.blogspot.com/2009/01/social-media-vs-recession.html>.

- The Fab Lab, a digitally assisted manufacturing facility to make almost anything[1]

As we saw in Chapter Six, the Fab Lab already exists in the form of commercial workshop space (for example TechShop); it also exists, in forms ranging from non-profit to commercial, in the "hacker space" movement. Regarding this latter, according to *Wired* magazine there are 96 hacker spaces worldwide—29 of them in the United States—including the Noisebridge hacker space profiled in the article.

> Located in rented studios, lofts or semi-commercial spaces, hacker spaces tend to be loosely organized, governed by consensus, and infused with an almost utopian spirit of cooperation and sharing.
>
> "It's almost a Fight Club for nerds," says Nick Bilton of his hacker space, NYC Resistor in Brooklyn, New York. Bilton is an editor in The New York Times R&D lab and a board member of NYC Resistor. Bilton says NYC Resistor has attracted "a pretty wide variety of people, but definitely all geeks. Not Dungeons & Dragons-type geeks, but more professional, working-type geeks." . . .
>
> Since it was formed last November, Noisebridge has attracted 56 members, who each pay $80 per month (or $40 per month on the "starving hacker rate") to cover the space's rent and insurance. In return, they have a place to work on whatever they're interested in, from vests with embedded sonar proximity sensors to web-optimized database software. . . .
>
> Noisebridge is located behind a nondescript black door on a filthy alley in San Francisco's Mission District. It is a small space, only about 1,000 square feet, consisting primarily of one big room and a loft. But members have crammed it with an impressive variety of tools, furniture and sub-spaces, including kitchen, darkroom, bike rack, bathroom (with shower), circuit-building and testing area, a small "chill space" with couches and whiteboard, and machine shop.
>
> The main part of the room is dominated by a battered work table. A pair of ethernet cables snakes down into the middle of the table, suspended overhead by a plastic track. Cheap metal shelves stand against the walls, crowded with spare parts and projects in progress.
>
> The drawers of a parts cabinet carry labels reflecting the eclecticism of the space: Altoids Tins, Crapulence, Actuators, DVDs, Straps/Buckles, Anchors/Hoisting, and Fasteners.
>
> Almost everything in the room has been donated or built by members—including a drill press, oscilloscopes, logic testers and a sack of stick-on googly eyes.
>
> While many movements begin in obscurity, hackers are unanimous about the birth of U.S. hacker spaces: August, 2007 when U.S. hackers Bre Pettis, Nicholas Farr, Mitch Altman and others visited Germany on a geeky field trip called Hackers on a Plane.
>
> German and Austrian hackers have been organizing into hacker collectives for years, including Metalab in Vienna, c-base in Berlin and the Chaos Computer Club in Hannover, Germany. Hackers on a Plane was a delegation of American hackers who visited the Chaos Communications Camp—"Burning Man for hackers," says Metalab founder Paul "Enki" Boehm—and their trip included a tour of these hacker spaces. They were immediately inspired, Altman says.
>
> On returning to the United States, Pettis quickly recruited others to the idea and set up NYC Resistor in New York, while Farr instigated a hacker space called HacDC in Washington, D.C. Both were open by late 2007. Noisebridge followed some months later, opening its doors in fall 2008.
>
> It couldn't have happened at a better time. Make magazine, which started in January, 2005, had found an eager audience of do-it-yourself enthusiasts. (The magazine's circulation now numbers 125,000.) Projects involving complex circuitry and microcontrollers were easier than ever for nonexperts to undertake,

[1]Nathan Cravens, "The Triple Alliance," *Appropedia: The sustainability wiki* <http://www.appropedia.org/ The_Triple_Alliance> (accessed July 3, 2009).

> thanks to open source platforms like Arduino and the easy availability of how-to guides on the internet.
>
> The idea spread quickly to other cities as visitors came to existing hacker spaces and saw how cool they were.
>
> "People just have this wide-eyed look of, 'I want this in my city.' It's almost primal," says Rose White, a sociology graduate student and NYC Resistor member. . . .
>
> Hacker spaces aren't just growing up in isolation: They're forming networks and linking up with one another in a decentralized, worldwide network. The hackerspaces.org website collects information about current and emerging hacker spaces, and provides information about creating and managing new spaces.[1]

Cravens specified that his model of Fab Labs was based on Open Source Ecology (for rural areas) and hacker spaces like NYC Resistor[2] (for urban areas).[3]

In discussion on the Open Manufacturing email list, I suggested that Cravens' three-legged stool needed a fourth leg: housing. Open-source housing would fill a big gap in the overall resiliency strategy. It might be some kind of cheap, bare bones cohousing project associated with the Cafe (water taps, cots, hotplates, etc) that would house people at minimal cost on the YMCA model. It might be an intentional community or urban commune, with cheap rental housing adapted to a large number of lodgers (probably in violation of laws restricting the number of unrelated persons living under one roof). Another model might be the commercial campground, with space for tents, water taps, etc., on cheap land outside the city, in connection with a ride-sharing arrangement of some sort to get to Alliance facilities in town. The government-run migrant worker camps, as depicted in *The Grapes of Wrath*, are an example of the kind of cheap and efficient, yet comfortable, bare bones projects that are possible based on a combination of prefab housing with common bathrooms. And finally, Vinay Gupta's work in the Hexayurt project on emergency life-support technology for refugees is also relevant to the housing problem: offering cheap LED lighting, solar cookers, water purifiers, etc., to those living in tent cities and Hoovervilles. Cravens replied:

> In an urban area, one large multi-level building could provide all basic needs. A floor for hydroponicly [sic] grown food, the fab, and cafe. The remaining space can be used for housing. The more sophisticated the fabs and availibility of materials, the better conditions may rival or exceed present middle class standards.[4]

Such large multi-level buildings resemble what actually exists in the networked manufacturing economies of Emilia-Romagna (as described by Sabel and Piore) and Shenzhen (as described by Bunnie Huang), which we examined in Chapter Six: publicly accessible retail space on the ground floor, a small factory upstairs, and worker housing above that.

This would probably fall afoul of local zoning laws and housing codes in the United States, in most cases. But as Dmitry Orlov points out, massive decreases in formal home ownership and increases in unemployment in coming years, coupled with increasingly hollowed-out local governments with limits on resources available for enforcement, will quite plausibly lead to a situation in which squatting on

[1]Dylan Tweney, "DIY Freaks Flock to 'Hacker Spaces' Worldwide," *Wired*, March29, 2009 <http://www.wired.com/gadgetlab/2009/03/hackerspaces/>.

[2]<http://www.nycresistor.com>.

[3]Nathan Cravens, "important appeal: social media and p2p tools against the meltdown," *Open Manufacturing* (Google Groups), March 13, 2009 <http://groups.google.com/group/openmanufacturing/msg/771617d04e45cd63>.

[4]*Ibid*.

(de facto) abandoned residential and commercial real estate is the norm, and local authorities turn a blind eye to it. Squats in abandoned/public buildings, and building with scavenged materials on vacant lots, etc. (a la Colin Ward), might be a black market version of what Cravens proposes. According to Gifford Hartman, although tent cities and squatter communities often receive hostile receptions, they're increasingly getting de facto acceptance from the local authorities in many parts of the country:

> In many places people creating tent encampments are met with hostility, and are blamed for their own condition. New York City, with a reputation for intolerance towards the homeless, recently shut down a tent city in East Harlem. Homeowners near a tent city of 200 in Tampa, Florida organised to close it down, saying it would 'devalue' their homes. In Seattle, police have removed several tent cities, each named 'Nickelsville' after the Mayor who ordered the evictions.
>
> Yet in some places, like Nashville, Tennessee, tent cities are tolerated by local police and politicians. Church groups are even allowed to build showers and provide services. Other cities that have allowed these encampments are: Champaign, Illinois; St. Petersburg, Florida; Lacey, Washington; Chattanooga, Tennessee; Reno, Nevada; Columbus, Ohio; Portland, Oregon. Ventura, California recently changed its laws to allow the homeless to sleep in cars and nearby Santa Barbara has made similar allowances. In San Diego, California a tent city appears every night in front of the main public library downtown.
>
> California seems to be where most new tent cities are appearing, although many are covert and try to avoid detection. One that attracted overflowing crowds is in the Los Angeles exurb of Ontario. The region is called the 'Inland Empire' and had been booming until recently; it's been hit extremely hard by the wave of foreclosures and mass layoffs. Ontario is a city of 175,000 residents, so when the homeless population in the tent city exploded past 400, a residency requirement was created. Only those born or recently residing in Ontario could stay. The city provides guards and basic services for those who can legally live there.[1]

Even squatting one's own residence after foreclosure has worked out fairly well in a surprising number of cases. A member of the Open Manufacturing email list

> Foreclosure is a double-edged sword. Dear friends of mine, a couple with two daughters, were really struggling two years ago, as the economy tanked, to pay their rent and feed their family from the same meager, erratic paychecks.
>
> When they heard that the owners of their rental unit had foreclosed, they saw it as the final blow. But unlike the other six residents they chose not to move out. It's been eighteen months since this happened: they have an ongoing relationship with corporations that provide heat, power and internet service but no rent is paid, while the former masters sue each other. It is probable that the so-called 'owners' of the property are themselves bankrupt and bought out, at this point; who knows when the situation might resolve itself.
>
> In the US this is called "Adverse possession", the legal term for squatting, and should they keep it up for seven years, they would own their apartment free and clear.
>
> This is in a dense neighborhood of Chicago, for perspective. It's happening all over the place, and with more foreclosure on the horizon, it's only going to

[1]Gifford Hartman, "Crisis in California: Everything Touched by Capital Turns Toxic," *Turbulence* 5 (2010) <http://turbulence.org.uk/turbulence-5/california/>.

> get more common. Single families aren't he only ones going bankrupt, it's happening to a lot of landlords and mortgage interests also.[1]

In addition, the proliferation of mortgage-based securities means the holder of a mortgage is several change-of-hands removed from the original lender, and may well lack any documentation of the original mortgage agreement. Some courts have failed to enforce eviction orders in such cases.

Another promising expedient for victims of foreclosure is to turn to firms like Boston Community Capital that specialize in buying up foreclosed mortgages, and then selling the property (with principle reduced to current market value) back to the original occupants. BCC's bargaining power is aided, in cutting a deal with foreclosing lenders, by embarrassing demonstrations by neighbors demanding they sell to BCC at market value rather than evict.[2]

In general, the resale value of foreclosed residences is so much lower, they are so difficult to resell, and managing the properties in the meantime is so inconvenient and costly, that—especially when the growing volume of defaults increases the difficulty of handling them—the bargaining power of defaulting home-owners is growing against lenders with an incentive to cut a deal rather than become real estate holding companies.

Although Cravens expressed some interest in the technical possibilities for social housing, he objected to my proposal to include housing as a fourth leg of an expanded Quadruple Alliance.

> I disagree with the name, Quadruple Alliance, as these three organizations I consider community ventures outside the home environment. Because the home I prefer to keep in the personal realm, I do not consider that an official community space.[3]

To the extent that my proposed housing "fourth leg" is a departure from Cravens' schema, it may be a closer approximation to Hine's original vision. Hine's original post addressed the basic question, from the individual in need of subsistence: "What do I do now that I'm unemployed." Housing is an integral part of such considerations. From the perspective of the sizable fraction of the general population that may soon be unemployed or unemployed, and consequently homeless, access to shelter falls in the same general class of pressing self-support needs as work in the Fab Lab and feeding oneself via the CSA farm. Although Cravens chose to focus on social production to the exclusion of private subsistence, if we revert to Hine's original concern, P2P housing projects are very much part of an overall resilient community package—analogous to the Roman villas of the Fifth Century—for weathering the Great Recession or Great Depression 2.0

[1]Sam Putman, "Walkable Community Networks for Spontaneous Gift Economy Development and Happiness," *Open Manufacturing*, March 20, 2010 <http://groups.google.com/group/openmanufacturing/browse_thread/thread/ 373013b9d631a374/78ba19a52d25e144>.

[2]John Leland, "Finding in Foreclosure a Beginning, Not an End," *New York Times*, March 21, 2010 <http://www.nytimes.com/2010/03/22/us/22foreclose.html>.

[3]Nathan Cravens, "[p2p-research] simpler way wiki," *P2P Research*, April 20, 2009 <http://listcultures.org/pipermail/p2presearch_listcultures.org/2009-April/002083.html>.

7

The Alternative Economy as a Singularity

We have seen the burdens of high overhead that the conventional, hierarchical enterprise and mass-production industry carry with them, their tendency to confuse the expenditure of inputs with productive output, and their culture of cost-plus markup. Running throughout this book, as a central theme, has been the superior efficiency of the alternative economy: its lower burdens of overhead, its more intensive use of inputs, and its avoidance of idle capacity.

Two economies are fighting to the death: one of them a highly-capitalized, high-overhead, and bureaucratically ossified conventional economy, the subsidized and protected product of one and a half century's collusion between big government and big business; the other a low capital, low-overhead, agile and resilient alternative economy, outperforming the state capitalist economy despite being hobbled and driven underground.

The alternative economy is developing within the interstices of the old one, preparing to supplant it. The Wobbly phrase "building the structure of the new society within the shell of the old" is one of the most fitting phrases ever conceived for summing up the concept.

A. Networked Production and the Bypassing of Corporate Nodes

One of the beauties of networked production, for subcontractors ranging from the garage shop to the small factory, is that it transforms the old corporate headquarters into a node to be bypassed.

Johan Soderberg suggests that the current model of outsourcing and networked production makes capital vulnerable to being cut out of the production process by labor. He begins with an anecdote about Toyota subcontractor Aisin Seiki, "the only manufacturer of a component critical to the whole Toyota network," whose factory was destroyed in a fire:

> The whole conglomerate was in jeopardy of grinding to a halt. In two months Toyota would run out of supplies of the parts produced by Aisin Seiki. Faced with looming disaster, the network of subcontractors fervently cooperated and created provisory means for substituting the factory. In a stunningly short time, Toyota subsidiaries had restructured themselves and could carry on unaffected by the incident. Duncan Watt attributes the swift response by the Toyota conglomerate to its networked mode of organisation. The relevance of this story for labour theory becomes apparent if we stipulate that the factory was not destroyed in an accident but was held-up in a labour conflict. Networked capital turns every point of production, from the firm down to the individual work as-

signment, into a node subject to circumvention. . . . [I]t is capital's ambition to route around labour strongholds that has brought capitalism into network production. . . . Nations, factories, natural resources, and positions within the social and technical division of labour, are all made subject to redundancy. Thus has capital annulled the threat of blockages against necks in the capitalist production chain, upon which the negotiating power of unions is based.

But this redundancy created by capital as a way of routing around blockages, Soderberg continues, threatens to make capital itself redundant:

> The fading strength of unions will continue for as long as organised labour is entrenched in past victories and outdated forms of resistance. But the networked mode of production opens up a "window of opportunity" for a renewed cycle of struggle, this time, however, of a different kind. *Since all points of production have been transformed into potentially redundant nodes of a network, capital as a factor of production in the network has itself become a node subject to redundancy.*[1]

(This was, in fact, what happened in the Third Italy: traditional mass-production firms attempted to evade the wave of strikes by outsourcing production to small shops, and were then blindsided when the shops began to federate among themselves.)[2]

Soderberg sees the growing importance of human relative to physical capital, and the rise of peer production in the informational realm, as reason for hope that independent and self-managed networks of laborers can route around capital. Hence the importance he attaches to the increasingly draconian "intellectual property" regime as a way of suppressing the open-source movement and maintaining control over the conditions of production.[3]

Dave Pollard, writing from the imaginary perspective of 2015, made a similar observation about the vulnerability of corporations that follow the Nike model of hollowing themselves out and outsourcing everything:

> In the early 2000s, large corporations that were once hierarchical end-to-end business enterprises began shedding everything that was not deemed 'core competency', in some cases to the point where the only things left were business acumen, market knowledge, experience, decision-making ability, brand name, and aggregation skills. This 'hollowing out' allowed multinationals to achieve enormous leverage and margin. It also made them enormously vulnerable and potentially dispensable.
>
> As outsourcing accelerated, some small companies discovered how to exploit this very vulnerability. When, for example, they identified North American manufacturers outsourcing domestic production to third world plants in the interest of 'increasing productivity', they went directly to the third world manufacturers, offered them a bit more, and then went directly to the North American retailers, and offered to charge them less. The expensive outsourcers quickly found themselves unnecessary middlemen. . . . The large corporations, having shed everything they thought was non 'core competency', learned to their chagrin that in the connected, information economy, the value of their core competency was much less than the inflated value of their stock, and they have lost

[1]Johan Soderberg, *Hacking Capitalism: The Free and Open Source Software Movement* (New York and London: Routledge, 2008), pp. 141-142.

[2]Michael J. Piore and Charles F. Sabel, *The Second Industrial Divide: Possibilities for Prosperity* (New York: HarperCollins, 1984), pp. 226-227.

[3]Soderberg, *Hacking Capitalism*, pp. 142-142

much of their market share to new federations of small entrepreneurial businesses.[1]

The worst nightmare of the corporate dinosaurs is that, in an economy where "imagination" or human capital is the main source of value, the imagination might take a walk: that is, the people who actually possess the imagination might figure out they no longer need the company's permission, and realize its "intellectual property" is unenforceable in an age of encryption and bittorrent (the same is becoming true in manufacturing, as the discovery and enforcement of patent rights against reverse-engineering efforts by hundreds of small shops serving small local markets becomes simply more costly than it's worth).

For example, Tom Peters gives the example of Oticon, which got rid of "the entire formal organization" and abolished departments, secretaries, and formal management titles. Employees put their personal belongings in "caddies, or personal carts, moving them to appropriate spots in the completely open space as their work with various colleagues requires."[2] The danger for the corporate gatekeepers, in sectors where outlays for physical capital cease to present significant entry barriers, is that one of these days knowledge workers may push their "personal carts" out of the organization altogether, and decide they can do everything just as well without the company.

B. The Advantages of Value Creation Outside the Cash Nexus

We already examined, in Chapters Three and Five, the tendencies toward a sharp reduction in the number of wage hours worked and increased production of value in the informal sector. From the standpoint of efficiency and bargaining power, this has many advantages.

On the individual level, a key advantage of the informal and household economy lies in its offer of an alternative to wage employment for meeting a major share of one's subsistence needs, and the increased bargaining power of labor in what wage employment remains.

> How much does the laborer increase his freedom if he happens to own a home, so that there is no landlord to evict him, and how much still greater is his freedom if he lives on a homestead where he can produce his own food?
>
> That the possession of capital makes a man independent in his dealings with his fellows is a self-evident fact. It makes him independent merely because it furnishes him actually or potentially means which he can use to produce support for himself without first securing the permission of other men.[3]

Ralph Borsodi demonstrated some eight decades ago—using statistics!—that the hourly "wage" from gardening and canning, and otherwise replacing external purchases with home production, is greater than the wages of most outside employment.[4]

Contra conventional finance gurus like Suze Orman, who recommend investments like lifetime cost averaging of stock purchases, contributing to a 401k up to the employer's maximum matching contribution, etc., the most sensible genu-

[1]David Pollard, "The Future of Business," *How to Save the World*, January 14, 2004 <http://blogs.salon.com/0002007/2004/01/14.html>.

[2]Tom Peters, *The Tom Peters Seminar: Crazy Times Call for Crazy Organizations* (New York: Vintage Books, 1994), pp. 29-30.

[3]Ralph Borsodi, *Prosperity and Security* (New York and London: Harper & Brothers, 1938), p. 241.

[4]Borsodi, *This Ugly Civilization* (Philadelphia: Porcupine Press, 1929, 1975), p. 99.

ine investment for the average person is capital investment in reducing his need for outside income. This includes building or purchasing the roof over his head as cheaply and paying it off as quickly as possible, and substituting home production for purchases with wage money whenever the first alternative is reasonably competitive. Compared to the fluctuation in value of financial investments, Borsodi writes,

> the acquisition of things which you can use to produce the essentials of comfort—houses and lands, machines and equipment—are not subject to these vicissitudes. . . . For their economic utility is dependent upon yourself and is not subject to change by markets, by laws or by corporations which you do not control.[1]

The home producer is free from "the insecurity which haunts the myriads who can buy the necessaries of life only so long as they hold their jobs."[2] A household with no mortgage payment, a large garden and a well-stocked pantry might survive indefinitely (if inconveniently) with only one part-time wage earner.

As we saw in Chapter Three, the evaporation of rents on artificial property rights like "intellectual property," and the rapid decline of capital outlays for physical production, mean a crisis in the ability to capture value from production. But, turning this on its head, it also means a collapse in the costs of living. As Bruce Sterling argued\ half facetiously (does he ever argue otherwise?), increased knowledge creates "poverty" in the sense that when everything is free, nothing is worth anything. But conversely, when nothing is worth everything, everything is free. And a world of free goods, while quite inconvenient for those who used to make their living selling those goods, is of a less unambiguously bad character for those who no longer need to make as much of a living to pay for stuff. When everything is free, the pressure to make a living in the first place is a lot less.

> - Waiting for the day of realization that Internet knowledge-richness actively MAKES people economically poor. "Gosh, Craigslist has such access to ultra-cheap everything now . . . hey wait a second, where did my job go?"
> - Someday the Internet will offer free food and shelter. At that point, hordes simply walk away. They abandon capitalism the way a real-estate bustee abandons an underwater building.[3]

C. More Efficient Extraction of Value from Inputs

John Robb uses STEMI compression, an engineering analysis template, as a tool for evaluating the comparative efficiency of his proposed Resilient Communities:

> In the evolution of technology, the next generation of a particular device/program often follows a well known pattern in the marketplace: its design makes it MUCH cheaper, faster, and more capable. This allows it to crowd out the former technology and eventually dominate the market (i.e. transistors replacing vacuum tubes in computation). A formalization of this developmental process is known as STEMI compression:
>
> - Space. Less volume/area used.
> - Time. Faster.
> - Energy. Less energy. Higher efficiency.

[1] *Ibid.*, p. 337.
[2] *Ibid.*, p. 352.
[3] Bruce Sterling, "The Power of Design in your exciting new world of abject poverty," *Wired: Beyond the Beyond*, February 21, 2010 <http://www.wired.com/beyond_the_ beyond/2010/02/the-power-of-design-in-your-exciting-new-world-of-abject-poverty/>.

- Mass. Less waste.
- Information. Higher efficiency. Less management overhead.

> So, the viability of a proposed new generation of a particular technology can often be evaluated based on whether it offers a substantial improvement in the compression of all aspects of STEMI without a major loss in system complexity or capability. This process of analysis also gives us an "arrow" of development that can be traced over the life of a given technology.

The relevance of the concept, he suggests, may go beyond new generations of technology. "Do Resilient Communities offer the promise of a generational improvement over the existing global system or not?"

> In other words: is the Resilient Community concept (as envisioned here) a viable self-organizing system that can rapidly and virally crowd out existing structures due to its systemic improvements? Using STEMI compression as a measure, there is reason to believe it is:
>
> - Space. Localization (or hyperlocalization) radically reduces the space needed to support any given unit of human activity. Turns useless space (residential, etc.) into productive space.
> - Time. Wasted time in global transport is washed away. JIT (just in time production) and place.
> - Energy. Wasted energy for global transport is eliminated. Energy production is tied to locality of use. More efficient use of solar energy (the only true exogenous energy input to our global system).
> - Mass. Less systemic wastage. Made to order vs. made for market.
> - Information. Radical simplification. Replaces hideously complex global management overhead with simple local management systems.[1]

The contrast between Robb's Resilient Communities and the current global system dovetails, more or less, with that between our two economies. And his STEMI compression template, as a model for analyzing the alternative economy's superiorities over corporate capitalism, overlaps with a wide range of conceptual models developed by other thinkers. Whether it be Buckminster Fuller's ephemeralization, or lean production's eliminating *muda* and "doing more and more with less and less," the same general idea has a very wide currency.

A good example is what Mamading Ceesay calls the "economies of agility." The emerging postindustrial age is a "network age where emerging Peer Production will be driven by the economies of agility."

> Economies of scale are about driving down costs of manufactured goods by producing them on a large scale. Economies of agility in contrast are about quickly being able to switch between producing different goods and services in response to demand.[2]

If the Toyota Production System is a quantum improvement on Sloanist mass-production in terms of STEMI compression and the economics of agility, and networked production on the Emilia-Romagna model is a similar advancement on the TPS, then the informal and household economy is an order of magnitude improvement on both of them.

[1]John Robb, "STEMI Compression," *Global Guerrillas* blog, November 12, 2008 <http://globalguerrillas.typepad.com/globalguerrillas/2008/11/stemi.html>.

[2]Mamading Ceesay, "The Economies of Agility and Disrupting the Nature of the Firm," *Confessions of an Autodidactic Engineer*, March 31, 2009 <http://evangineer.agoraworx.com/blog/2009-03-31-the-economies-of-agility-and-disrupting-the-nature-of-the-firm.html>.

Jeff Vail uses the term "Rhizome" for the forms of organization associated with Robb's Resilient Communities, and with the alternative economy in general: "an alternative mode of human organization consisting of a network of minimally self-sufficient nodes that leverage non-hierarchal coordination of economic activity."

> The two key concepts in my formulation of rhizome are 1) minimal self-sufficiency, which eliminates the dependencies that accrete [sic] hierarchy, and 2) loose and dynamic networking that uses the "small worlds" theory of network information processing to allow rhizome to overcome information processing burdens that normally overburden hierarchies.[1]

By these standards, the alternative economy that we saw emerging from the crises of state capitalism in previous chapters is capable of eating the corporate-state economy for lunch. Its great virtue is its superior efficiency in using limited resources intensively, as opposed to mass-production capitalist industry's practice of adding subsidized inputs extensively. The alternative economy reduces waste and inefficiency through the greater efficiency with which it extracts use-value from a given amount of land or capital.

An important concept for understanding the alternative economy's more efficient use of inputs is "productive recursion," which Nathan Cravens uses to refer to the order of magnitude reduction in labor required to obtain a good when it is produced in the social economy, without the artificial levels of overhead and waste associated with the corporate-state nexus.[2] Savings in productive recursion include (say) laboring to produce a design in a fraction of the time it would take to earn the money to pay for a proprietary design, or simply using an open source design; or reforging scrap metal at a tenth the cost of using virgin metal.[3]

> Production methods lower the cost of products when simplified for rapid replication. That is called productive recursion. Understanding productive recursion is the first step to understanding how we need to restructure Industrial economic systems in response to this form of technological change. If Industrial systems are not reconfigured for productive recursion, they will collapse before reaching anywhere near full automation. I hope this writing helps divert a kink in the proliferation of personal desktop fabrication and full productive automation generally.[4]

He cites, from Neil Gershenfeld's *Fab*, a series of "cases that prove the theory of productive recursion in practice." One example is the greatly reduced cost for cable service in rural Indian villages, "due to reverse engineered satellite receivers by means of distributed production." Quoting from *Fab*:

> A typical village cable system might have a hundred subscribers, who pay one hundred rupees (about two dollars) per month. Payment is prompt, because the "cable-wallahs" stop by each of their subscribers personally and rather persuasively make sure that they pay. Visiting one of these cable operators, I was intrigued by the technology that makes these systems possible and financially viable.
>
> A handmade satellite antenna on his roof fed the village's cable network. Instead of a roomful of electronics, the head end of his cable network was just a

[1]Jeff Vail, "What is Rhizome?" *JeffVail.Net*, January 28, 2008 <http://www.jeffvail.net/2007/01/what-is-rhizome.html>.

[2]Nathan Cravens, "Productive Recursion Proven," *Open Manufacturing* (Google Groups), March 8, 2009 <http://groups.google.com/group/openmanufacturing/browse_thread/thread/f819aab7683b93ac?pli=1>.

[3]Cravens, "Productive Recursion," *Open Source Ecology* Wiki <http://openfarmtech.org/index.php?title=Productive_Recursion>.

[4]Cravens, "Productive Recursion Proven."

> shelf at the foot of his bed. A sensitive receiver there detects and interprets the weak signal from the satellite, then the signal is amplified and fed into the cable for distribution around the village. The heart of all this is the satellite receiver, which sells for a few hundred dollars in the United States. He reported that the cost of his was one thousand rupees, about twenty dollars.[1]

The cheap satellite receiver was built by Sharp, which after some legwork Gershenfeld found to be "an entirely independent domestic brand" run out of a room full of workbenches in a district of furniture workshops in Delhi.

> They produced all of their own products, although not in that room—done there, it would cost too much. The assembly work was farmed out to homes in the community, where the parts were put together. Sharp operated like a farm market or grain elevator, paying a market-based per-piece price on what was brought in. The job of the Sharp employees was to test the final products.
>
> The heart of the business was in a back room, where an engineer was busy taking apart last-generation video products from developed markets. Just as the students in my fab class would learn from their predecessors' designs and use them as the starting point for their own, this engineer was getting a hands-on education in satellite reception from the handiwork of unknown engineers elsewhere. He would reverse engineer their designs to understand them, then redo the designs so that they could be made more simply and cheaply with locally available components and processes. And just as my students weren't guilty of plagiarism because of the value they added to the earlier projects, this engineer's inspiration by product designs that had long since become obsolete was not likely to be a concern to the original satellite-receiver manufacturers.
>
> The engineer at the apex of the Sharp pyramid was good at his job, but also frustrated. Their business model started with existing product designs. The company saw a business opportunity to branch out from cable television to cable Internet access, but there weren't yet available obsolete cable modems using commodity parts that they could reverse-engineer. Because cable modems are so recent, they use highly integrated state-of-the-art components that can't be understood by external inspection, and that aren't amenable to assembly in a home. But there no technological reason that data networks couldn't be produced in just this way, providing rural India with Internet access along with Bollywood soap operas. . . .
>
> . . . There isn't even a single entity with which to partner on a joint venture; the whole operation is fundamentally distributed.[2]

Another example of productive recursion, also from Gershenfeld's experiences in India, is the reverse engineering of ground resistance meters.

> For example, the ground resistance meters that were used for locating water in the area cost 25,000 rupees (about $500). At Vigyan Ashram they bought one, stripped it apart, and from studying it figured out how to make them for just 5,000 rupees. . . . Another example arose because they needed a tractor on the farm at Vigyan Ashram, but could not afford to buy a new one. Instead, they developed their own "MechBull" made out of spare jeep parts for 60,000 rupees ($1,200). This proved to be so popular that a Vigyan Ashram alum built a business making and selling these tractors.[3]

Yet another is a walk-behind tractor, developed from a modified motorcycle within Anil Gupta's "Honeybee Network" (an Indian alternative technology group).

> Modeled on how honeybees work—collecting pollen without harming the flowers and connecting flowers by sharing the pollen—the Honeybee Network col-

[1]Neil Gershenfeld, *Fab: The Coming Revolution on Your Desktop—from Personal Computers to Personal Fabrication* (New York: Basic Books, 2005), p. 182.

[2]*Ibid.*, pp. 185-187.

[3]*Ibid.* p. 164.

> lects and helps develop ideas from grassroots inventors, sharing rather than taking their ideas. At last count they had a database of ten thousand inventions.
>
> One Indian inventor couldn't afford or justify buying a large tractor for his small farm; it cost the equivalent of $2,500. But he could afford a motorcycle for about $800. So he came up with a $400 kit to convert a motorcycle into a three-wheeled tractor (removable of course, so that it's still useful as transportation). Another agricultural inventor was faced with a similar problem in applying fertilizer; his solution was to modify a bicycle.[1]

According to Marcin Jakubowski of Open Source Ecology, the effects of productive recursion are cumulative. "Cascading Factor 10 cost reduction occurs when the availability of one product decreases the cost of the next product."[2] We already saw, in Chapter Five, the specific case of the CEB Press, which can be produced for around 20% of the cost of purchasing a competing commercial model.

Amory Lovins and his coauthors, in *Natural Capitalism*, described the cascading cost savings ("Tunneling Through the Cost Barrier") that result when the efficiencies of one stage of design reduce costs in later stages. Incremental increases in efficiency may increase costs, but large-scale efficiency improvements in entire designs may actually result in major cost reductions. Improving the efficiency of individual components in isolation can be expensive, but improving the efficiency of systems can reduce costs by orders of magnitude.[3]

> Much of the art of engineering for advanced resource efficiency involves harnessing helpful interactions between specific measures so that, like loaves and fishes, the savings keep on multiplying. The most basic way to do this is to "think backward," from downstream to upstream in a system. A typical industrial pumping system, for example . . . , contains so many compounding losses that about a hundred units of fossil fuel at a typical power station will deliver enough electricity to the controls and motor to deliver enough torque to the pump to deliver only ten units of flow out of the pipe—a loss factor of about tenfold.
>
> But turn those ten-to-one compounding losses around backward . . . , and they generate a one-to-ten compounding *saving*. That is, saving one unit of energy furthest downstream (such as by reducing flow or friction in pipes) avoids enough compounding losses from power plant to end use to save about ten units of fuel, cost, and pollution back at the power plant.[4]

To take another example, both power steering and V-8 engines resulted from Detroit's massive increases in automobile weight in the 1930s, along with marketing-oriented decisions to add horsepower that would be idle except during rapid acceleration. The introduction of lightweight frames, conversely, makes possible the use of much lighter internal combustion engines or even electric motors, which in turn eliminate the need for power steering.

Most of the order-of-magnitude efficiencies of whole-system design that Lovins et all describe result, not from new technology, but from more conscious use of existing technology: what Edwin Land called "the sudden cessation of stupidity" or "stopping having an old idea."[5] Simply combining existing technological elements in the most effective way can result in efficiency increases of Factor Four,

[1]*Ibid.*, p. 88.

[2]Marcin Jakubowski, "OSE Proposal—Towards a World Class Open Source Research and Development Facility," v0.12, January 16, 2008 <http://openfarmtech.org/OSE_Proposal.doc>.

[3]Paul Hawken, Amory Lovins, and L. Hunter Lovins, *Natural Capitalism: Creating the Next Industrial Revolution* (Boston, New York, and London: Little, Brown and Company, 1999), pp. 113-124.

[4]*Ibid.*, p. 121.

[5]*Ibid.*, pp. 65, 117.

Factor Eight, or more. The overall designs are generally the kinds of mashups of off-the-shelf technology that Cory Doctorow and Murray Bookchin comment on below.

The increased efficiencies result from a design process like Eric Raymond's Bazaar: designers operate intelligently, with constant feedback.[1] The number of steps and the transaction costs involved in aggregating user feedback with the design process are reduced. The inefficiencies that result from an inability to "think backward" are far more likely to occur in a stovepiped organizational framework, where each step or part is designed in isolation by a designer whose relation to the overall process is mediated by a bureaucratic hierarchy. For example, in building design:

> Conventional buildings are typically designed by having each design specialist "toss the drawings over the transom" to the next specialist. Eventually, all the contributing specialists' recommendations are integrated, sometimes simply by using a stapler.[2]

This approach inevitably results in higher costs, because increased efficiencies of a single step taken in isolation generally *are* governed by a law of increased costs and diminishing returns. Thicker insulation, better windows, etc., cost more than their conventional counterparts. Lighter materials and more efficient engines for a car, similarly, cost more than conventional components. So optimizing the efficiency of each step in isolation follows a rising cost curve, with each marginal improvement in efficiency of the step costing more than the last. But by approaching design from the perspective of a whole system, it becomes possible to "tunnel through the cost barrier":

> When intelligent engineering and design are brought into play, big savings often cost less *up front* than small or zero savings. Thick enough insulation and good enough windows can eliminate the need for a furnace, which represents an investment of more capital than those efficiency measures cost. Better appliances help eliminate the cooling system, too, saving even more capital cost. Similarly, a lighter, more aerodynamic car and a more efficient drive system work together to launch a spiral of decreasing weight, complexity and cost. The only moderately more efficient house and car do cost more to build, but when designed as whole systems, the *super*efficient house and car often cost less than the original, unimproved versions.[3]

While added insulation and tighter windows increase the cost of insulation or windows, taken in isolation, if integrated into overall building design they may *reduce* total costs up front by reducing the required capacity—and hence outlays on capital equipment—of heating and cooling systems. A more energy-efficient air conditioner, *given* unchanged cooling requirements, will cost more; but energy-efficient windows, office equipment, etc., can reduce the cooling load by 85%, and thus make it possible to replace the cooling system with one three-fourths smaller than the original—thereby not only reducing the energy bill by 75%, but enormously reducing capital expenditures on the air conditioner.[4] The trick is to "do the right things in the right order":

> . . . if you're going to retrofit your lights and air conditioner, do the lights first so you can make the air conditioner smaller. If you did the opposite, you'd pay for

[1]Eric S. Raymond, *The Cathedral and the Bazaar* <http://catb.org/~esr/writings/homesteading>.

[2]Hawken et al, *Natural Capitalism*, p. 90.

[3]*Ibid.*, p. 114.

[4]*Ibid.*, pp. 119-120.

> more cooling capacity than you'd need after the lighting retrofit, and you'd also make the air conditioner less efficient because it would either run at part-load or cycle on and off too much.[1]

This is also a basic principle of lean production: most costs come from five percent of point consumption needs, and from scaling the capacity of the load-bearing infrastructure to cover that extra five percent instead of just handling the first ninety-five percent. It ties in, as well, with another lean principle: getting production out of sync with demand (including the downstream demand for the output of one step in a process), either spatially or temporally, creates inefficiencies. Optimizing one stage without regard to production flow and downstream demand usually involves expensive infrastructure to get an in-process input from one stage to another, often with intermediate storage while it is awaiting a need. The total resulting infrastructure cost greatly exceeds the saving at individual steps. Inefficient synchronization of sequential steps in any process results in bloated overhead costs from additional storage and handling infrastructure.

A good example of the cost-tunneling phenomenon was engineer Jan Schilham's work at the Interface carpet factory in Shanghai, which reduced horsepower requirements for pumping in one process twelvefold—while *reducing* capital costs. In conventional design, the factory layout and system of pipes are assumed as given, and the pumps chosen against that background.

> . . . First, Schilham chose to deploy big pipes and small pumps instead of the original design's small pipes and big pumps. Friction falls as nearly the fifth power of pipe diameter, so making the pipes 50 percent fatter reduces their friction by 86 percent. The system needs less pumping energy—*and* smaller pumps and motors to push against the friction. If the solution is this easy, why weren't the pipes originally specified to be big enough? . . . Traditional optimization compares the cost of fatter pipe with only the value of the saved *pumping energy*. This comparison ignores the size, and hence the capital cost, of the [pumping] *equipment* needed to combat the pipe friction. Schilham found he needn't calculate how quickly the savings would repay the extra up-front cost of the fatter pipe, because capital cost would fall more for the pumping and drive equipment than it would rise for the pipe, making the efficient system as a whole cheaper to construct.
>
> Second, Schilham laid out the pipes first and *then* installed the equipment, in reverse order from how pumping systems are conventionally installed. Normally, equipment is put in some convenient and arbitrary spot, and the pipe fitter is then instructed to connect point A to point B. the pipe often has to go through all sorts of twists and turns to hook up equipment that's too far apart, turned the wrong way, mounted at the wrong height, and separated by other devices installed in between. . . .
>
> By laying out the pipes before placing the equipment that the pipes connect, Schilham was able to make the pipes short and straight rather than long and crooked. That enabled him to exploit their lower friction by making the pumps, motors, inverters and electricals even smaller and cheaper.[2]

Vinay Gupta described some of the specific efficiencies involved in productive recursion, that combine to reduce the alternative economy's costs by an order of magnitude.[3] The most important efficiency comes from distributed infrastructure which provides

[1]*Ibid*., p. 122.

[2]*Ibid*., pp. 116-117.

[3]Vinay Gupta, "The Global Village Development Bank: financing infrastructure at the individual, household and village level worldwide" Draft 2 (March 12, 2009)

the same class of services that are provided by centralized systems like the water and power grids, but without the massive centralized investments in physical plant. For example, dry toilets and solar panels can provide high quality services household by household without a grid.

The digital revolution and network organization interact with distributed infrastructure to remove most of the administrative and other transaction costs involved in getting the technologies to the people who can benefit from them. It is, in other words, governed by the rules of Raymond's Bazaar, which Robb made the basis of his "open source insurgency."

Distributed infrastructure also benefits from "economies of agility," as opposed to the enormous capital outlays in conventional blockbuster investments that must frequently be abandoned as "sunk costs" when the situation changes or funding stops. " . . . [H]alf a dam is no dam at all, but 500 of 1000 small projects is half way to the goal." And distributed infrastructure projects manage to do without the enormous administrative and overhead costs of conventional organizations, which we saw described by Paul Goodman in Chapter Two; most of the organization and planning are done by those with the technical knowledge and sweat equity, who are directly engaged in the project and reacting to the situation on the ground.

And finally, Gupta argues, distributed finance—microcredit—interacts with distributed infrastructure and network organization to heighten the advantages of agility and low overhead still further.

We also saw, in Chapter Five, the ways that modular design and the forms of stigmergic organization facilitated by open-source design contribute to lower costs. Modular design is a way of getting more bang for the R&D buck by maximizing use of a given innovation across an entire product ecology, and at the same time building increased redundancy into the system through interchangeable parts.[1] And stigmergic organization with open-source designs eliminates barriers to widespread use of the most efficient existing designs.

Malcolm Gladwell's "David vs. Goliath" analysis of military history is an excellent illustration of the economies of agility. Victory goes to the bigger battalions about seven times out of ten—when Goliath outnumbers David ten to one, that is. But when the smaller army, outnumbered ten to one, acknowledges the fact and deliberately chooses unconventional tactics that target Goliath's weaknesses, it actually wins about six times out of ten. "When underdogs choose not to play by Goliath's rules, they win . . . " Guerrilla fighters from J.E.B. Stuart to T. E. Lawrence to Ho Chi Minh have learned, as General Maurice de Saxe put it, that victory is about legs rather than arms. As Lawrence wrote, "Our largest available resources were the tribesmen, men quite unused to formal warfare, whose assets were movement, endurance, individual intelligence, knowledge of the country, courage."[2] Another good example is what the U.S. military (analyzing Chinese asymmetric warfare capabilities) calls "Assassin's Maces": "anything which provides a

<http://vinay.howtolivewiki.com/blog/hexayurt/my-latest-piece-the-global-village-development-bank-1348>.

[1]Jonathan Dugan, for example, stresses Redundancy and Modularity as two of the central principles of resilience. Chris Pinchen, "Resilience: Patterns for thriving in an uncertain world," *P2P Foundation Blog*, April 17, 2010 <http://blog.p2pfoundation.net/resilience-patterns-for-thriving-in-an-uncertain-world/2010/04/17>.

[2]Malcolm Gladwell, "How David Beats Goliath," *The New Yorker*, May 11, 2009 <http://www.newyorker.com/reporting/2009/05/11/090511fa_fact_gladwell?currentPage=all>.

cheap means of countering an expensive weapon." A good example is the black box that transmits ten thousand signals on the same frequency used by SAM missiles, and thus overwhelms American air-to-surface missiles which target SAM radio signals. The Chinese, apparently, work from the assumption that the U.S. develops countermeasures to "Assassin's Mace" weapons, and deliberately make it easier for American intelligence to acquire older such weapons as a form of disinformation; there's good reason to believe the Chinese military can work around American countermeasures much more quickly, and cheaply, than the U.S. can develop them.[1]

A recent example of "Assassin's Mace" technology is Skygrabber, an off-the-shelf software product that costs $26. Insurgents in Afghanistan use it to capture video feeds from U.S. military drones. The Pentagon has known about the problem since the Balkan wars, but—get this—didn't bother spending the money to encrypt the feed because they "assumed local adversaries wouldn't know how to exploit it."[2] In our discussion of networked resistance in Chapter Three, if you recall, we saw that the music industry assumed its DRM only had to be good enough to thwart the average user, because the geeks who could crack it would be too few to have a significant economic impact. But as Cory Doctorow pointed out, it takes only one geek to figure it out and then explain it to everybody else. It's called "stigmergic organization." Well, here's Dat Ole Debbil stigmergy again, and the Pentagon's having about as much fun with it as the record companies. John Robb describes the clash of organizational cultures:

> This event isn't an aberration. It is an inevitable development, one that will only occur more and more often. Why? Military cycles of development and deployment take decades due to the dominance of a lethargic, bureaucratic, and bloated military industrial complex. Agility isn't in the DNA of the system nor will it ever be (my recent experience with a breakthrough and inexpensive information warfare system my team built, is yet another example of how FAIL the military acquisition system is).
>
> In contrast, vast quantities of cheap/open/easy technologies (commercial and open source) are undergoing rapid rates of improvement. Combined with tinkering networks that can repurpose them to a plethora of unintended needs (like warfare), this development path becomes an inexorable force. The delta (a deficit from the perspective of the status quo, an advantage for revisionists) between the formal and the informal will only increase as early stage networks that focus specifically on weapons/warfare quickly become larger, richer, etc. (this will happen as they are combined with the economic systems of more complex tribal/community "Darknets").[3]

In theory, it's fairly obvious what the U.S. national security establishment needs to do. All the assorted "Fourth Generation Warfare" doctrines are pretty much agreed on that. It has to reconfigure itself as a network, more decentralized and agile than the network it's fighting, so that it can respond quickly to intelligence and small autonomous units can "swarm" enemy targets from many direc-

[1]David Hambling, "China Looks to Undermine U.S. Power, With 'Assassin's Mace'." *Wired*, July 2 <http://www.wired.com/dangerroom/2009/07/china-looks-to-undermine-us-power-with-assassins-mace/>.

[2]Siobhan Gorman, Yochi J. Dreazen and August Cole, "Insurgents Hack U.S. Drones," *Wall Street Journal*, December 17, 2009 <http://online.wsj.com/article/SB126102247889095011.html>.

[3]John Robb, "SUPER EMPOWERMENT: Hack a Predator Drone," *Global Guerrillas*, December 17, 2009 <http://globalguerrillas.typepad.com/globalguerrillas/2009/12/super-empowerment-hack-a-predator-drone.html>.

tions at once.[1] The problem is, it's easier said than done. Al Qaeda had one huge advantage over the U.S. national security establishment: Osama bin Laden is simply *unable* to interfere with the operations of local Al Qaeda cells in the way that American military bureaucracies interfere with the operations of military units. No matter what 4GW doctrine calls for, no matter what the slogans and buzzwords at the academies and staff colleges say, it will be impossible to *do* any of it so long as the military bureaucracy exists because military bureaucracies are constitutionally *incapable* of restraining themselves from interference. Robb describes the problem. He quotes Jonathan Vaccaro's op-ed from the *New York Times*:

> In my experience, decisions move through the process of risk mitigation like molasses. When the Taliban arrive in a village, I discovered, it takes 96 hours for an Army commander to obtain necessary approvals to act. In the first half of 2009, the Army Special Forces company I was with repeatedly tried to interdict Taliban. By our informal count, however, we (and the Afghan commandos we worked with) were stopped on 70 percent of our attempts because we could not achieve the requisite 11 approvals in time.
>
> For some units, ground movement to dislodge the Taliban requires a colonel's oversight. In eastern Afghanistan, traveling in anything other than a 20-ton mine-resistant ambush-protected vehicle requires a written justification, a risk assessment and approval from a colonel, a lieutenant colonel and sometimes a major. These vehicles are so large that they can drive to fewer than half the villages in Afghanistan. They sink into wet roads, crush dry ones and require wide berth on mountain roads intended for donkeys. The Taliban walk to these villages or drive pickup trucks.
>
> The red tape isn't just on the battlefield. Combat commanders are required to submit reports in PowerPoint with proper fonts, line widths and colors so that the filing system is not derailed. Small aid projects lag because of multimonth authorization procedures. A United States-financed health clinic in Khost Province was built last year, but its opening was delayed for more than eight months while paperwork for erecting its protective fence waited in the approval queue.
>
> Communication with the population also undergoes thorough oversight. When a suicide bomber detonates, the Afghan streets are abuzz with Taliban propaganda about the glories of the war against America. Meanwhile, our messages have to inch through a press release approval pipeline, emerging 24 to 48 hours after the event, like a debutante too late for the ball.[2]

Robb adds his own comments on just how badly the agility-enhancing potential of network technology is sabotaged:

- Risk mitigation trumps initiative every time. Careers are more important than victory. Risk evaluation moves upward in the hierarchy. Evaluation of risk takes time, particularly with the paucity of information that can be accessed at positions removed from the conflict. . . .
- New communications technology isn't being used for what it is designed to do (enable decentralized operation due to better informed people on the ground). Instead it is being used to enable more complicated and hierarchical approval processes—more sign offs/approvals, more required processes, and higher level oversight. For example: a general, and his staff, directly commanding a small strike team remotely.[3]

[1]John Arquilla and David Ronfeldt, "Fighting the Network War," *Wired*, December 2001 <http://www.wired.com/ wired/archive/9.12/netwar.html>.

[2]Jonathan J. Vaccaro, "The Next Surge—Counterbureaucracy," *New York Times*, December 7, 2009 <http://www.nytimes.com/2009/12/08/opinion/08vaccaro.html>.

[3]Robb, "Fighting an Automated Bureaucracy," *Global Guerrillas*, December 8, 2009 <http://globalguerrillas.typepad.com/globalguerrillas/2009/12/journal-fighting-an-automated-bureaucracy.html>.

So long as the military bureaucracy exists, it will be impossible to put 4GW ideas into practice without interference from the pointy-haired bosses.

Another example of the same phenomenon is the way the Transportation Security Administration deals with security threats: as the saying goes, by "always planning for the last war."

> First they attacked us with box cutters, so the TSA took away anything even vaguely sharp or pointy. Then they tried (and failed) to hurt us with stuff hidden in their shoes. So the TSA made us take off our shoes at the checkpoint. Then there was a rumor of a planned (but never executed) attack involving liquids, so the TSA decided to take away our liquids.[1]

Distributed infrastructure benefits, as well, from what Robb calls "scale invariance"[2]: the ability of the part, in cases of system disruption, to replicate the whole. Each part conserves the features that define the whole, on the same principle as a hologram. Projects like Open-Source Ecology,[3] once the major components of a local site are in place, can duplicate any of the individual components or duplicate them all to create a second site. The Fab Lab can produce the parts for a steam engine, CEB press, tractor, sawmill, etc., or even the machine tools for another Fab Lab.

Distributist writer John Medaille pointed out, by private email, that the Israelites under the Judges were a good example of superior extraction of value from inputs. At a time when the "more civilized" Philistines dominated most of the fertile valleys of Palestine, the Israelite confederacy stuck to the central highlands. But their "alternative technology," focused on extracting more productivity from marginal land, enabled them to make more intensive use of what was unusable to the Philistines.

> The tribes clung to the hilltops because the valleys were "owned" by the townies (Philistines) and the law of rents was in full operation. The Hebrews were free in the hills, and increasingly prosperous, both because of their freedom and because of new technologies, namely contoured plowing and waterproof cement, which allowed the construction of cisterns to put them through the dry season.[4]

In other words, a new technological regime supplanted a more privileged form of society through superior efficiency, despite being disadvantaged in access to productive inputs. The Hebrews were able to outcompete the dominant social system by making more efficient and intensive use of inputs that were "unusable" with conventional methods of economic organization.

The alternative economy, likewise, has taken for its cornerstone the stone which the builders refused. As I put it in a blog post (in an admittedly grandiose yet nevertheless eminently satisfying passage):

> . . . [T]he owning classes use less efficient forms of production precisely because the state gives them preferential access to large tracts of land and subsidizes the inefficiency costs of large-scale production. Those engaged in the alternative economy, on the other hand, will be making the most intensive and efficient use of the land and capital available to them. So the balance of forces between the al-

[1]Thoreau, "More on the swarthy threat to our precious carry-on fluids," *Unqualified Offerings*, December 26, 2009 <http://highclearing.com/index.php/archives/2009/12/26/10438>.

[2]Robb, "Resilient Communities and Scale Invariance," *Global Guerrillas*, April 16, 2009 <http://globalguerrillas.typepad.com/globalguerrillas/2009/04/resilient-communities-and-scale-invariance.html>.

[3]See Chapter Five.

[4]John Medaille, personal email to author, January 28, 2009.

> ternative and capitalist economy will not be anywhere near as uneven as the distribution of property might indicate.
>
> If everyone capable of benefiting from the alternative economy participates in it, and it makes full and efficient use of the resources already available to them, eventually we'll have a society where most of what the average person consumes is produced in a network of self-employed or worker-owned production, and the owning classes are left with large tracts of land and understaffed factories that are almost useless to them because it's so hard to hire labor except at an unprofitable price. At that point, the correlation of forces will have shifted until the capitalists and landlords are islands in a mutualist sea—and their land and factories will be the last thing to fall, just like the U.S Embassy in Saigon.[1]

Soderberg refers to the possibility that increasing numbers of workers will "defect from the labour market" and "establish means of non-waged subsistence," through efficient use of the waste products of capitalism.[2] The "freegan" lifestyle (less charitably called "dumpster diving") is one end of a spectrum of such possibilities. At the other end is low-cost recycling and upgrading of used and discarded electronic equipment: for example, the rapid depreciation of computers makes it possible to add RAM to a model a few years old at a small fraction of the cost of a new computer, with almost identical performance.

Reason's Brian Doherty, in a display of rather convoluted logic, attempted to depict freeganism as proof of capitalism's virtues:

> It's nice of capitalism to provide such an overflowing cornucopia that the [freegans] of the world can opt out. Wouldn't it be gracious of them to show some love to the system that manages to keep them alive and thriving without even trying?[3]

To take Doherty's argument and stand it on its head, consider the amount of waste resulting from the perverse incentives under the Soviet planned economy. In some cases, new refrigerators and other appliances were badly damaged by being roughly thrown off the train and onto a pile at the point of delivery, because the factory got credit simply for manufacturing them, and the railroad got credit for delivering them, under the metrics of the Five Year Plan. Whether they actually worked, or arrived at the retailer in a condition such that someone was willing to buy them, was beside the point. Now, imagine if some handy fellow in the Soviet alternative economy movement had bought up those fridges as factory rejects for a ruble apiece, or just bought them for scrap prices from a junkyard, and then got them in working order at little or no cost. Would Doherty be praising Soviet socialism for its efficiency in producing such a surplus that the Russian freegan could live off the waste?

When the alternative economy is able to make more efficient use of the waste byproducts of state capitalism—waste byproducts that result from the latter's inefficient use of subsidized inputs—and thereby supplant state capitalism from within by the superior use of its underutilized resources and waste, it is rather perverse to dismiss the alternative economy as just another hobby or lifestyle choice enabled by the enormous efficiencies of corporate capitalism. And the alternative economy is utilizing inputs that would otherwise be waste, and thereby establish-

[1]Kevin Carson, "'Building the Structure of the New Society Within the Shell of the Old,'" *Mutualist Blog: Free Market Anti-Capitalism*, March 22, 2005 <http://mutualist.blogspot.com/2005/03/building-structure-of-new-society.html>.

[2]Soderberg, *Hacking Capitalism*, p. 172.

[3]Brian Doherty, "The Glories of Quasi-Capitalist Modernity, Dumpster Diving Division," *Reason Hit & Run* Blog, September 12, 2007 <http://reason.com/blog/show/122450.html>.

ing an ecological niche based on the difference between capitalism's actual and potential efficiencies; so to treat capitalism's inefficiencies as a mark of efficiency—i.e., how inefficient it can afford to be—is a display of Looking Glass logic.

The alternative economy's superior extraction of value from waste inputs extends, ultimately, to the entire economy.

> If these isolated nodes of self-sufficiency connect, communicate, and interact, then they will enjoy an improve position relative to hierarchal structures. . . .
>
> Additionally, from the perspective of the diagonal, the Diagonal Economy will begin as a complementary structure that is coextensive but out of phase with our current system. However, it will be precisely because it leverages a more efficient information processing structure that it will be able to eventually supplant the substrate hierarchies as the dominant system.[1]

One example of how the alternative economy permits the increasingly efficient extraction of value from waste material, by the way, is the way in which network technology facilitates repair even within the limits of proprietary design and the planned obsolescence model. In Chapter Two, we considered Julian Sanchez's account of how Apple's design practices serve to thwart cheap repair. iFixit is an answer to that problem:

> Kyle Wiens and Luke Soules started iFixit (ifixit.com) out of their dorms at Cal Poly in San Luis Obispo, Calif. That was six years ago. Today they have a self-funded business that sells the parts and tools you need to repair Apple equipment. One of their innovations is creating online repair manuals for free that show you how to make the repairs.
>
> "Our biggest source of referrals is Apple employees, particularly folks at the Genius Bar," Wien says. They refer customers who complain when Apple won't let them fix an out-of-warranty product. (Apple: "Just buy a new one!")
>
> iFixit will also buy your old Mac and harvest the reusable parts to resell. . . . If it's starting to sound like an auto parts franchise, well, Wiens and Soules have been thinking about someday doing for cars what they do for computers and handhelds today.[2]

In other words, the same open-source insurgency model that governs the file-sharing movement is spreading to encompass the development of all kinds of measures for routing around planned obsolescence and the other irrationalities of corporate capitalism. The reason for the quick adaptability of fourth generation warfare organizations, as described by John Robb, is that any innovation developed by a particular cell becomes available to the entire network. And by the same token, in the file-sharing world, it's not enough that DRM be sufficiently hard to circumvent to deter the average user. The average user need only use Google to benefit from the superior know-how of the geek who has already figured out how to circumvent it. Likewise, once anyone figures out how to circumvent any instance of planned obsolescence, their hardware hack becomes part of a universally accessible repository of knowledge.

As Cory Doctorow notes, cheap technologies which can be modularized and mixed-and-matched for any purpose are just lying around. " . . . [T]he market for facts has crashed. The Web has reduced the marginal cost of discovering a fact to $0.00." He cites Robb's notion that "[o]pen source insurgencies don't run on detailed instructional manuals that describe tactics and techniques." Rather,they just

[1]Jeff Vail, "The Diagonal Economy 5: The Power of Networks," *Rhizome*, December 21, 2009 <http://www.jeffvail.net/2009/12/diagonal-economy-5-power-of-networks.html>.

[2]Dale Dougherty, "What's in Your Garage?" *Make*, vol. 18 <http://www.make-digital.com/make/vol18/?pg=39>.

run on "plausible premises." You just put out the plausible premise—i.e., the suggestion based on your gut intuition, based on current technical possibilities, that something can be done—that IED's can kill enemy soldiers, and then anyone can find out *how* to do it via the networked marketplace of ideas, with virtually zero transaction costs.

> But this doesn't just work for insurgents—it works for anyone working to effect change or take control of her life. Tell someone that her car has a chip-based controller that can be hacked to improve gas mileage, and you give her the keywords to feed into Google to find out how to do this, where to find the equipment to do it—even the firms that specialize in doing it for you.
>
> In the age of cheap facts, we now inhabit a world where knowing something is possible is practically the same as knowing how to do it.
>
> This means that invention is now a lot more like collage than like discovery.

Doctorow mentions Bruce Sterling's reaction to the innovations developed by the protagonists of his (Doctorow's) *Makers*: "There's hardly any engineering. Almost all of this is mash-up tinkering." Or as Doctorow puts it, it "assembles rather than invents."

> It's not that every invention has been invented, but we sure have a lot of basic parts just hanging around, waiting to be configured. Pick up a $200 FPGA chip-toaster and you can burn your own microchips. Drag and drop some code-objects around and you can generate some software to run on it. None of this will be as efficient or effective as a bespoke solution, but it's all close enough for rock-n-roll.[1]

Murray Bookchin anticipated something like this back in the 1970s, writing in *Post-Scarcity Anarchism*:

> Suppose, fifty years ago, that someone had proposed a device which would cause an automobile to follow a white line down the middle of the road, automatically and even if the driver fell asleep. . . . He would have been laughed at, and his idea would have been called preposterous. . . . But suppose someone called for such a device today, and was willing to pay for it, leaving aside the question of whether it would actually be of any genuine use whatever. Any number of concerns would stand ready to contract and build it. No real invention would be required. There are thousands of young men in the country to whom the design of such a device would be a pleasure. They would simply take off the shelf some photocells, thermionic tubes, servo-mechanisms, relays, and, if urged, they would build what they call a breadboard model, and it would work. The point is that the presence of a host of versatile, reliable, cheap gadgets, and the presence of men who understand all their cheap ways, has rendered the building of automatic devices almost straightforward and routine. It is no longer a question of whether they can be built, it is a question of whether they are worth building.[2]

D. Seeing Like a Boss

The contrast in agility and learning ability between stigmergic organizations and hierarchies is beautifully brought out by David Pollard:

[1]Cory Doctorow, "Cheap Facts and the Plausible Premise," *Locus Online*, July 5, 2009 <http://www.locusmag.com/Perspectives/2009/07/cory-doctorow-cheap-facts-and-plausible.html>.

[2]Murray Bookchin, "Toward a Liberatory Technology," in *Post-Scarcity Anarchism* (Berkeley, Calif.: The Ramparts Press, 1971), pp. 49-50.

So Management by SMART Objective [Specific, Measurable, Achievable, Realistic, and Time-Based—Peter Drucker] leads to this ludicrous and dysfunctional dance:

• Leaders hire 'expert' consultants, or huddle among themselves, or decide by fiat, what the SMART objectives should be for their organization: "increase revenues by 10% and profits by 20% next year by introducing 'improved' versions of 15 selected products that can be sold for an average price 25% higher than the old version, and which, through internal efficiencies, cost 15% less per unit to produce"

• These leaders then 'cascade down' these objectives and command subordinates to come up with SMART business unit plans that will, if successful, collectively achieve these top-level objectives.

• The subordinates understand that their success depends on ratcheting up profits, and that the objectives set by the leaders are ridiculous, magical thinking. So they come up with alternative plans to increase profits by 20% through a series of difficult, but realistic, moves. These entail offshoring everything to China, layoffs, pressuring staff to work longer hours for no more money, and, if all else fails, firing people or leaving vacancies unfilled.

• The good people in the organization all leave, because they know this short-range thinking is dysfunctional, damaging to the organizations in the longer term, unsustainable, and a recipe for a miserable workplace. Their departure creates more vacancies that aren't filled, which in the short term reduces costs.

• The clueless and the losers, who are left, attempt to pick up the slack. They work harder, find workarounds for the dumbest management decrees, and do their best to achieve these objectives. Those fortunate enough to be in the right market areas in the right economies get promoted into some of the vacant spots left by the good people, but without the commensurate salary increase.

• The leaders, as a result, achieve their short-run objectives, award themselves huge bonuses, profit from increases in the value of their stock options, and repeat the whole cycle the next year.

• At some point the utter sustainability of this "management process" becomes apparent. There is a really bad year. The economy is blamed, perhaps. Or the top leaders are fired, and rehired in other organizations suffering from really bad years. Or the company is bought out, or 'reorganized' so that all the old objectives and measures no longer apply, and a completely new set is established.

The byproduct is a blizzard of plans, budgets and strategies, which are substantially meaningless. Everyone does ad hoc things to protect their ass and try to make the best of impossible targets and incompetent, arrogant leaders self-deluded about their own brilliance and about their ability to control what is really happening in the organization and the marketplace.

There are, however, some things of real value happening in these organizations. None of them are 'SMART' so none is recognized or rewarded, and most of these things are actively discouraged. Nevertheless, because most people take pride in what they do, these valuable things happen. They include:

• Learning: People learn by making mistakes (that they don't admit to), and this makes them better at doing their jobs.

• Conversations: People share, peer-to-peer, what works and doesn't work, through mostly informal conversations, and this too makes them better at doing their jobs. These conversations are often surreptitious, since they are not considered 'productive' work.

• Practice: The more people work at doing a particular task, the better they get at it. Most such practices are substantially workarounds, self-developed ways to do their particular specialized work optimally, despite instructions to the contrary from leaders and published manuals, and despite the burden of reporting SMART data up the hierarchy, which has to be creatively invented and explained so that the practices aren't disrupted by new orders from the leaders.

> • Judgement: Through the above improved learning, conversations and practice, people develop good judgement. They make better decisions. The leaders get all the credit for these decision, but it doesn't matter.
>
> • Trust Relationships: Through peer-to-peer conversations, trust relationships develop. When people trust each other, whole layers of bureaucracy are stripped away. People are left to do what they do well. Unfortunately leaders in large organizations almost never trust their subordinates, so these trust relationships are almost always horizontal, not vertical. Despite this, these relationships profoundly improve productivity.
>
> • Professionalism: The net result of all of the above is increased professionalism. People just become more competent.
>
> This is why, in all my years as a manager, I always saw my role as listening and clearing away obstacles my staff were facing, identifying and getting rid of the small percentage who could not be trusted (too ambitious, too self-serving, uncollaborative, secretive or careless), and trusting the rest to do what they do best, and staying out of their way. In recent years I started to lose the heart to do this, but I still tried.
>
> The ideal organization is therefore not SMART, but self-organized, trusting (no need to measure results, just practice your craft and the results will inevitably be good), highly conversational, and ultimately collaborative (impossible in large organizations because performance is measured individually not collectively). It's one where the non-performers are collectively identified by their peers and self-select out by sheer peer pressure. It's one without hierarchy. It's agile, resilient and improvisational, because it runs on principles, not rules, and because when issues arise they're dealt with by the self-organized group immediately, not shelved until someone brings them to the attention of the 'leaders'. It's designed for complexity. It's organic, natural.
>
> In my experience, such an organizational model can be replicated, but it doesn't scale.[1]

Eric Raymond sees the phase transition between forms of social organization as a response to insupportable complexity. The professionalized meritocracies that managed the centralized state and large corporation through the middle of the 20th century were an attempt to manage complexity by applying Weberian and Taylorist rules. And they did a passable job of managing the system competently for most of that time, he says. But in recent years we've reached a level of complexity beyond their capacity to deal with.

> The "educated classes" are adrift, lurching from blunder to blunder in a world that has out-complexified their ability to impose a unifying narrative on it, or even a small collection of rival but commensurable narratives. They're in the exact position of old Soviet central planners, systemically locked into grinding out products nobody wants to buy.

The answer, under these conditions, is to "[a]dapt, decentralize, and harden"—i.e., to reconfigure the system along the stigmergic lines he described earlier in "The Cathedral and the Bazaar":

> Levels of environmental complexity that defeat planning are readily handled by complex adaptive systems. A CAS doesn't try to plan against the future; instead, the agents in it try lots of adaptive strategies and the successful ones propagate. This is true whether the CAS we're speaking of is a human immune system, a free market, or an ecology.
>
> Since we can no longer count on being able to plan, we must adapt. When planning doesn't work, centralization of authority is at best useless and usually

[1] David Pollard, "Replicating (Instead of Growing) Natural Small Organizations," *how to save the world*, January 14, 2009 <http://howtosavetheworld.ca/2010/01/14/not-so-smart-replicating-instead-of-growing-natural-small-organizations/>.

> harmful. And we must harden: that is, we need to build robustness and the capacity to self-heal and self-defend at every level of the system. I think the rising popular sense of this accounts for the prepper phenomenon. Unlike old-school survivalists, the preppers aren't gearing up for apocalypse; they're hedging against the sort of relatively transient failures in the power grid, food distribution, and even civil order that we can expect during the lag time between planning failures and CAS responses.
>
> CAS hardening of the financial system is, comparatively speaking, much easier. Almost trivial, actually. About all it requires is that we re-stigmatize the carrying of debt at more than a very small proportion of assets. By anybody. With that pressure, there would tend to be enough reserve at all levels of the financial system that it would avoid cascade failures in response to unpredictable shocks.
>
> Cycling back to terrorism, the elite planner's response to threats like underwear bombs is to build elaborate but increasingly brittle security systems in which airline passengers are involved only as victims. The CAS response would be to arm the passengers, concentrate on fielding bomb-sniffers so cheap that hundreds of thousands of civilians can carry one, and pay bounties on dead terrorists.[1]

Compared to the stigmergic organization, a bureaucratic hierarchy is systematically stupid. This was the subject of a recent debate between Roderick Long and Bryan Caplan. Here's what Long wrote:

> Rand describes a "pyramid of ability" operating within capitalism, wherein the dull masses are carried along by the intelligent and enterprising few. "The man at the top," Rand assures us, "contributes the most to all those below him," while the "man at the bottom who, left to himself, would starve in his hopeless ineptitude, contributes nothing to those above him, but receives the bonus of all of their brains." Rand doesn't say that the top and the bottom always correspond to employers and employees respectively, but she clearly takes that to be the usual situation. And that simply does *not* correspond with the reality of most people's everyday experience.
>
> If you've spent any time at all in the business world, you've almost certainly discovered that the reality on the ground resembles the comic-strip Dilbert a lot more than it resembles Rand's pyramid of ability. In Kevin Carson's words: as in government, so likewise in business, the "people who regulate what you do, in most cases, know less about what you're doing than you do," and businesses generally get things done only to the extent that "rules imposed by people not directly involved in the situation" are treated as "an obstacle to be routed around by the people actually doing the work." To a considerable extent, then, in the real world we see the people at the "bottom" carrying the people at the "top" rather than vice versa.[2]

Caplan, in challenging this assessment, missed the point. He treated Long's critique as an attack on the intelligence of the average manager:

> But what about the "tons of empirical evidence" that Rand's pyramid of ability is real? The Bell Curve is a good place to start. Intelligence is one of the strongest—if not the strongest—predictors of income, occupation, and social status. More to the point, simple pencil-and-paper tests of intelligence are the single best predictor of independently measured job performance and trainability. If you want to dig deeper, check out the large literature on why income runs in families.

[1]Eric Raymond, "Escalating Complexity and the Collapse of Elite Authority," *Armed and Dangerous*, January 5, 2010 <http://esr.ibiblio.org/?p=1551>.

[2]Roderick Long, "The Winnowing of Ayn Rand," *Cato Unbound*, January 20, 2010 <http://www.cato-unbound.org/2010/01/20/roderick-long/the-winnowing-of-ayn-rand/>.

> How then can we reconcile first-hand observation with economic theory and statistical fact? It's easier than it seems. Lots of people think their bosses are stupid because:
>
> 1. The market doesn't measure merit perfectly, so success is partly luck. As a result, some bosses are unimpressive. (Though almost all of them are smarter than the average rank-and-file worker).
>
> 2. There's a big contrast effect: If you expect bosses to be in the 99th percentile of ability, but they're only in the 90th, it's natural to misperceive them as "stupid." (Similarly, if someone scores in the 99th percentile on the SAT in math, and the 80th in English, many people will perceive him as "terrible in English.")
>
> 3. Bosses are much more visible than regular workers, so their flaws and mistakes—even if minor—are quickly noticed. When normal people screw up, there's usually no one paying attention.
>
> 4. Perhaps most importantly, people over-rate themselves. We like to imagine that we're so great that we intellectually tower over our so-called "superiors." Only a small percentage of us are right.
>
> If Rod Long's point is merely that markets would be even more meritocratic under laissez-faire, I agree. But to deny that actually-existing capitalism is highly meriocratic is misguided. To suggest that the pyramid of ability is actually inverted is just silly.[1]

But the point, as I argued with Caplan, is not that managers are inherently less intelligent or capable as individuals. Rather, it's that hierarchical organizations are—to borrow that wonderful phrase from Feldman and March—*systematically* stupid. For all the same Hayekian reasons that make a planned economy unsustainable, *no* individual is "smart" enough to manage a large, hierarchical organization. *Nobody*–not Einstein, not John Galt–possesses the qualities to make a bureaucratic hierarchy function rationally. Nobody's that smart, any more than anybody's smart enough to run Gosplan efficiently–that's the whole point. No matter how insightful and resourceful they are, no matter how prudent, as human beings in dealing with actual reality, nevertheless by their very nature hierarchies insulate those at the top from the reality of what's going on below, and *force* them to operate in imaginary worlds where all their intelligence becomes useless. No matter how intelligent managers are as individuals, a bureaucratic hierarchy makes their intelligence less *usable*.

In the case of network organization, just the opposite is the case: networked, stigmergic organization promotes *maximum* usability of intelligence.

The fundamental reason for agility, in a self-managed peer network, is the lack of a bureaucratic hierarchy separating the worker from the end-user. The main metric of quality is direct end-user feedback. And in a self-managed peer network, "employee education" follows directly from what workers actually learn by doing their jobs.

In a corporate hierarchy, in contrast, most quality metrics are developed to inform bureaucratic intermediaries who are neither providers nor end-users of the company's services.

And, much like management metrics of quality, their metrics of employee skill and competence are utterly divorced from reality. At just about every job where I've ever worked, for example, "employee education" credits were utterly worthless busy work that had nothing to do with what I actually did.

Steve Herrick, commenting under a blog post of mine, confirmed my impression of the (lack of) value of most "in-service meetings" and "employee education hours," based on his own experience working in hospitals:

[1]Bryan Caplan, "Pyramid Power," *EconLog*, January 21, 2010 <http://econlog.econlib.org/archives/2010/01/pyramid_power.html>.

> . . . I work as a medical interpreter. According to the rules, I can't touch patients (let alone provide care) or computers. However, according to other rules, I have [to] pass tests on sharps disposal, pathogen transmission, proper use of portable computers, etc.[1]

Such nonsense results, of necessity, from a situation in which a bureaucratic hierarchy must develop some metric for assessing the skills or work quality of a labor force whose actual work they know nothing about. When management doesn't know (in Paul Goodman's words) "what a good job of work is," they are forced to rely on arbitrary metrics. Blogger Atrios describes his experience with the phenomenon.

> During my summers doing temp office work I was always astounded by the culture of "face time"—the need to be at your desk early and stay late even when there was no work to be done and doing so in no way furthered any company goals. Doing your work and doing it adequately was entirely secondary to looking like you were working hard as demonstrated by your desire to stay at work longer than strictly necessary.[2]

One of his commenters, in considerably more pointed language, added: "If you are a manager who is too stupid to figure out that what you should actually measure is real output then the next best thing is to measure how much time people spend pretending to produce that output." But in fairness, again, establishing a satisfactory measure of real output that can convey information to those outside the production process, without being gamed by those engaged in the process, in a situation where the interests of the two diverge, is a lot easier said than done.

Most of the constantly rising burden of paperwork exists to give an illusion of transparency and control to a bureaucracy that is out of touch with the actual production process. Most new paperwork is added to compensate for the fact that existing paperwork reflects poorly designed metrics that poorly convey the information they're supposed to measure. "If we can only design the perfect form, we'll finally know what's going on."

Weberian work rules result of necessity when performance and quality metrics are not tied to direct feedback from the work process itself. It is a metric *of* work *for* someone who is neither a creator/provider not an end user.

In a self-managed process, if we may recur to the terminology of James Scott cited in the previous chapter, work quality is horizontally legible to those directly engaged in it. In a hierarchy, managers are forced to see "in a glass darkly" a process which is necessarily opaque to them because they are not directly engaged in it. They are forced to carry out the impossible task of developing accurate metrics for evaluating the behavior of subordinates, based on the self-reporting of people with whom they have a fundamental conflict of interest. All of the paperwork burden that management imposes on workers reflects an attempt to render legible a set of social relationships that by its nature must be opaque and closed to them, because they are outside of it. Each new form is intended to remedy the heretofore imperfect self-reporting of subordinates. The need for new paperwork is predicated on the assumption that compliance must be verified because those being monitored have a fundamental conflict of interest with those making the policy, and hence cannot be trusted; but at the same time, that paperwork relies on their self-reporting as the main source of information. Every time new evidence is presented

[1]Comment under Carson, "The People Making 'The Rules' are Dumber than You," Center for a Stateless Society, January 11, 2010 <http://c4ss.org/content/1687>.

[2]Atrios, "Face Time," *Eschaton*, July 9, 2005 <http://atrios.blogspot.com/2005_07-03_atrios_archive.html>.

that this or that task isn't being performed to management's satisfaction, or this or that policy isn't being followed, despite the existing reams of paperwork, management's response is to design yet *another* form. "If you don't trust me to do the job right without filling out all these forms, why do you trust me to fill out the forms truthfully?"

The difficulties are inherent in the agency problem. Human agency is inalienable. When someone agrees to work under someone else's direction for a period of time, the situation is comparable to selling a car but remaining in the driver's seat. There is no magical set of compliance paperwork or quality/performance metrics that will enable management to sit in the driver's seat of the worker's consciousness, to exercise direct control over his hands, or to look out through his eyes.

The only solution is to build incentives into the work itself, and into the direct relationships between the worker and customer, so that it is legible to them It is necessary to create a situation in which creators/providers and end-users are the only parties directly involved in the provision of goods and services, so that metrics of quality are *for* them as well as *of* them. Michel Bauwens writes:

> The capacity to cooperate is verified in the process of cooperation itself. Thus, projects are open to all comers provided they have the necessary skills to contribute to a project. These skills are verified, and communally validated, in the process of production itself. This is apparent in open publishing projects such as citizen journalism: anyone can post and anyone can verify the veracity of the articles. Reputation systems are used for communal validation. The filtering is a posteriori, not a priori. Anti-credentialism is therefore to be contrasted to traditional peer review, where credentials are an essential prerequisite to participate.
>
> P2P projects are characterized by holoptism. Holoptism is the implied capacity and design of peer to [peer] processes that allows participants free access to all the information about the other participants; not in terms of privacy, but in terms of their existence and contributions (i.e. horizontal information) and access to the aims, metrics and documentation of the project as a whole (i.e. the vertical dimension). This can be contrasted to the panoptism which is characteristic of hierarchical projects: processes are designed to reserve 'total' knowledge for an elite, while participants only have access on a 'need to know' basis. However, with P2P projects, communication is not top-down and based on strictly defined reporting rules, but feedback is systemic, integrated in the protocol of the cooperative system.[1]

When you make a sandwich for yourself, or for a member of your family, you don't need a third-party inspection regime to guarantee that the sandwich is up to snuff, because there is a fundamental unity of interest between you as sandwich maker and sandwich eater, or between you and the person you're making food for. And if the quality of the sandwich is substandard, you or your family know it because it tastes bad when they take a bite of it. In other words, the process is run directly for the benefit of those engaged in it, and the quality feedback is built directly into the process itself.

It's only when people are engaged in work with no intrinsic value or meaning to themselves, with which they don't identify, which they don't control, and which is for the benefit of people whose interests are fundamentally opposed to their own, that a complicated system of compliance and quality metrics are required to vouch for its quality to third parties removed from the immediate situation. And in such circumstances, because the managerial hierarchy lacks the job-related tacit

[1]Michel Bauwens, "The Political Economy of Peer Production," *Ctheory.net*, December 1, 2005 <http://www.ctheory.net/articles.aspx?id=499>.

knowledge required to formulate meaningful metrics or evaluate incoming data, the function of the metrics and data is at best largely symbolic: e.g., elaborate exercises in shining it on, like JCAHO inspections and ISO-9000. At worst, they *reduce* quality when people who don't understand the work interfere with those who do. So you wind up with a 300-page manual for making the sandwich, along with numerous other 300-page manuals for vendor specifications—and it still tastes like crap.

A classic example of the counterproductivity of using bureaucratic rules to obstruct the initiative of those directly involved in a situation is the story of a train fire which was widely circulated on the Internet (which, according to Snopes.Com, it turns out was legitimate). A faulty bearing caused a wheel on one of the cars to overheat and melt down. The crew, spotting the smoke, stopped the train in compliance with the rules. Unfortunately, it came to rest on a wooden bridge with creosote ties. Still more unfortunately, the management geniuses directing the crew from afar refused to budge on the rules, which prohibited moving the train. As a result, the bridge burned and six burning coal cars dropped into the creek below.[1]

The same principle was illustrated by an anecdote from the Soviet Great Patriotic War (I'm afraid I can't track down the original source, but it's too good a story not to relate). A division commander was denied permission to pull his divisional artillery back far enough to be in effective range of a road, and thus to be able to target German armor moving along that road, because he couldn't convince the political officer that backward movement didn't constitute "retreat."

And then there's the old saw about how the Egyptians lost the 1967 Arab-Israeli War because they literally obeyed the instructions in their Russian field manuals: "retreat into the interior and wait for the first snowfall."

Rigid hierarchies and rigid work rules only work in a predictable environment. When the environment is unpredictable, the key to success lies with empowerment and autonomy for those in direct contact with the situation. A good example is the Transportation Safety Administration's response to the threat of Al Qaeda attacks. As Matthew Yglesias has argued, "the key point about identifying al-Qaeda operatives is that there are extremely few al-Qaeda operatives so (by Bayes' theorem) any method you employ of identifying al-Qaeda operatives is going to mostly reveal false positives."[2] So (this is me talking) when your system for anticipating attacks upstream is virtually worthless, the "last mile" becomes monumentally important: having people downstream capable of recognizing and thwarting the attempt, and with the freedom to use their own discretion in stopping it, when it is actually made.

An almost universal problem, when bureaucratic, stovepiped industrial design processes isolate designers from user feedback, is the "gold plated turd." Whenever a product is designed by one bureaucracy, for sale to procurement officers in another bureaucracy who are buying it for someone else's use, a gold-plated turd is almost invariably the result.

A good example from my experience as a hospital worker is the kind of toilet paper dispenser sold to large institutional clients. If you've ever used a public restroom or patient restroom in a hospital, you've almost certainly encountered one of those Georgia-Pacific monstrosities: a plastic housing that makes it almost impos-

[1]"A Bridge Too Far: Train Sets Bridge on Fire," Snopes.Com <http://www.snopes.com/photos/accident/trainfire.asp>.

[2]Matthew Yglesias, "Too Much Information," *Matthew Yglesias*, December 28, 2009 <http://yglesias.thinkprogress.org/ archives/2009/12/too-much-information.php>.

sible to manipulate the roll without breaking your wrist, and so much resistance that you tear the paper rather than turning the spool more often than not. And these toilet paper dispensers, seemingly engineered at great effort to perform their functions as badly as possible, sell for $20 or more. On the other hand, an ordinary toilet paper spool—one that actually turns easily and is convenient to use—can probably be bought at Lowe's or Home Depot for a dollar.

I've had similar experiences as a consumer of goods and services, outside of my job. A good example is my experience with the IT officer at the local public library, which I described earlier in the book. I emailed the library on how poorly the newly installed Word 2007 software, and whatever Windows desktop upgrade they'd bought, performed compared to the earlier version of Windows and the Word 2003 they replaced. As Windows products go, Word 2003 is about the best word processing software you can get. It's got a user interface pretty much the same as that of Open Office, in terms of complexity. In fact, I'd go so far as to say it was as good as Open Office, aside from the $200 price tag and the forced upgrades that open source software is mercifully free of. Word 2007, on the other hand, is a classic gold-plated turd. Its user interface is so complicated and busy that the dashboard actually has to be tabbed to accommodate all the bells and whistles. I told the IT officer that it was a good idea, whenever she found a Windows product that worked acceptably, to hold onto it like grim death and run like hell when offered anything "new and improved" from Redmond. Her response: Word 2007 is the standard "productivity software" choice of major public libraries and corporations all across America. In my follow-up, I told her the very fact that something worked worse than what it replaced, despite being the "standard choice" of pointy-haired bosses all across the country, was an object lesson in the wisdom of basing one's software choice on corporate bureaucrats' "best practices" rather than on feedback from user communities. Never heard back from her, for some reason. Nice lady, though.

Niall Cook, in *Enterprise 2.0*, describes the comparative efficiencies of social software outside the enterprise to the "enterprise software" in common use by employers. Self-managed peer networks, and individuals meeting their own needs in the outside economy, organize their efforts through social software chosen by the users themselves based on its superior usability for their purposes. And they are free to do so without corporate bureaucracies and their officially defined procedural rules acting as a ball and chain. Enterprise software, in contrast, is chosen by non-users for use by other people of whose needs they know little (at best). Hence enterprise software is frequently a gold-plated turd. Blogs and wikis, and the free, browser-based platforms offered by Google and Mozilla, are a quantum improvement on the proprietary enterprise software that management typically forces on its employees. The kinds of productivity software and social software freely available to individuals in their private lives is far better than the enterprise software that corporate bureaucrats buy for a captive clientele of users—consumer software capabilities amount to "a fully functioning, alternative IT department."[1] Corporate IT departments, in contrast, "prefer to invest in a suite of tools 'offered by a major incumbent vendor like Microsoft or IBM'." System specs are driven by management's top-down requirements rather than by user needs.

> . . . a small group of people at the top of the organization identify a problem, spend 12 months identifying and implementing a solution, and a huge amount of

[1]Niall Cook, *Enterprise 2.0: How Social Software Will Change the Future of Work* (Burlington, Vt.: Gower, 2008), p. 91.

resources launching it, only then to find that employees don't or won't use it because they don't buy in to the original problem.[1]

Management is inclined "to conduct a detailed requirements analysis with the gestation period of an elephant simply in order to chose a $1,000 social software application."[2] Employees often wind up using their company credit cards to purchase needed tools online rather than "wait for [the] IT department to build a business case and secure funding."[3] This is the direct opposite of agility.

As a result of all this, people are more productive away from work than they are at work.

Corporate IT departments are a lot like the IT department at my public library, as recounted above. They are obsessed with security and control, and see the free exchange of information between employees as a threat to that security and control. They also have an affinity for doing business with other bureaucracies like themselves, which means a preference for buying proprietary enterprise software from giant corporations. They select software on pretty much the same basis as a Grandma buying a gift for her granddaughter just entering college: "I just knew it had to be the best, dear, because it's the latest thing from Microsoft!"

Nascent "Enterprise 2.0" organization within a traditional firm is often forced to fight obstruction from top-down management styles, even in areas where human capital is the main source of value. With corporate cultures based on obsession with security and control, management instinctively fights workers' attempts to choose their own platforms based on usability. Attempts to facilitate information sharing between employees falls afoul of this culture, because employees obviously wouldn't desire access to information unless they were up to no good. On the outside, peer networks are free to self-organize without interference from hierarchy. As a result, in forms of production where the main source of value is human capital, and human relationships for sharing knowledge, autonomous outside peer networks have a leg up on corporate hierarchies.

The parallels between Enterprise 2.0 and the military's doctrines for Fourth Generation Warfare are striking. The military's Fourth Generation Warfare doctrines are an attempt to take advantage of network communications technology and cybernetic information processing capabilities in order replicate, within a conventional military force, the agility and resilience of networked organizations like Al Qaeda. The problem, as we saw earlier in this chapter, is that interference from the military's old bureaucratic hierarchies systematically impede all the possibilities offered by network technology. The basic idea behind the new doctrines is, through the use of networked communications technology, to increase the autonomy and reduce the reaction time of the "boots on the ground" directly engaged in a situation. But as John Robb suggested, military hierarchies wind up seeing the new communications technologies instead as a way of increasing mid-level commanders' realtime control over operations, and increasing the number of sign-offs required to approve any proposed operation. By the time those engaged in combat operations get the required eleven approvals of higher-ups, and the staff officers have had time to process the information into some kind of unrecognizable scrapple (PowerPoint presentations and all), the immediate situation has changed to the point that their original plan is meaningless anyway.

[1]*Ibid.*, p. 93.
[2]*Ibid.*, p. 95.
[3]*Ibid.*, p. 96.

So the real thing—genuinely independent, self-managed networked resistance movements unimpeded by bureaucratic interference with the natural feedback and reaction mechanisms of a stigmergic organization—is incomparably better than the military hierarchy's pallid imitations.

Similarly, Enterprise 2.0 is an attempt to replicate, within the boundaries of a corporation, the kinds of networked, stigmergic organization that Raymond wrote about in "The Cathedral and the Bazaar." But networked producers inside the corporation find themselves thwarted, at every hand, by bureaucratic impediments to their putting immediately into practice their own judgment of what's necessary based on direct experience of the situation.

What actually happens, when management attempts to "empower" employees by adopting a networked organization within corporate boundaries, is suggested by an anecdote from an HR blog. Management came up with a brilliant idea for reducing the number of round-robin emails selling extra concert tickets and used cars, soliciting rides, etc.: to put an official bulletin board at one convenient central location! But rather than simply mounting a square of corkboard and leaving employees to their own devices in posting notices, management had to come up with an official procedure for advance submission of notices for approval, followed—a week later, if they were lucky and the notice was successfully vetted for all conceivable violations of company policy—by a manager unlocking the glass case with his magic set of keys and posting the ad. Believe it or not, management was puzzled as to why the round-robin emails continued and the bulletin board wasn't more popular.[1]

This sort of thing is the currency of one school of organization theorists, who as Charles Sabel describes them, assert that

> So bounded is the rationality of organizations that they are incapable of learning in the sense of improving decisions by deliberation on experience. Thus the assumption that decision makers 'survey' only the first feasible choices immediately accessible to them at the moment of decision, and 'prefer' that choice to any other or inaction, yields 'garbage-can' models of organizations, in which decisions result from collisions between decision makers and solutions. . . . The assumption that decision makers can compare only a few current solutions to their problem, and prefer the one that best meets their needs, but cannot draw from this decision any analytic conclusions regarding subsequent choices, turns organized decision making into muddling through. . . . [2]

To take just one example: Martha Feldman and James March found little relationship between the gathering of information and the policies that were ostensibly based on it. In corporate legitimizing rhetoric, of course, management decisions are always based on a rational assessment of the best available information.[3] They did case studies of three organizations, and found an almost total disconnect between policies and the information they were supposedly based on.

Feldman and March did their best to provide a charitable explanation—an explanation, that is, other than "organizations are systematically stupid."[4] "Systematically stupid" probably comes closest to satisfying Occam's Razor, and I'd have

[1]Chloe, "Important People," *Corporate Whore*, September 21, 2007 <http://web.archive.org/web/20071014221728/http://corporatewhore.us/important-people/>.

[2]Charles F. Sabel, "A Real-Time Revolution in Routines," in Charles Hecksher and Paul S. Adler, *The Firm as a Collaborative Community: Reconstructing Trust in the Knowledge Economy* (New York: Oxford University Press, 2006), pp. 110-111.

[3]Martha S. Feldman and James G. March, "Information in Organizations as Signal and Symbol," *Administrative Science Quarterly* 26 (April 1981).

[4]*Ibid.*, p. 174.

happily stuck with that explanation. But Feldman and March struggled to find some adaptive purpose in the observed use of information.

The interesting thing, from my perspective, is that most of the "adaptive purposes" they describe reflect *precisely* what I'd call "systematic stupidity." They began by surveying more conventional assessments of organizational inefficiency as an explanation for the observed pattern. First, organizations are "unable . . . to process the information they have. They experience an explanation glut as a shortage. Indeed, it is possible that the overload contributes to the breakdown in processing capabilities. . . . " Second, " . . . the information available to organizations is systematically the wrong kind of information. Limits of analytical skill or coordination lead decision makers to collect information that cannot be used."[1]

Then they made three observations of their own on how organizational structure affects the use of information:

> First, ordinary organizational procedures provide positive incentives for underestimating the costs of information relative to its benefits. Second, much of the information in an organization is gathered in a surveillance mode rather than in a decision mode. Third, much of the information used in organizational life is subject to strategic misrepresentations.
>
> Organizations provide incentives for gathering more information than is optimal from a strict decision perspective. . . . First, the costs and benefits of information are not all incurred at the same place in the organization. Decisions about information are often made in parts of the organization that can transfer the costs to other parts of the organization while retaining the benefits. . . .
>
> Second, post hoc accountability is often required of both individual decision makers and organizations . . .
>
> Most information that is generated and processed in an organization is subject to misrepresentation. . . .

The decision maker, in other words, must gather excess information in anticipated defense against the possibility that his decision will be second-guessed.[2] By "surveillance mode," the authors mean that the organization seeks out information not for any specific decision, but rather to monitor the environment for surprises. The lead time for information gathering is longer than the lead time for decisions. Information must therefore be gathered and processed without clear regard to the specific decisions that may be made.[3]

All the incentives mentioned so far seem to result mainly from large size and hierarchy—i.e., to result (again) from "systematic stupidity." The problem of non-internalization of the costs and benefits of information-gathering by the same actor, of course, falls into the inefficiency costs of large size. The problem of post hoc accountability results from hierarchy. At least part of the problem of surveillance mode is another example of poor internalization: the people gathering the information are different from the ones using it, and are therefore gathering it with a second-hand set of goals which does not coincide with their own intrinsic motives. The strategic distortion of information, as an agency problem, is (again) the result of hierarchy and the poor internalization of costs and benefits in the same responsible actors. In other words, the large, hierarchical organization is "systematically stupid."

The authors' most significant contribution in this article is their fourth observation: that the gathering of information serves a legitimizing function in the organization.

[1] *Ibid.*, p. 175.
[2] *Ibid.*, pp. 175-176.
[3] *Ibid.*, p. 176.

Bureaucratic organizations are edifices built on ideas of rationality. The cornerstones of rationality are values regarding decision making. . . .

The gathering of information provides a ritualistic assurance that appropriate attitudes about decision making exist. Within such a scenario of performance, information is not simply a basis for action. It is a representation of competence and a reaffirmation of social virtue. Command of information and information sources enhances perceived competence and inspires confidence. The belief that more information characterizes better decisions engenders a belief that having information, in itself, is good and that a person or organization with more information is better than a person or organization with less. Thus the gathering and use of information in an organization is part of the performance of a decision maker or an organization trying to make decisions intelligently in a situation in which the verification of intelligence is heavily procedural and normative. . . .

Observable features of information use become particularly important in this scenario. When there is no reliable alternative for asserting a decision maker's knowledge, visible aspects of information gathering and storage are used as implicit measures of the quality and quantity of information possessed and used. . . .[1]

In other words, when an organization gets too big to have any clear idea how well it is performing the function for which it officially exists, it creates a metric for "success" defined—as we saw in our study of Sloanist organizational pathologies—in terms of the processing of inputs.

But in fairness to management, it's not the stupidity of the individual; to repeat my point above contra Caplan, it's the stupidity of the organization. Large, hierarchical organizations are *systematically* stupid, regardless of how intelligent and competent the people running them are. By definition, *nobody* is smart enough to run a large, hierarchical organization, just as nobody's smart enough to centrally plan an economy.

The reality of corporate life is apt to bear a depressing resemblance to the Ministry of Central Services in *Brazil*, or to "The Feds" in Neal Stephenson's *Snow Crash*. "The Feds" in the latter example are the direct successor to the United States government, claiming continued sovereign jurisdiction over the territory of the former U.S., but in fact functioning as simply one of many competing franchise "governments" or networked civil societies in the panarchy that exists following the collapse of most territorial states. Mainly occupying the federal office buildings on what used to be federal property, its primary activity is designing enterprise software for sale to corporations. Its internal governance seems to reflect, in equal parts, the bureaucratic world of *Brazil* and the typical IT department's idealized vision of a corporate intranet (not that there's much difference).

One employee of the Feds shows up for work and logs on, after negotiating the endless series of biometric scans, only to receive a long and excruciatingly detailed memo on the policies governing the unauthorized bringing in of toilet paper from home, sparked by toilet paper shortages in the latest austerity drive.

The memo includes an announcement that "Estimated reading time for this document is 15.62 minutes (and don't think we won't check)." Her supervisor's standard template, in checking up on memo reading times, is something like this:

Less than 10 min. Time for an employee conference and possible attitude counseling.

10-14 min. Keep an eye on this employee; may be developing slipshod attitude.

[1] *Ibid.*, pp. 177-178.

> 14-15.61 min. Employee is an efficient worker, may sometimes miss important details.
>
> Exactly 15.62 min. Smartass. Needs attitude counseling.
>
> 15.63-16 min. Asswipe. Not to be trusted.
>
> 16-18 min. Employee is a methodical worker, may sometimes get hung up on minor details.
>
> More than 18 min. Check the security videotape, see just what this employee was up to (e.g., possible unauthorized restroom break).

The employee decides, accordingly, to spend between fourteen and fifteen minutes reading the memo. "It's better for younger workers to spend too long, to show that they're careful, not cocky. It's better for older workers to go a little fast, to show good management potential."

Their actual work is similarly micromanaged:

> She is an applications programmer for the Feds. In the old days, she would have written computer programs for a living. Nowadays, she writes fragments of computer programs. These programs are designed by Marietta and Marietta's superiors in massive week-long meetings on the top floor. Once they get the design down, they start breaking up the problem into tinier and tinier segments, assigning them to group managers, who break them down even more and feed little bits of work to the individual programmers. In order to keep the work done by the individual coders from colliding, it all has to be done according to a set of rules and regulations even bigger and more fluid than the Government procedure manual [even bigger than the rules for reading a toilet paper memo?].
>
> So the first thing [she] does, having read the new subchapter on bathroom tissue pools, is to sign on to a subsystem of the main computer system that handles the particular programming project she's working on. She doesn't know what the project is—that's classified—or what it's called. She shares it with a few hundred other programmers, she's not sure exactly who. And every day when she signs on to it, there's a stack of memos waiting for her, containing new regulations and changes to the rules that they all have to follow when writing code for the project. These regulations make the business with the bathroom tissue seem as simple and elegant as the Ten Commandments.
>
> So she spends until about eleven A.M. reading, rereading, and understanding the new changes in the Project [presumably with recommended reading times, carefully monitored, for each one]. . . .
>
> Then she starts going back over all the code she has previously written for the Project and making a list of all the stuff that will have to be rewritten in order to make it compatible with the new specifications. Basically, she's going to have to rewrite all of her material from the ground up. For the third time in as many months.

But hey, it's a job.[1]

If you think that's a joke, go back and reread the material in the last section on the rules governing PowerPoint presentations in the U.S. military command in Afghanistan.

E. The Implications of Reduced Physical Capital Costs

The informal and household economy reduces waste by its reliance on "spare cycles" of ordinary capital goods that most people already own. It makes productive use of idle capital assets the average person owns anyway, provides a productive outlet for the surplus labor of the unemployed, and transforms the small surpluses of household production into a ready source of exchange value.

[1]Neal Stephenson, *Snow Crash* (Westminster, Md.: Bantam Dell Pub Group, 2000).

Let's consider again our example of the home-based microenterprise—the microbrewery or restaurant—from Chapter Five. Buying a brewing vat and a few small fermenters for your basement, using a few tables in an extra room as a public restaurant area, etc., would require a small bank loan for at most a few thousand dollars. And with that capital outlay, you could probably make payments on the debt with the margin from one customer a day. A few customers evenings and weekends, probably found mainly among your existing circle of acquaintances, would enable you to initially shift some of your working hours from wage labor to work in the restaurant, with the possibility of gradually phasing out wage labor altogether or scaling back to part time, as you built up a customer base. In this and many other lines of business (for example a part-time gypsy cab service using a car and cell phone you own anyway), the minimal entry costs and capital outlay mean that the minimum turnover required to pay the overhead and stay in business would be quite modest. In that case, a lot more people would be able to start small businesses for supplementary income and incrementally shift some of their wage work to self employment, with minimal risk or sunk costs.

The lower the initial capital outlays, and the lower the resulting overhead that must be serviced, the larger the percentage of its income stream belongs to the microenterprise without encumbrance—regardless of how much business it is able to do. It is under no pressure to "go big or not go at all," to "get big or get out," or to engage in large batch production to minimize unit costs from overhead, because it has virtually no overhead costs. So the microenterprise can ride out prolonged periods of slow business. If the microenterprise is based in a household which owns its living space free and clear and has a garden and well-stocked pantry, the household may be able to afford to go without income during slow spells and live off its savings from busy periods. Even if the household is dependent on some wage labor, the microenterprise in good times can be used as a supplemental source of income with no real cost or risk of the kind that would exist were there overhead to be serviced, and therefore enable a smaller wage income to go further in a household income-pooling unit.

That's why, as we saw in Chapter Two, one of the central functions of so-called "health" and "safety" codes, and occupational licensing is to prevent people from using idle capacity (or "spare cycles") of what they already own anyway, and thereby transforming them into capital goods for productive use. In general, state regulatory measures that increase the minimum level of overhead needed to engage in production will increase the rate of failure for small businesses, with pressure to intensified "cutthroat competition." In the specific case of high burdens of interest-bearing debt, and the pressure to earn a sufficient revenue stream to repay the interest as well as the principal, Tom Greco writes,

> As borrowers compete with one another to try to meet their debt obligations in this game of financial "musical chairs," they are forced to expand their production, sales, and profits. . . .
>
> . . . Thus, debt continually mounts up, and businesses and individuals are forced to compete for markets and scarce money in a futile attempt to avoid defaulting on their debts. The system makes it certain that some *must* fail. [1]

Because the household economy and the microenterprise require few or no capital outlays, their burden of overhead is miniscule. This removes the pressure to large-batch production. It removes the pressure to get out of business altogether

[1]Thomas Greco, *The End of Money and the Future of Civilization* (White River Junction, Vt.: Chelsea Green Publishing, 2009), p. 55.

and liquidate one's assets when business is slow, because there is no overhead to service. Reduced overhead costs reduce the failure rate; they reduce the cost of staying in business indefinitely, enjoying revenue free and clear in good periods and riding out slow ones with virtually no loss. As Borsodi wrote,

> Only in the home can the owner of a machine afford the luxury of using it only when he has need of it. The housewife uses her washing machine only an hour or two per week. The laundry has to operate its washing machine continuously. Whether operating or not operating all of its machines, the factory has to earn enough to cover depreciation and obsolescence on them. Office overhead, too, must be earned, whether the factory operates on full time or only on part time.[1]

And a housewife who uses her washing machine to full capacity in a household micro-laundry, with no additional marginal cost besides the price of soap, water, and power, will eat the commercial laundry alive.

F. Strong Incentives and Reduced Agency Costs

We already saw, above, Eric Raymond's description of how self-selection and incentives work in the Linux "Bazaar" model of open-source development. As Michel Bauwens put it,

> the permissionless self-aggregation afforded by the internet, allowed humans to congregate around their passionate pursuits. . . . It was discovered that when people are motivated by intrinsic positive motivation, they are hyperproductive. . . .
>
> . . . [W]hile barely one in five of corporate workers are passionately motivated, one hundred percent of peer producers are, since the system filters out those lacking it![2]

And Johan Soderberg, likewise:

> To a hired programmer, the code he is writing is a means to get a pay check at the end of the month. Any shortcut when getting to the end of the month will do. For a hacker, on the other hand, writing code is an end in itself. He will always pay full attention to his endeavour, or else he will be doing something else.[3]

The alternative economy reduces waste by eliminating all the waste of time involved in the "face time" paradigm. Wage labor and hierarchy are characterized by high degrees of "presenteeism." Because the management is so divorced from the actual production process, it has insufficient knowledge of the work to develop a reliable metric of actual work accomplished. So it is required to rely on proxies for work accomplished, like the amount of time spent in the office and whether people "look busy." Workers, who have no intrinsic interest in the work and who get paid for just being there, have no incentive to use their time efficiently.

Matthew Yglesias describes this as "the office illusion": the equation of "being in the office" to "working."

> Thus, minor questions like *am I getting any work done?* can tend to slip away. Similarly, when I came into an office every day, I felt like I couldn't just leave the office just because I didn't want to do anymore work, so I would kind of foot-drag on things to make sure whatever task I had stretched out to fill the entire working day. If I'm not in an office, by contrast, I'm acutely aware that I have a

[1]Borsodi, *This Ugly Civilization*, p. 126.

[2]Michel Bauwens, "The three revolutions in human productivity," *P2P Foundation Blog*, November 29, 2009 <http://blog.p2pfoundation.net/the-three-revolutions-in-human-productivity/2009/11/29>.

[3]Johan Soderberg, *Hacking Capitalism*, p. 26.

> budget of *tasks* that need to be accomplished, that "working" means finishing some of those tasks, and that when the tasks are done, I can go to the gym or go see a movie or watch TV. Thus, I tend to work in a relatively focused, disciplined manner and then go do something other than work rather than slack off.[1]

Under the "face time" paradigm of wage employment at a workplace away from home, there is no trade-off between work and leisure. Anything done at work is "work," for which one gets paid. There is no opportunity cost to slacking off on the job. In home employment, on the other hand, the trade-off between effort and consumption is clear. The self-employed worker knows how much productive labor is required to support his desired level of consumption, and gets it done so he can enjoy the rest of his life. If his work itself is a consumption good, he still balances it with the rest of his activities in a rational, utility-maximizing manner, because he is the conscious master of his time, and has no incentive to waste time because "I'm here anyway." Any "work" he does which is comparatively unproductive or unrewarding comes at the expense of more productive or enjoyable ways of spending his time.

At work, on the other hand, all time belongs to the boss. A shift of work is an eight-hour chunk of one's life, cut off and flushed down the toilet for the money it will bring. And as a general rule, people do not make very efficient use of what belongs to someone else.

J.E. Meade contrasts the utility-maximizing behavior of a self-employed individual to that of a wage employee:

> A worker hired at a given hourly wage in an Entrepreneurial firm will have to observe the minimum standard of work and effort in order to keep his job; but he will have no immediate personal financial motive . . . to behave in a way that will promote the profitability of the enterprise. . . . [A]ny extra profit due to his extra effort will in the first place accrue to the entrepreneur. . . .
>
> Let us go to the other extreme and consider a one-man Cooperative, i.e. a single self-employed worker who hires his equipment. He can balance money income against leisure and other amenities by pleasing himself over hours of work, holidays, the pace and concentration of work, tea-breaks or the choice of equipment and methods of work which will make his work more pleasant at the cost of profitability. Any innovative ideas which he has, he can apply at once and reap the whole benefit himself.[2]

This is true not only of self-employment in the household sector and of self-managed peer networks, but of self-managed cooperatives in the money economy as well. The latter require far less in the way of front-line managers than do conventional capitalist enterprises. Edward Greenberg contrasts the morale and engagement with work, among the employees of a capitalist enterprise, with that of workers who own and manage their place of employment:

> Rather than seeing themselves as a group acting in mutuality to advance their collective interests and happiness, workers in conventional plants perceive their work existence, quite correctly, as one in which they are almost powerless, being used for the advancement and purposes of others, subject to the decisions of higher and more distant authority, and driven by a production process that is relentless. . . .

[1]Matthew Yglesias, "The Office Illusion," *Matthew Yglesias*, September 1, 2007 <http://matthewyglesias.theatlantic.com/archives/2007/09/the_office_illusion.php>.

[2]J.E. Meade, "The Theory of Labour-Managed Firms and Profit Sharing," in Jaroslav Vanek, ed., *Self-Management: Economic Liberation of Man* (Hammondsworth, Middlesex, England: Penguin Education, 1975), p. 395.

> The general mood of these two alternative types of work settings could not be more sharply contrasting. To people who find themselves in conventional, hierarchically structured work environments, the work experience is not humanly rewarding or enhancing. This seems to be a product of the all-too-familiar combination of repetitious and monotonous labor . . . and the structural position of powerlessness, one in which workers are part of the raw material that is manipulated, channeled, and directed by an only partly visible managerial hierarchy. Workers in such settings conceive of themselves, quite explicitly, as objects rather than subjects of the production process, and come to approach the entire situation, quite correctly, since they are responding to an objective situation of subordination, as one of a simple exchange of labor for wages. Work, done without a great deal of enthusiasm, is conceived of as intrinsically meaningless, yet necessary for the income that contributes to a decent life away from the workplace.[1]

Greenberg notes a "striking" fact: "the vast difference in the number of supervisors and foremen found in conventional plants as compared with the plywood cooperatives."

> While the latter were quite easily able to manage production with no more than two per shift, and often with only one, the former often requires six or seven. Such a disparity is not uncommon. I discovered in one mill that had recently been converted from a worker-owned to a conventional, privately owned firm that the very first action taken by the new management team was to quadruple the number of line supervisors and foremen. In the words of the general manager of this mill who had also been manager of the mill prior to its conversion,
>
> We need more foremen because, in the old days, the shareholders supervised themselves.... They cared for the machinery, kept their areas picked up, helped break up production bottlenecks all by themselves. That's not true anymore. We've got to pretty much keep on them all of the time.[2]

Workers in a cooperative enterprise put more of themselves into their work, and feel free to share their private knowledge—knowledge that would be exploited far more ruthlessly as a source of information rent in a conventional enterprise. Greenberg quotes a comment by a worker in a plywood co-op that speaks volumes on wage labor's inefficiency at aggregating distributed knowledge, compared to self-managed labor:

> If the people grading off the end of the dryer do not use reasonable prudence and they start mixing the grades too much, I get hold of somebody and I say, now look, this came over to me as face stock and it wouldn't even make decent back. What the hell's goin' on here?
>
> [Interviewer: That wouldn't happen if it were a regular mill?]
>
> That wouldn't happen. [In a regular mill] . . . he has absolutely no money invested in the product that's being manufactured. . . . He's selling nothing but his time. *Any knowledge he has on the side, he is not committed or he is not required to share that.* [emphasis added]
>
> It took me a little while to get used to this because where I worked before . . . there was a union and you did your job and you didn't go out and do something else. Here you get in and do anything to help. . . . I see somebody needs help, why you just go help them.

[1]Edward S. Greenberg. "Producer Cooperatives and Democratic Theory" in Robert Jackall and Henry M. Levin, eds., *Worker Cooperatives in America* (Berkeley, Los Angeles, London: University of California Press, 1984), p. 185.

[2]*Ibid.*, p. 193.

> I also tend to . . . look around and make sure things are working right a little more than . . . if I didn't have anything invested in the company. . . . I would probably never say anything when I saw something wrong.[1]

G. Reduced Costs from Supporting Rentiers and Other Useless Eaters

The alternative economy reduces waste and increases efficiency by eliminating the burden of supporting a class of absentee investors. By lowering the threshold of capital investment required to enter production, and easing the skids for self-employment at the expense of wage employment, the informal economy increases efficiency. Because producer-owned property must support only the laborer and his family, the rate of return required to make the employment of land and capital worthwhile is reduced. As a result, fewer productive resources are held out of use and there are more opportunities for productive labor.

The absentee ownership of capital skews investment in a different direction from what it would be in an economy of labor-owned capital, and reduces investment to lower levels. Investments that would be justified by the bare fact of making labor less onerous and increasing productivity, in an economy of worker-owned capital,[2] must produce an additional return on the capital to be considered worth making in an economy of rentiers. It is directly analogous to the holding of vacant land out of use that might enable laborers to subsist comfortably, because it will not in addition produce a rent over and above the laborer's subsistence. As Thomas Hodgskin observed in *Popular Political Economy*,

> It is maintained . . . that labour is not productive, and, in fact, the labourer is not allowed to work, unless, in addition to replacing whatever he uses or consumes, and comfortably subsisting himself, his labour also gives a profit to the capitalist . . . ; or unless his labour produces a great deal more . . . than will suffice for his own comfortable subsistence. Capitalists becoming the proprietors of all the wealth of the society . . . act on this principle, and never . . . will they suffer labourers to have the means of subsistence, unless they have a confident expectation that their labour will produce a profit over and above their own subsistence. This . . . is so completely the principle of slavery, to starve the labourer, unless his labour will feed his master as well as himself, that we must not be surprised if we should find it one of the chief causes . . . of the poverty and wretchedness of the labouring classes.[3]

When capital equipment is owned by the same people who make and use it, or made and used by different groups of people who divide the entire product according to their respective labor and costs, it is productive. But when capital equipment is owned by a class of rentiers separate from those who make it or use it, the owners may be said more accurately to impede production rather than "contribute" to it.

> If there were only the makers and users of capital to share between them the produce of their co-operating labour, the only limit to productive labour would be, that it should obtain for them and their families a comfortable subsistence. But when in addition to this . . . , they must also produce as much more as satisfies the capitalist, this limit is much sooner reached. When the capitalist . . . will allow labourers neither to make nor use instruments, unless he obtains a profit

[1]*Ibid.*, p. 191.

[2]Thomas Hodgskin, *Popular Political Economy: Four Lectures Delivered at the London Mechanics' Institution* (New York: Augustus M. Kelley, 1966 [1827]) , pp. 255-256.

[3]*Ibid.*, pp. 51-52.

over and above the subsistence of the labourer, it is plain that bounds are set to productive labour much within what Nature prescribes. In proportion as capital in the hands of a third party is accumulated, so the whole amount of profit required by the capitalist increases, and so there arises an artificial check to production and population. The impossibility of the labourer producing all which the capitalist requires prevents numberless operations, such as draining marshes, and clearing and cultivating waste lands; to do which would amply repay the labourer, by providing him with the means of subsistence, though they will not, in addition, give a large profit to the capitalist. In the present state of society, the labourers being in no case the owners of capital, every accumulation of it adds to the amount of profit demanded from them, and extinguishes all that labour which would only procure the labourer his comfortable subsistence.[1]

Hodgskin developed this same theme, as it applied to land, in *The Natural and Artificial Right of Property Contrasted*:

> It is, however, evident, that the labour which would be amply rewarded in cultivating all our waste lands, till every foot of the country became like the garden grounds about London, were all the produce of labour on those lands to be the reward of the labourer, cannot obtain from them a sufficiency to pay profit, tithes, rent, and taxes. . . .
>
> In the same manner as the cultivation of waste lands is checked, so are commercial enterprise and manufacturing industry arrested. Infinite are the undertakings which would amply reward the labour necessary for their success, but which will not pay the additional sums required for rent, profits, tithes, and taxes. These, and no want of soil, no want of adequate means for industry to employ itself, are the causes which impede the exertions of the labourer and clog the progress of society.[2]

The administrative and tranaction costs of conventional commercial economy have a similar effect to that of rentier incomes: they increase the number of people the laborer must support, in addition to himself, and thereby increase the minimum scale of output required for entering the market. The social economy enables its participants to evade the overhead costs of conventional organization (of the kind we saw skewered by Paul Goodman in Chapter Two), as described by Scott Burns in *The Household Economy*. The most enthusiastic celebrations of increased efficiencies from division of labor—like those at Mises.Org—tend to rely on illustrations in which, as Burns puts it, "labor can be directly purchased," or be made the object of direct exchange between the laborers themselves. But in fact,

> [m]arketplace labor must not only bear the institutional burden of taxation, it must also carry the overhead costs of organization and the cost of distribution. Even the most direct service organizations charge two and one-half times the cost of labor. The accountant who is paid ten dollars an hour is billed out to clients at twenty-five dollars an hour. . . . When both the general and the specific overhead burdens are considered, it becomes clear that any productivity that accrues to specialization is vitiated by the overhead burdens it must carry.
>
> Consider, for example, what happens when an eight-dollar-an-hour accountant hires an eight-dollar-an-hour service repairman, and vice versa. The repairman is billed out by his company at two and one-half times his hourly wage, or twenty dollars; to earn this money, the accountant must work three hours and twenty minutes, because 25 per cent of his wages are absorbed by

[1] *Ibid.*, pp. 243-244

[2] Hodgskin, "Letter the Eighth: Evils of the Artificial Right of Property," *The Natural and Artificial Right of Property Contrasted. A Series of Letters, addressed without permission to H. Brougham, Esq. M.P. F.R.S.* (London: B. Steil, 1832).
<http://oll.libertyfund.org/index.php?option=com_staticxt&staticfile=show.php%3Ftitle=323&layout=html>

taxes. Thus, to be truly economically efficient, the service repairman must be at least three and one-third times as efficient as the accountant at repairing things.[1]

The same principle applies to exchange, with household and informal arrangements requiring far less in the way of administrative overhead than conventional retailers. Food buying clubs run out of people's homes, barter bazaars[2] and freecycling networks, the imploding transaction costs of aggregating information and putting buyer and seller together on Craigslist, etc., all involve little or no overhead cost. Projects like FreeCycle, in fact, kill two birds with one stone: they simultaneously provide a low-overhead alternative to conventional retail, and maximize the efficiency with which the alternative economy extracts the last drop of value from the waste byproducts of capitalism.

To take just one example, consider the enormous cost of factoring in the apparel industry. Because most large retailers don't pay their apparel suppliers on time (delays of as much as six months are common), apparel producers must rely on factors to buy their accounts receivable at a heavy discount ("loan shark rates," in the words of Eric Husman, an engineer who blogs on lean manufacturing issues—typically 15-20%).[3] The requirement either to absorb several months' expenses while awaiting payment, or to get timely payment only at a steep discount, is an enormous source of added cost which exerts pressure to make it up on volume through large batch size. Now the large retailers, helpfully, are introducing a new "Supplier Alliance Program," which amounts to bringing the factoring operation in-house.[4] That's right: they actually "lend you the money they owe you" (in Husman's words). Technically, the retailers aren't actually lending the money, but rather extending their credit rating to cover your dealings with independent banks. The program is a response to the bankruptcy of several major factors in the recent financial crisis, and the danger that hundreds of vendors would go out of business in the absence of factoring. (Of course actually paying for orders on receipt would be beyond the meager resources of the poor big box chains.)

For the small apparel producer, in contrast, producing directly for an independent local retailer, for a local barter network, or for networked operations like Etsy, carries little or no overhead. Consider also the number of other industries in which something like the factoring system prevails (i.e., selling you, on credit, the rope to hang yourself with). A good example is the relationship Cargill and ADM have with family farmers: essentially a recreation of the 18th century putting-out system. Kathleen Fasanella, a consultant to the small apparel industry who specializes—among other things—in applying lean principles to apparel manufacturing, is for this reason an enthusiastic supporter of pull distribution networks (farmers selling at farmers' markets, craft producers selling on Etsy, etc.).[5]

[1]Scott Burns, *The Household Economy: Its Shape, Origins, & Future* (Boston: The Beacon Press, 1975), pp. 163-164.

[2]Gul Tuysuz, "An ancient tradition makes a little comeback," *Hurriyet DailyNews*, January 23, 2009 <http://www.hurriyet.com.tr/english/domestic/10826653.asp?scr=1>.

[3]Eric Husman, private email, November 18, 2009; Kathleen Fasanella, "Selling to Department Stores pt. 1," *Fashion Incubator*, August 11, 2009 <http://www.fashion-incubator.com/archive/selling-to-department-stores-pt1/>.

[4]"Supply Chain News: Walmart Joins Kohl's in Offering Factoring Program to Apparel Suppliers," *Supply Chain Digest*, November 17, 2009 <http://www.scdigest.com/ASSETS/ON_TARGET/09-11-17-2.PHP?cid=2954&ctype=conte>.

[5]Kathleen Fasanella, private email, November 19, 2009. Fasanella wrote the best-known book in the industry on how to start an apparel company: *The Entrepreneur's Guide to Sewn Product Manufacturing* (Apparel Technical Svcs, 1998). Eric Husman also happens to be her husband.

The shift to dispersed production in countless micro-enterprises also makes the alternative economy far less vulnerable to state taxation and imposition of artificial levels of overhead. In an economy of large-scale, conventional production, the required scale of capital outlays and resulting visibility of enterprises provides a physical hostage for the state's enforcement of overhead-raising regulations and "intellectual property" laws.

The conventional enterprise also provides a much larger target for taxation, with much lower costs for enforcement.[1] But as required physical capital outlays implode, and conventional manufacturing melts into a network of small machine shops and informal/household "hobby" shops, the targets become too small and dispersed to bother with.

This effect of rentier income, by the way, is just another example of a broader phenomenon we have been observing in various guises throughout this book: the effect of any increase in the minimum capital outlay, overhead, etc., to carry out a function, is to increase the scale of production necessary to service fixed costs. Overhead is a baffle that disrupts the flow from effort to output, and has an effect on the productive economy comparable to that of constipation or edema on the human body.

On the Open Manufacturing list, Eric Hunting argues that one of the side-effects of the kind of relocalized flexible manufacturing we examined in Chapter Five is that increasing competition, easy diffusion of new technology and technique, and increasing transparency of cost structure will—between them—arbitrage the rate of profit to virtually zero and squeeze artificial scarcity rents and spot-market profits from price almost entirely.

> What Open Manufacturing is doing is on the bleeding-edge of a general tend in industrial automation for progressively increasing productivity and production flexibility (mass customization/demand-driven flex production) with systems of decreasing scale and up-front cost. At the same time the economics of manufacturing has used-up the potential of Globalization as a means to exploit geographic spot-market bargains in materials and labor costs and is now dealing in a world of increasingly homogenous materials costs and expensive energy—and therefore transportation—costs. The efficiency of manufacturing logistics really matters now. It no longer makes economic sense to manufacture whole goods in far away places no matter how cheap the labor is. And—though the executive class remains slow on the uptake as usual—the trend is toward localization of production with increasing flexibility. So, ironically driven by the profit motive, commercial manufacturing is on a parallel track to the same goal as Open Manufacturing; progressive localization and diversification of production.
>
> I foresee this producing a progressive 'commoditization' of global economics. In other words, global trade will increasingly be trade of commodities materials and components, because it no longer makes economic sense to move finished goods around when their transportation is so inefficient. Commodities trade is highly automated because commodities production is highly automated, produces uniform products, and deals in large volumes relative to the number of workers. Production costs are highly quantifiable when the amortized cost of equipment supersedes the human labor overhead and that tends to factor out the variability in that only remaining (and deliberately) 'fuzzy' valued commodity. The result is that there is increasing global price capitulation in the value of commodities—largely because its increasingly difficult to hide costs, find exclu-

[1]See, for example, Benjamin Darrington, "Government Created Economies of Scale and Capital Specificity" (Austrian Student Scholars' Conference, 2007) pp. 6-7 <http://agorism.info/_media/government_created_economies_of_scale_and_capital_specificity.pdf>.

sive geographical spot-market bargains, or maintain exclusive distribution hegemonies. Trading systems have a very high and steadily increasing quantitative awareness of the costs of everything and the projected demand and production capacity for everything. At a certain point they can algorithmically factor out profit and can start trading commodities for commodities without cash indexed to projected demand/production. Profit in trade is based on divergence in the perception of value between buyer and seller. Scarcity is often a perception created by hiding data—and that's increasingly hard to do in a world where quantitative analysis trading knows more about an industry than the CEOs do. When everybody knows ahead of time what the concrete values of everything is and you have an actual open market where everyone has alternate sources for just about everything, profit becomes impossible.[1]

H. The Stigmergic Non-Revolution

Kim Stanley Robinson, in the second volume of his Mars trilogy, made some interesting comments (through the mouth of one of his characters) on the drawbacks of traditional models of revolution:

> " . . . [R]evolution has to be rethought. Look, even when revolutions have been successful, they have caused so much destruction and hatred that there is always some kind of horrible backlash. It's inherent in the method. If you choose violence, then you create enemies who will resist you forever. And ruthless men become your revolutionary leaders, so that when the war is over they're in power, and likely to be as bad as what they replaced."[2]

Arthur Silber, in similar vein, wrote that "with no exception in history that I can think of, violent revolutions on any scale lead to a state of affairs which is no better and frequently worse than that which the rebels seek to replace."[3]

A political movement is useful mainly for running interference, defending safe spaces in which we can build the real revolution—the revolution that *matters*. To the extent that violence is used, it should not be perceived by the public at large as a way of conquering anything, but as defensive force that raises the cost of government attacks on the counter-economy in a situation where the government is clearly the aggressor. The movement should avoid, at all costs, being seen as an attempt to impose a new "alternative" way of life on the "conventional" public, but instead strive to be seen as a fight to enable everyone to live their own lives the way they want. And even in such cases, non-cooperation and civil disobedience—while taking advantage of the possibilities of exposure that networked culture provide—are likely to be more effective than violent defense.

Rather than focusing on ways to shift the correlation of forces between the state's capabilities for violence and ours, it makes far more sense to focus on ways to increase our capabilities of living how we want below the state's radar. The networked forms of organization we've examined in Chapter Three and in this chapter are key to that process.

The focus on securing liberty primarily through political organization—organizing "one big movement" to make sure everybody is on the same page, be-

[1]Eric Hunting, "Re: Roadmap to Post-Scarcity," *Open Manufacturing*, January 12, 2010 <http://groups.google.com/group/openmanufacturing/msg/96166c44e39d3e6e?hl=en>.

[2]Kim Stanley Robinson, *Green Mars* (New York, Toronto, London, Sydney, Auckland: Bantam Books, 1994), p. 309.

[3]Arthur Silber, "An Evil Monstrosity: Thoughts on the Death State," *Once Upon a Time*, April 20, 2010 <http://powerofnarrative.blogspot.com/2010/04/evil-monstrosity-thoughts-on-death.html>.

fore anyone can put one foot in front of the other—embodies all the worst faults of 20th century organizational culture. What we need, instead, is to capitalize on the capabilities of network culture.

Network culture, in its essence, is stigmergic: that is, an "invisible hand" effect results from the several efforts of individuals and small groups working independently. Such independent actors may have a view to coordinating their efforts with a larger movement, and take the actions of other actors into account, but they do so without any single coordinating apparatus set over and above their independent authority.

In other words, we need a movement that works like Wikipedia at its best (without the deletionazis), or like open-source developers who independently tailor modular products to a common platform.

The best way to change "the laws," in practical terms, is to make them irrelevant and unenforceable through counter-institution building and through counter-economic activity outside the state's control. States claim all sorts of powers that they are utterly unable to enforce. It doesn't matter what tax laws are on the books if most commerce is in encrypted currency of some kind and invisible to the state. It doesn't matter how industrial patents enforce planned obsolescence, when a garage factory produces generic replacements and modular accessories for proprietary corporate platforms, and sells to such a small market that the costs of detecting and punishing infringement are prohibitive. It doesn't matter that local zoning regulations prohibit people doing business out of their homes, when their clientele is so small they can't be effectively monitored.

Without the ability of governments to enforce their claimed powers, the claimed powers themselves are about as relevant as the edicts of the Emperor Norton. That's why Charles Johnson argues that it's far more cost-effective to go directly after the state's enforcement capabilities than to try to change the law.

> In point of fact, if options other than electoral politics are allowed onto the table, then it might very well be the case that exactly the opposite course would be more effective: if you can establish effective means for individual people, or better yet large groups of people, to evade or bypass government enforcement and government taxation, then that might very well provide a much more effective route to getting rid of particular bad policies than getting rid of particular bad policies provides to getting rid of the government enforcement and government taxation.
>
> To take one example, consider immigration. If the government has a tyrannical immigration law in place . . . , then there are two ways you could go about trying to get rid the tyranny. You could start with the worst aspects of the law, build a coalition, do the usual stuff, get the worst aspects removed or perhaps ameliorated, fight off the backlash, then, a couple election cycles later, start talking about the almost-as-bad aspects of the law, build another coalition, fight some more, and so on, and so forth, progressively whittling the provisions of the immigration law down until finally you have whittled it down to nothing, or as close to nothing as you might realistically hope for. Then, if you have gotten it down to nothing, you can now turn around and say, "Well, since we have basically no restrictions on immigration any more, why keep paying for a border control or internal immigration cops? Let's go ahead and get rid of that stuff." And then you're done.
>
> The other way is the reverse strategy: to get rid of the tyranny by first aiming at the enforcement, rather than aiming at the law, by making the border control and internal immigration cops as irrelevant as you can make them. What you would do, then, is to work on building up more or less loose networks of black-market and grey-market operators, who can help illegal immigrants get into the country without being caught out by the Border Guard, who provide

> safe houses for them to stay on during their journey, who can help them get the papers that they need to skirt surveillance by La Migra, who can hook them up with work and places to live under the table, etc. etc. etc. To the extent that you can succeed in doing this, you've made immigration enforcement irrelevant. And without effective immigration enforcement, the state can bluster on as much as it wants about the Evil Alien Invasion; as a matter of real-world policy, the immigration law will become a dead letter.[1]

It's a principle anticipated over twenty years ago by Chuck Hammill, in an early celebration of the liberatory potential of network technology:

> While I certainly do not disparage the concept of political action, I don't believe that it is the only, nor even necessarily the most cost-effective path toward increasing freedom in our time. Consider that, for a fraction of the investment in time, money and effort I might expend in trying to convince the state to abolish wiretapping and all forms of censorship—I can teach every libertarian who's interested how to use cryptography to abolish them unilaterally. . . .
>
> . . . . Suppose this hungry Eskimo never learned to fish because the ruler of his nation-state had decreed fishing illegal. . . .
>
> . . . . However, it is here that technology—and in particular information technology—can multiply your efficacy literally a hundredfold. I say "literally," because for a fraction of the effort (and virtually none of the risk) attendant to smuggling in a hundred fish, you can quite readily produce a hundred Xerox copies of fishing instructions. . . .
>
> And that's where I'm trying to take The LiberTech Project. Rather than beseeching the state to please not enslave, plunder or constrain us, I propose a libertarian network spreading the technologies by which we may seize freedom for ourselves. . . .
>
> So, the next time you look at the political scene and despair, thinking, "Well, if 51% of the nation and 51% of this State, and 51% of this city have to turn Libertarian before I'll be free, then somebody might as well cut my goddamn throat now, and put me out of my misery"—recognize that such is not the case. There exist ways to make yourself free.[2]

This coincides to a large extent with what Dave Pollard calls "incapacitation": "rendering the old order unable to function by sapping what it needs to survive."[3]

> But suppose if, instead of waiting for the collapse of the market economy and the crumbling of the power elite, we brought about that collapse, guerrilla-style, by making information free, by making local communities energy self-sufficient, and by taking the lead in biotech away from government and corporatists (the power elite) by working collaboratively, using the Power of Many, Open Source, unconstrained by corporate allegiance, patents and 'shareholder expectations'?[4]

In short, we undermine the old corporate order, not by the people we elect to Washington, or the policies those people make, but by how we do things where we live. A character in Marge Piercy's *Woman on the Edge of Time*, describing the revolution that led to her future decentralist utopia, summed it up perfectly. Revolution, she said, was not uniformed parties, slogans, and mass-meetings. "It's the people who worked out the labor-and-land intensive farming we do. It's all the

[1]Charles Johnson, "In which I fail to be reassured," *Rad Geek People's Daily*, January 26, 2008 <http://radgeek.com/gt/2008/01/26/in_which/>.

[2]Chuck Hammill, "From Crossbows to Cryptography: Techno-Thwarting the State" (Given at the Future of Freedom Conference, November 1987) <www.csua.berkeley.edu/~ranga/papers/crossbows2crypto/crossbows2crypto.pdf>.

[3]David Pollard, "All About Power and the Three Ways to Topple It (Part 1)," *How to Save the World*, February 18, 2005 <http://blogs.salon.com/0002007/2005/02/18.html>.

[4]Pollard, "All About Power—Part Two," *How to Save the World*," February 21, 2005 <http://blogs.salon.com/0002007///2005/02/21.html>.

people who changed how people bought food, raised children, went to school!. . . . Who made new unions, withheld rent, refused to go to wars, wrote and educated and made speeches."[1]

One of the benefits of stigmergic organization, as we saw in earlier discussions of it, is that individual problems are tackled by the self-selected individuals and groups best suited to deal with them—and that their solutions are then passed on, via the network, to everyone who can benefit from them. DRM may be so hard to crack that only a handful of geeks can do it; but that doesn't mean, as the music and movie industries had hoped, that that would make "piracy" economically irrelevant. When a handful of geeks figure out how to crack DRM today, thanks to stigmergic organization, grandmas will be downloading DRM-free "pirated" music and movies at torrent sites next week.

Each individual innovation in ways of living outside the control of the corporate-state nexus, of the kind mentioned by Pollard and Piercy, creates a demonstration effect: You can do this too! Every time someone figures out a way to produce "pirated" knockoff goods in a microfactory in defiance of a mass-production corporation's patents, or build a cheap and livable house in defiance of the contractor-written building code, or run a microbakery or unlicensed hair salon out of their home with virtually zero overhead in defiance of local zoning and licensing regulations, they're creating another hack to the system, and adding it to the shared culture of freedom. And the more they're able to do business with each other through encrypted currencies and organize the kind of darknet economy described by John Robb, the more the counter-economy becomes a coherent whole opaque to the corporate state.

Statism will ultimately end, not as the result of any sudden and dramatic failure, but as the cumulative effect of a long series of little things. The costs of enculturing individuals to the state's view of the world, and of dissuading a large enough majority of people from disobeying when they're pretty sure they're not being watched, will result in a death of a thousand cuts. More and more of the state's activities, from the perspective of those running things, will just cost more (in terms not only of money but of just plain mental aggravation) than they're worth. The decay of ideological hegemony and the decreased feasibility of enforcement will do the same thing to the state that file-sharing is now doing to the RIAA.

One especially important variant of the stigmergic principle is educational and propaganda effort. Even though organized, issue-oriented advocacy groups arguably can have a significant effect on the state, in pressuring the state to cease or reduce suppression of the alternative economy, the best way to maximize bang for the buck in such efforts is simply to capitalize on the potential of network culture: that is, put maximum effort into just getting the information out there, giving the government lots and lots of negative publicity, and then "letting a thousand flowers bloom" when it comes to efforts to leverage it into political action. That being done, the political pressure itself will be organized by many different individuals and groups operating independently, spurred by their own outrage, without even sharing any common antistatist ideology.

In the case of any particular state abuse of power or intervention into the economy, there are likely to be countless subgroups of people who oppose it for any number of idiosyncratic reasons of their own, and not from any single dogmatic principle. If we simply expose the nature of the state action and all its unjust particular effects, it will be leveraged into action by people in numbers many times

[1]Marge Piercy, *Woman on the Edge of Time* (New York: Fawcett Columbine, 1976), p. 190.

larger than those of the particular alternative economic movement we are involved in. Even people who do not particularly sympathize with the aims of a counter-economic movement may be moved to outrage if the state's enforcers can be put in a position of looking like Bull Connor. As John Robb says: "The use of the media to communicate intent and to share innovation with other insurgent groups is a staple of open source insurgency. . . . "[1] The state and the large corporations are a bunch of cows floundering around in the Amazon. Just get the information out there, and the individual toothy little critters in the school of piranha, acting independently, will take care of the skeletonizing on their own.

A good example, in the field of civil liberties, is what Radley Balko does every day, just through his own efforts at exposing the cockroaches of law enforcement to the kitchen light, or the CNN series about gross civil forfeiture abuses in that town in Texas. When Woodward and Bernstein uncovered Watergate, they didn't start trying to organize a political movement to capitalize on it. They just published the info and a firestorm resulted.

This is an example of what Robb calls "self-replication": "create socially engineered copies of your organization through the use of social media. Basically, this means providing the motivation, knowledge, and focus necessary for an unknown person (external and totally unconnected to your group) to conduct operations that advance your group's specific goals (or the general goals of the open source insurgency)."[2]

It's because of increased levels of general education and the diffusion of more advanced moral standards that countries around the world have had to rename their ministries of war "ministries of defense." It's for the same reason that, in the twentieth and twenty-first centuries, governments could no longer launch wars for reasons of naked realpolitik on the model of the dynastic wars of two centuries earlier; rather, they had to manufacture pretexts based on self-defense. Hence pretexts like the mistreatment of ethnic Germans in Danzig as a pretext for Hitler's invasion of Poland, and the Tonkin Gulf incident and Kuwaiti incubator babies as pretexts for American aggressions. That's not to say that the pretexts had to be very good to fool the general public; but network culture is changing that as well, as witnessed by the contrasting levels of anti-war mobilization in the first and second Gulf wars.

More than one thinker on network culture has argued that network technology and the global justice movements piggybacked on it are diffusing more advanced global moral norms and putting increasing pressure on governments that violate those norms.[3] Global activism and condemnation of violations of human rights in countries like China and Iran—like American nationwide exposure and boycotts of measures like Arizona's "papers, please" law—are an increasing source of embarrassment and pressure. NGOs and global civil society are emerging as a powerful countervailing force against both national governments and global corporations. As we saw in the subsection on networked resistance in Chapter Three, governments and corporations frequently can find themselves isolated and ex-

[1]John Robb, "Links: 2 APR 2010," *Global Guerrillas*, April 2, 2010 <http://globalguerrillas.typepad.com/globalguerrillas/2010/04/links-2-apr-2010.html>.

[2]John Robb, "STANDING ORDER 8: Self-replicate," *Global Guerrillas*, June 3, 2009 <http://globalguerrillas.typepad.com/globalguerrillas/2009/06/standing-order-8-selfreplicate.html>.

[3]Paul Hartzog, "Panarchy: Governance in the Network Age," <http://www.panarchy.com/Members/PaulBHartzog/Papers/Panarchy%20-%20Governance%20in%20the%20Network%20Age.htm#_ftn>.

posed in the face of an intensely hostile global public opinion quite suddenly, thanks to networked global actors.

In light of all this, the most cost-effective "political" effort is simply making people understand that they don't need anyone's permission to be free. Start telling them right now that the law is unenforceable, and disseminating knowledge as widely as possible on the most effective ways of evading it. Publicize examples of ways we can live our lives the way we want, with institutions of our own making, under the radar of the state's enforcement apparatus: local currency systems, free clinics, ways to protect squatter communities from harassment, and so on. Educational efforts to undermine the state's moral legitimacy, educational campaigns to demonstrate the unenforceability of the law, and efforts to develop and circulate means of circumventing state control, are all things best done on a stigmergic basis.

Critics of "digital communism" like Jaron Lanier and Mark Helprin, who condemn network culture for submerging "individual authorial voice" in the "collective," are missing the point. Stigmergy synthesizes the highest realization of both individualism and collectivism, and represents the most absolute form of each of them, without either being limited or qualified in any way.

Stigmergy is not "collectivist" in the traditional sense, as it was understood in the days when a common effort on any significant scale required a large organization to represent the collective, and the coordination of individual efforts through a hierarchy. But it is the ultimate realization of collectivism, in that it removes the transaction cost of free collective action by many individuals.

It is the ultimate in individualism because all actions are the free actions of individuals, and the "collective" is simply the sum total of several individual actions. Every individual is free to formulate any innovation he sees fit, without any need for permission from the collective. Every individual or voluntary association of individuals is free to adopt the innovation, or not, as they see fit. The extent of adoption of any innovation is based entirely on the unanimous consent of every voluntary grouping that adopts it. Each innovation is modular, and may be adopted into any number of larger projects where it is found useful. Any grouping where there is disagreement over adoption may fork and replicate their project with or without the innovation.

Group action is facilitated with greater ease and lower transaction costs than ever before, but all "group actions" are the unanimous actions of individuals.

I. The Singularity

The cumulative effect of all these superior efficiencies of peer production, and of the informal and household economy, is to create a singularity.

The problem, for capital, is that—as we saw in previous chapters—the miniaturization and cheapness of physical capital, and the emergence of networked means of aggregating investment capital, are rendering capital increasingly superfluous.

The resulting crisis of realization is fundamentally threatening. Not only is capital superfluous in the immaterial realm, but the distinction between the immaterial and material realms is becoming increasingly porous. Material production, more and more, is taking on the same characteristics that caused the desktop computer to revolutionize production in the material realm.

The technological singularity means that labor is ceasing to depend on capital, and on wage employment by capital, for its material support.

For over two centuries, as Immanuel Wallerstein observed, the system of capitalist production based on wage labor has depended on the ability to externalize many of its reproduction functions on the non-monetized informal and household economies, and on organic social institutions like the family which were outside the cash nexus.

Historically, capital has relied upon its superior bargaining power to set the boundary between the money and social economies to its own advantage. The household and informal economies have been allowed to function to the extent that they bear reproduction costs that would otherwise have to be internalized in wages; but they have been suppressed (as in the Enclosures) when they threaten to increase in size and importance to the point of offering a basis for independence *from* wage labor.

The employing classes' fear of the subsistence economy made perfect sense. For as Kropotkin asked:

> If every peasant-farmer had a piece of land, free from rent and taxes, if he had in addition the tools and the stock necessary for farm labour—Who would plough the lands of the baron? Everyone would look after his own. . . .
>
> If all the men and women in the countryside had their daily bread assured, and their daily needs already satisfied, who would work for our capitalist at a wage of half a crown a day, while the commodities one produces in a day sell in the market for a crown or more?[1]

"The household as an income-pooling unit," Wallerstein writes, "can be seen as a fortress both of accommodation to and resistance to the patterns of labor-force allocation favored by accumulators." Capital has tended to favor severing the nuclear family household from the larger territorial community or extended kin network, and to promote an intermediate-sized income-pooling household. The reason is that too small a household falls so far short as a basis for income pooling that the capitalist is forced to commodify too large a portion of the means of subsistence, i.e. to internalize the cost in wages.[2] It is in the interest of the employer not to render the worker *totally* dependent on wage income, because without the ability to carry out some reproduction functions through the production of use value within the household subsistence economy, the worker will be "compelled to demand higher real wages. . . . "[3] On the other hand, too large a household meant that "the level of work output required to ensure survival was too low," and "diminished pressure to enter the wage-labor market."[4]

It's only common sense that when there are multiple wage-earners in a household, their dependence on any one job is reduced, and the ability of each member to walk away from especially onerous conditions is increased: "While a family with two or more wage-earners is no less dependent on the sale of labor power in general, it is significantly shielded from the effects of particular unem-

[1]Peter Kropotkin, *The Conquest of Bread* (New York: Vanguard Press, 1926), pp. 36-37.

[2]Immanuel Wallerstein, "Household Structures and Labor-Force Formation in the Capitalist World Economy," in Joan Smith, Immanuel Wallerstein, Hans-Dieter Evers, eds., *Households and the World Economy* (Beverly Hills, London, New Delhi: Sage Publications, 1984), pp. 20-21.

[3]Wallerstein and Joan Smith, "Households as an institution of the world-economy," in Smith and Wallerstein, eds., *Creating and Transforming Households: The constraints of the world-economy* (Cambridge; New York; Oakleigh, Victoria; Paris: Cambridge University Press, 1992), p. 16.

[4]Wallerstein, "Household Structures," p. 20.

ployment . . . "[1] And in fact it is less dependent on the sale of labor power in general, to the extent that the per capita overhead of fixed expenses to be serviced falls as household size increases. And the absolute level of fixed expenses can also be reduced by substituting the household economy for wage employment, in part, as the locus of value creation. As we saw Borsodi put it in the previous chapter, "[a] little money, where wages are joined to the produce of the soil, will go a long way. . . . "

The new factor today is a revolutionary shift in competitive advantage from wage labor to the informal economy. The rapid growth of technologies for home production, based on small-scale electrically powered machinery and new forms of intensive cultivation, has radically altered the comparative efficiencies of large- and small-scale production. This was pointed out by Borsodi almost eighty years ago, and the trend has continued since. The current explosion in low-cost manufacturing technology promises to shift competitive advantage in the next decade much more than in the entire previous century.

The practical choice presented to labor by this shift of comparative advantage was ably stated by Marcin Jakubowski, whose Factor E Farm is one of the most notable attempts to integrate open manufacturing and digital fabrication with an open design repository:

> Friends and family still harass me. They still keep telling me to 'get a real job.' I've got a good response now. It is:
>
> 1. Take a look at the last post on the soil pulverizer
>
> 2. Consider 'getting a real job at $100k,' a well-paid gig in The System. Tax and expense take it down to $50k, saved, if you're frugal.
>
> Ok. I can 'get a real job', work for 6 months, and then buy a Soil Pulverizer for $25k. Or, I make my own in 2 weeks at $200 cost, and save the world while I'm at it.
>
> Which one makes more sense to you? You can see which one makes more sense to me. It's just economics.[2]

In other words, how ya gonna keep 'em down in the factory, when the cost of getting your own garage factory has fallen to two months' wages?

As James O'Connor described the phenomenon in the 1980s, "the accumulation of stocks of means and objects of reproduction within the household and community took the edge off the need for alienated labor."

> Labor-power was hoarded through absenteeism, sick leaves, early retirement, the struggle to reduce days worked per year, among other ways. Conserved labor-power was then expended in subsistence production. . . . The living economy based on non- and anti-capitalist concepts of time and space went underground: in the reconstituted household; the commune; cooperatives; the single-issue organization; the self-help clinic; the solidarity group. Hurrying along the development of the alternative and underground economies was the growth of underemployment . . . and mass unemployment associated with the crisis of the 1980s. "Regular" employment and union-scale work contracted, which became an incentive to develop alternative, localized modes of production. . . .
>
> . . . New social relationships of production and alternative employment, including the informal and underground economies, threatened not only labor discipline, but also capitalist markets. . . . Alternative technologies threatened capital's monopoly on technological development . . . Hoarding of labor-power

[1]Samuel Bowles and Herbert Gintis. "The Crisis of Liberal Democratic Capitalism: The Case of the United States," *Politics and Society* 11:1 (1982), p. 83.

[2]Marcin Jakubowski, "Get a Real Job!" *Factor E Farm Weblog*, September 7, 2009 <http://openfarmtech.org/weblog/?p=1067>.

> threatened capital's domination of production. Withdrawal of labor-power undermined basic social disciplinary mechanisms. . . .[1]

More recently, "Eleutheros," of *How Many Miles from Babylon?* blog, described the sense of freedom that results from a capacity for independent subsistence:

> . . . if we padlocked the gate to this farmstead and never had any trafficking with Babylon ever again, we could still grow corn and beans in perpetuity. . . .
>
> What is this low tech, low input, subsistence economy all about, what does it mean to us? It is much like Jack Sparrow's remark to Elizabeth Swann when . . . he told her what the Black Pearl really was, it was freedom. Like that to us our centuries old agriculture represents for us a choice. And having a choice is the very essence and foundation of our escape from Babylon.
>
> . . . To walk away from Babylon, you must have choices. . . . Babylon, as with any exploitative and controlling system, can only exist by limiting and eliminating your choices. After all, if you actually have choices, you may in fact choose the things that benefit and enhance you and your family rather than things that benefit Babylon.
>
> Babylon must eliminate your ability to choose. . . .
>
> So I bring up my corn field in way of illustration of what a real choice looks like. We produce . . . our staple bread with no input at all from Babylon. So we always have the choice to eat that instead of what Babylon offers. We also buy wheat in bulk and make wheat bread sometimes, but if (when, as it happened this year) the transportation cost or scarcity of wheat makes the price beyond the pale, we can look at it and say, "No, not going there, we will just go home and have our cornbread and beans." Likewise we sometimes buy food from stands and stores, and on a few occasions we eat out. But we always have the choice, and if we need to, we can enforce that choice for months on end. . . .
>
> Your escape from Babylon begins when you can say, "No, I have a choice. Oh, I can dine around Babylon's table if I choose, but if the Babyonian terms and conditions are odious, then I don't have to."[2]

And the payoff doesn't require a total economic implosion. This is a winning strategy even if the money economy and division of labor persist indefinitely to a large extent—as I think they almost surely will—and most people continue to get a considerable portion of their consumption needs through money purchases. The end-state, after Peak Oil and the other terminal crises of state capitalism have run their course, is apt to bear a closer resemblance to Warren Johnson's *Muddling Toward Frugality* and Brian Kaller's "Return to Mayberry" than Jim Kunstler's *World Made by Hand*. The knowledge that you are debt-free and own your living space free and clear, and that you could keep a roof over your head and food on the table without wage labor indefinitely, if you had to, has an incalculable effect on your bargaining power here and now, even while capitalism persists.

As Ralph Borsodi observed almost eighty years ago, his ability to "retire" on the household economy for prolonged periods of time—and potential employers' knowledge that he could do so—enabled him to negotiate far better terms for what outside work he did decide to accept. He described, from his own personal experience, the greatly increased bargaining power of labor when the worker has the ability to walk away from the table:

> . . . Eventually income began to go up as I cut down the time I devoted to earning money, or perhaps it would be more accurate to say I was able to secure more for my time as I became less and less dependent upon those to whom I

[1] James O'Connor, *Accumulation Crisis* (New York: Basil Blackwell, 1984), pp. 184-186.

[2] Eleutheros, "Choice, the Best Sauce," *How Many Miles from Babylon*, October 15, 2008 <http://milesfrombabylon.blogspot.com/2008/10/choice-best-sauce.html>.

> sold my services. . . . This possibility of earning more, by needing to work less, is cumulative and is open to an immense number of professional workers. It is remarkable how much more appreciative of one's work employers and patrons become when they know that one is independent enough to decline unattractive commissions. And of course, if the wage-earning classes were generally to develop this sort of independence, employers would have to compete and bid up wages to secure workers instead of workers competing by cutting wages in order to get jobs.[1]
>
> . . . . Economic independence immeasurably improves your position as a seller of services. It replaces the present "buyer's market" for your services, in which the buyer dictates terms with a "seller's market," in which you dictate terms. It enables you to pick and choose the jobs you wish to perform and to refuse to work if the terms, conditions, and the purposes do not suit you. The next time you have your services to sell, see if you cannot command a better price for them if you can make the prospective buyer believe that you are under no compulsion to deal with him.[2]
>
> . . . [T]*he terms upon which an exchange is made between two parties are determined by the relative extent to which each is free to refuse to make the exchange*. . . . The one who was "free" (to refuse the exchange), dictated the terms of the sale, and the one who was "not free" to refuse, had to pay whatever price was exacted from him.[3]

Colin Ward, in "Anarchism and the informal economy," envisioned a major shift from wage labor to the household economy:

> [Jonathan Gershuny of the Science Policy Research Unit at Sussex University] sees the decline of the service economy as accompanied by the emergence of a self-service economy in the way that the automatic washing machine in the home can be said to supersede the laundry industry. His American equivalent is Scott Burns, author of *The Household Economy*, with his claim that 'America is going to be transformed by nothing more or less than the inevitable maturation and decline of the market economy. The instrument for this positive change will be the household—the family—revitalized as a powerful and relatively autonomous productive unit'.
>
> The only way to banish the spectre of unemployment is to break free from our enslavement to the idea of employment. . . .
>
> The first distinction we have to make then is between work and employment. The world is certainly short of jobs, but it has never been, and never will be, short of work. . . . The second distinction is between the regular, formal, visible and official economy, and the economy of work which is not employment. . . .
>
> . . . Victor Keegan remarks that 'the most seductive theory of all is that what we are experiencing now is nothing less than a movement back towards an informal economy after a brief flirtation of 200 years or so with a formal one'.
>
> We are talking about the movement of work back into the domestic economy. . . .[4]

Burns, whom Ward cited above, saw the formation of communes, the buying of rural homesteads, and other aspects of the back to the land movement, as an attempt

[1]Borsodi, *Flight From the City: An Experiment in Creative Living on the Land* (New York, Evanston, San Francisco, London: Harper & Row, 1933, 1972), p. 100.

[2]Borsodi, p. 335.

[3]*Ibid*., p. 403.

[4]Colin Ward, "Anarchism and the informal economy," *The Raven* No. 1 (1987), pp. 27-28.

> to supplant the marketplace entirely. By building their own homes and constructing them to minimize energy consumption, by recycling old cars or avoiding the automobile altogether, by building their own furniture, sewing their own clothes, and growing their own food, they are minimizing their need to offer their labor in the marketplace. They pool it, instead, in the extended household. . . . [T]he new homesteader can internalize 70-80 per cent of all his needs in the household; his money work is intermittent when it can't be avoided altogether.[1]

To reiterate: we're experiencing a singularity in which it is becoming impossible for capital to prevent a shift in the supply of an increasing proportion of the necessities of life from mass produced goods purchased with wages, to small-scale production in the informal and household sector. The upshot is likely to be something like Vinay Gupta's "Unplugged" movement, in which the possibilities for low-cost, comfortable subsistence off the grid result in exactly the same situation, the fear of which motivated the propertied classes in carrying out the Enclosures: a situation in which the majority of the public can take wage labor or leave it, if it takes it at all, the average person works only on his own terms when he needs supplemental income for luxury goods and the like, and (even if he considers supplemental income necessary in the long run for an optimal standard of living) can afford in the short run to quit work and live off his own resources for prolonged periods of time, while negotiating for employment on the most favorable terms. It will be a society in which workers, not employers, have the greater ability to walk away from the table. It will, in short, be the kind of society Wakefield lamented in the colonial world of cheap and abundant land: a society in which labor is hard to get on any terms, and almost impossible to hire at a low enough wage to produce significant profit.

Gupta's short story "The Unplugged"[2] related his vision of how such a singularity would affect life in the West.

> To "get off at the top" requires millions and millions of dollars of stored wealth. Exactly how much depends on your lifestyle and rate of return, but it's a lot of money, and it's volatile depending on economic conditions. A crash can wipe out your capital base and leave you helpless, because all you had was shares in a machine.
>
> So we Unpluggers found a new way to unplug: an independent life-support infrastructure and financial architecture—a society within society—which allowed anybody who wanted to "buy out" to "buy out at the bottom" rather than "buying out at the top."
>
> If you are willing to live as an Unplugger does, your cost to buy out is only around three months of wages for a factory worker, the price of a used car. You never need to "work" again—that is, for money which you spend to meet your basic needs.

The more technical advances lower the capital outlays and overhead for production in the informal and household economy, the more the economic calculus is shifted in the way described by Jakubowski above.

The basic principle of Unplugging was to combine "Gandhi's Goals" ("self-sufficiency," or "the freedom that comes from owning your own life support system") with "Fuller's Methods" (getting more from less). Such freedom

> allows us to disconnect from the national economy as a way of solving the problems of our planet one human at a time. But Gandhi's goals don't scale past the

[1]Burns, *The Household Economy*, p. 47.

[2]Vinay Gupta, "The Unplugged," How to Live Wiki, February 20, 2006 <http:// howtolivewiki.com/en/The_Unplugged>.

lifestyle of a peasant farmer and many westerners view that way of life as unsustainable for them personally. . . .

Fuller's "do more with less" was a method we could use to attain self-sufficiency with a much lower capital cost than "buy out at the top." An integrated, whole-systems-thinking approach to a sustainable lifestyle—the houses, the gardening tools, the monitoring systems—all of that stuff was designed using inspiration from Fuller and later thinkers inspired by efficiency. The slack—the waste—in our old ways of life were consuming 90% of our productive labor to maintain.

A thousand dollar a month combined fuel bill is your life energy going down the drain because the place you live sucks your life way [sic] in waste heat, which is waste money, which is waste time. Your car, your house, the portion of your taxes which the Government spends on fuel, on electricity, on waste heat . . . all of the time you spent to earn that money is wasted to the degree those systems are inefficient systems, behind best practices!

James L. Wilson, in a vignette of family life in the mid-21st century, writes of ordinary people seceding from the wage system and meeting as many of their needs as possible locally, primarily as a response to the price increases from Peak Oil—but in so doing, also regaining control of their lives and ending their dependence on the corporation and the state.

"Well, you see all these people working on their gardens? They used to not be here. People had grass lawns, and would compete with each other for having the greenest, nicest grass. But your gramma came home from the supermarket one day, sat down, and said, 'That's it. We're going to grow our own food.' And the next spring, she planted a vegetable garden where the grass used to be.

"And boy, were some of the neighbors mad. The Homeowners Association sued her. They said the garden was unsightly. They said that property values would fall. But then, the next year, more people started planting their own gardens.

"And not just their lawns. People started making improvements on their homes, to make them more energy-efficient. They didn't do it to help the environment, but to save money. People in the neighborhood started sharing ideas and working together, when before they barely ever spoke to each other. . . .

"And people also started buying from farmer's markets, buying milk, meat, eggs and produce straight from nearby farmers. This was fresher and healthier than processed food. They realized they were better off if the profits stayed within the community than if they went to big corporations far away.

"This is when your gramma, my Mom, quit her job and started a bakery from home. It was actually in violation of the zoning laws, but the people sided with gramma against the government. When the government realized it was powerless to crack down on this new way of life, and the people realized they didn't have to fear the government, they became free. And so more and more people started working from home. Mommies and Daddies used to have different jobs in different places, but now more and more of them are in business together in their own home, where they're close to their children instead of putting them in day care.". . . .[1]

Conclusion

We have seen throughout this chapter the superiority of the alternative economy, in terms of a number of different conceptual models—Robb's STEMI compression, Ceesay's economies of agility, Gupta's distributed infrastructure, and Cravens' productive recursion—to the corporate capitalist economy. All these su-

[1]James L. Wilson, "Standard of Living vs. Quality of Life," *The Partial Observer*, May 29, 2008 <http://www.partialobserver.com/article.cfm?id=2955&RSS=1>.

periorities can be summarized as the ability to make better use of material inputs than capitalism, and the ability to make use of the waste inputs of capitalism.

Localized, small-scale economies are the rats in the dinosaurs' nests. The informal and household economy operates more efficiently than the capitalist economy, and can function on the waste byproducts of capitalism. It is resilient and replicates virally. In an environment in which resources for technological development have been almost entirely diverted toward corporate capitalism, it takes technologies that were developed to serve corporate capitalism, adapts them to small-scale production, and uses them to destroy corporate capitalism. In fact, it's almost as though the dinosaurs themselves had funded a genetic research lab to breed mammals: "Let's reconfigure the teeth so they're better for sucking eggs, and ramp up the metabolism to survive a major catastrophe—like, say, an asteroid collision. Nah, I *don't* really know what it would be good for—but what the fuck, the Pangean Ministry of Defense is paying for it!"

To repeat, there are two economies competing: their old economy of bureaucracy, high overhead, enormous capital outlays, and cost-plus markup, and our new economy of agility and low overhead. And in the end . . . we will bury them.

APPENDIX
THE SINGULARITY IN THE THIRD WORLD

If the coming singularity will enable the producing classes in the industrialized West to defect from the wage system, in the Third World it may enable them to skip that stage of development altogether. Gupta concluded "The Unplugged" with a hint about how the principle might be applied in the Third World: "We encourage the developing world to Unplug as the ultimate form of Leapfrogging: skip hypercapitalism and anarchocapitalism and democratic socialism entirely and jump directly to Unplugging."

Gupta envisions a corresponding singularity in the Third World when the cost of an Internet connection, through cell phones and other mobile devices, falls low enough to be affordable by impoverished villagers. At that point, the transaction costs which hampered previous attempts at disseminating affordable intermediate technologies in the Third World, like Village Earth's Appropriate Technology Library or Schumacher's Intermediate Technology Development Group, will finally be overcome by digital network technology.

> It is inevitable that the network will spread everywhere across the planet, or very nearly so. Already the cell phone has reached 50% of the humans on the planet. As technological innovation transforms the ordinary cell phone into a little computer, and ordinary cell services into connections to the Internet, the population of the internet is going to change from being predominantly educated westerners to being mainly people in poorer countries, and shortly after that, to being predominantly people living on a few dollars a day. . . .
>
> . . . Most people are very poor, and as the price of a connection to the Internet falls to a level they can afford, as they can afford cell phones now, we're going to get a chance to really help these people get a better life by finding them the information resources they need to grow and prosper.
>
> Imagine that you are a poor single mother in South America who lives in a village without a clean water source. Your child gets sick now and again from the dirty water, and you feel there is nothing you can do, and worry about their survival. Then one of your more prosperous neighbors gets a new telephone, and there's a video which describes how to purify water [with a solar purifier made from a two-liter soda bottle]. It's simple, in your language, and describes all the basic steps without showing anything which requires schooling to understand. After a while, you master the basic practical skills—the year or two of high school you caught before having the child and having to work helps. But then you teach your sisters, and none of the kids get sick as often as they used to . . . life has improved because of the network.
>
> Then comes solar cookers, and improved stoves, and preventative medicine, and better agriculture [earlier Gupta mentions improved green manuring techniques], and diagnosis of conditions which require a doctor's attention, with a GPS map and calendar of when the visiting doctors will be in town again.[1]

The revolution is already here, according to a *New York Times* story. Cell phones, with service plans averaging $5 a month, have already spread to a third of the population of India. That means that mobile phones, with Internet service, have "seeped down the social strata, into slums and small towns and villages, becoming that rare Indian possession to traverse the walls of caste and region and class; a majority of subscribers are now outside the major cities and wealthiest

[1]Vinay Gupta, "What's Going to Happen in the Future," *The Bucky-Gandhi Design Institution*, June 1, 2008 <http://vinay.howtolivewiki.com/blog/global/whats-going-to-happen-in-the-future-670>.

states." And the mushrooming growth of cell phone connections, 15 million in March 2009, amounts to something like a 45% annual growth rate over the 400 million currently in use—a rate which, if it continues, will mean universal cell phone ownership within five years.[1]

Interestingly, Jeff Vail predicts that this increased connectivity, combined with especially exacerbated trends toward the hollowing out of the nation-state (see Chapter Three), will cause India to be a pioneer in the early development of the Diagonal Economy (see Chapter Six).

> . . . I do want to give one location-specific example of where I think this trend toward the degradation of the Nation-State construct will be especially severe: India. No, I don't think India will collapse (though there will be plenty of stories of woe), nor that the state government that occupies most of the geographic territory of "India" will collapse (note that careful wording). Rather, I think that the trend for a disconnection between any abstract notion of "Nation" and a unitary state in India will become particularly pronounced over the course of 2010. This is already largely apparent in India, but look for it to become more so. While Indian business and economy will fare decently well in 2010 from an international trade perspective, the real story will be a rising failure of this success to be effectively distributed by the government outside of a narrow class of urban middle class. It will instead be a rising connectivity and self-awareness of their situation among India's rural poor, resulting in an increasing push for localized self-sufficiency and resiliency of food production (especially the "tipping" of food forests and perennial polycultures), that will most begin to tear at the relevancy of India's central state government. In India there is a great potential for the beginnings of the Diagonal Economy to emerge in 2010.[2]

[1]Anand Giridhardas, "A Pocket-Size Leveler in an Outsized Land," *New York Times*, May 9, 2009 <http://www.nytimes.com/2009/05/10/weekinreview/10giridharadas.html?ref=world>.

[2]Jeff Vail, "2010—Predictions and Catabolic Collapse," *Rhizome*, January 4, 2010 <http://www.jeffvail.net/2010/01/2010-predictions-and-catabolic-collapse.html>.

Bibliography

"270-day libel case goes on and on . . . ," June 28 1996, *Daily Telegraph* (UK) <http://www.mcspotlight.org/media/thisweek/jul3.html>.

"100kGarages is Building a MakerBot." *100kGarages*, October 17, 2009 <http://blog.100kgarages.com/2009/10/17/100kgarages-is-building-a-makerbot/>.

Scott Adams. "Ridesharing in the Future." *Scott Adams Blog*, January 21, 2009 <http://dilbert.com/blog/entry/ridesharing_in_the_future/>.

Gar Alperowitz, Ted Howard, and Thad Williamson. "The Cleveland Model." *The Nation*, February 11, 2010 <http://www.thenation.com/doc/20100301/alperowitz_et_al/single>.

Lloyd Alter. "Ponoko + ShopBot = 100kGarages: This Changes Everything in Downloadable Design." *Treehugger*, September 16, 2009 <http://www.treehugger.com/files/2009/09/ponoko-shopbot.php>.

Oscar Ameriger. "Socialism for the Farmer Who Farms the Farm." Rip-Saw Series No. 15 (Saint Louis: The National Rip-Saw Publishing Co., 1912).

Beatrice Anarow, Catherine Greener, Vinay Gupta, Michael Kinsley, Joanie Henderson, Chris Page and Kate Parrot, Rocky Mountain Institute. "Whole-Systems Framework for Sustainable Consumption and Production." Environmental Project No. 807 (Danish Environmental Protection Agency, Ministry of the Environment, 2003), p. 24. <http://files.howtolivewiki.com/A%20Whole%20Systems%20Framework%20for%20Sustainable%20Production%20and%20Consumption.pdf>.

Chris Anderson. *Free: The Future of a Radical Price* (New York: Hyperion, 2009).

Anderson. "In the Next Industrial Revolution, Atoms Are the New Bits." *Wired*, January 25, 2010 <http://www.wired.com/magazine/2010/01/ff_newrevolution/all/1>.

Poul Anderson. *Orion Shall Rise* (New York: Pocket Books, 1983).

Massimo de Angelis. "Branding + Mingas + Coops = Salinas." *the editor's blog*, March 26, 2010 <http://www.commoner.org.uk/blog/?p=239>.

John Arquilla and David Ronfeldt. *The Advent of Netwar* MR-789 (Santa Monica, CA: RAND, 1996) <http://www.rand.org/pubs/monograph_reports/MR789/>.

Arquilla and Ronfeldt. "Fighting the Network War," Wired, December 2001 <http://www.wired.com/wired/archive/9.12/netwar.html>.

Arquilla and Ronfeldt. "Introduction," in Arquilla and Ronfeldt, eds., "Networks and Netwars: The Future of Terror, Crime, and Militancy" MR-1382-OSD (Santa Monica: Rand, 2001) <http://www.rand.org/pubs/monograph_reports/MR1382/>.

Arquilla and Ronfeldt. *Swarming & the Future of Conflict* DB-311 (Santa Monica, CA: RAND, 2000), iii <http://www.rand.org/pubs/documented_briefings/DB311/>.

Arquilla, Ronfeldt, Graham Fuller, and Melissa Fuller. The Zapatista "Social Netwar" in Mexico MR-994-A (Santa Monica: Rand, 1998) <http://www.rand.org/pubs/monograph_reports/MR994/index.html>.

Adam Arvidsson. "Review: Cory Doctorow, The Makers." *P2P Foundation Blog*, February 24, 2010 <http://blog.p2pfoundation.net/review-cory-doctorow-the-makers/2010/02/24>.

Arvidsson. "The Makers—again: or the need for keynesian management of abundance," *P2P Foundation Blog*, February 25, 2010 <http://blog.p2pfoundation.net/the-makers-again-or-the-need-for-keynesian-management-of-abundance/2010/02/25>.

Associated Press. "Retail sales fall after Cash for Clunkers ends," MSNBC, October 14, 2009 <http://www.msnbc.msn.com/id/33306465/ns/business-retail/>.

Associated Press. "U.S. government fights to keep meatpackers from testing all slaughtered cattle for mad cow," *International Herald-Tribune*, May 29, 2007 <http://www.iht.com/articles/ap/2007/05/29/america/NA-GEN-US-Mad-Cow.php>.

Atrios. "Face Time." *Eschaton*, July 9, 2005 <http://atrios.blogspot.com/2005_07 - 03_atrios_archive.html>.

Ronald Bailey. "Post-Scarcity Prophet: Economist Paul Romer on growth, technological change, and an unlimited human future," *Reason*, December 2001 <http://reason.com/archives/2001/12/01/post-scarcity-prophet/>.

Gopal Balakrishnan. "Speculations on the Stationary State," *New Left Review*, September-October 2009 <http://www.newleftreview.org/A2799>.

Paul Baran and Paul Sweezy. *Monopoly Capitalism: An Essay in the American Economic and* Social Order (New York: Monthly Review Press, 1966).

David Barboza. "In China, Knockoff Cellphones are a Hit," *New York Times*, April 27, 2009 <http://www.nytimes.com/2009/04/28/technology/28cell.html>.

Taylor Barnes. "America's 'shadow economy' is bigger than you think—and growing." *Christian Science Monitor*, November 12, 2009 <http://features.csmonitor.com/economyrebuild/2009/11/12/americas-shadow-economy-is-bigger-than-you-think-and-growing>.

Albert Bates. "Ecovillage Roots (and Branches): When, where, and how we re-invented this ancient village concept." *Communities Magazine* No. 117 (2003).

Michel Bauwens. "Asia needs a Social Innovation Stimulus plan." *P2P Foundation Blog*, March 23, 2009 <http://blog.p2pfoundation.net/asia-needs-a-social-innovation-stimulus-plan/2009/03/23>.

Bauwens. "Can the experience economy be capitalist?" *P2P Foundation Blog*, September 27, 2007 <http://blog.p2pfoundation.net/can-the-experience-economy-be-capitalist/2007/09/27>.

Bauwens. "Conditions for the Next Long Wave." *P2P Foundation Blog*, May 28, 2009 <http://blog.p2pfoundation.net/conditions-for-the-next-long-wave/2009/05/28>.

Bauwens. "Contract manufacturing as distributed manufacturing." *P2P Foundation Blog*, September 11, 2008 <http://blog.p2pfoundation.net/contract-manufacturing-as-distributed-manufacturing/2008/09/11>.

Bauwens. "The Emergence of Open Design and Open Manufacturing." *We Magazine*, vol. 2 <http://www.we-magazine.net/we-volume-02/the-emergence-of-open-design-and-open-manufacturing>.

Bauwens. "The great internet/p2p deflation." *P2P Foundation Blog*, November 11, 2009 <http://blog.p2pfoundation.net/the-great-internetp2p-deflation/2009/11/11>.

Bauwens. "A milestone for distributed manufacturing: 100kGarages." *P2P Foundation Blog*, September 19, 2009 <http://blog.p2pfoundation.net/a-milestone-for-distributed-manufacturing-100k-garages/2009/09/19>.

Bauwens. *P2P and Human Evolution*. Draft 1.994 (Foundation for P2P Alternatives, June 15, 2005) <http://integralvisioning.org/article.php?story=p2ptheory1>.

Bauwens. "Phases for implementing peer production: Towards a Manifesto for Mutually Assured Production." P2P Foundation *Forum*, August 30, 2008 <http://p2pfoundation.ning.com/forum/topics/ 2003008:Topic:6275>.

Bauwens. "The Political Economy of Peer Production." *CTheory*, December 1, 2005 <http://www.ctheory.net/articles.aspx?id=499>.

Bauwens. "Strategic Support for Factor e Farm and Open Source Ecology." *P2P Foundation Blog*, June 19, 2009 <http://blog.p2pfoundation.net/strategic-support-for-factor-e-farm-and-open-source-ecology/2009/06/19>.

Bauwens. "The three revolutions in human productivity." *P2P Foundation Blog*, November 29, 2009 <http://blog.p2pfoundation.net/the-three-revolutions-in-human-productivity/2009/11/29>.

Bauwens. "Three Times Exodus, Three Phase Transitions." *P2P Foundation Blog*, May 2, 2010 <http://blog.p2pfoundation.net/three-times-exodus-three-phase-transitions/2010/05/02>.

Bauwens. "What kind of economy are we moving to? 3. A hierarchy of engagement between companies and communities." *P2P Foundation Blog*, October 5, 2007 <http://blog.p2pfoundation.net/what-kind-of-economy-are-we-moving-to-3-a-hierarchy-of-engagement-between-companies-and-communities/2007/10/05>.

Robert Begg, Poli Roukova, John Pickles, and Adrian Smith. "Industrial Districts and Commodity Chains: The Garage Firms of Emilia-Romagna (Italy) and Haskovo (Bulgaria)." *Problems of Geography* (Sofia, Bulgarian Academy of Sciences), 1-2 (2005).

Walden Bello. "Asia: The Coming Fury." *Asia Times Online*, February 11, 2009 <http://www.atimes.com/atimes/Asian_Economy/KB11Dk01.html>.

Bello. "Can China Save the World from Depression?" *Counterpunch*, May 27, 2009 <http://www.counterpunch.org/bello05272009.html>.

Bello. "Keynes: A Man for This Season?" *Share the World's Resources*, July 9, 2009 <http://www.stwr.org/globalization/keynes-a-man-for-this-season.html>.

Bello. "A Primer on Wall Street Meltdown." *MR Zine*, October 3, 2008 <http://mrzine.monthlyreview.org/bello031008.html>.

James C. Bennett. "The End of Capitalism and the Triumph of the Market Economy." from *Network Commonwealth: The Future of Nations in the Internet Era* (1998, 1999) <http://www.pattern.com/bennettj-endcap.html>.

Yochai Benkler. *The Wealth of Networks: How Social Production Transforms Markets and Freedom* (New Haven and London: Yale University Press, 2006), pp. 220-223, 227-231.

Edwin Black. "Hitler's Carmaker: How Will Posterity Remember General Motors' Conduct? (Part 4)." *History News Network*, May 14, 2007 <http://hnn.us/articles/38829.html>.

"Black Mountain College." *Wikipedia* <http://en.wikipedia.org/wiki/Black_Mountain_College> (captured March 30, 2009).

David G. Blanchflower and Andrew J. Oswald. "What Makes an Entrepreneur?" <http://www2.warwick.ac.uk/fac/soc/economics/staff/faculty/oswald/entrepre.pdf >. Later appeared in *Journal of Labor Economics*, 16:1 (1998), pp. 26-60.

Murray Bookchin. *Post-Scarcity Anarchism* (Berkeley, Ca.: The Ramparts Press, 1971).
Ralph Borsodi. *The Distribution Age* (New York and London: D. Appleton and Company, 1929).
Borsodi. *Flight From the City: An Experiment in Creative Living on the Land* (New York, Evanston, San Francisco, London: Harper & Row, 1933, 1972).
Borsodi. *Prosperity and Security: A Study in Realistic Economics* (New York and London: Harper & Brothers Publishers, 1938).
Borsodi. *This Ugly Civilization* (Philadelphia: Porcupine Press, 1929, 1975).
Kenneth Boulding. *Beyond Economics* (Ann Arbor: University of Michigan Press, 1968).
Samuel Bowles and Herbert Gintis. "The Crisis of Liberal Democratic Capitalism: The Case of the United States." *Politics and Society* 11:1 (1982).
Ben Brangwyn and Rob Hopkins. *Transition Initiatives Primer: becoming a Transition Town, City, District, Village, Community or even Island* (Version 26—August 12, 2008) <http://transitionnetwork.org/Primer/ TransitionInitiativesPrimer.pdf>.
Brad Branan. "Police: Twitter used to avoid DUI checkpoints." *Seattle Times*, December 28, 2009 <http://seattletimes.nwsource.com/html/nationworld/2010618380_ twitterdui29.html>.
Gareth Branwyn. "ShopBot Open-Sources Their Code." *Makezine*, April 13, 2009 <http://blog.makezine.com/archive/2009/04/shopbot_open-sources_their_code.html>.
John Brummett. "Delta Solution: Move." *The Morning News of Northwest Arkansas*, June 14, 2009 <http://arkansasnews.com/2009/06/14/delta-solution-move/>.
Stewart Burgess. "Living on a Surplus." *The Survey* 68 (January 1933).
Scott Burns. *The Household Economy: Its Shape, Origins, & Future* (Boston: The Beacon Press, 1975).
Bryan Caplan. "Pyramid Power." *EconLog*, January 21, 2010 <http://econlog.econlib.org/archives/ 2010/01/pyramid_power.html>.
Kevin Carey. "College for $99 a Month." *Washington Monthly*, September/October 2009 <http://www.washingtonmonthly.com/college_guide/feature/college_for_99_a_month.php>.
Kevin Carson. "Abundance Creates Utility but Destroys Exchange Value." *P2P Foundation Blog*, February 2, 2010 <http://blog.p2pfoundation.net/abundance-creates-utility-but-destroys-exchange-value/2010/02/02>.
Carson. "'Building the Structure of the New Society Within the Shell of the Old.'" *Mutualist Blog: Free Market Anti-Capitalism*, March 22, 2005 <http:// mutualist.blogspot.com/2005/ 03/ building-structure-of-new-society.html>.
Carson. "The Cleveland Model and Micromanufacturing." *P2P Foundation Blog*, April 2, 2010 <http://blog.p2pfoundation.net/the-cleveland-model-and-micromanufacturing-an-opportunity-for-collaboration/2010/04/06>.
Carson. "Cory Doctorow. Makers." *P2P Foundation Blog*, October 25, 2009 <http://blog.p2pfoundation.net/cory-doctorow-makers/2009/10/25>.
Carson. "Daniel Suarez. Daemon and Freedom(TM)." *P2P Foundation Blog*, April 26, 2010 <http://blog.p2pfoundation.net/daniel-suarez-daemon-and-freedom/2010/04/26>.
Carson. "The People Making 'The Rules' are Dumber than You." Center for a Stateless Society, January 11, 2010 <http://c4ss.org/content/1687>.
Carson. *Studies in Mutualist Political Economy* (Blitzprint, 2004).

Carson. "Three Works on Abundance and Technological Unemployment." *Mutualist Blog*, March 30, 2010 <http://mutualist.blogspot.com/2010/03/three-works-on-abundance-and.html>.

"Carter Doctrine." *Wikipedia*, accessed December 23, 2009 <http://en.wikipedia.org/wiki/ Carter_Doctrine>.

"Doug Casey on Unemployment." *LewRockwell.Com*, January 22, 2010. Interviewed by Louis James, editor, *International Speculator* <http://www.lewrockwell.com/casey/casey38.1.html>.

Cassander. "It's Hard Being a Bear (Part Three): Good Economic History." *Steve Keen's Debtwatch*, September 5, 2009 <http://www.debtdeflation.com/blogs/2009/09/05/it%E2%80%99s-hard-being-a-bear-part-three-good-economic-history/>.

Mamading Ceesay. "The Economies of Agility and Disrupting the Nature of the Firm." *Confessions of an Autodidactic Engineer*, March 31, 2009 <http://evangineer.agoraworx.com/blog/2009-03-31-the-economies-of-agility-and-disrupting-the-nature-of-the-firm.html>.

Alfred D. Chandler, Jr. *Inventing the Electronic Century* (New York: The Free Press, 2001).

Chandler. *Scale and Scope: The Dynamics of Industrial Capitalism* (Cambridge and London: The Belknap Press of Harvard University Press, 1990).

Chandler. *The Visible Hand: The Managerial Revolution in American Business* (Cambridge and London: The Belknap Press of Harvard University Press, 1977).

Aimin Chen. "The structure of Chinese industry and the impact from China's WTO entry." *Comparative Economic Studies* (Spring 2002) <http://www.entrepreneur.com/tradejournals/article/ print/86234198.html>.

Chloe. "Important People." *Corporate Whore*, September 21, 2007 <http://web.archive.org/web/20071014221728/http://corporatewhore.us/important-people/>.

"The CloudFab Manifesto." Ponoko Blog, September 28, 2009 <http://blog.ponoko.com/ 2009/09/28/the-cloudfab-manifesto/>.

"CNC machine v2.0—aka 'Valkyrie'." *Let's Make Robots*, July 14, 2009 <http://letsmakerobots.com/node/9006>.

Moses Coady. *Masters of Their Own Destiny: The Story of the Antigonish Movement of Adult Education Through Economic Cooperation* (New York, Evanston, and London: Harper & Row, 1939).

Coalition of Immokalee Workers. "Burger King Corp. and Coalition of Immokalee Workers to Work Together." May 23, 2008 <http://www.ciw-online.org/BK_CIW_joint_release.html>.

Tom Coates. "(Weblogs and) The Mass Amateurisation of (Nearly) Everything . . ." Plasticbag.org, September 3, 2003 <http://www.plasticbag.org/archives/2003/09/weblogs_and_the_mass_ amateurisation_of_nearly_everything>.

G.D.H. Cole. *A Short History of the British Working Class Movement (1789-1947)* (London: George Allen & Unwin, 1948).

John R. Commons. *Institutional Economics* (New York: Macmillan, 1934).

"Community Wealth Building Conference in Cleveland, OH." *GVPT News*, February 2007, p. 14 <http://www.bsos.umd.edu/gvpt/newsletter/February07.news.pdf>.

Abe Connally. "Open Source Self-Replicator." *MAKE Magazine*, No. 21 <http://www.make-digital.com/make/vol21/?pg=69>.

Niall Cook. *Enterprise 2.0: How Social Software Will Change the Future of Work* (Burlington, Vt.: Gower, 2008).

Alan Cooper. *The Inmates are Running the Asylum: Why High-Tech Products Drive Us Crazy and How to Restore the Sanity* (Indianapolis: Sams, 1999).

James Coston, Amtrak Reform Council, 2001. In "America's long history of subsidizing transportation." <http://www.trainweb.org/moksrail/advocacy/ resources/subsidies/transport.htm>.

Tyler Cowen. "Was recent productivity growth an illusion?" *Marginal Revolution*, March 3, 2009 <http://www.marginalrevolution.com/marginalrevolution/2009/03/was-recent-productivity-growth-an-illusion.html>.

Nathan Cravens. "important appeal: social media and p2p tools against the meltdown." *Open Manufacturing* (Google Groups), March 13, 2009 <http://groups.google.com/group/openmanufacturing/msg/771617d04e45cd63>.

Cravens. "[p2p-research] simpler way wiki." *P2P Research*, April 20, 2009 <http://listcultures.org/pipermail/p2presearch_listcultures.org/2009-April/002083.html>.

Cravens. "Productive Recursion." *Open Source Ecology* Wiki <http://openfarmtech.org/index.php?title=Productive_Recursion>.

Cravens. "Productive Recursion Proven." *Open Manufacturing* (Google Groups), March 8, 2009 <http://groups.google.com/group/openmanufacturing/browse_thread/thread/f819aab7683b93ac?pli=1>.

Cravens. "The Triple Alliance." *Appropedia: The sustainability wiki* <http://www.appropedia.org/The_Triple_Alliance> (accessed July 3, 2009).

Matthew B. Crawford. "Shop Class as Soulcraft." *The New Atlantis*, Number 13, Summer 2006, pp. 7-24 <http://www.thenewatlantis.com/publications/shop-class-as-soulcraft>.

"CubeSpawn, An open source, Flexible Manufacturing System (FMS)." <http://www.kickstarter.com/projects/1689465850/cubespawn-an-open-source-flexible-manufacturing>.

John Curl. *For All the People: Uncovering the Hidden History of Cooperation, Cooperative Movements, and Communalism in America* (Oakland, CA: PM Press, 2009).

Fred Curtis. "Peak Globalization: Climate change, oil depletion and global trade." *Ecological Economics* Volume 69, Issue 2 (December 15, 2009).

Benjamin Darrington. "Government Created Economies of Scale and Capital Specificity." (Austrian Student Scholars' Conference, 2007).

Craig DeLancey. "Openshot." *Analog* (December 2006).

Brad DeLong. "Another Bad Employment Report (I-Wish-We-Had-a-Ripcord-to-Pull Department)." *Grasping Reality with All Eight Tentacles*, October 2, 2009 <http://delong.typepad.com/sdj/2009/10/another-bad-employment-report-i-wish-we-had-a-ripcord-to-pull-department.html>.

DeLong. "Jobless Recovery: Quiddity Misses the Point." *J. Bradford DeLong's Grasping Reality with All Eight Tentacles*, October 25, 2009 <http://delong.typepad.com/sdj/2009/10/jobless-recovery-quiddity-misses-the-point.html>.

Karl Denninger. "GDP: Uuuuggghhhh—UPDATED." *The Market Ticker*, July 31, 2009 <http://market-ticker.denninger.net/archives/1276-GDP-Uuuuggghhhh.html>.

Chris Dillow. "Negative Credibility." *Stumbling and Mumbling*, October 12, 2007 <http://stumblingandmumbling.typepad.com/stumbling_and_mumbling/2007/10/negative-credib.html>.

Maurice Dobb. *Political Economy and Capitalism: Some Essays in Economic Tradition*, 2nd rev. ed. (London: Routledge & Kegan Paul Ltd, 1940, 1960).

Cory Doctorow. "Australian seniors ask Pirate Party for help in accessing right-to-die sites." *Boing Boing*, April 9, 2010 <http://www.boingboing.net/2010/04/09/australian-seniors-a.html>.

Doctorow. "Cheap Facts and the Plausible Premise." *Locus Online*, July 5, 2009 <http://www.locusmag.com/Perspectives/2009/07/cory-doctorow-cheap-facts-and-plausible.html>.

Doctorow. *Content: Selected Essays on Technology, Creativity, Copyright, and the Future of the Future* (San Francisco: Tachyon Publications, 2008).

Doctorow. "The criticism that Ralph Lauren doesn't want you to see!" *BoingBoing*, October 6, 2009 <http://www.boingboing.net/2009/10/06/the-criticism-that-r.html>.

Brian Doherty. "The Glories of Quasi-Capitalist Modernity, Dumpster Diving Division." *Reason Hit & Run* Blog, September 12, 2007 <http://reason.com/blog/show/122450.html>.

Dale Dougherty. "What's in Your Garage?" *Make*, vol. 18 <http://www.make-digital.com/make/ vol18/?pg=39>.

Steve Dubb, Senior Research Associate, The Democracy Collaborative. "A Report on the Cleveland Community Wealth Building Roundtable December 7—8, 2007." <http://www.community-wealth.org/_pdfs/innovations/article-dubb-cleve.pdf>.

Deborah Durham-Vichr. "Focus on the DeCSS trial." CNN.Com, July 27, 2000 <http://archives.cnn.com/2000/TECH/computing/07/27/decss.trial.p1.idg/index.html>.

Barry Eichengreen and Kevin H. O'Rourke. "A Tale of Two Depressions." *VoxEU.Org*, June 4, 2009 <http://www.voxeu.org/index.php?q=node/3421>.

Eleutheros. "Choice, the Best Sauce." *How Many Miles from Babylon*, October 15, 2008 <http://milesfrombabylon.blogspot.com/2008/10/choice-best-sauce.html>.

Mark Elliott. "Some General Off-the-Cuff Reflections on Stigmergy." *Stigmergic Collaboration*, May 21, 2006 <http://stigmergiccollaboration.blogspot.com/2006/05/some-general-off-cuff-reflections-on.html>.

Elliott. "Stigmergic Collaboration: The Evolution of Group Work." *M/C Journal*, May 2006 <http://journal.media-culture.org.au/0605/03-elliott.php>.

Elliott. *Stigmergic Collaboration: A Theoretical Framework for Mass Collaboration*. Doctoral Dissertation, Centre for Ideas, Victorian College of the Arts, University of Melbourne (October 2007).

Ralph Estes. *Tyranny of the Bottom Line: Why Corporations Make Good People Do Bad Things* (San Francisco: Berrett-Koehler Publishers, 1996).

Stuart Ewen. *Captains of Consciousness: Advertising and the Social Roots of Consumer Culture* (New York: McGraw-Hill, 1976).

Kathleen Fasanella. "IP Update: DPPA & Fashion Law Blog." *Fashion Incubator*, March 10, 2010 <http://www.fashion-incubator.com/archive/ip-update-dppa-fashion-law-blog/>.

Fasanella. "Selling to Department Stores pt. 1." *Fashion Incubator*, August 11, 2009 <http://www.fashion-incubator.com/archive/selling-to-department-stores-pt1/>.

Martha Feldman and James G. March. "Information in Organizations as Signal and Symbol." *Administrative Science Quarterly* 26 (April 1981).

Mike Ferner. "Taken for a Ride on the Interstate Highway System." MRZine (Monthly Review) June 28, 2006 <http://mrzine.monthlyreview.org/ferner 280606.html>.

Ken Fisher. "Darknets live on after P2P ban at Ohio U." *Ars Technica*, May 9, 2007 <http://arstechnica.com/tech-policy/news/2007/05/darknets-live-on-after-p2p-ban-at-ohio-u.ars>.

Joseph Flaherty. "Desktop Injection Molding." February 1, 2010 <http://replicatorinc.com/blog/ 2010/02/desktop-injection-molding/>.

Richard Florida. "Are Bailouts Saving the U.S. from a New Great Depression." *Creative Class*, March 18, 2009 <http://www.creativeclass.com/creative_class/ 2009/03/18/are-the-bailouts-saving-us-from-a-new-great-depression/>.

Florida. *The Rise of the Creative Class* (New York: Basic Books, 2002).

Martin Ford. *The Lights in the Tunnel: Automation, Accelerating Technology and the Economy of the Future* (CreateSpace, 2009).

John Bellamy Foster and Fred Magdoff. "Financial Implosion and Stagnation: Back to the Real Economy." *Monthly Review*, December 2008 <http://www .monthlyreview.org/081201foster-magdoff.php>.

Justin Fox. "The Great Paving How the Interstate Highway System helped create the modern economy—and reshaped the FORTUNE 500." Reprinted from *Fortune*. CNNMoney.Com, January 26, 2004 <http://money.cnn.com/ magazines/fortune/fortune_archive/2004/01/26/358835/index.htm>.

John Kenneth Galbraith. *The New Industrial State* (New York: Signet Books, 1967).

John Gall. *Systemantics: How Systems Work and Especially How They Fail* (New York: Pocket Books, 1975).

Priya Ganapati. "Open Source Hardware Hackers Start P2P Bank." *Wired*, March 18, 2009 <http://www.wired.com/gadgetlab/2009/03/open-source-har/>.

Neil Gershenfeld. *Fab: The Coming Revolution on Your Desktop—from Personal Computers to Personal Fabrication* (New York: Basic Books, 2005), p. 182.

Kathryn Geurin. "Toybox Outlaws." *Metroland Online*, January 29, 2009 <http://www.metroland.net/back_issues/vol32_no05/features.html>.

Bruno Giussani. "Open Source at 90 MPH." *Business Week*, December 8, 2006 <http://www.businessweek.com/innovate/content/dec2006/id20061208_5090 41.htm?>. See also the OS Car website, <http://www.theoscarproject.org/>.

Malcolm Gladwell. "How David Beats Goliath." *The New Yorker*, May 11, 2009 <http://www.newyorker.com/reporting/2009/05/11/090511fa_fact_gladwell?cu rrentPage=all>.

Paul Goodman. *Compulsory Miseducation* and *The Community of Scholars* (New York: Vintage books, 1964, 1966).

Goodman. *People or Personnel* and *Like a Conquered Province* (New York: Vintage Books, 1964, 1966).

Paul and Percival Goodman. *Communitas: Means of Livelihood and Ways of Life* (New York: Vintage Books, 1947, 1960).

David Gordon. "Stages of Accumulation and Long Economic Cycles." in Terence K. Hopkins and Immanuel Wallerstein, eds., *Processes of the World-System* (Beverly Hills, Calif.: Sage, 1980), pp. 9-45.

Siobhan Gorman, Yochi J. Dreazen and August Cole. "Insurgents Hack U.S. Drones." *Wall Street Journal*, December 17, 2009 <http://online.wsj.com/ article/SB126102247889095011.html>.

Thomas Greco. *The End of Money and the Future of Civilization* (White River Junction, Vermont: Chelsea Green Publishing, 2009).

Greco. *Money and Debt: A Solution to the Global Crisis* (1990), Part III: Segregated Monetary Functions and an Objective, Global, Standard Unit of Account <http://circ2.home.mindspring.com/Money_and_Debt_Part3_lo.PDF>.

Edward S. Greenberg. "Producer Cooperatives and Democratic Theory." in Robert Jackall and Henry M. Levin, eds., *Worker Cooperatives in America* (Berkeley, Los Angeles, London: University of California Press, 1984).

Anand Giridhardas. "A Pocket-Size Leveler in an Outsized Land." *New York Times*, May 9, 2009 <http://www.nytimes.com/2009/05/10/weekinreview/10giridharadas.html?ref=world>.

Vinay Gupta. "The Global Village Development Bank: financing infrastructure at the individual, household and village level worldwide." Draft 2 (March 12, 2009) <http://vinay.howtolivewiki.com/ blog/hexayurt/my-latest-piece-the-global-village-development-bank-1348>.

Gupta. "The Unplugged." How to Live Wiki, February 20, 2006 <http://howtolivewiki.com/en/ The_Unplugged>.

Gupta. "What's Going to Happen in the Future." *The Bucky-Gandhi Design Institution*, June 1, 2008 <http://vinay.howtolivewiki.com/blog/global/whats-going-to-happen-in-the-future-670>.

Ted Hall. "100kGarages is Open: A Place to Get Stuff Made." Open Manufacturing email list, September 15, 2009 <http://groups.google.com/group/openmanufacturing/browse_thread/thread/ ae45b45de1d055a7?hl=en#>.

Hall (ShopBot) and Derek Kelley (Ponoko). "Ponoko and ShopBot announce partnership: More than 20,000 online creators meet over 6,000 digital fabricators." joint press release, September 16, 2009. Posted on Open Manufacturing email list, September 16, 2009 <http://groups.google.com/group/openmanufacturing/browse_thread/thread/fdb7b4d562f5e59d?hl=en>.

David Hambling. "China Looks to Undermine U.S. Power, With 'Assassin's Mace'." *Wired*, July 2 <http://www.wired.com/dangerroom/2009/07/china-looks-to-undermine-us-power-with-assassins-mace/>.

Chuck Hammill. "From Crossbows to Cryptography: Techno-Thwarting the State" (Given at the Future of Freedom Conference, November 1987) <www.csua.berkeley.edu/~ranga/papers/crossbows2crypto/crossbows2crypto.pdf>.

Bascha Harris. "A very long talk with Cory Doctorow, part 1." redhat.com, January 2006 <http://www.redhat.com/magazine/015jan06/features/doctorow/>.

Jed Harris. "Capitalists vs. Entrepreneurs." *Anomalous Presumptions*, February 26, 2007 <http://jed.jive.com/?p=23>.

Gifford Hartman. "Crisis in California: Everything Touched by Capital Becomes Toxic." *Turbulence* 5 (2010) <http://turbulence.org.uk/turbulence-5/california/>.

Paul Hartzog. "Panarchy: Governance in the Network Age." <http://www.panarchy.com/Members/PaulBHartzog/Papers/Panarchy%20-%20Governance%20in%20the%20Network%20Age.htm>.

Paul Hawken, Amory Lovins, and L. Hunter Lovins. *Natural Capitalism: Creating the Next Industrial Revolution* (Boston, New York, London: Little, Brown, and Company, 1999).

Richard Heinberg. *Peak Everything: Waking Up to the Century of Declines* (Gabriola Island, B.C.: New Society Publishers, 2007).

Richard Heinberg. *Powerdown* (Gabriola Island, British Columbia: New Society Publishers, 2004).

Martin Hellwig. "On the Economics and Politics of Corporate Finance and Corporate Control." in Xavier Vives, ed., *Corporate Governance: Theoretical and Empirical Perspectives* (Cambridge: Cambridge University Press, 2000).

Doug Henwood. *Wall Street: How it Works and for Whom* (London and New York: Verso, 1997).

Karl Hess. *Community Technology* (New York, Cambridge, Hagerstown, Philadelphia, San Francisco, London, Mexico City, Sao Paulo, Sydney: Harper & Row, Publishers, 1979).

Hess and David Morris. *Neighborhood Power: The New Localism* (Boston: Beacon Press, 1975).

Dougald Hine. "Social Media vs the Recession." *Changing the World*, January 28, 2009 <http://otherexcuses.blogspot.com/2009/01/social-media-vs-recession.html>.

Thomas Hodgskin. *Labour Defended Against the Claims of Capital* (New York: Augustus M. Kelley, 1969 [1825]).

Hodgskin. *The Natural and Artificial Right of Property Contrasted. A Series of Letters, addressed without permission to H. Brougham, Esq. M.P. F.R.S.* (London: B. Steil, 1832).

Hodgskin. *Popular Political Economy: Four Lectures Delivered at the London Mechanics' Institution* (London: Printed for Charles and William Tait, Edinburgh, 1827).

Joshua Holland. "Let the Banks Fail: Why a Few of the Financial Giants Should Crash." *Alternet*, December 15, 2008 <http://www.alternet.org/workplace/112166/let_the_banks_fail%3A_why_a_few_of_the_financial_giants_should_crash_/>.

Holland. "The Spectacular, Sudden Crash of the Global Economy." *Alternet*, February 24, 2009 <http://www.alternet.org/module/printversion/128412/the_spectacular%2C_sudden_crash_of_the_global_economy/>.

Lisa Hoover. "Riversimple to Unveil Open Source Car in London This Month." *Ostatic*, June 11, 2009 <http://ostatic.com/blog/riversimple-to-unveil-open-source-car-in-london-this-month>.

Rob Hopkins. *The Transition Handbook: From Oil Dependency to Local Resilience* (Totnes: Green Books, 2008).

"How to Fire Your Boss: A Worker's Guide to Direct Action." <http://www.iww.org/organize/ strategy/strikes.shtml> (originally a Wobbly Pamphlet, it is reproduced in all its essentials at the I.W.W. Website under the heading of "Effective Strikes and Economic Actions"—although the Wobblies no longer endorse it in its entirety).

Ebenezer Howard. *To-Morrow: A Peaceful Path to Real Reform*. Facsimile of original 1898 edition, with introduction and commentary by Peter Hall, Dennis Hardy and Colin Ward (London and New York: Routledge, 2003).

Bunnie Huang. "Copycat Corolla?" *bunnie's blog*, December 13, 2009 <http://www.bunniestudios.com/blog/?p=749>.

Huang. "Tech Trend: Shanzhai." *Bunnie's Blog*, February 26, 2009 <http://www.bunniestudios.com/blog/?p=284>.

Michael Hudson. "What Wall Street Wants." *Counterpunch*, February 11, 2009 <http://www.counterpunch.org/hudson02112009.html>.

Eric Hunting. "On Defining a Post-Industrial Style (1): from Industrial blobjects to post-industrial spimes." *P2P Foundation Blog*, November 2, 2009 <http://blog.p2pfoundation.net/on-defining-a-post-industrial-style-1-from-industrial-blobjects-to-post-industrial-spimes/2009/11/02>.

Hunting. "On Defining a Post-Industrial Style (2): some precepts for industrial design." *P2P Foundation Blog*, November 3, 2009 <http://blog.p2pfoundation.net/on-defining-a-post-industrial-style-2-some-precepts-for-industrial-design/2009/11/03>.

Hunting. "On Defining a Post-Industrial Style (3): Emerging examples." *P2P Foundation Blog*, November 4, 2009 <http://blog.p2pfoundation.net/on-defining-a-post-industrial-style-3-emerging-examples/2009/11/04>.

Hunting. "[Open Manufacturing] Re: Roadmap to Post-Scarcity." *Open Manufacturing*, January 12, 2010 <http://groups.google.com/group/openmanufacturing/msg/96166c44e39d3e6e?hl=en>.

Hunting. "[Open Manufacturing] Re:Vivarium." *Open Manufacturing*, March 28, 2009 <http://groups.google.com/group/openmanufacturing/browse_thread/thread/a891d6f72243436d/e58d837ac4022484?hl=en&q=vivarium+hunting#>.

Hunting. "[Open Manufacturing] Re: Why automate? and opinions on Energy Descent?" *Open Manufacturing*, September 22, 2008 <http://groups.google.com/group/openmanufacturing/browse_thread/thread/1f40d031453b94eb>.

Hunting. "Toolbook and the Missing Link." *Open Manufacturing*, January 30, 2009 <http://groups.google.com/group/openmanufacturing/msg/2fccdde02f402a5b>.

Samuel P. Huntington, Michael J. Crozier, and Joji Watanuki. *The Crisis of Democracy*. Report on the Governability of Democracies to the Trilateral Commission: Triangle Paper 8 (New York: New York University Press, 1975).

Jon Husband. "How Hard is This to Understand?" *Wirearchy*, June 22, 2007 <http://blog.wirearchy.com/blog/ _archives/2007/6/22/3040833.html>.

Eric Husman. "Human Scale Part II—Mass Production." *Grim Reader* blog, September 26, 2006 <http://www.zianet.com/ehusman/weblog/2006/09/human-scale-part-ii-mass-production.html>.

Husman. "Human Scale Part III—Self-Sufficiency" *GrimReader* blog, October 2, 2006 <http://www.zianet.com/ehusman/weblog/2006/10/human-scale-part-iii-self-sufficiency.html>.

Husman. "Open Source Automobile," *GrimReader*, March 3, 2005 <http://www.zianet.com/ehusman/weblog/2005/03/open-source-automobile.html>.

Tom Igoe. "Idle speculation on the shan zhai and open fabrication" *hello* blog, September 4, 2009 <http://www.tigoe.net/blog/category/environment/295/>.

Ivan Illich. *Deschooling Society* (New York, Evanston, San Francisco, London: Harper & Row, 1973).

Illich. *Disabling Professions* (New York and London: Marion Boyars, 1977).

Illich. "The Three Dimensions of Public Opinion," in *The Mirror of the Past: Lectures and Addresses, 1978-1990* (New York and London: Marion Boyars, 1992).

Illich. *Tools for Conviviality* (New York, Evanston, San Francisco, London: Harper & Row, 1973).

Illich. *Vernacular Values* (1980). Online edition courtesy of The Preservation Institute <http://www.preservenet.com/theory/Illich/Vernacular.html>.

"Ironworkers." Open Source Ecology Wiki <http://openfarmtech.org/index.php?title=Ironworkers>.

Neil Irwin. "Economic data don't point to boom times just yet." *Washington Post*, April 13, 2010 <http://www.washingtonpost.com/wp-dyn/content/article/2010/04/12/AR2010041204236.html?sid=ST2010041204452>.

Andrew Jackson. "Recession Far From Over." *The Progressive Economics Forum*, August 7, 2009 <http://www.progressive-economics.ca/2009/08/07/recession-far-from-over/>.

Ross Jackson. "The Ecovillage Movement." *Permaculture Magazine* No. 40 (Summer 2004).

Jane Jacobs. *Cities and the Wealth of Nations: Principles of Economic* Life (New York: Vintage Books, 1984).

Jacobs. *The Economy of Cities* (New York: Vintage Books, 1969, 1970).

Marcin Jakubowski. "CEB Proposal—Community Supported Manufacturing." *Factor e Farm* weblog, October 23, 2008 <http://openfarmtech.org/weblog/?p =379>.

Jakubowski. "CEB Prototype II Finished." *Factor e Farm Weblog*, August20, 2009 <http://openfarmtech.org/weblog/?p=1025>.

Jakubowski. "CEB Sales: Rocket Fuel for Post-Scarcity Economic Development?" *Factor e Farm Weblog*, November 28 2009 <http://openfarmtech.org/weblog/?p=1331>.

Jakubowski. "Clarifying OSE Vision." *Factor E Farm Weblog*, September 8, 2008 <http://openfarmtech.org/weblog/?p=325>.

Jakubowski. "Exciting Times: Nearing Product Release." *Factor e Farm Weblog*, October 10, 2009 <http://openfarmtech.org/weblog/?p=1168>.

Jakubowski. "Factor e Live Distillations—Part 8—Solar Power Generator," *Factor e Farm Weblog*, February 3, 2009 <http://openfarmtech.org/weblog/?p=507>.

Jakubowski. "Get a Real Job!" *Factor E Farm Weblog*, September 7, 2009 <http://openfarmtech.org/weblog/?p=1067>.

Jakubowski. "Initial Steps to the Open Source Multimachine." *Factor e Farm Weblog*, January 26, 2010 <http://openfarmtech.org/weblog/?p=1408>.

Jakubowski. "MicroTrac Completed." Factor e Farm Weblog, July 7, 2009 <http://openfarmtech.org/weblog/?p=852>.

Jakubowski. "Moving Forward." Factor e Farm Weblog, August 20, 2009 <http://openfarmtech.org/weblog/?p=1020>

Jakubowski. "Open Source Induction Furnace." *Factor e Farm Weblog*, December 15, 2009 <http://openfarmtech.org/weblog/?p=1373>.

Jakubowski. "OSE Proposal—Towards a World Class Open Source Research and Development Facility." vo.12, January 16, 2008 <http://openfarmtech.org/OSE _Proposal.doc> (accessed August 25, 2009).

Jakubowski. "Power Cube Completed." *Factor e Farm Weblog*, June 29, 2009 <http://openfarmtech.org/weblog/?p=814>.

Jakubowski. "PowerCube on LifeTrak." *Factor e Farm Weblog* , April 26, 2010 <http://openfarmtech.org/weblog/?p=1761>.

Jakubowski. "Product." *Factor e Farm Weblog*, November 4, 2009 <http://openfarmtech.org/weblog/?p=1224>.

Jakubowski. "Rapid Prototyping for Industrial Swadeshi." *Factor E Farm Weblog*, August 10, 2008 <http://openfarmtech.org/weblog/?p=293>.

Jakubowski. "Soil Pulverizer Annihilates Soil Handling Limits." *Factor e Farm Weblog*, September 7, 2009 <http://openfarmtech.org/weblog/?p=1063>.

Jakubowski. ""TED Fellows." *Factor e Farm Weblog*, September 22, 2009 <http://openfarmtech.org/weblog/?p=1121>.

Jakubowski. "The Thousandth Brick: CEB Field Testing Report." *Factor e Farm Weblog*, Nov. 16, 2008 <http://openfarmtech.org/weblog/?p=422>.

Jeff Jarvis. "When innovation yields efficiency." BuzzMachine, June 12, 2009 <http://www.buzzmachine.com/2009/06/12/when-innovation-yields-efficiency/>.

"Jay Rogers: I Challenge You to Make Cool Cars," Alphachimp Studio Inc., November 10, 2009 <http://www.alphachimp.com/poptech-art/2009/11/10/jay-rogers-

i-challenge-you-to-make-cool-cars.html>; Local Motors website at <http://www.local-motors.com>.

Charles Johnson. "Coalition of Imolakee Workers marches in Miami," *Rad Geek People's Daily*, November 30, 2007 <http://radgeek.com/gt/2007/11/30/coalition_of/>.

Johnson. "Dump the rentiers off your back" *Rad Geek People's Daily*, May 29, 2008 <http://radgeek.com/gt/2008/05/29/dump_the/>.

Johnson. "In which I fail to be reassured." *Rad Geek People's Daily*, January 26, 2008 <http://radgeek.com/gt/2008/01/26/in_which/>.

Johnson. "Liberty, Equality, Solidarity: Toward a Dialectical Anarchism." in Roderick T. Long and Tibor R. Machan, eds., *Anarchism/Minarchism: Is a Government Part of a Free Country?* (Hampshire, UK, and Burlington, Vt.: Ashgate Publishing Limited, 2008).

Johnson. "¡Sí, Se Puede! Victory for the Coalition of Imolakee Workers in the Burger King penny-per-pound campaign." *Rad Geek People's Daily*, May 23, 2008 <http://radgeek.com/gt/2008/05/ 23/ si_se/>.

H. Thomas Johnson. "Foreword." William H. Waddell and Norman Bodek, *Rebirth of American Industry: A Study of Lean Management* (Vancouver, WA: PCS Press, 2005).

Warren Johnson. *Muddling Toward Frugality* (San Francisco: Sierra Club Books, 1978).

Linda Joseph and Albert Bates. "What Is an 'Ecovillage'?" *Communities Magazine* No. 117 (2003).

Matthew Josephson. *The Robber Barons: The Great American Capitalists 1861-1901* (New York: Harcourt, Brace & World, Inc., 1934, 1962).

Brian Kaller. "Future Perfect: the future is Mayberry, not Mad Max." *Energy Bulletin*, February 27, 2009 (from *The American* Conservative, August 2008) <http://www.energybulletin.net/node/ 48209>.

Jeffrey Kaplan. "The Gospel of Consumption: And the better future we left behind." *Orion*, May/June 2008 <http://www.orionmagazine.org/index.php/articles/article/2962>.

Raphael Kaplinsky. "From Mass Production to Flexible Specialization: A Case Study of Microeconomic Change in a Semi-Industrialized Economy." *World Development* 22:3 (March 1994).

Kevin Kelly. "Better Than Free." *The Technium*, January 31, 2008 <http://www.kk.org/thetechnium/archives/2008/01/ better_than_fre.php>.

Marjorie Kelly. "The Corporation as Feudal Estate." (an excerpt from *The Divine Right of Capital*) *Business Ethics*, Summer 2001. Quoted in *GreenMoney Journal*, Fall 2008 <http://greenmoneyjournal.com/article.mpl?articleid=60&newsletterid=15>.

Paul T. Kidd. *Agile Manufacturing: Forging New Frontiers* (Addison-Wesley Publishing Company, 1994).

Lawrence Kincheloe. "First Dedicated Project Visit Comes to a Close." *Factor e Farm Weblog*, October 25, 2009 <http://openfarmtech.org/weblog/?p=1187>

Kincheloe. "One Month Project Visit: Take Two." *Factor e Farm Weblog*, October 4, 2009 <http://openfarmtech.org/weblog/?p=1146>.

Mark Kinney. "In Whose Interest?" (n.d) <http://www.appropriate-economics.org/materials/ in_whose_interest.pdf>.

Kinsale 2021: An Energy Descent Action Plan. Version.1. 2005. By Students of Kinsale Further Education College. Edited by Rob Hopkins <http:// transitionculture.org/wp-content/uploads/members/KinsaleEnergyDescentActionPlan.pdf>.

Peter Kirwan. “Bad News: What if the money’s not coming back?” *Wired.Co.Uk*, August 7, 2009 <http://www.wired.co.uk/news/archive/2009-08/07/bad-news-what-if-the-money%27s-not-coming-back.aspx>.

Ezra Klein. “A Fast Recovery? Or a Slow One?” *Washington Post*, April 14, 2010 <http://voices.washingtonpost.com/ezra-klein/2010/04/a_fast_recovery_or_a_slow_one.html>.

Ezra Klein. “Why Labor Matters.” *The American Prospect*, November 14, 2007 <http://www.prospect.org/csnc/blogs/ezraklein_archive?month=11&year=2007&base_name=why_labor_matters>.

Naomi Klein. *No Logo* (New York: Picador, 1999).

Keith Kleiner. “3D Printing and Self-Replicating Machines in Your Living Room—Seriously.” *Singularity Hub*, April 9, 2009 <http://singularityhub.com/2009/04/09/3d-printing-and-self-replicating-machines-in-your-living-room-seriously/>.

Thomas L. Knapp. “The Revolution Will Not Be Tweeted,” *Center for a Stateless Society*, October 5, 2009 <http://c4ss.org/content/1179>.

Jennifer Kock. “Employee Sabotage: Don’t Be a Target!” <http://www.workforce.com/archive/ features/22/20/88/mdex-printer.php>.

Frank Kofsky. *Harry Truman and the War Scare of 1948*, (New York: St. Martin’s Press, 1993).

Leopold Kohr. *The Overdeveloped Nations: The Diseconomies of Scale* (New York: Schocken Books, 1978, 1979).

Gabriel Kolko. *Confronting the Third World: United States Foreign Policy 1945-1980* (New York: Pantheon Books, 1988).

Kolko. *The Triumph of Conservatism: A Reinterpretation of American History 1900-1916* (New York: The Free Press of Glencoe, 1963).

Sam Kornell. “Will PeakOil Turn Flying into Something Only Rich People Can Afford?” *Alternet*, May 7, 2010 <http://www.alternet.org/economy/146769/will_peak_oil_turn_flying_into_something_only_rich_people_can_afford>.

Peter Kropotkin. *The Conquest of Bread* (New York: Vanguard Press, 1926).

Kropotkin. *Fields, Factories and Workshops: or Industry Combined with Agriculture and Brain Work with Manual Work* (New York: Greenwood Press, Publishers, 1968 [1898]).

Paul Krugman. “Averting the Worst.” *New York Times*, August 9, 2009 <http://www.nytimes.com/2009/08/10/opinion/10krugman.html>.

Krugman. “Double dip warning.” Paul Krugman Blog, *New York Times*, Dec. 1, 2009 <http://krugman.blogs.nytimes.com/2009/12/01/double-dip-warning/>.

Krugman. “Life Without Bubbles.” New York Times, January 6, 2009 <http://www.nytimes.com/ 2008/12/22/opinion/22krugman.html?ref=opinion>.

Krugman. “Use, Delay, and Obsolescence.” *The Conscience of a Liberal*, February 13, 2009 <http://krugman.blogs.nytimes.com/2009/02/13/use-delay-and-obsolescence/>.

James Howard Kunstler. “Lagging Recognition.” *Clusterfuck Nation*, June 8, 2009 <http://kunstler.com/blog/2009/06/lagging-recognition.html>.

Kunstler. *The Long Emergency: Surviving the End of Oil, Climate Change, and Other Converging Catastrophes of the Twenty-First Century* (Grove Press, 2006).

Kunstler. “Note: Hope = Truth.” *Clusterfuck Nation*, April 20, 2009 <http://jameshowardkunstler.typepad.com/clusterfuck_nation/2009/04/note-hope-truth.html>.

Kunstler. *World Made by Hand* (Grove Press, 2009).

Karim Lakhana. “Communities Driving Manufacturers Out of the Design Space.” *The Future of Communities Blog*, March 25, 2007 <http://www. futureofcommunities.com/2007/03/25/communities-driving-manufacturers-out-of-the-design-space/>.

"Lawrence Kincheloe Contract." OSE Wiki <http://openfarmtech.org/index.php?title=Lawrence_Kincheloe_Contract>.

Eli Lake. "Hacking the Regime." *The New Republic*, September 3, 2009 <http://www.tnr.com/ article/politics/hacking-the-regime>.

Steve Lawson. "The Future of Music is . . . Indie!" *Agit8*, September 10, 2009 <http://agit8.org.uk/?p=336>.

David S. Lawyer. "Are Roads and Highways Subsidized?" March 2004 <http://www.lafn.org/ ~dave/trans/econ/highway_subsidy.html>.

William Lazonick. *Business Organization and the Myth of the Market Economy* (Cambridge, 1991).

John Leland. "Finding in Foreclosure a Beginning, Not an End." *New York Times*, March 21, 2010 <http://www.nytimes.com/2010/03/22/us/22foreclose.html>.

Jay Leno. "Jay Leno's 3-D Printer Replaces Rusty Old Parts." *Popular Mechanics*, July 2009 <http://www.popularmechanics.com/automotive/jay_leno_garage/4320759.html?page=1>.

Daniel S. Levine. *Disgruntled: The Darker Side of the World of Work* (New York: Berkley Boulevard Books, 1998).

Rick Levine, Christopher Locke, Doc Searls and David Weinberger. *The Cluetrain Manifesto: The End of Business as Usual* (Perseus Books Group, 2001) <http://www.cluetrain.com/book/index.html>.

Claude Lewenz. *How to Build a Village* (Auckland, New Zealand: Village Forum Press and Jackson House Publishing Company, 2007).

Bernard Lietaer. *The Future of Money: A New Way to Create Wealth, Work and a Wiser World* (London: Century, 2001).

"LifeTrac." Open Source Ecology wiki <http://openfarmtech.org/index.php?title=LifeTrac>.

Roderick Long. "Free Market Firms: Smaller, Flatter, and More Crowded." *Cato Unbound*, Nov. 25, 2008 <http://www.cato-unbound.org/2008/11/25/roderick-long/free-market-firms-smaller-flatter-and-more-crowded>.

Long. "The Winnowing of Ayn Rand." *Cato Unbound*, January 20, 2010 <http://www.cato-unbound.org/2010/01/20/roderick-long/the-winnowing-of-ayn-rand/>.

"Long-Term Unemployment." *Economist's View*, November 9, 2009 <http:// economistsview.typepad.com/economistsview/2009/10/longterm-unemployment.html>.

Luca. "TeleKommunisten." (interview with Dmytri Kleiner), ecopolis, May 21, 2007 <http://www.ecopolis.org/telekommunisten/>.

Spencer H. MacCallum. "E. C. Riegel on Money." (January 2008) <http://www.newapproachtofreedom.info/documents/AboutRiegel.pdf>.

Andrew MacLeod. "Mondragon—Cleveland—Sacramento." *Cooperate and No One Gets Hurt*, October 10, 2009 <http://coopgeek.wordpress.com/2009/10/10/mondragon-cleveland-sacramento/>.

"McDonald's Restaurants v Morris & Steele." Wikipedia <http://en.wikipedia.org/wiki/McLibel_case> (accessed December 26, 2009).

Bill McKibben. *Deep Economy: The Wealth of Communities and the Durable Future* (New York: Times Books, 2007).

Karl Marx. *The Poverty of Philosophy*. Marx and Engels *Collected Works*, vol. 6 (New York: International Publishers, 1976).

Harry Magdoff and Paul Sweezy. *The End of Prosperity: The American Economy in the 1970s* (New York and London: Monthly Review Press, 1977).

Magdoff and Sweezy. *The Irreversible Crisis: Five Essays by Harry Magdoff and Paul M. Sweezy* (New York: Monthly Review Press, 1988).

"Mahatma Gandhi on Mass Production." (1936). *TinyTech Plants* <http://www.tinytechindia.com/ gandhiji2.html>.

Katherine Mangu-Ward. "The Sheriff is Coming! The Sheriff is Coming!" *Reason Hit & Run*, January 6, 2010 <http://reason.com/blog/2010/01/06/the-sheriff-is-coming-the-sher>.

"Manufacture Goods, Not Needs." E. F. Schumacher Society Blog, October 11, 2009 <http://efssociety.blogspot.com/2009/10/manufacture-goods-not-needs_11.html>.

Mike Masnick. "Artificial Scarcity is Subject to Massive Deflation." *Techdirt*, <http://techdirt.com/articles/ 20090624/ 0253385345.shtml>.

Masnick. "How Automakers Abuse Intellectual Property Laws to Force You to Pay More For Repairs." *Techdirt*, December 29, 2009 <http://techdirt.com/articles/20091228/0345127515.shtml>.

Masnick. "Yet Another High School Newspaper Goes Online to Avoid District Censorship." *Techdirt*, January 15, 200 <http://www.techdirt.com/articles/20090112/1334043381.shtml>.

Jeremy Mason. "Sawmill Development." *Factor e Farm Weblog*, January 22, 2009 <http://openfarmtech.org/weblog/?p=498>.

Jeremy Mason. "What is Open Source Ecology?" *Factor e Farm Weblog*, March 20, 2009 <http://openfarmtech.org/weblog/?p=595>.

Race Matthews. *Jobs of Our Own: Building a Stakeholder Society—Alternatives to the Market & the State* (Annandale, NSW, Australia: Pluto Press, 1999).

Paul Mattick. "The Economics of War and Peace." *Dissent* (Fall 1956).

J.E. Meade. "The Theory of Labour-Managed Firms and Profit Sharing." in Jaroslav Vanek, ed., *Self-Management: Economic Liberation of Man* (Hammondsworth, Middlesex, England: Penguin Education, 1975).

Seymour Melman. *The Permanent War Economy: American Capitalism in Decline* (New York: Simon and Schuster, 1974).

Richard Milne. "Crisis and climate force supply chain shift." *Financial Times*, August 9, 2009 <http://www.ft.com/cms/s/0/65a709ec-850b-11de-9a64-00144feabdc0.html>.

MIT Center for Bits and Atoms. "Fab Lab FAQ." <http://fab.cba.mit.edu/about/faq/> (accessed August 31, 2009).

"Monsanto Declares War on 'rBGH-free' Dairies." April 3, 2007 (reprint of Monsanto press release by Organic Consumers Association) <http://www.organicconsumers.org/articles/article_4698.cfm>.

Dante-Gabryell Monson. "[p2p-research] trends? : "Corporate Dropouts." towards Open diy ? . . ." *P2P Research*, October 13, 2009 <http://listcultures.org/pipermail/p2presearch_listcultures.org/2009-October/005128.html>.

William Morris. *News From Nowhere: or, An Epoch of Rest* (1890). Marxists.Org online text <http://www.marxists.org/archive/morris/works/1890/nowhere/nowhere.htm>.

Jim Motavalli. "Getting Out of Gridlock: Thanks to the Highway Lobby, Now We're Stuck in Traffic. How Do We Escape?" *E Magazine*, March/April 2002 <http://www.emagazine.com/view/?534>.

"Multimachine." *Wikipedia* <http://en.wikipedia.org/wiki/Multimachine> (accessed August 31, 2009>; <http://groups.yahoo.com/group/multimachine/>.

"Multimachine & Flex Fab—Open Source Ecology." <http://openfarmtech.org/index.php?title=Multimachine_%26_Flex_Fab>.

Lewis Mumford. *The City in History: Its Transformations, and Its Prospects* (New York: Harcourt, Brace, & World, Inc., 1961).

Mumford. *Technics and Civilization* (New York: Harcourt, Brace, and Company, 1934).

Charles Nathanson. "The Militarization of the American Economy." in David Horowitz, ed., *Corporations and the Cold War* (New York and London: Monthly Review Press, 1969).

Daisy Nguyen. "High tech vehicles pose trouble for some mechanics." *North County Times*, December 26, 2009 <http://nctimes.com/news/state-and-regional/article_4ea03fd6-090d-5c2e-bd91-dfb5508495ef.html>.

David F. Noble. *America by Design: Science, Technology, and the Rise of Corporate Capitalism* (New York: Alfred A. Knopf, 1977).

Noble. *Forces of Production: A Social History of American Automation* (New York: Alfred A. Knopf, 1984).

James O'Connor. *Accumulation Crisis* (New York: Basil Blackwell, 1984).

O'Connor. *The Fiscal Crisis of the State* (New York: St. Martin's Press, 1973).

"October 30 2009: An interview with Stoneleigh—the case for deflation." *The Automatic Earth* <http://theautomaticearth.blogspot.com/2009/10/october-30-2009-interview-with.html>.

Ohio Employee Ownership Center. "Cleveland Goes to Mondragon." *Owners at Work* (Winter 2008-2009) <http://dept.kent.edu/oeoc/OEOCLibrary/OaW_Winter08_Cleveland_Goes_to_ Mondragon.pdf>.

"Open Source Hardware." *P2P Foundation Wiki* <http://www.p2pfoundation.net/Open_Source_Hardware>.

"Open Source Fab Lab." Open Source Ecology wiki (accessed August 22, 2009) <http://openfarmtech.org/index.php?title=Open_Source_Fab_Lab>.

"Organizational Strategy." Open Source Ecology wiki, February 11, 2009 <http://openfarmtech.org/index.php?title=Organizational_Strategy> (accessed August 28, 2009).

Franz Oppenheimer. "A Post Mortem on Cambridge Economics (Part Three)." *The American Journal of Economics and Sociology*, vol. 3, no. 1 (1944), pp, 122-123, [115-124].

George Orwell. *1984*. Signet Classics reprint (New York: Harcourt Brace Jovanovich, 1949, 1981).

"Our Big Idea!" 100kGarages site <http://100kgarages.com/our_big_idea.html>.

"Pa. bars hormone-free milk labels." *USA Today*, November 13, 2007 <http://www.usatoday.com/ news/nation/2007-11-13-milk-labels_N.htm>.

Keith Paton. *The Right to Work or the Fight to Live?* (Stoke-on-Trent, 1972).

Michael Parenti. "Capitalism's Self-Inflicted Apocalypse." *Common Dreams*, January 21, 2009 <http://www.commondreams.org/view/2009/01/20-9>.

David Parkinson. "A coming world that's 'a whole lot smaller.'" *The Globe and Mail*, May 19, 2009 <http://docs.google.com/Doc?id=dg5dgmrv_79hjb66vc3>.

Michael Perelman. "The Political Economy of Intellectual Property." *Monthly Review*, January 2003 <http://www.monthlyreview.org/0103perelman.htm>.

Tom Peters. *The Tom Peters Seminar: Crazy Times Call for Crazy Organizations* (New York: Vantage Books, 1999).

Diane Pfeiffer. "Digital Tools, Distributed Making and Design." Thesis submitted to the faculty of the Virginia Polytechnic Institute and State University in partial fulfillment of the requirements for Master of Science in Architecture, 2009.

"PhysicalDesignCo teams up with 100kGarages." *100kGarages News*, October 4, 2009 <http://blog.100kgarages.com/2009/10/04/physicaldesignco-teams-up-with-100kgarages/>.

Marge Piercy. *Woman on the Edge of Time* (New York: Fawcett Columbine, 1976).

Chris Pinchen. "Resilience: Patterns for thriving in an uncertain world." *P2P Foundation Blog*, April 17, 2010 <http://blog.p2pfoundation.net/resilience-patterns-for-thriving-in-an-uncertain-world/2010/04/17>.

Michael J. Piore and Charles F. Sabel. "Italian Small Business Development: Lessons for U.S. Industrial Policy." in John Zysman and Laura Tyson, eds., *American Industry in International Competition: Government Policies and Corporate Strategies* (Ithaca and London: Cornell University Press, 1983).

Piore and Sabel. "Italy's High-Technology Cottage Industry." *Transatlantic Perspectives* 7 (December 1982).

Piore and Sabel. *The Second Industrial Divide: Possibilities for Prosperity* (New York: HarperCollins, 1984).

"Plowboy Interview." (Ralph Borsodi), *Mother Earth News*, March-April 1974 <http://www.soilandhealth.org/03sov/0303critic/Brsdi.intrvw/The%20Plowboy-Borsodi%20Interview.htm>.

David Pollard, "All About Power and the Three Ways to Topple It (Part 1)." *How to Save the World*, February 18, 2005 <http://blogs.salon.com/0002007/2005/02/18.html>.

Pollard. "All About Power—Part Two," *How to Save the World.*" February 21, 2005 <http://blogs.salon.com/0002007///2005/02/21.html>.

Pollard. "The Future of Business." *How to Save the World*, January 14, 2004 <http://blogs.salon.com/0002007/2004/01/14.html>.

Pollard. "Peer Production." *How to Save the World*, October 28, 2005 <http://blogs.salon.com/0002007/2005/10/28.html#a1322>.

Pollard. "Replicating (Instead of Growing) Natural Small Organizations." *how to save the world*, January 14, 2009 <http://howtosavetheworld.ca/2010/01/14/not-so-smart-replicating-instead-of-growing-natural-small-organizations>.

Pollard. "Ten Important Business Trends." *How to Save the World*, May 12, 2009 <http://blogs.salon.com/0002007/2009/05/12.html#a2377>.

J.A. Pouwelse, P. Garbacki, D.H.J. Epema, and H.J. Sips. "Pirates and Samaritans: a Decade of Measurements on Peer Production and their Implications for Net Neutrality and Copyright." (The Netherlands: Delft University of Technology, 2008) <http://www.tribler.org/trac/wiki/ PiratesSamaritans>.

"PR disaster, Wikileaks and the Streisand Effect." PRdisasters.com, March 3, 2007 <http://prdisasters.com/pr-disaster-via-wikileaks-and-the-streisand-effect/>.

David L Prychitko. *Marxism and Workers' Self-Management: The Essential Tension* (New York; London; Westport, Conn.: Greenwood Press, 1991).

"Public Service Announcement—Craig Murray, Tim Ireland, Boris Johnson, Bob Piper and Alisher Usmanov . . ." *Chicken Yoghurt*, September 20, 2007 <http://www.chickyog.net/2007/09/20/public-service-announcement/>.

Jeff Quackenbush and Jessica Puchala. "Middleville woman threatened with fines for watching neighbors' kids," WZZM13.Com, September 24, 2009 <http://www.wzzm13.com/news/news_story.aspx?storyid=114016&catid=14#>.

Quiddity. "Job-loss recovery," *uggabugga*, October 25, 2009 <http://uggabugga.blogspot.com/2009/10/job-loss-recovery-experts-see.html#comments>.

John Quiggin. "The End of the Cash Nexus." *Crooked Timber*, March 5, 2009 <http://crookedtimber.org/2009/03/05/the-end-of-the-cash-nexus>.

Nick Raaum. "Steam Dreams." *Factor e Farm Weblog*, January 22, 2009 <http://openfarmtech.org/weblog/?p=499>.

Raghuram Rajan and Luigi Zingales. "The Governance of the New Enterprise," in Xavier Vives, ed., *Corporate Governance: Theoretical and Empirical Perspectives* (Cambridge: Cambridge University Press, 2000).

Joshua Cooper Ramo. "Jobless in America: Is Double-Digit Unemployment Here to Stay?" *Time*, September 11, 2009 <http://www.time.com/time/printout/0,8816,1921439,00.html>.

JP Rangaswami. "Thinking about predictability: More musings about Push and Pull." *Confused of Calcutta*, May 4, 2010 <http://confusedofcalcutta.com/2010/05/04/thinking-about-predictability-more-musings-about-push-and-pull/>.

Eric S. Raymond. *The Cathedral and the Bazaar* <http://catb.org/~esr/writings/homesteading>.

Raymond. "Escalating Complexity and the Collapse of Elite Authority." *Armed and Dangerous*, January 5, 2010 <http://esr.ibiblio.org/?p=1551>.

Eric Reasons. "Does Intellectual Property Law Foster Innovation?" *The Tinker's Mind*, June 14, 2009 <http://blog.ericreasons.com/2009/06/does-intellectual-property-law-foster.html>.

Reasons. "The Economic Reset Button." *The Tinker's Mind*, July 2, 2009 <http://blog.ericreasons.com/2009/07/economic-reset-button.html>.

Reasons. "Innovative Deflation." *The Tinker's Mind*, July 5, 2009 <http://blog.ericreasons.com/ 2009/07/innovative-deflation.html>.

Reasons, "Intellectual Property and Deflation of the Knowledge Economy." *The Tinker's Mind*, June 21, 2009. <http://blog.ericreasons.com/2009/06/ intellectual-property-and-deflation-of.html>.

Lawrence W. Reed. "A Tribute to the Polish People," *The Freeman: Ideas on Liberty*, October 2009 <http://www.thefreemanonline.org/columns/ideas-and-consequences/a-tribute-to-the-polish-people/>.

George Reisman. "Answer to Paul Krugman on Economic Inequality." *The Webzine*, March 3, 2006 <http://thewebzine.com/articles/030306Reisman Answer.html>.

"RepRap Project." Wikipedia <http://en.wikipedia.org/wiki/RepRap_Project> (accessed August 31, 2009).

E. C. Riegel. *The New Approach to Freedom: together with Essays on the Separation of Money and State.* Edited by Spencer Heath MacCallum (San Pedro, California: The Heather Foundation, 1976) <http://www.newapproachtofreedom.info/naf/>.

Riegel. *Private Enterprise Money: A Non-Political Money System* (1944) <http://www.newapproachtofreedom.info/pem/>.

John Robb. "THE BAZAAR'S OPEN SOURCE PLATFORM." *Global Guerrillas*, Sept3ember 24, 2004 <http://globalguerrillas.typepad.com/globalguerrillas/2004/09/bazaar_dynamics.html>.

Robb. "Below Replacement Level." *Global Guerrillas*, February 20, 2009 <http://globalguerrillas.typepad.com/johnrobb/2009/02/below-replacement-level.html>.

Robb. "An Entrepreneur's Approach to Resilient Communities." *Global Guerrillas*, February 22, 2010 <http://globalguerrillas.typepad.com/globalguerrillas/2010/02/turning-resilient-communities-into-a-business-opportunity.html>.

Robb. "Fighting an Automated Bureaucracy." *Global Guerrillas*, December 8, 2009 <http://globalguerrillas.typepad.com/globalguerrillas/2009/12/journal-fighting-an-automated-bureaucracy.html>.

Robb. "HOLLOW STATES vs. FAILED STATES." *Global Guerrillas*, March 24, 2009 <http://globalguerrillas.typepad.com/globalguerrillas/2009/03/hollow-states-vs-failed-states.html>.

Robb. "INFOWAR vs. CORPORATIONS." *Global Guerrillas*, October 1, 2009 <http://globalguerrillas.typepad.com/globalguerrillas/2009/10/infowar-vs-corporations.html>.

Robb. "Onward to a Hollow State." *Global Guerrillas*, September 22, 2009 <http://globalguerrillas.typepad.com/globalguerrillas/2008/09/onward-to-a-hol.html>.

Robb. "Resilient Communities and Scale Invariance." *Global Guerrillas*, April 16, 2009 <http://globalguerrillas.typepad.com/globalguerrillas/2009/04/resilient-communities-and-scale-invariance.html>.

Robb. "Resilient Communities: Transition Towns." *Global Guerrillas*, April 7, 2008 <http://globalguerrillas.typepad.com/globalguerrillas/2008/04/transition-town.html>.

Robb. "STANDING ORDER 8: Self-replicate." *Global Guerrillas*, June 3, 2009 <http://globalguerrillas.typepad.com/globalguerrillas/2009/06/standing-order-8-selfreplicate.html>.

Robb. "STEMI Compression." *Global Guerrillas*, November 12, 2008 <http://globalguerrillas.typepad.com/globalguerrillas/2008/11/stemi.html>.

Robb. "Stigmergic Leaning and Global Guerrillas." *Global Guerrillas*, July 14, 2004 <http://globalguerrillas.typepad.com/globalguerrillas/2004/07/stigmergic_sys t.html>.

Robb. "SUPER EMPOWERMENT: Hack a Predator Drone." *Global Guerrillas*, December 17, 2009 <http://globalguerrillas.typepad.com/globalguerrillas/2009/12/super-empowerment-hack-a-predator-drone.html>.

Robb. "The Switch to Local Manufacturing." *Global Guerrillas*, July 8, 2009 <http://globalguerrillas.typepad.com/globalguerrillas/2009/07/journal-the-switch-to-local-manufacturing.html>.

Robb. "Viral Resilience." *Global Guerrillas*, January 12, 2009 <http://globalguerrillas.typepad.com/globalguerrillas/2009/01/journal-phase-t.html>.

Robb. "You Are in Control." *Global Guerrillas*, January 3, 2010 <http://globalguerrillas .typepad.com/globalguerrillas/2010/01/you-are-in-control.html>.

Andy Robinson. "[p2p research] Berardi essay." P2P Research email list, May 25, 2009 <http://listcultures.org/pipermail/p2presearch_listcultures.org/2009-May/003079.html>.

Kim Stanley Robinson. *Green Mars* (New York, Toronto, London, Sydney, Auckland: Bantam Books, 1994).

Nick Robinson. "Even Without a Union, Florida Wal-Mart Workers Use Collective Action to Enforce Rights." *Labor Notes*, January 2006. Reproduced at Infoshop, January 3, 2006 <http://www.infoshop.org/inews/article.php?story =20060103065054461>.

Janko Roettgers. "The Pirate Bay: Distributing the World's Entertainment for $3,000 a Month." *NewTeeVee.Com*, July 19, 2009 <http://newteevee.com/2009/07/19/the-pirate-bay-distributing-the-worlds-entertainment-for-3000-a-month/>.

Paul M. Romer. "Endogenous Technological Change." (December 1989). NBER Working Paper No. W3210.

Joseph Romm. "McCain's Cruel Offshore Drilling Hoax." *CommonDreams.Org*, July 11, 2008 <http://www.commondreams.org/archive/2008/07/11/10301>.

David F. Ronfeldt. *Tribes, Institutions, Markets, Networks* P-7967 (Santa Monica: RAND, 1996) <http://www.rand.org/pubs/papers/P7967/>.

Ronfeldt and Armando Martinez. "A Comment on the Zapatista Netwar." in Ronfeldt and Arquilla, *In Athena's Camp: Preparing for Conflict in th Information Age* (Santa Monica: Rand, 1997),

Murray N. Rothbard. *Power and Market: Government and the Economy* (Menlo Park, Calif.: Institute for Humane Studies, Inc., 1970).

Jonathan Rowe. "Entrepreneurs of Cooperation," *Yes!*, Spring 2006 <http://www.yesmagazine.org/article.asp?ID=1464>.

Jeffrey Rubin. "The New Inflation," *StrategEcon* (CIBC World Markets), May 27, 2008 <http://research.cibcwm.com/economic_public/download/smay08pdf>.

Rubin. *Why Your World is About to Get a Whole Lot Smaller: Oil and the End of Globalization* (Random House, 2009).

Rubin and Benjamin Tal. "Will Soaring Transport Costs Reverse Globalization?" *StrategEcon*, May 27, 2008.

Eric Rumble, "Toxic Shocker." *Up! Magazine*, January 1, 2007 <http://www.up-magazine.com/magazine/exclusives/Toxic_Shocker_3.shtml>.

Alan Rusbridge. "First Read: The Mutualized Future is Bright," *Columbia Journalism Review*, October 19, 2009 <http://www.cjr.org/reconstruction/the_mutualized_future_is_brigh.php>.

Douglas Rushkoff. "How the Tech Boom Terminated California's Economy," *Fast Company*, July 10, 2009 <http://www.fastcompany.com/article/how-tech-boom-terminated-californias-economy?page=0%2C1>.

Charles F. Sabel. "A Real-Time Revolution in Routines." Charles Hecksher and Paul S. Adler, *The Firm as a Collaborative Community: Reconstructing Trust in the Knowledge Economy* (New York: Oxford University Press, 2006).

Reihan Salam. "The Dropout Economy." *Time*, March 10, 2010 <http://www.time.com/time/specials/packages/printout/0,29239,1971133_1971110_1971126,00.html>.

Kirkpatrick Sale. *Human Scale* (New York: Coward, McCann, & Geoghegan, 1980).

Julian Sanchez. "Dammit, Apple," *Notes from the Lounge*, June 2, 2008 <http://www.juliansanchez.com/2008/06/02/dammit-apple/>.

"Say No to Schultz Mansion Purchase" Starbucks Union <http://www.starbucksunion.org/node/1903>.

F.M. Scherer and David Ross. *Industrial Market Structure and Economic Performance*. 3rd ed (Boston: Houghton Mifflin, 1990).

Ron Scherer. "Number of long-term unemployed hits highest rate since 1946." *Christian Science Monitor*, January 8, 2010 <http://www.csmonitor.com/USA/2010/0108/Number-of-long-term-unemployed-hits-highest-rate-since-1948>.

E. F. Schumacher. *Good Work* (New York, Hagerstown, San Fransisco, London: Harper & Row, 1979).

Schumacher. *Small is Beautiful: Economics as if People Mattered* (New York, Hagerstown, San Francisco, London: Harper & Row, Publishers, 1973).

Joseph Schumpeter. *History of Economic Analysis*. Edited from manuscript by Elizabeth Boody Schumpeter (New York: Oxford University Press, 1954).

Schumpeter. "Imperialism," in *Imperialism, Social Classes: Two Essays* by Joseph Schumpeter. Translated by Heinz Norden. Introduction by Hert Hoselitz (New York: Meridian Books, 1955).

Tom Scotney. "Birmingham Wragge team to focus on online comment defamation." *Birmingham Post*, October 28, 2009 <http://www.birminghampost.net/birmingham-business/birmingham-business-news/legal-business/2009/10/28/birmingham-wragge-team-to-focus-on-online-comment-defamation-65233-25030203/>.

James Scott. *Seeing Like a State* (New Haven and London: Yale University Press, 1998).

Butler Shaffer. *Calculated Chaos: Institutional Threats to Peace and Human Survival* (San Francisco: Alchemy Books, 1985).

Shaffer. *In Restraint of Trade: The Business Campaign Against Competition, 1918-1938* (Lewisburg: Bucknell University Press, 1997).

Laurence H. Shoup and William Minter. "Shaping a New World Order: The Council on Foreign Relations' Blueprint for World Hegemony, 1939-1945," in Holly Sklar, ed., *Trilateralism: The Trilateral Commission and Elite Planning for World Management* (Boston: South End Press, 1980).

Christian Siefkes. *From Exchange to Contributions: Generalizing Peer Production into the Physical World* Version 1.01 (Berlin, October 2007).

Siefkes. "[p2p-research] Fwd: Launch of Abundance: The Journal of Post-Scarcity Studies, preliminary plans," Peer to Peer Research List, February 25, 2009 <http://listcultures.org/pipermail/p2presearch_listcultures.org/2009-February/ 001555.html>.

Arthur Silber. "An Evil Monstrosity: Thoughts Upon the Death State." *Once Upon a Time*, April 20, 2010 <http://powerofnarrative.blogspot.com/2010/04/evil-monstrosity-thoughts-on-death.html>.

Charles Hugh Smith. "End of Work, End of Affluence III: The Rise of Informal Businesses." *Of Two Minds*, December 10, 2009 <http://www.oftwominds.com/blogdec08/informal12-08.html>.

Smith. "The Future of Manufacturing in the U.S." oftwominds, February 5, 2010 <http://charleshughsmith.blogspot.com/2010/02/future-of-manufacturing-in-us.html>.

Smith. "Globalization and China: Neoliberal Capitalism's Last 'Fix'," *Of Two Minds*, June 29, 2009 <http://www.oftwominds.com/blogjune09/globalization06-09.html>.

Smith. "The Travails of Small Business Doom the U.S. Economy," *Of Two Minds*, August 17, 2009 <http://charleshughsmith.blogspot.com/2009/08/he-travails-of-small-business-doom-us.html>.

Smith. "Trends for 2009: The Rise of Informal Work." *Of Two Minds*, December 30, 2009 <http://www.oftwominds.com/blogdec08/rise-of-informal12-08.html>.

Smith. "Unemployment: The Gathering Storm," *Of Two Minds*, September 26, 2009 <http://charleshughsmith.blogspot.com/2009/09/unemployment-gathering-storm.html>.

Smith. "Welcome to America's Lost Decade(s)," *Of Two Minds*, September 18, 2009 <http://charleshughsmith.blogspot.com/2009/09/welcome-to-americas-lost-decades.html>.

Smith. "What if the (Debt Based) Economy Never Comes Back?" *Of Two Minds*, July 2, 2009 <http://www.oftwominds.com/blogjuly09/what-if07-09.html>.

Johan Soderberg. *Hacking Capitalism: The Free and Open Source Software Movement* (New York and London: Routledge, 2008).

"Solar Turbine—Open Source Ecology." <http://openfarmtech.org/index.php?title=Solar_Turbine> Accessed January 5, 2009.

Donna St. George. "Pew report shows 50-year high point for multi-generational family households." *Washington Post*, March 18, 2010 <http://www.washingtonpost.com/wp-dyn/content/article/2010/03/18/AR2010031804510.html>.

L. S. Stavrianos. *The Promise of the Coming Dark Age* (San Francisco: W. H. Freeman and Co. 1976).

Barry Stein. *Size, Efficiency, and Community Enterprise* (Cambridge: Center for Community Economic Development, 1974).

Anton Steinpilz. "Destructive Creation: BuzzMachine's Jeff Jarvis on Internet Disintermediation and the Rise of Efficiency." *Generation Bubble*, June 12, 2009 <http://generationbubble.com/2009/06/12/destructive-creation-buzzmachines-jeff-jarvis-on-internet-disintermediation-and-the-rise-of-efficiency/>.

Neal Stephenson. *Snow Crash* (Westminster, Md.: Bantam Dell Pub Group, 2000).

Bruce Sterling. "The Power of Design in your exciting new world of abject poverty." *Wired: Beyond the Beyond*, February 21, 2010 <http://www.wired.com/beyond_the_beyond/2010/02/the-power-of-design-in-your-exciting-new-world-of-abject-poverty/>.

"Stigmergy." *Wikipedia* <http://en.wikipedia.org/wiki/Stigmergy> (accessed September 29, 2009).

Carin Stillstrom and Mats Jackson. "The Concept of Mobile Manufacturing." *Journal of Manufacturing Systems* 26:3-4 (July 2007) <http://www.sciencedirect.com/science?ob=ArticleURL&_udi=B6VJD-4TK3FG8-6&_user=108429&_rdoc=1&_fmt=&_orig=search&_sort=d&view=c&_version=1&_urlVersion=0&_userid=108429&md5=bf6e603b5de29cdfd026d5d003379877c>.

David Streitman. "Rock Bottom for Decades, but Showing Signs of Life." *New York Times*, February 1, 2009 <http://www.nytimes.com/2009/02/01/us/ 01braddock.html>.

Dan Strumpf. "Exec Says Toyota Prepared for GM Bankruptcy." Associated Press, April 8, 2009 <http://abcnews.go.com/Business/wireStory?id=7288650>.

Daniel Suarez. *Daemon* (Signet, 2009).

Suarez. *Freedom(TM)* (Dutton 2010).

Kevin Sullivan. "As Economy Plummets, Cashless Bartering Soars on the Internet." *Washington Post*, March 14, 2009 <http://www.washingtonpost.com/wp-dyn/content/article/2009/03/13/ AR2009031303035_pf.html>.

"Supply Chain News: Walmart Joins Kohl's in Offering Factoring Program to Apparel Suppliers." *Supply Chain Digest*, November 17, 2009 <http://www.scdigest.com/ASSETS/ON_TARGET/ 09-11-17-2.PHP?cid=2954&ctype=conte>.

Vin Suprynowicz. "Schools guarantee there can be no new Washingtons." *Review Journal*, February 10, 2008 <http://www.lvrj.com/opinion/15490456.html>.

Paul Sweezy. "Competition and Monopoly." *Monthly Review* (May 1981).

Joseph Tainter. *The Collapse of Complex Societies* (Cambridge, New York, New Rochelle, Melbourne, Sydney: Cambridge University Press, 1988).

Don Tapscott and Anthony D. Williams. *Wikinomics: How Mass Collaboration Changes Everything* (New York: Portfolio, 2006).

"Telekommunisten: The Revolution is Coming." <http://telekommunisten.net/about> Accessed October 19, 2009.

Clive Thompson. "The Dream Factory." *Wired*, September 2005 <http://www.wired.com/wired/archive/13.09/fablab_pr.html>.

E. P. Thompson. *The Making of the English Working Class* (New York: Vintage Books, 1963, 1966).

Thoreau. "More on the swarthy threat to our precious carry-on fluids." *Unqualified Offerings*, December 26, 2009 <http://highclearing.com/index.php/archives/2009/12/26/10438>.

"Torch Table Build." *Open Source Ecology* wiki (accessed August 22, 2009 <http://openfarmtech.org/index.php?title=Torch_Table_Build>.

Ted Trainer. "Local Currencies." (September 4, 2008), The Simpler Way <http://ssis.arts.unsw.edu.au/tsw/localcurrency.html>.

Trainer. "The Transition Towns Movement; its huge significance and a friendly criticism." *(We) can do better*, July 30, 2009 <http://candobetter.org/node/1439>.

Trainer. "We Need More Than LETS." The Simpler Way <http://ssis.arts.unsw.edu.au/tsw/ D11WeNdMreThLETS2p.html>.

Gul Tuysuz. "An ancient tradition makes a little comeback." *Hurriyet DailyNews*, January 23, 2009 <http://www.hurriyet.com.tr/english/domestic/0826653.asp ?scr=1>.

Dylan Tweney. "DIY Freaks Flock to 'Hacker Spaces' Worldwide." *Wired*, March29, 2009 <http://www.wired.com/gadgetlab/2009/03/hackerspaces/>.

"Uh, oh, higher jobless rates could be the new normal." *New York Daily News*, October 23, 2009 <http://www.nydailynews.com/money/work_career/2009/10/19/2009-10-19_uh_oh_higher_jobless_rates_could_be_the_new_normal.html>.

United States Participation in the Multilateral Development Banks in the 1980s. Department of the Treasury (Washingon, DC: 1982).

Bob Unruh. "Food co-op hit by SWAT raid fights back." *WorldNetDaily*, December 24, 2008 <http://www.wnd.com/index.php?fa=PAGE.view&pageId=84445>.

Unruh. "SWAT raid on food co-op called 'entrapment'." *WorldNetDaily*, December 26, 2008 <http://www.wnd.com/index.php?fa=PAGE.view&pageId=84594>.

"U.S. Suffering Permanent Destruction of Jobs." *Washington's Blog*, October 5, 2009 <http://www.washingtonsblog.com/2009/10/us-suffering-permanent-destruction-of.html>

Jonathan J. Vaccaro. "The Next Surge—Counterbureaucracy." *New York Times*, December 7, 2009 <http://www.nytimes.com/2009/12/08/opinion/08vaccaro. html>.

Jeff Vail. "2010—Predictions and Catabolic Collapse." *Rhizome*, January 4, 2010 <http://www.jeffvail.net/2010/01/2010-predictions-and-catabolic-collapse.html>.

Vail. "The Design Imperative." JeffVail.Net, April 8, 2007 <http://www.jeffvail.net/2007/04/design-imperative.html>.

Vail. "Diagonal Economy 1: Overview." JeffVail.Net, August 24, 2009 <http://www .jeffvail.net/ 2009/08/diagonal-economy-1-overview.html>.

Vail. "The Diagonal Economy 5: The Power of Networks." *Rhizome*, December 21, 2009 <http://www.jeffvail.net/2009/12/diagonal-economy-5-power-of-networks.html>.

Vail. "Five Geopolitical Feedback-Loops in Peak Oil." *JeffVail.Net*, April 23, 2007 <http://www.jeffvail.net/2007/04/five-geopolitical-feedback-loops-in.html>.

Vail. "Re-Post: Hamlet Economy." *Rhizome*, July 28, 2008 <http://www.jeffvail .net/2008/07/re-post-hamlet-economy.html>.

Vail. *A Theory of Power* (iUniverse, 2004) <http://www.jeffvail .net/atheoryofpower.pdf>.

Vail. "What is Rhizome?" *JeffVail.Net*, January 28, 2008 <http://www.jeffvail.net/2007/01/what-is-rhizome.html>.

Lyman P. van Slyke. "Rural Small-Scale Industry in China." in Richard C. Dorf and Yvonne L. Hunter, eds., *Appropriate Visions: Technology the Environment and the Individual* (San Francisco: Boyd & Fraser Publishing Company, 1978).

"Venture Communism." *P2P Foundation Wiki* <http://p2pfoundation.net/Venture _Communism> (accessed August 8, 2009.

Chris Vernon. "Peak Coal—Coming Soon?" *The Oil Drum: Europe*, April 5, 2007 <http://europe.theoildrum.com/node/2396>.

William Waddell. "But You Can't Fool All the People All the Time." *Evolving Excellence*, August 25, 2009 <http://www.evolvingexcellence.com/blog/2009/08/but-you-cant-foool-all-the-people-all-the-time.html>.

Waddell. "The Irrelevance of the Economists." *Evolving Excellence*, May 6, 2009 <http://www.evolvingexcellence.com/blog/2009/05/the-irrelevance-of-the-economists.html>.

Waddell and Norman Bodek. *The Rebirth of American Industry: A Study of Lean Management* (Vancouver, WA: PCS Press, 2005).

"Wal-Mart Nixes 'Open Availability' Policy." *Business & Labor Reports* (Human Resources section), June 16, 2005 <http://hr.blr.com/news.aspx?id=15666>.

Jesse Walker. "The Satellite Radio Blues: Why is XM Sirius on the verge of bankruptcy?" *Reason*, February 27, 2009 <http://reason.com/news/show/131905.html>.

Tom Walker. "The Doppelganger Effect." *EconoSpeak*, January 2, 2010 <http://econospeak.blogspot.com/2010/01/ doppelg-effect.html>.

Todd Wallack. "Beware if your blog is related to work." San Francisco Chronicle, January 25, 2005 <http://www.sfgate.com/cgi-bin.article.cgi?f=/c/a/2005/01/24/BIGCEAT1l01.DTL>.

Immanuel Wallerstein. "Household Structures and Labor-Force Formation in the Capitalist World Economy." in Joan Smith, Immanuel Wallerstein, Hans-Dieter Evers, eds., *Households and the World Economy* (Beverly Hills, London, New Delhi: Sage Publications, 1984).

Wallerstein and Joan Smith. "Households as an institution of the world-economy." in Smith and Wallerstein, eds., *Creating and Transforming Households: The constraints of the world-economy* (Cambridge; New York; Oakleigh, Victoria; Paris: Cambridge University Press, 1992).

Colin Ward. "Anarchism and the informal economy." *The Raven* No. 1 (1987).

Ward. *Anarchy in Action* (London: Freedom Press, 1982).

"What are we working on?" *100kGarages*, January 8, 2010 <http://blog.100kgarages.com/2010/01/08/what-are-we-working-on/>.

"What is an Ecovillage?" Gaia Trust website <http://www.gaia.org/gaia/ecovillage/whatis/>.

"What is an Ecovillage?" (sidebar), Agnieszka Komoch, "Ecovillage Enterprise." *Permaculture Magazine* No. 32 (Summer 2002).

"What is the relationship between RepRap and Makerbot?" *Hacker News* <http://news.ycombinator.com/item?id=696785>.

"What's Digital Fabrication?" 100kGarages website <http://100kgarages.com/ digital_fabrication.html>.

"What's Next for 100kGarages?" *100kGarages News*, February10, 2010 <http://blog.100kgarages.com/2010/02/10/whats-next-for-100kgarages/>.

Shawn Wilbur, "Re: [Anarchy-List] Turnin' rebellion into money (or not . . . your choice)." *Anarchy* list, July 17, 2009 <http://lists.anarchylist.org/private.cgi/anarchy-list-anarchylist.org/2009-July/003406.html>.

Wilbur. "Taking Wing: Corvus Editions." *In the Libertarian Labyrinth*, July 1, 2009 <http://libertarian-labyrinth.blogspot.com/2009/07/taking-wing-corvus-editions.html>; Corvus Distribution website <http://www.corvusdistribution .org/shop/>.

Wilbur. "Who benefits most economically from state centralization." *In the Libertarian Labyrinth*, December 9, 2008 <http://libertarian-labyrinth.blogspot.com/2008/12/who-benefits-most-economically-from.html>.

Chris Williams. "Blogosphere shouts 'I'm Spartacus' in Usmanov-Murray case: Uzbek billionaire prompts Blog solidarity." *The Register*, September 24, 2007 <http://www.theregister.co.uk/2007/09/24/usmanov_vs_the_internet/>.

William Appleman Williams. *The Contours of American History* (Cleveland and New York: The World Publishing Company, 1961).

Williams. *The Tragedy of American Diplomacy* (New York: Dell Publishing Company, 1959, 1962).

Frank N. Wilner. "Give truckers an inch, they'll take a ton-mile: every liberalization has been a launching pad for further increases—trucking wants long combination vehicle restrictions dropped." *Railway Age*, May 1997 <http://findarticles.com/p/articles/mi_m1215/is_n5_v198/ai_19460645>.

James L. Wilson. "Standard of Living vs. Quality of Life." *The Partial Observer*, May 29, 2008 <http://www.partialobserver.com/article.cfm?id=2955&RSS=1>.

James P. Womack and Daniel T. Jones. *Lean Thinking: Banish Waste and Create Wealth in Your Corporation* (Simon & Schuster, 1996).

Womack, Jones, and Daniel Roos. *The Machine That Changed the World* (New York: Macmillian Publishing Company, 1990).

Nicholas Wood. "The 'Family Firm'—Base of Japan's Growing Economy." *The American Journal of Economics and Sociology*, vol. 23 no. 3 (1964).

Matthew Yglesias. "The Elusive Post-Bubble Economy," *Yglesias/ThinkProgress.Org*, December 22, 2008 <http://yglesias.thinkprogress.org/archives/2008/12/the_elusive_post_bubble_economy.php>.

Yglesias. "The Office Illusion." *Matthew Yglesias*, September 1, 2007 <http://matthewyglesias.theatlantic.com/archives/2007/09/the_office_illusion.php>.

Yglesias. "Too Much Information." *Matthew* Yglesias, December 28, 2009 <http://yglesias.thinkprogress.org/archives/2009/12/too-much-information.php>.

Andrea Zippay. "Organic food co-op raid sparks case against health department, ODA." FarmAndDairy.Com, December 19, 2008 <http://www.farmanddairy.com/news/organic-food-co-op-raid-sparks-court-case-against-health-department-oda/10752.html.

Luigi Zingales. "In Search of New Foundations." *The Journal of Finance*, vol. lv, no. 4 (August 2000).

Ethan Zuckerman. "Samuel Bowles introduces Kudunomics." *My Heart's in Accra* November 17, 2009 <http://www.ethanzuckerman.com/blog/2009/11/17/ samuel-bowles-introduces-kudunomics/>.

Index

About the Author

KEVIN A. CARSON is Research Associate at the Center for a Stateless Society. He is the author of *Organization Theory: A Libertarian Perspective, Studies in Mutualist Political Economy*—the focus of a symposium published in the *Journal of Libertarian Studies*—as well as of the pamphlets *Austrian and Marxist Theories of Monopoly-Capital* and *Contract Feudalism: A Critique of Employer Power Over Employees* (Libertarian Alliance); *The Ethics of Labor Struggle* (Alliance of the Libertarian Left); and *The Iron Fist behind the Invisible Hand: Corporate Capitalism As a State-Guaranteed System of Privilege* (Red Lion Press). His writing has also appeared in *Just Things, Any Time Now, The Freeman: Ideas on Liberty,* and *Land and Liberty,* as well as on the P2P Foundation blog. A member of the Industrial Workers of the World, the Voluntary Cooperation Movement, and the Alliance of the Libertarian Left, and a leader in the contemporary revival of Proudhonian mutualist anarchism, he maintains the blog *Free Market Anti-Capitalism* at http://Mutualist.BlogSpot.Com.